Medicinal Plants of Sabah, North Borneo

Researchers in the field of natural product chemistry and pharmacology focus on discovering new drugs from medicinal plants. The medicinal plants of Sabah (North Borneo), many of which are found in all of Asia and the Pacific, represent a vast array of resources that could be used for the discovery of drugs and the development of cosmetics as well as functional food. No other book focuses on the medicinal plants from this part of the globe in such depth. The author is well known in natural products and this text will appeal to a global audience in presenting the botanical description, chemical constituents, pharmacology, pharmaceutical, cosmetic, and functional food potentials of these plants.

Features:

1. Describes in a phylogenetic order the habitats, distributions, botanical descriptions, local names, uses, chemical constituents, pharmacological activities, and toxic effects of more than 250 species of medicinal plants of Sabah.
2. Provides a selection of botanical plates of medicinal plants endemic to Sabah, hand-drawn by the author.
3. Provides comment sections to invite further research on the topic of the development of drugs, dietary products, and cosmetics from Sabah medicinal plants.

Natural Products Chemistry of Global Plants

Series Editor: Clara Bik-san Lau
Founding Editor: Raymond Cooper

This unique book series focuses on the natural products chemistry of botanical medicines from different countries such as Bangladesh, Borneo, Brazil, Cameroon, China, Ecuador, India, Iran, Laos, Romania, Sri Lanka, Turkey, etc. These fascinating volumes are written by experts from their respective countries. The series will focus on the pharmacognosy, covering recognized areas rich in folklore as well as botanical medicinal uses as a platform to present the natural products and organic chemistry. Where possible, the authors will link these molecules to pharmacological modes of action, reflecting the ethnopharmacological uses. The series intends to trace a route through history from ancient civilizations to the modern day showing the importance to man of natural products in medicines, in foods and a variety of other ways. With special emphasis on plant parts for medicinal uses, phytochemistry and biological activities, this book series will be of useful reference to scientists/pharmacognosists/pharmacists/chemists/graduates/undergraduates/researchers in the fields of natural products, herbal medicines, ethnobotany, pharmacology, chemistry, and biology. Furthermore, pharmaceutical companies may also found valuable information on potential herbals and lead compounds for the future development of health supplements and western medicines.

Recent Titles in this Series:

Traditional Herbal Remedies of Sri Lanka
Viduranga Y. Waisundara

Medicinal Plants of Ecuador
Pablo A. Chong Aguirre, Patricia Manzano Santana, Migdalia Miranda Martínez (Eds)

Medicinal Plants of Laos
Djaja Djendoel Soejarto, Bethany Gwen Elkington and Kongmany Sydara

Edible and Medicinal Mushrooms of the Himalayas
Climate Change, Critically Endangered Species and the Call for Sustainable Development
Edited by Ajay Sharma, Garima Bhardwaj and Gulzar Ahmad Nayik

Medicinal Plants of Turkey
Ufuk Koca-Caliskan and Esra Akkol

Natural Products and Medicinal Properties of Carpathian (Romanian) Plants
Adina-Elena Segneanu

Medicinal Plants of Sabah, North Borneo
Christophe Wiart

For more information about this series, please visit: www.routledge.com/Natural-Products-Chemistry-of-Global-Plants/book-series/CRCNPCGP

Medicinal Plants of Sabah, North Borneo

Christophe Wiart

CRC Press is an imprint of the
Taylor & Francis Group, an **informa** business

Designed cover credit: Fogs Mist Over Dipterocarp Rain Forest, Stock Photo 493080355 | Shutterstock

First edition published 2024
by CRC Press

2385 Executive Center Drive, Suite 320, Boca Raton, FL 33431
and by CRC Press

4 Park Square, Milton Park, Abingdon, Oxon, OX14 4RN

CRC Press is an imprint of Taylor & Francis Group, LLC

Library of Congress Cataloging-in-Publication Data
Names: Wiart, Christophe, author.
Title: Medicinal Plants of Sabah, North Borneo / by Christophe Wiart, University Malaysia Sabah, Malaysia.
Description: First edition. | Boca Raton : CRC Press, 2024. | Series: Natural products chemistry of global plants | Includes bibliographical references and index.
Identifiers: LCCN 2023055882 (print) | LCCN 2023055883 (ebook) | ISBN 9781032515762 (hardback) | ISBN 9781032514314 (paperback) | ISBN 9781003402886 (ebook)
Subjects: LCSH: Medicinal plants—Malaysia—Sabah. | Materia medica, Vegetable—Malaysia—Sabah.
Classification: LCC QK99.M282 W53 20224 (print) | LCC QK99.M282 (ebook) | DDC 581.6/34095953—dc23/eng/20240213
LC record available at https://lccn.loc.gov/2023055882
LC ebook record available at https://lccn.loc.gov/2023055883

ISBN: 978-1-032-51576-2 (hbk)
ISBN: 978-1-032-51431-4 (pbk)
ISBN: 978-1-003-40288-6 (ebk)

DOI: 10.1201/ 9781003402886

Typeset in Times
by Apex CoVantage, LLC

A ma grand-mère, Madame Renée Monllor
A ma mère, Madame Flora Monllor
A ma famille
A mes amis et maîtres, les arbres

Contents

Foreword xix
Preface xxi
Author Biography xxiii

Chapter 1. The Medicinal Plants of Sabah (North Borneo): The Lycophytes 1

1.1 Subclass Lycopodiidae Bek. (1862) 1
1.1.1 Order Lycopodiales DC. ex Bercht. & J. Presl (1820) 1
1.1.1.1 Family Lycopodiaceae P. Beauv. ex Mirb. (1802), the Club Moss Family 1
Lycopodium cernuum L. 1
Lycopodium phlegmaria L. 4
1.1.2 Order Selaginellales Prantl (1854) 6
1.1.2.1 Family Selaginellaceae Willk. (1854), the Spike Moss Family 6
Selaginella argentea (Wall. ex Hook. & Grev.) Spring 6
Selaginella plana (Desv. ex Poir.) Hieron. 7

Chapter 2. The Medicinal Plants of Sabah (North Borneo): The Monilophytes 9

2.1 Subclass Ophioglissidae Klinge (1832) 9
2.1.1 Order Ophioglossales Link (1833) 9
2.1.1.1 Family Ophioglossaceae Martinov (1820), the Adder's Tongue Family 9
Helminthostachys zeylanica (L.) Hook. 9
2.2 Subclass Polypodiidae Cronquist, Takht. & W. Zimm. (1966) 12
2.2.1 Order Blechnales Pic. Sem. ex Reveal (1993) 12
2.2.1.1 Family Blechnaceae Newman (1844) the Chain Fern family 12
Blechnum orientale L. 12
Stenochlaena palustris (Burm. f.) Bedd. 14
2.2.2 Order Gleicheniales Link (1825) 16
2.2.2.1 Family Gleicheniaceae C. Presl (1825), the Forking Fern Family 16
Gleichenia truncata (Willd.) Sprain 16
2.2.3 Order Polypodiales Link (1833) 18
2.2.3.1 Family Athyriaceae Alston (1956), the Lady Fern Family 18
Diplazium cordifolium Bl. 18
2.2.3.2 Family Polypodiaceae Link J. Presl. & C. Presl. (1822), the Polypody Family 19
Drymoglossum piloselloides (L.) C. Presl 19
Drynaria sparsisora (Desv.) T. Moore 21
Pyrrosia lanceolata (L.) Fawr. 23
2.2.4 Order Schizaeales Schimp. (1869) 25
2.2.4.1 Family Lygodiaceae M. Roem. (1840), the Climbing Ferns family 25
Lygodium circinnatum Sw. 25

Lygodium salicifolium C. Presl. 27
Lygodium flexuosum (L.) Sw. 27
2.2.4.2 Family Thelypteridaceae Ching ex Pic. Serm. (1970), the Marsh Fern Family 29
Pronephrium asperum (C. Presl) Holttum 29

Chapter 3. The Medicinal Plants of Sabah (North Borneo): The Gymnosperms 31

3.1 Subclass Pinidae Cronquist, Takht. & W. Zimm. (1966) 31
3.1.1 Order Auraucariales Gorozh. (1904) 31
3.1.1.1 Family Auraucariaceae Henkel & W. Hochst. (1865), the Araucaria Family 31
Agathis borneensis Warb. 31
3.2 Subclass Gnetidae Pax (1894) 33
3.2.1 Order Gnetales Blume (1835) 33
3.2.1.1 Family Gnetaceae Blume (1833), the Gnetum Family 33
Gnetum macrostachyum Hook.f. 33

Chapter 4. The Medicinal Plants of Sabah (North Borneo): The Angiosperms 35

4.1 Subclass Magnoliidae Novák ex Takht. (1967) 35
4.1.1 Superorder Austrobaileyanae Doweld ex M.W. Chase & Reveal (2009), the Protomagnoliids 35
4.1.1.1 Order Austrobaileyales Takht. ex Reveal (1992) 35
4.1.1.1.1 Family Schisandraceae Blume (1830), the Star Vine family 35
Kadsura borneensis A.C. Sm 35
4.1.2 Superorder Magnolianae Takht. (1967), the Magnoliids 37
4.1.2.1 Order Piperales Bercht. & C. Presl. (1820) 37
4.1.2.1.1 Family Aristolochiaceae Juss. (1789), the Dutchman's Pipe Family 37
Aristolochia minutiflora Ridl. ex Gamble 37
Aristolochia papillifolia Ding Hou 37
4.1.2.1.2 Family Piperaceae Giseke (1792), the Pepper Family 39
Peperomia pellucida (L.) Kunth 39
Piper betle L. 41
Piper caducibracteum C. DC 44
Piper caninum Bl. 46
Piper sarmentosum Roxb. 47
Piper umbellatum L. 48
4.1.2.2 Order Laurales Juss. ex Bercht. & J. Presl (1820) 49
4.1.2.2.1 Family Lauraceae Juss. (1789), the Laurel Family 49
Cinnamomum iners Reinw. ex Blume 49
Lindera pipericarpa Boerl. 50
Litsea accedens (Bl.) Boerl. 51
Litsea garciae Vidal. 51
Litsea odorifera Valeton 53
Litsea umbellata (Lour.) Merr. 54

4.1.2.2.2 Family Annonaceae Juss. (1789), the Custard Apple Family 55
Annona muricata L. 55
Artabotrys roseus Boerl. 57
Desmos teysmannii (Boerl.) Merr. 58
Fissistigma fulgens Merr. 58
Fissistigma latifolium (Dunal) Merr. 59
Fissistigma manubriatum Merr. 60
Goniothalamus roseus Stapf 61
Phaeanthus ophthalmicus (Roxb. ex G. Don) J. Sinclair 62
Polyalthia bullata King 64
Polyalthia insignis (Hook.f.) Airy Shaw 65
Polyalthia sumatrana (Miq.) Kurz 66
Uvaria cuneifolia (Hook.f. & Thomson) L.L. Zhou, Y.C.F Su, & R.M.K. Saunders 67
Uvaria grandiflora Roxb. ex Hornem 67
Uvaria sorzogonensis C. Presl. 68
Xylopia dehiscens (Blanco) Merr. 70
4.1.3 Superorder Lilianae Takht. (1967), the Monocots 70
4.1.3.1 Order Acorales Link (1835) 70
4.1.3.1.1 Family Acoraceae Martinov (1820), the Sweet Flag family 70
Acorus calamus L. 70
4.1.3.1.2 Family Araceae Juss. (1789), the Arum Family 72
Aglaonema oblongifolium (Roxb.) Kunth 72
Alocasia macrorrhizos (L.) G. Don 73
Amydrium medium Nicolson 74
Homalomena propinqua Schott 74
Rhaphidophora korthalsii Schott 76
Scindapsus perakensis Hook.f. 77
4.1.3.2 Order Asparagales Link (1829) 77
4.1.3.2.1 Family Asparagaceae Juss. (1789), the Asparagus Family 77
Cordyline fruticosa (L.) A. Chev. 77
Dracaena elliptica Thunb. 79
Dracaena umbratica Ridl. 80
Sansevieria trifasciata Prain 81
4.1.3.2.2 Family Hypoxidaceae R.Br. (1814), the Star-Grass family 82
Curculigo latifolia Dryand. ex W.T. Aiton 82
4.1.3.2.3 Family Aspholedaceae Juss. (1789), the Asphodel Family 84
Aloe vera (L.) Burm.f. 84
Dianella ensifolia (L.) Redouté 85
4.1.3.2.4 Family Orchidaceae Juss. (1789), the Orchid Family 86
Dendrobium umbellatum Rchb. f. 86

4.1.3.3 Order Liliales Pelerb (1826) ... 88
4.1.3.3.1 Family Smilacaceae Vent. (1799), the Catbrier Family ... 88
Smilax odoratissima Bl. ... 88
4.1.3.4 Order Pandanales R.Br. ex Bercht. & J. Presl (1820) ... 90
4.1.3.4.1 Family Pandanaceae R.Br. (1810), the Screwpine Family ... 90
Pandanus amaryllifolius Roxb. ... 90
4.1.4 Superorder Lilianae Takht. (1967), the Commelinids ... 91
4.1.4.1 Order Arecales Bromhead (1840) ... 91
4.1.4.1.1 Family Arecaceae Bercht. & J. Presl (1820), the Palm Family ... 91
Areca catechu L. ... 91
Caryota mitis Lour. ... 93
Licuala spinosa Wurmb. ... 94
Metroxylon sagu Rottb. ... 95
Plectocomiopsis geminiflora (Griff.) Becc. ... 96
Salacca zalacca (Gaertn.) Voss ... 96
4.1.4.2 Order Commelinales Mirb. ex Bercht. & J. Presl (1820) ... 98
4.1.4.2.1 Family Commelinaceae Mirbel (1804), the Spiderwort Family ... 98
Commelina communis L. ... 98
Commelina nudiflora L. ... 99
Forrestia griffithii C.B. Clarke ... 100
4.1.4.2.2 Family Hanguanaceae Airy Shaw (1965) ... 101
Hanguana malayana (Jack.) Merr. ... 101
4.1.4.2.3 Family Pontederiaceae Kunth (1816), the Pickerel-Weed Family ... 102
Eichhornia crassipes (Mart.) Solms ... 102
4.1.4.3 Order Poales Small (1903) ... 104
4.1.4.3.1 Family Bromeliaceae Juss. (1789), the Pineapple Family ... 104
Ananas comosus (L.) Merr. ... 104
4.1.4.3.2 Family Cyperaceae Juss. (1789), the Sedge Family ... 105
Cyperus rotundus L. ... 105
Hypolytrum nemorum (Vahl) Spreng. ... 107
4.1.4.3.3 Family Flagellariaceae Dumort. (1829), the Bushcane Family ... 107
Flagellaria indica L. ... 107
Scleria bancana Miq. ... 109
4.1.4.3.4 Family Poaceae Barnhart (1895), the Grass Family ... 109
Bambusa vulgaris Schrad. ex J.C. Wendl. ... 109
Coix lacryma-jobi L. ... 112
Cymbopogon citratus (DC.) Stapf. ... 114
Dinochloa scandens (Blume) Kuntze ... 116
Dinochloa scabrida S. Dransf. ... 117
Dinochloa sublaevigata S. Dransf. ... 117
Dinochloa trichogona S. Dransf. ... 117
Eleusine indica (L.) Gaertn. ... 119
Garnotia acutigluma (Steud.) Ohwi. ... 120

Imperata cylindrica (L.) Raeusch. 120
Lophatherum gracile Brongn. 122
Miscanthus floridulus (Labill.) Warb. ex K. Schum. & Lauterb. 123
Panicum palmifolium J. Koenig 123
Paspalum conjugatum P.J. Bergius 124
Thysanolaena latifolia (Roxb. ex Hornem.) Honda.. 125
4.1.4.4. Order Zingiberales Griseb. (1854) 127
4.1.4.4.1 Family Cannaceae Juss. (1789), the Canna Family... 127
Canna indica L. 127
4.1.4.4.2 Family Costaceae Nakai (1941), the Costus Family 129
Costus speciosus (J. Koenig ex Retz.) Sm. 129
4.1.4.4.3 Family Marantaceae R.Br. (1814), the Arrowroot Family....................................... 131
Donax canniformis (G. Forst.) K. Schum. 131
4.1.4.4.4 Family Musaceae Juss. (1789), the Banana Family... 133
Musa paradisiaca L. 133
4.1.4.4.5 Family Zingiberaceae Martinov (1820), the Ginger Family... 135
Alpinia galanga (L.) Sw. 135
Boesenbergia pulchella (Ridl.) Merr. 137
Boesenbergia rotunda (L.) Mansf. 137
Boesenbergia stenophylla R.M Sm. 139
Curcuma caesia Roxb. 141
Curcuma longa L. 142
Curcuma xanthorrhiza Roxb. 143
Etlingera brevilabrum (Valeton) R.M. Sm. 145
Etlingera coccinea (Blume) S. Sakai & Nagam. 146
Etlingera elatior (Jack) R.M. Sm. 147
Etlingera littoralis (J. Koenig) Giseke 149
Etlingera punicea (Roxb.) R.M. Sm. 150
Globba francisci Ridl. 151
Globba propinqua Ridl. 151
Hedychium longicornutum Griff. ex Baker....... 151
Kaempferia galanga L. 152
Zingiber officinale Roscoe 154
Zingiber purpureum Roscoe............................. 156
Zingiber zerumbet (L.) Roscoe ex Sm. 158
4.1.5 Superorder Ranunculanae Takht. ex Reveal (1992), the Eudicots.. 159
4.1.5.1 Order Ranunculales Juss. ex Bercht. & J. Presl (1820) 159
4.1.5.1.1 Family Menispermaceae Juss. (1789), the Moonseed Family 159
Coscinium fenestratum (Gaertn.) Colebr. 159
Fibraurea tinctoria Lour. 161
Pycnarrhena tumefacta Miers. 163

Tinospora crispa (L.) Hook.f. & Thoms. 164
Stephania corymbosa Walp. 165
4.1.6 Superorder Proteanae Takht. (1967), the Eudicots 166
4.1.6.1 Order Proteales Juss. ex Bercht. & J. Presl (1820) 166
4.1.6.1.1 Family Proteaceae Juss. (1789), the Protea Family 166
Helicia serrata Bl. 166
4.1.7 Superorder Dillenianae Takht. ex Doweld (2001), the Core Eudicots 168
4.1.7.1 Order Dilleniales DC. ex Bercht. & J. Presl (1820) 168
4.1.7.1.1 Family Dilleniaceae Salisb. (1807), the Elephant Apple Family 168
Dillenia excelsa (Jack) Martelli 168
Dillenia grandifolia Wall. ex Hook. f. & Thomson 169
Dillenia indica L. 169
Tetracera akara Merr. 171
Tetracera indica (Christm. & Panz.) Merr. 172
Tetracera scandens (L.) Merr. 174
4.1.8 Superorder Myrothamnanae Takht. (1997), the Core Eudicot 175
4.1.8.1 Order Saxifragales Bercht. & J. Presl (1820) 175
4.1.8.1.1 Family Crassulaceae J.St.-Hil. (1805), the Stonecrop Family 175
Kalanchoe pinnata (Lam.) Pers. 175
4.1.9 Superorder Rosanae Takht. (1967), the Rosids 177
4.1.9.1 Order Vitales Juss. ex Bercht. & J. Presl (1820)................. 177
4.1.9.1.1 Family Vitaceae Juss. (1789), the Grape Family 177
Leea indica (Burm. f.) Merr. 177
Tetrastigma leucostaphylum (Dennst.) Alston... 178
Tetrastigma diepenhorstii (Miq.) Latiff 179
Vitis trifolia L. 179
4.1.10 Superorder Rosanae Takht. (1967), the Fabids 181
4.1.10.1 Order Cucurbitales Juss. ex Bercht. & J. Presl (1820)......... 181
4.1.10.1.1 Family Anisophylleaceae Ridl. (1922), the Anisophyllea Family.................................. 181
Anisophyllea disticha Baill. 181
4.1.10.1.2 Family Cucurbitaceae Juss. (1789), the Gourd Family 182
Benincasa hispida (Thunb.) Cogn. 182
Luffa cylindrica M. Roem. 183
Momordica charantia L. 184
Trichosanthes cucumerina L. 186
4.1.10.1.3 Family Tetramelaceae Airy Shaw (1965), the Tetramela Family 188
Octomeles sumatrana Miq. 188
4.1.10.2 Order Fabales Bromhead (1838).. 189

4.1.10.2.1 Family Fabaceae Lindley (1836), the Pea Family ... 189
Airyantha borneensis (Oliv.) Brummitt ... 189
Cassia alata L. ... 190
Millettia nieuwenhuis J.J. Smith ... 192
Mimosa pudica L. ... 192
Vigna unguiculata (L.) Walp. ... 194
4.1.10.2.2 Family Polygalaceae Hoffmanns. & Link (1809), the Milkwort Family ... 196
Polygala paniculata L. ... 196
Xanthophyllum excelsum (Blume) Miq. ... 198
4.1.10.3 Order Fagales Engl. (1892) ... 200
4.1.10.3.1 Family Casuarinaceae R.Br. (1814), the Casuarina Family ... 200
Casuarina sumatrana Jungh. ex de Vriese. ... 200
4.1.10.4 Order Malpighiales Juss, ex Bercht. & J. Presl (1820) ... 200
4.1.10.4.1 Family Clusiaceae Lindley (1836), the Garcinia Family ... 200
Garcinia mangostana L. ... 200
4.1.10.4.2 Family Dichapetalaceae Baill. (1886), the Dichapetalum Family ... 203
Dichapetalum gelonioides (Roxb.) Engl. ... 203
4.1.10.4.3 Family Euphorbiaceae Juss. (1789), the Spurge Family ... 205
Antidesma montanum Bl. ... 205
Baccaurea lanceolata (Miq.) Müll. Arg. ... 206
Bischofia javanica Bl. ... 207
Bridelia stipularis (L.) Bl. ... 209
Euphorbia hirta L. ... 210
Euphorbia prostrata Aiton ... 212
Glochidion macrostigma Hook.f. ... 213
Homalanthus populneus (Geiseler) Pax & Prantl ... 213
Jatropha curcas L. ... 215
Jatropha podagrica Hook. ... 217
Macaranga gigantea (Zoll.) Müll.Arg. ... 218
Macaranga gigantifolia Merr. ... 220
Macaranga tanarius (L.) Müll. Arg. ... 221
Mallotus paniculatus (Lam.) Müll. Arg. ... 222
Manihot esculenta Crantz ... 223
Phyllanthus niruri L. ... 224
Sauropus androgynus (L.) Merr. ... 226
4.1.10.4.4 Family Flacourtiaceae Rich ex DC. (1824), the Flacourtia Family ... 228
Casearia grewiifolia Vent. ... 228
4.1.10.4.5 Family Rhizophoraceae Pers. (1806), the Rhizophora Family ... 231
Bruguiera parviflora Wight ... 231
Rhizophora apiculata Bl. ... 232

4.1.10.5 Order Oxalidales Bercht. & J. Presl. (1820) ... 234
4.1.10.5.1 Family Connaraceae R.Br. (1818), the Kana Family ... 234
Cnestis platantha Griff. ... 234
4.1.10.5.2 Family Elaeocarpaceae Juss. ex DC. (1816), the Elaeocarpus Family ... 236
Elaeocarpus clementis Merr. ... 236
4.1.10.6 Order Rosales Bercht. & J. Presl. (1820) ... 237
4.1.10.6.1 Family Moraceae Link (1831), the Mulberry Family ... 237
Artocarpus elasticus Reinw ex Bl. ... 237
Artocarpus tamaran Becc. ... 238
Ficus deltoidea Jack ... 238
Ficus elliptica Hook ex Miq. ... 240
Ficus lepicarpa Bl. ... 240
Ficus septica Burm.f. ... 241
4.1.10.6.2 Family Rhamnaceae Juss. (1789), the Buckthorn Family ... 243
Alphitonia incana (Roxb.) Teijsm. & Binn. ex Kurz ... 243
Ziziphus horsfieldii Miq. ... 245
4.1.10.6.3 Family Rosaceae Juss. (1789), the Rose Family ... 245
Prunus arborea (Bl.) Kalkman ... 245
4.1.10.6.4 Family Urticaceae Juss. (1789), the Nettle Family ... 247
Dendrocnide elliptica (Merr.) Chew ... 247
Poikilospermum cordifolium (Barg.-Petr.) Merr. ... 248
Poikilospermum suaveolens (Bl.) Merr. ... 249
4.1.11 Superorder Rosanae Takht. (1967), the Malvids ... 250
4.1.11.1 Order Brassicales Bromhead (1838) ... 250
4.1.11.1.1 Family Caricaceae Dumort. (1829), the Papaya Family ... 250
Carica papaya L. ... 250
4.1.11.1.2 Family Cleomaceae Bercht. & J. Presl (1820), the Spider Flower Family ... 252
Cleome chelidonii L.f. ... 252
4.1.11.2 Order Malvales Juss. ex Bercht. & J. Presl (1820) ... 253
4.1.11.2.1 Family Bixaceae Kunth (1822), the Achiote Family ... 253
Bixa orellana L. ... 253
4.1.11.2.2 Family Bombacaceae Kunth (1822), the Bombax Family ... 254
Bombax ceiba L. ... 254
Ceiba pentandra (L.) Gaertn. ... 256
Durio zibethinus Rumph. ex Murray ... 257
4.1.11.2.3 Family Dipterocarpaceae Blume (1825), the Dipterocarp Family ... 258
Shorea macroptera Dyer ... 258
Shorea parvistipulata F. Heim ... 259

4.1.11.2.4 Family Malvaceae Juss. (1789), the Mallow Family 260
Abelmoschus esculentus L. 260
Urena lobata L. 261
4.1.11.2.5 Family Tiliaceae Juss. (1789), the Jute Family 263
Microcos antidesmifolia (King) Burret 263
4.1.11.2.6 Family Sterculiaceae Vent. (1807), the Cocoa Family 264
Theobroma cacao L. 264
4.1.11.2.7 Family Thymeleaceae Juss. (1789), the Eaglewood Family 266
Wikstroemia androsaemifolia Hand.-Mazz. 266
Wikstroemia ridleyi Gamble 267
4.1.11.3 Order Myrtales Juss. ex Bercht. & J. Presl (1820) 268
4.1.11.3.1 Family Combretaceae R.Br. (1810), the Indian Almond Family 268
Combretum nigrescens King 268
4.1.11.3.2 Family Lythraceae J. St.-Hil. (1805), the Loosestrife Family 268
Lawsonia inermis L. 268
4.1.11.3.3 Family Melastomataceae Juss. (1789), the Melastomes Family 270
Clidemia hirta (L.) D. Don 270
Dissochaeta monticola Bl. 271
Melastoma malabathricum L. 272
4.1.11.3.4 Family Myrtaceae Juss. (1789), the Myrtle Family 274
Psidium guajava L. 274
4.1.11.4 Order Sapindales Juss. ex Bercht. & J. Presl (1820) 276
4.1.11.4.1 Family Anacardiaceae R.Br. (1818), the Cashew Family 276
Mangifera caesia Jack 276
Mangifera pajang Kosterm. 277
Pegia sarmentosa (Lecomte) Hand.-Mazz. 278
Semecarpus cuneiformis Blanco 278
4.1.11.4.2 Family Rutaceae Juss. (1789), the Citrus Family 280
Clausena excavata Burm.f. 280
4.1.11.4.3 Family Sapindaceae Juss. (1789), the Soapberry Family 282
Guioa pleuropteris (Blume) Radlk. 282
Lepisanthes amoena (Hassk.) Leenh. 282
Mischocarpus pentapetalus (Roxb.) Radlk. 284
4.1.11.4.4 Family Simaroubaceae DC. (1811), the Quassia Family 285
Brucea javanica (L.) Merr. 285
Eurycoma longifolia Jack 287

4.1.12 Superorder Caryophyllanae Takht. (1967), the Malvids.....289
4.1.12.1 Order Caryophyllales Juss. ex Bercht. & J. Presl (1820).....289
4.1.12.1.1 Family Amaranthaceae Juss. (1789), the Amaranth Family.....289
Alternanthera sessilis (L.) R.Br. ex DC. 289
Amaranthus spinosus L. 290
Cyathula prostrata (L.) Bl. 292
4.1.12.1.2 Family Polygonaceae Juss. (1789), the Buckwheat Family.....294
Polygonum odoratum Lour. 294
Polygonum orientale L. 295
4.1.13 Superorder Asteranae Takht. (1967), the Asterids.....297
4.1.13.1 Order Ericales Bercht. & J. Presl (1820).....297
4.1.13.1.1 Family Actinidiaceae Engl. & Gilg. (1824), the Kiwi Fruit Family.....297
Saurauia fragrans Hoogland.....297
4.1.13.1.2 Family Balsaminaceae Bercht. & J. Presl (1820), the Touch-me-not Family.....297
Impatiens balsamina L. 298
4.1.13.1.3 Family Ebenaceae Gürke (1891), the Ebony Family.....300
Diospyros elliptifolia Merr. 300
4.1.13.1.4 Family Primulaceae Batsch ex Borkh (1797), the Primrose Family.....301
Embelia philippinensis A. DC.....301
4.1.14 Superorder Asteranae Takht. (1967), the Lamiids.....303
4.1.14.1 Order Gentianales Juss. ex Bercht. & J. Presl (1820).....303
4.1.14.1.1 Family Apocynaceae Juss. (1789), the Dogbane Family.....303
Alstonia angustifolia Wall. 303
Alstonia angustiloba Miq. 304
Alstonia macrophylla Wall. ex G.Don.....305
Alstonia scholaris (L.) R.Br. 306
Alstonia spatulata Bl. 308
Kopsia dasyrachis Ridl. 309
Plumieria acuminata W.T. Aiton.....311
Tabernaemontana macrocarpa Jack.....312
Tabernaemontana sphaerocarpa Bl. 313
4.1.14.1.2 Family Loganiaceae R.Br ex Mart (1827), the Logania Family.....315
Fagraea cuspidata Bl. 315
4.1.14.1.3 Family Rubiaceae Juss. (1789), the Madder Family.....315
Gardenia tubifera Wall. 315
Hedyotis auricularia L. 317
Hydnophytum formicarum Jack.....319
Ixora blumei (Bl.) Zoll. & Moritzi.....321
Morinda citrifolia L. 321
Mussaenda frondosa L. 323

Myrmecodia platytyrea Becc. 324
Neonauclea gigantea (Valeton) Merr. 325
Nauclea officinalis (Pierre ex Pitard) Merr. & Chun. 326
Oxyceros bispinosus (Griff.) Tirveng. 327
Paedaria verticillata Bl. 327
Psychotria gyrulosa Stapf 328
Ridsdalea pseudoternifolia (Valeton) J.T.Pereira 329
Uncaria acida (W. Hunter) Roxb. 329
4.1.14.2 Order Lamiales Bromhead (1838) 331
4.1.14.2.1 Family Acanthaceae Juss. (1789), the Acanthus Family 331
Clinacanthus nutans (Burm. f.) Lindau 331
Justicia gendarussa Burm.f. 333
4.1.14.2.2 Family Avicenniaceae Miq. (1845) 335
Avicennia marina (Forssk.) Vierh. 335
4.1.14.2.3 Family Bignoniaceae Juss. (1789), the Trumpet-creeper Family 337
Oroxylum indicum (L.) Vent. 337
4.1.14.2.4 Family Lamiaceae Martinov (1820), the Mint Family 339
Hyptis capitata Jacq. 339
Orthosiphon stamineus Benth. 340
4.1.14.2.5 Family Gesneriaceae Rich. & Juss. (1816), the Gerneriad Family 342
Cyrtandra areolata (Stapf) B.L. Burtt 342
4.1.14.2.6 Family Scrophulariaceae Juss. (1789), the Figwort Family 343
Scoparia dulcis L. 343
4.1.14.2.7 Family Verbenaceae Martinov (1820), the Vervain Family 345
Callicarpa longifolia Lam. 345
Clerodendrum philippinum Schauer 346
Lantana camara L. 347
Stachytarpheta jamaicensis (L.) Vahl 349
Vitex pubescens (L.) Vahl 350
4.1.14.3 Order Solanales Juss. ex Bercht. & J.Presl (1820) 351
4.1.14.3.1 Family Convolvulaceae Juss. (1789), the Morning Glory Family 351
Ipomoea aquatica Forssk. 351
Ipomoea batatas (L.) Lam. 353
Merremia peltata (L.) Merr. 354
4.1.14.3.2 Family Solanaceae Juss. (1789), the Nightshade Family 355
Capsicum frutescens L. 355
Nicotiana tabacum L. 357
Physalis minima L. 358
Solanum erianthum D.Don 360
Solanum ferox L. 361
Solanum melongena L. 362

Solanum nigrum L. 363
Solanum torvum Sw. 364
4.1.15 Superorder Asteranae Takht. (1967), the Campanulids 366
4.1.15.1 Order Asterales Link (1829) 366
4.1.15.1.1 Family Asteraceae Martinov (1820), the Aster Family 366
Ageratum conyzoides L. 366
Blumea balsamifera DC. 367
Blumea riparia DC. 368
Crassocephalum crepidioides (Benth.) S. Moore 370
Elephantopus mollis Kunth 371
Eupatorium odoratum L. 373
Synedrella nodiflora (L.) Gaertn. 374
4.1.15.2 Order Apiales Nakai (1930) 376
4.1.15.2.1 Family Apiaceae Lindley (1836), the Parsley Family 376
Centella asiatica (L.) Urb. 376
Eryngium foetidum L. 377
4.1.15.2.2 Family Araliaceae Juss. (1789), the Ginseng Family 378
Schefflera nervosa (King) R. Vig. 378
4.1.15.2.3 Family Pittosporaceae R.Br. (1814), the Cheesewood Family 379
Pittosporum ferrugineum W.T. Aiton 379

Bibliography 381

Index 387

Foreword

It is with great pleasure that I introduce this remarkable volume on the medicinal plants of Sabah (North Borneo) by Professor Christophe Wiart. This comprehensive work offers an unparalleled exploration of the botanical, ethnopharmacological, and pharmacological aspects of over 250 plants used in Sabah's traditional medicinal practices.

Christophe Wiart's dedication to this field is evident in the meticulous detail provided for each plant, including botanical classifications, traditional uses, pharmacological properties, and phytochemical analyses. His work not only consolidates existing knowledge but also sheds new light on the potential of these plants in modern medicine.

This book is a testament to Christophe Wiart's expertise and his invaluable contributions to the field of herbal medicine. I hope that his work will inspire more researchers to explore the rich botanical resources of Sabah, ultimately leading to the development of new herbal remedies and therapeutic molecules.

I congratulate Christophe Wiart on this outstanding achievement and thank him for his significant advancements in our understanding of the medicinal plants of Sabah. May this book serve as a beacon for future research in this fascinating field.

Edwin Liew
President of the Sabah Society

Preface

The northern part of Borneo encompassing the territory of Sabah stands at latitude 5° north, 117° east, with an area almost equivalent to that of Ireland. It used to be covered with a tropical primary rainforest of considerable extent, grace, and beauty, and possessed a biological richness that words cannot describe. Unfortunately, the precious timbers and resins did not escape the gaze of traders from commercially inclined powers, which explains the waves of foreign colonization of this little paradise, especially from about the 7th century AD. It is said by some that North Borneo has been associated with the Buddhist Srivijaya Empire since the 7th century and with the Hindu Javanese Empire of Majapahit from the 13th century, when trade and commercial dealings between North Borneo and China were established.

More direct commercialization of North Borneo's products became possible when neighboring sultanates backed by the Ming Dynasty shared various portions of its land from the 16th century. In the 19th century, the North Borneo Company acquired what is now known as the territory of Sabah from one Sultanate, legitimized according to British law for their purposes of acquisition and ran it as an independent enclave under British Protection until 1941, when this land was for a short time part of the Empire of Japan.

After WW2, its pre-war independent status like Sarawak was not recognized, but they became colonies of Britain, which engineered their becoming part of Malaysia in 1963. Sabah is treated operationally as a mere state, like the other states in Malaya, although legally its status is markedly different.

All of these changes and exploitations of local resources, especially after 1963, were witnessed and endured by the many indigenous native tribes of North Borneo, which include the Dusun, Kadazan, Murut, and Rungus. The most painful and despicable events that these indigenous people had to courageously witness and endure was the almost complete deforestation of their sacred tropical rainforest to make way for oil palm and agro plantations in the 1980s and 1990s.

Another catastrophe of ethnobotanic significance was the simultaneous destruction of their demography by the importation of illegal and alien migrants from the surrounding lands which culturally displaced these people and their rights to the land. The native ethnic groups, unlike the migrants, have a very deep attachment and great respect for the trees, the rivers, the majestic and mystical Mount Kinabalu, and nature in general, and they remain the main source of ethnobotanic knowledge and information.

While walking and talking to these friendly and kindly people in their villages, one may be surprised to see men and women over 60 years old without gray hair, in good health, strong, and without apparent chronic illnesses. The Dusun and Kadazan or "people of the orchards" used to live on the slope of Mount Kinabalu and rely on plants on a daily basis, and we can see among them elderly men and women climbing hills quickly while carrying on their backs very heavy loads with little effort. When enquiring about the source of their strength and good health, we often hear them say, "my grandmother was a shaman and knew about medicinal plants."

For the last decade, Sabah has, and still attracts, the migration of an enormous population of Filipinos and Indonesians, and combined with globalization, would all contribute to the displacement and extinction of the culture and knowledge of the historical indigenous peoples and their contribution to the science of pharmacognosy and ethnobotany.

Since the field of medicinal plants for maintaining the health of the elderly is of worldwide interest, and since the medicinal knowledge of the ethnic groups of Sabah is vanishing at an alarming rate, the writing of *Medicinal Plants of Sabah, North Borneo* seems both timely and necessary. This book presents the medicinal plants used by natives as well as non-natives in a phylogenetic manner, within their respective orders and families, and for each plant, the scientific names are given, as well as local names, common names, habitat, distribution, and detailed botanical descriptions. For each plant, medicinal uses are listed along with what is known about their pharmacological properties and possible adverse effects. Botanical plates are provided, as well as comments, carefully selected references, and a list of consulted literature for post-graduate students and academics.

Writing this book took about 2 years of daily work, during which I realized that some medicinal plants of Sabah have the potential to be used for the development of medecines and herbal remedies for better health in aging, and also for diabetes, obesity, cancer, and other diseases. However, the development of such drugs might never be realized because of administrative red tape, the absence of local research funds, and the impossibility of working with large pharmaceutical laboratories and foreign universities because of the Nagoya Protocol. We must understand that these senseless obstacles will result in the inability to develop drugs that will allow at least the poor among us to survive the global health upheavals that sooner or later will shake up a world that is rapidly changing. COVID-19 is a toddler compared to the novel bacteria, fungi, and viruses that are anticipated to appear, due in no small part to deforestation and ecological violations. These malpractices may be compared, contrasted, and gauged with the active experimentation and intervention of natives over the flora and fauna of their habitat.

I would like to express my gratitude to Universiti Malaysia Sabah for providing favorable conditions for the writing of this book and I would like to particularly thank the indigenous villagers of Sabah for their warm welcome, kindness, honesty, and guidance.

Christophe Wiart PharmD, PhD
Universiti Malaysia Sabah
Kota Kinabalu, Sabah, Malaysia
November 30, 2023

Author Biography

Christophe Wiart was born on 12th of August, 1967 in Saint Malo, France. After his A-levels, he completed his Pharm.D. at the Facultée des Sciences Pharmaceutiques et Biologiques, Université Rennes 2 (France) and earned his Ph.D. in natural products chemistry at the Universiti Pertanian Malaysia. He has taught pharmacognosy at the University of Malaya, and elsewhere. Dr. Wiart is the author of *Medicinal Plants of the Asia-Pacific: Drugs from the Future?* (2006), *Medicinal Plants of Asia and the Pacific* (2006), *Ethnopharmacology of Medicinal Plants: Asia and the Pacific* (2006), *Medicinal Plants from the East* (2010), *Medicinal Plants from China, Korea and Japan: Bioresource for Tomorrow's Drug and Cosmetic Discovery* (2012), *Lead Compounds from Medicinal Plants for the Treatment of Cancer* (2012), *Lead Compounds from Medicinal Plants for the Treatment of Neurodegenerative Diseases* (2013), *Medicinal Plants in Asia for Metabolic Syndrome* (2018), *Medicinal Plants from West Bengal and Bangladesh* (2019), *Medicinal Plants in Asia and Pacific for Parasitic Infections: Botany, Ethnopharmacology, Molecular Basis, and Future Prospect* (2020), *Medicinal Plants in Asia and the Pacific for Zoonotic Pandemics* (2021), and *Handbook of Medicinal Plants of the World for Aging* (2023).

He has published numerous articles. Other current research interests include the ethnopharmacological study of the medicinal plants of Southeast Asia for the development of herbal remedies and lead therapeutic compounds.

1 The Medicinal Plants of Sabah (North Borneo)

The Lycophytes

1.1 SUBCLASS LYCOPODIIDAE BEK. (1862)

1.1.1 Order Lycopodiales DC. ex Bercht. & J. Presl (1820)

1.1.1.1 Family Lycopodiaceae P. Beauv. ex Mirb. (1802), the Club Moss Family

The family Lycopodiaceae consists of about 5 genera and 400 species of club mosses growing from rhizomes or corms. Stems: dichotomously branched, somewhat like those of algae. Leaves: simple, tiny, sessile, spiral, or whorled. Ligules: none. Sporangia: reniform at the axis of sporophylls. Spores: trilete. Gametophytes: subterranean.

Lycopodium cernuum **L.**

[From Greek *lycos* = wolf, *pous* = foot, and Latin *cernuum* = bowing]

Published in *Species Plantarum* 2: 1103. 1753

Synonyms: *Lepidotis cernua* P. Beauv.; *Lycopodiella cernua* (L.) Pic. Serm.; *Lycopodium sikkimense* Mull. Hal.; *Palhinhaea cernua* (L.) Franco & Vasc.; *Palhinhaea lufengensis* C.Y. Yang

Local names: *dapok dapokan, tapok tapokan, sembunyi, kerudop* (Kedayan), *gogor gogor, susut susut* (Dusun)

Common name: stag's-horn clubmoss

Habitat: hill slopes, roadsides, wet soils, at the edges of forests and swamps

Geographical distribution: subtropical, tropical

Botanical description: this club moss is erect and reaches a height of about 60 cm. This plant somewhat has the appearance of a miniature fir. Stems: dichotomously branched. Leaves: simple, linear, sessile, spirally arranged, about 5 mm long, subulate, rounded at base, and acuminate at apex. Strobili: solitary and terminal, pendulous, and about 1 cm long. Sporophylls: rhombic, acuminate, serrate, and tiny. Sporangia: globose, enclosed (Figure 1.1).

Traditional therapeutic indications: asthma, chest pain, coughs (Kedayan); fever, hypertension, stroke, canker sores (Dusun)

Pharmacology and phytochemistry: serratane-type triterpenes (Liu et al. 2022; Zhang et al. 2022). Other constituents are lycopodium alkaloids including cernuine (acetylcholinesterase, $IC_{50:}$ 32.7 μM) and lycopodine (toxic to cervical cancer cells) (Ma & Gang 2004; Morita et al. 2004; Mandal et al. 2010; Konrath et al. 2013).

Toxicity, side effects, and drug interaction: poisonous (Jaspersen-Schib et al. 1996).

DOI:10.1201/9781003402886-1

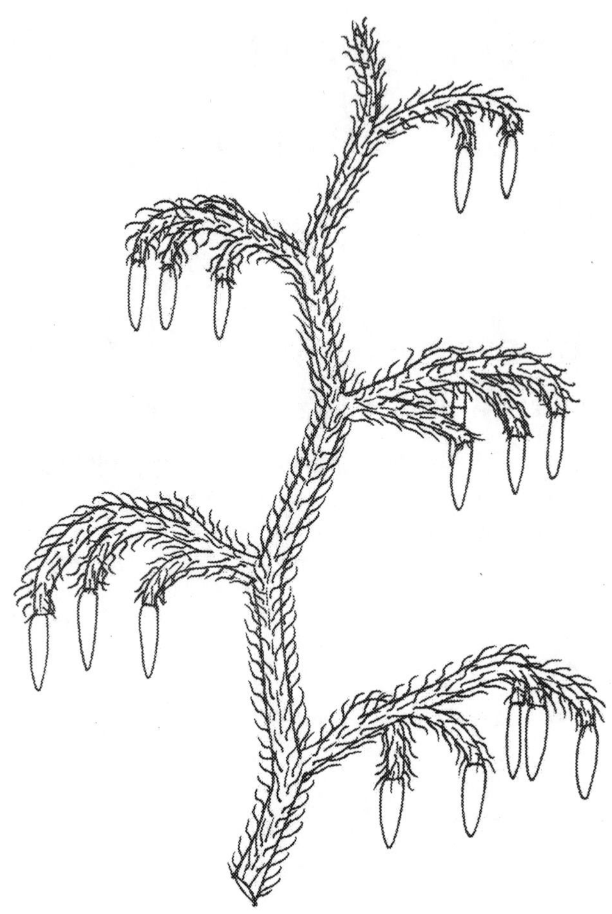

FIGURE 1.1 *Lycopodium cernuum* L.

Comment: Someone might say that the name of this plant has recently been changed to *Lycopodiella cernua* (L.) Pic. Serm. The incessant modification of the scientific names of plants by committees of botanists is a source of confusion, and we therefore retain in this volume the most commonly used scientific names of plants. Also, today's plant taxonomists have an unpleasant habit, without global consultation, of creating catch-all families. This was done with the merger of Apocynaceae and Asclepadiaceae or Lamiaceae with Verbenaceae, despite obvious morphological differences.

REFERENCES

Jaspersen-Schib, R., Theus, L., Guirguis-Oeschger, M., Gossweiler, B. and Meier-Abt, P.J. 1996. Serious plant poisonings in Switzerland 1966–1994. Case analysis from the Swiss toxicology information center. *Schweizerische Medizinische Wochenschrift*. 126(25): 1085–1098.

Konrath, E.L., Passos, C.D.S., Klein-Júnior, L.C. and Henriques, A.T. 2013. Alkaloids as a source of potential anticholinesterase inhibitors for the treatment of Alzheimer's disease. *Journal of Pharmacy and Pharmacology*. 65(12): 1701–1725.

Liu, B.R., Zheng, H.R., Jiang, X.J., Zhang, P.Z. and Wei, G.Z. 2022. Serratene triterpenoids from *Lycopodium cernuum* L. as α-glucosidase inhibitors: Identification, structure—activity relationship and molecular docking studies. *Phytochemistry*. 195: 113056.

Ma, X. and Gang, D.R. 2004. The lycopodium alkaloids. *Natural Product Reports.* 21(6): 752–772.

Mandal, S.K., Biswas, R., Bhattacharyya, S.S., Paul, S., Dutta, S., Pathak, S. and Khuda-Bukhsh, A.R. 2010. Lycopodine from *Lycopodium clavatum* extract inhibits proliferation of HeLa cells through induction of apoptosis via caspase-3 activation. *European Journal of Pharmacology.* 626(2–3): 115–122.

Morita, H., Hirasawa, Y., Shinzato, T. and Kobayashi, J.I. 2004. New phlegmarane-type, cernuane-type, and quinolizidine alkaloids from two species of *Lycopodium. Tetrahedron.* 60(33): 7015–7023.

Zhang, Z., ElSohly, H.N., Jacob, M.R., Pasco, D.S., Walker, L.A. and Clark, A.M. 2002. Natural products inhibiting *Candida albicans* secreted aspartic proteases from *Lycopodium cernuum. Journal of Natural Products.* 65(7): 979–985.

***Lycopodium phlegmaria* L.**

[From Greek *lycos* = wolf, pous = foot, *phlegma* = flame, and *oura* = tail]

Published in *Species Plantarum* 2: 1101. 1753

Synonyms: *Huperzia phlegmaria* (L.) Rothm.; *Lepidotis phlegmaria* (L.) P. Beauv.; *Phlegmariurus phlegmaria* (L.) Holub; *Urostachys phlegmaria* (L.) Herter ex Nessel
Local name: *sari gading bini* (Kedayan)
Common name: common tassel fern
Habitat: mountainsides
Geographical distribution: subtropical and tropical Asia and Africa
Botanical description: this epiphytic club moss is pendulous and grows up to a length of about 60 cm. Stems: dichotomously branched. Leaves: simple, subsessile, spiral, up to about 1.5 cm long, ovate to lanceolate, somewhat knifelike, glossy, fleshy, rounded to truncate at base, asymmetrical, sharply triangular at apex. Strobili: terminal, pendulous, dichotomously branched, up to about 15 cm long. Sporophylls: ovate to somewhat deltoid, acuminate, serrate, tiny. Sporangia: reniform, grooved. Spores: trilete (Figure 1.2).
Traditional therapeutic indications: hair wash, alopecia (Kedayan)

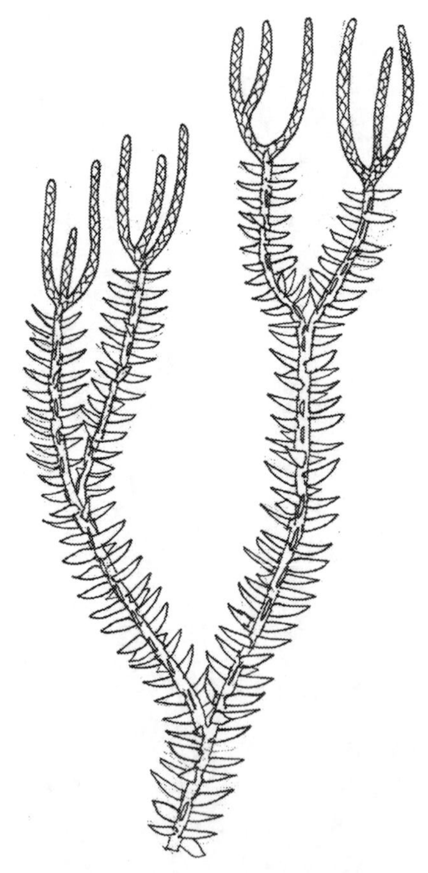

FIGURE 1.2 *Lycopodium phlegmaria* L.

Pharmacology and phytochemistry: serratane-type triterpenes (lycophlegmariol B) toxic to acute lymphoblastic leukemia cells (Shi et al. 2005; Wittayalai et al. 2012). Lycopodium alkaloid with anti-acetylcholinesterase activity (huperphlegmine A) (Nguyen et al. 2018).
Toxicity, side effects, and drug interaction: poisonous (Jaspersen-Schib et al. 1996).

Comment: Plants of the genus *Lycopodium* L. (1753) are medicinal for the Dusun and Kadazan: "*rongilut*" for feverish colds and "*pogou*" for burns.

REFERENCES

Jaspersen-Schib, R., Theus, L., Guirguis-Oeschger, M., Gossweiler, B. and Meier-Abt, P.J. 1996. Serious plant poisonings in Switzerland 1966–1994. Case analysis from the Swiss toxicology information center. *Schweizerische Medizinische Wochenschrift*. 126(25): 1085–1098.

Nguyen, H.T., Doan, H.T., Ho, D.V., Pham, K.T., Raal, A. and Morita, H. 2018. Huperphlegmines A and B, two novel *Lycopodium* alkaloids with an unprecedented skeleton from *Huperzia phlegmaria*, and their acetylcholinesterase inhibitory activities. *Fitoterapia*. 129: 267–271.

Shi, H., Li, Z.Y. and Guo, Y.W. 2005. A new serratane-type triterpene from Lycopodium phlegmaria. *Natural Product Research*. 19(8): 777–781.

Wittayalai, S., Sathalalai, S., Thorroad, S., Worawittayanon, P., Ruchirawat, S. and Thasana, N. 2012. Lycophlegmariols A—D: Cytotoxic serratene triterpenoids from the club moss *Lycopodium phlegmaria* L. *Phytochemistry*. 76: 117–123.

1.1.2 Order Selaginellales Prantl (1854)

1.1.2.1 Family Selaginellaceae Willk. (1854), the Spike Moss Family

The family Selaginellaceae consists of the single genus *Selaginella* P. Beauv. (1804).

***Selaginella argentea* (Wall. ex Hook. & Grev.) Spring**

[From Latin *selago* = club moss and *argentea* = silvery]

Published in *Bulletin de l'Academie Royale des Sciences et Belles-lettres de Bruxelles* 10(1): 137. 1843

Synonym: *Lycopodium argenteum* Wall. ex Hook. & Grev.
Local name: *sondotnulogo* (Murut)
Habitat: forests
Geographical distribution: Southeast Asia
Botanical description: this spike moss grows up to a length of about 30 cm. Stems: stiff, somewhat dichotomous, quadrangular, and develop aerial roots (rhizophores). Leaves: simple, alternate, lanceolate, scalelike, up to 3 mm long, ciliate, pellucid at margin. Strobili: terminal, squarish. Sporophylls: ovate, acute, pellucid at margin, serrulate (Figure 1.3).
Traditional therapeutic indications: asthma, body pains, fever, headaches (Murut)
Pharmacology and phytochemistry: not known
Toxicity, side effects, and drug interaction: not known

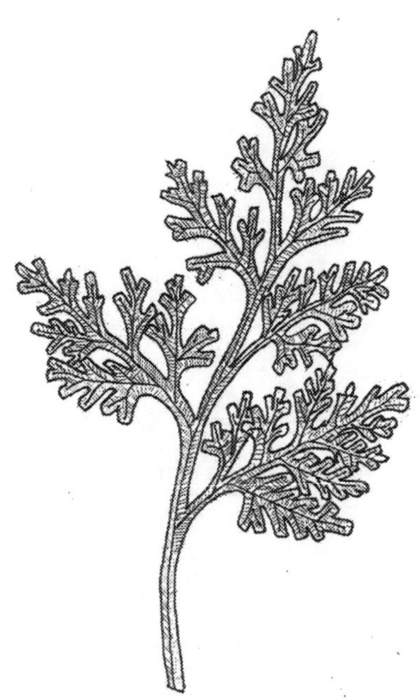

FIGURE 1.3 *Selaginella argentea* (Wall. ex Hook. & Grev.) Spring.

Selaginella plana (Desv. ex Poir.) Hieron.

[From Latin *selago* = club moss and *planus* = flat]

Published in *Die Natürlichen Pflanzenfamilien* 1(4): 703. 1901

Synonym: *Lycopodium planum* Desv. ex Poir.
Common name: Asian spikemoss
Habitat: forests
Geographical distribution: from Malaysia to Pacific Islands
Botanical description: this spike moss grows up to a length of about 50 cm. Stems: stiff, somewhat dichotomous. Leaves: simple, alternate, lanceolate to oblong, up to about 2 mm × 5 mm, and scalelike. Strobili: terminal, elongated, light green, up to about 5 cm long. Sporophylls: ovate to acuminate, acute, pellucid at margin, serrulate (Figure 1.4).
Traditional therapeutic indications: fever, headaches (Murut)
Pharmacology and phytochemistry: antibacterial (Novita et al. 2020), toxic to breast cancer cells (Handayani et al. 2013).
Toxicity, side effects, and drug interaction: not known

FIGURE 1.4 *Selaginella plana* (Desv. ex Poir.) Hieron.

REFERENCES

Handayani, S., Hermawan, A., Meiyanto, E. and Udin, Z. 2013. Induction of apoptosis on MCF-7 cells by selaginella fractions. *Journal of Applied Pharmaceutical Science*. 3(4): 31–34.

Novita, H., Novitaningrum, A.D., Madusari, B.D., Nugraha, M.F.I., Rajamuddin, M.A., Yunita, R. and Enggarini, W. 2002. Antibacterial activity of *Anadendrum microstachyum* and *Selaginella plana* against two pathogenic bacteria (*Edwardsiella ictaluri* and *Streptococcus agalactiae*) causing freshwater fish diseases. In *IOP Conference Series: Earth and Environmental Science* (535, No. 1: 012025). IOP Publishing.

2 The Medicinal Plants of Sabah (North Borneo)
The Monilophytes

2.1 SUBCLASS OPHIOGLISSIDAE KLINGE (1832)

2.1.1 ORDER OPHIOGLOSSALES LINK (1833)

2.1.1.1 Family Ophioglossaceae Martinov (1820), the Adder's Tongue Family

The family Ophioglossaceae consists of about 4 genera and 80 species of ferns. Rhizomes: erect. The frond consists of a stipe, a simple or compound blade (trophophore), and a spike containing sporangia (sporophore). Sporangia: exposed, or spores: trilete.

***Helminthostachys zeylanica* (L.) Hook.**

[From Greek *helmin* = worm, *stachys* = a spike, and Latin *zeylanica* = from Sri Lanka]

Published in *Genera Filicum*, pl. 47, B. 1842

Synonyms: *Botrychium zeylanicum* (L.) Sw.; *Helminthostachys dulcis* Kaulf.; *Ophiala zeylanica* (L.) Desv.; *Osmunda zeylanica* L.
Local name: *pajerok* (Lundayeh)
Habitat: moist and shady soils, on the banks of streams
Geographical distribution: from India to Pacific Islands
Botanical description: this fern reaches a height of about 60 cm. Rhizomes: erect. Stipes: 10–50 cm long, blackish. Blades: somewhat palmate, each foliole narrowly lanceolate, 1.5–3 cm × 6–20 cm, membranous, wavy, irregularly serrulate at margin, glossy, asymmetrical at base, with numerous thin veinations. Spikes: light yellowish–green, 4–20 cm long, containing sporangia (sporophore). Sporangia: exposed. Spores: trilete (Figure 2.1).
Traditional therapeutic indications: cancer (Lundayeh)
Pharmacology and phytochemistry: anti-inflammatory (Liou et al. 2017), hepatoprotective (Suja et al. 2004), insulin resistance (Chang et al. 2019), gastric cancer (Tsai et al. 2021), osteogenesis (Su et al. 2022). Cytotoxic prenylated stilbenes (ugonstilbene A) (Chen et al. 2003; Lin et al. 2023), prenylated flavonols with anti-inflammatory activity (ugonin M) (Huang et al. 2009; Wu et al. 2017), and osteogenic properties (ugonin K) (Lee et al. 2011).
Toxicity, side effects, and drug interaction: not known

Comments:

(i) In the Subclass Marattiidae Klinge (1882), order Marattiales Link (1833), family Marattiaceae Kaulf. (1824). A plant of the genus *Angiopteris* (1753). is sold as medicine in the wet markets of Sabah.
(ii) The Dusun use *Angiopteris evecta* (G. Forst.) Hoffm. for diarrhea and swellings (local names: *paku tiou, radou*).
(iii) In the Subclass Equisetidae Warm. (1883), order Equisetales DC. ex Bercht. & J. Presl

DOI: 10.1201/ 9781003402886-2

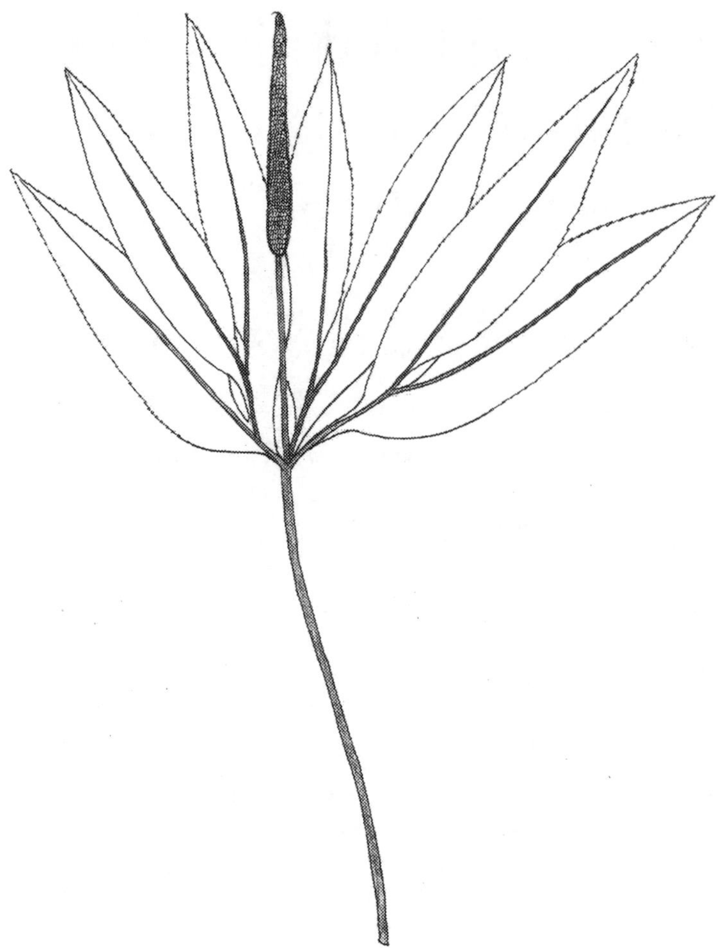

FIGURE 2.1 *Helminthostachys zeylanica* (L.) Hook.

(1820), family Equisetaceae Michx. ex DC. (1804), *Equisetum ramosissimum* Desf. is used for fever by the Dusun (local name*: langod langod*).

REFERENCES

Chang, T.C., Chiang, H., Lai, Y.H., Huang, Y.L., Huang, H.C., Liang, Y.C., Liu, H.K. and Huang, C. 2019. *Helminthostachys zeylanica* alleviates hepatic steatosis and insulin resistance in diet-induced obese mice. *BMC Complementary and Alternative Medicine*. 19: 1–11.

Chen, C.C., Huang, Y.L., Yeh, P.Y. and Ou, J.C. 2003. Cyclized geranyl stilbenes from the rhizomes of *Helminthostachys zeylanica. Planta Medica*. 69(10): 964–967.

Huang, Y.C., Hwang, T.L., Chang, C.S., Yang, Y.L., Shen, C.N., Liao, W.Y., Chen, S.C. and Liaw, C.C. 2009. Anti-inflammatory flavonoids from the rhizomes of *Helminthostachys zeylanica. Journal of Natural Products*. 72(7): 1273–1278.

Lee, C.H., Huang, Y.L., Liao, J.F. and Chiou, W.F. 2011. Ugonin K promotes osteoblastic differentiation and mineralization by activation of p38 MAPK-and ERK-mediated expression of Runx2 and osterix. *European Journal of Pharmacology*. 668(3): 383–389.

Lin, C.T., Yang, Y.H., Cheng, J.J. and Don, M.J. 2023. Total syntheses, absolute configurations, and cytotoxicity evaluation of ugonstilbenes a, b, and c from the rhizomes of Helminthostachys zeylanica. *Journal of Natural Products*. 86(2): 307–316.

Liou, C.J., Huang, Y.L., Huang, W.C., Yeh, K.W., Huang, T.Y. and Lin, C.F. 2017. Water extract of *Helminthostachys zeylanica* attenuates LPS-induced acute lung injury in mice by modulating NF-κB and MAPK pathways. *Journal of Ethnopharmacology*. 199: 30–38.

Su, C.H., Chen, Y.C., Yang, Y.H., Wang, C.Y., Ko, P.W., Huang, P.J., Liaw, C.C., Liao, W.L., Cheng, T.L., Lee, D.Y. and Lo, L.C. 2022. Effect of the traditional Chinese herbs *Helminthostachys zeylanica* on post-surgical recovery in patients with ankle fracture: A double-blinded randomized controlled clinical trial. *Journal of Ethnopharmacology*. 295: 115435.

Suja, S.R., Latha, P.G., Pushpangadan, P. and Rajasekharan, S. 2004. Evaluation of hepatoprotective effects of *Helminthostachys zeylanica* (L.) Hook against carbon tetrachloride-induced liver damage in Wistar rats. *Journal of Ethnopharmacology*. 92(1): 61–66.

Tsai, M.M., Lin, H.C., Yu, M.C., Lin, W.J., Chu, M.Y., Tsai, C.C. and Cheng, C.Y. 2021. Anticancer effects of *Helminthostachys zeylanica* ethyl acetate extracts on human gastric cancer cells through downregulation of the TNF-α-activated COX-2-cPLA2-PGE2 pathway. *Journal of Cancer*. 12(23): 7052.

Wu, K.C., Huang, S.S., Kuo, Y.H., Ho, Y.L., Yang, C.S., Chang, Y.S. and Huang, G.J. 2017. Ugonin M, a *Helminthostachys zeylanica* constituent, prevents LPS-induced acute lung injury through TLR4-mediated MAPK and NF-κB signaling pathways. *Molecules*. 22(4): 573.

2.2 SUBCLASS POLYPODIIDAE CRONQUIST, TAKHT. & W. ZIMM. (1966)

2.2.1 Order Blechnales Pic. Sem. ex Reveal (1993)

2.2.1.1 Family Blechnaceae Newman (1844), the Chain Fern Family

The family Blechnaceae includes 14 genera and about 250 species of rhizomatous ferns. Stipes: thin, coriaceous. Blades: pinnate, pinnatifid, or bipinnatifid. Sori: elongated or continuous along vascular network. Spores: elliptic, bilateral, monolete.

***Blechnum orientale* L.**

[From Greek *blechnon* = a fern and Latin *orientale* = from the Orient]

Published in *Species Plantarum* 2: 1077. 1753

Synonym: *Blechnopsis orientalis* (L.) C. Presl
Local names: *dudugau, dungau, tungau* (Dusun, Kadazan); *garintik* (Dusun); *paku* (Murut)
Common name: centipede fern
Habitat: hillsides
Geographical distribution: tropical Asia, Pacific Islands
Botanical description: this rhizomatous fern reaches a height of about 2 m. Stipes: 10–60 cm long, scaly to hairy at base. Blades: imparipinnate, about 1 m long. Folioles: about 30 pairs, linear, reddish to green, subopposite, sessile, 0.8–1.8 cm × 10–30 cm, asymmetrical at base, pointed at apex, light red at first, somewhat fleshy, with innumerable thin nerves. Sori: linear. Spores: elliptic, smooth (Figure 2.2).

FIGURE 2.2 *Blechnum orientale* L.

Traditional therapeutic indications: abscesses, boils, insect stings, flatulence, headaches, wounds, ulcers (Dusun, Kadazan)

Pharmacology and phytochemistry: wound healing (Lai et al. 2011), anticandidal, anti-inflammatory (Amoroso et al. 2014). Antibacterial proanthocyanidin oligomers (Lai et al. 2017).

Toxicity, side effects, and drug interaction: not known

Comment: This plant is used as food in Indonesia.

REFERENCES

Amoroso, V.B., Antesa, D.A., Buenavista, D.P. and Coritico, F.P. 2014. Antimicrobial, antipyretic, and anti-inflammatory activities of selected Philippine medicinal pteridophytes. *Asian Journal of Biodiversity*. 5(1).

Lai, H.Y., Lim, Y.Y. and Kim, K.H. 2011. Potential dermal wound healing agent in *Blechnum orientale* Linn. *BMC Complementary and Alternative Medicine*. 11(1): 1–9.

Lai, H.Y., Lim, Y.Y. and Kim, K.H. 2017. Isolation and characterisation of a proanthocyanidin with antioxidative, antibacterial and anti-cancer properties from fern *Blechnum orientale*. *Pharmacognosy Magazine*. 13(49): 31.

***Stenochlaena palustris* (Burm.f.) Bedd.**

[From Latin *Stenochlaena* = narrow cloak and *palustris* = of the swamp]

Published in *Supplement to the Ferns of Southern India and British India* (l.): 26, pl. 201. 1876

Synonyms: *Acrostichum palustre* (Burm.f.) C.B. Clarke; *Acrostichum scandens* (Sw.) Hook.; *Chrysodium palustre* (Burm.f.) Luerss.; *Lomaria scandens* Willd.; *Lomariopsis palustris* (Burm.f.) Kuhn; *Lomariopsis scandens* Mett.; *Olfersia scandens* C. Presl; *Polypodium palustre* Burm.f.; *Pteris scandens* (Willd.) Roxb.; *Stenochlaena hainanensis* Ching & P.S. Chiu; *Stenochlaena scandens* J. Sm.

Local names: *lambiding, lembiding, lominding* (Dusun, Kadazan); *kuraunolot* (Murut)

Common name: climbing fern

Habitat: swamps, forests

Geographical distribution: Southeast Asia, Australia, Pacific Islands

Botanical description: this climbing fern grows up to a length of about 1.5 m. Rhizomes: stout, blackish. Both sterile and fertile fronds are present (dimorphic frond). Stipes: 10–80 cm long. Blades: 1–30 cm × 5–80 cm, pinnate. Folioles: 8–16 pairs, alternate, with numerous thin nervations, wavy, glossy, reddish–brown to somewhat light burgundy when young, 2–4.5 cm × 15–20 cm, attenuate at base, somewhat serrulate at margin, acuminate at apex. Sori: cover the whole beneath underside of fertile pinnae (acrostichoid). Spores: somewhat shaped like curry puffs (Figure 2.3).

Traditional therapeutic indications: postpartum, fever, skin diseases (Dusun, Kadazan)

FIGURE 2.3 *Stenochlaena palustris* (Burm.f.) Bedd.

Pharmacology and phytochemistry: toxic to liver cancer cells (Yanti et al. 2021). Prenylated glycoside (stenopaluside) (Liu et al. 1998), antibacterial flavonol glycosides (stenopalustroside A) (Liu et al. 1999).

Toxicity, side effects, and drug interaction: not known

Comment: The young leaves of this plant are used as food in Sabah.

REFERENCES

Liu, H., Orjala, J., Rali, T. and Sticher, O. 1998. Glycosides from *Stenochlaena palustris*. *Phytochemistry*. 49(8): 2403–2408.

Liu, H., Orjala, J., Sticher, O. and Rali, T. 1999. Acylated flavonol glycosides from leaves of *Stenochlaena palustris*. *Journal of Natural Products*. 62(1): 70–75.

Yanti, M.N., Rahmawati, I. and Herdwiani, W. 2021. Uji Aktivitas Sitotoksik herba Kelakai (*Stenochlaena palustris* (Burm.F.) Bedd.) terhadap Sel Kanker Hati HEPG2. *Jurnal Bioteknologi & Biosains Indonesia (JBBI)*. 8(2): 255–266.

2.2.2 Order Gleicheniales Link (1825)

2.2.2.1 Family Gleicheniaceae C. Presl (1825), the Forking Fern Family

The family Gleicheniaceae consists of about 5 genera and 15 species of rhizomatous ferns. Stipes: terete, soft. Blades: pinnate. The sori form 1–3 lines. Spores: angular to kidney-shaped, smooth.

Gleichenia truncata (Willd.) Sprain

[After the 18th century German botanist Wilhelm Friedrich von Gleichen and Latin *truncata* = mutilated]

Published in *Systema Vegetabilium, editio decima sexta* 4(1): 25. 1827

Synonyms: *Mertensia truncata* Willd.; *Sticherus truncatus* (Willd.) Nakai
Local names: *laputong, loputung* (Dusun)
Common name: creepy fingers fern
Habitat: highland roadsides, forests
Geographical distribution: from Malaysia to Papua New Guinea
Botanical description: this invasive fern grows up to a length of about 2 m. Rhizomes: creeping, scaly. Stipes: scandent and terete. Fronds: distant and erect. Blades: dichotomously divided, of a dull kind of green, pinnate. Folioles: up to about 20 cm long, numerous, straight, about 1 cm long, linear. Sori orbicular, exindusiate, in one line on each side of costule, with 4 or 5 sessile sporangia (Figure 2.4).

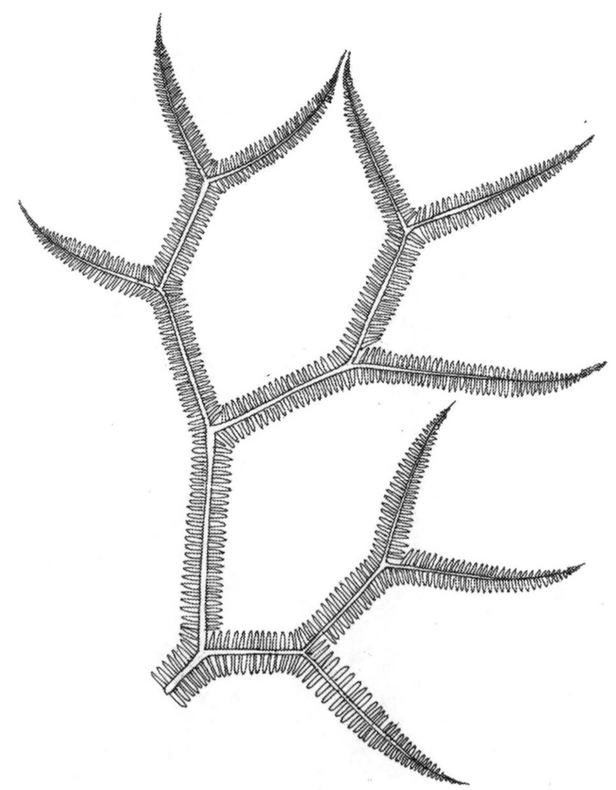

FIGURE 2.4 *Gleichenia truncata* (Willd.) Sprain.

Traditional therapeutic indications: medicinal (Dusun)
Pharmacology and phytochemistry: antibacterial, antimalarial, anti-inflammatory (Chai et al. 2013; Suhaini et al. 2015).
Toxicity, side effects, and drug interaction: not known

Comment: A plant of the genus *Dicranopteris* Bernh. (1805) is used by Dusun and Kadazan (local name: *kawang kawang*) for sore eyes.

REFERENCES

Chai, T.T., Elamparuthi, S., Yong, A.L., Quah, Y., Ong, H.C. and Wong, F.C. 2013. Antibacterial, anti-glucosidase, and antioxidant activities of selected highland ferns of Malaysia. *Botanical Studies*. 54: 1–7.

Suhaini, S., Liew, S., Norhaniza, J., Lee, P., Jualang, G., Embi, N. and Hasidah, M. 2015. Anti-malarial and anti-inflammatory effects of *Gleichenia truncata* mediated through inhibition of GSK3β. *Tropical Biomedicine*. 32(3): 419–433.

2.2.3 Order Polypodiales Link (1833)

2.2.3.1 Family Athyriaceae Alston (1956), the Lady Fern Family

The family Athyriaceae consists of about 5 genera and 600 species. Rhizomes: scaly. Stipes: hairy, scaly, or glabrous. Blades: simple or pinnate. Sori: variously shaped. Spores: ellipsoid, bilateral.

***Diplazium cordifolium* Bl.**

[From Greek *diplasios* = double and Latin *cor* = heart, and *folium* = leaf]

Published in *Enumeratio Plantarum Javae* 2: 190. 1828

Synonyms: *Anisogonium cordifolium* (Blume) Bedd.; *Asplenium cordifolium* (Blume) Mett.; *Athyrium cordifolium* (Blume) Copel.; *Callipteris cordifolia* (Blume) Copel.; *Oxygonium cordifolium* (Blume) J. Sm.
Local name: *giman* (Bajau)
Habitat: watery and shady places in forests
Geographical distribution: from Bangladesh to Pacific Islands
Botanical description: this fern reaches a height of about 30 cm. Rhizomes: scaly, stout. Stipes: 30–60 cm long, dark purplish, channeled above, somewhat scaly. Blades: simple or pinnate, cordate at base, somewhat fleshy, glossy, with numerous thin nerves, lanceolate to tongue-shaped, with a dark purplish midrib, wavy, acuminate at apex, up to about 13 cm × 30 cm. Sori: elongated, up to about 4 cm long (Figure 2.5).
Traditional therapeutic indications: colds, fever (Bajau)
Pharmacology and phytochemistry: not known
Toxicity, side effects, and drug interaction: not known

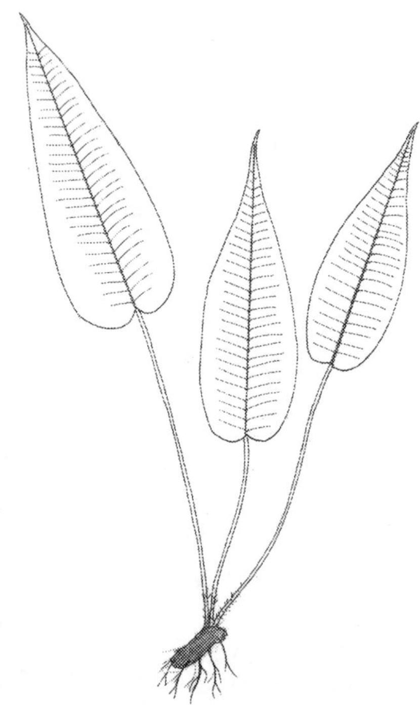

FIGURE 2.5 *Diplazium cordifolium* Bl.

2.2.3.2 Family Polypodiaceae Link J. Presl. & C. Presl. (1822), the Polypody Family

The family Polypodiaceae consists of about 50 genera and 1,200 species of ferns. Rhizomes: scaly. Stipes: articulate at base. Fronds: monomorphic or dimorphic. Blades: simple or pinnate. Sori: abaxial on venations and variously shaped. Spores: monolete, bilateral.

Drymoglossum piloselloides **(L.) C. Presl**

[From Greek *drymo* = forests, *glossa* = tongue, and Latin *pilus* =hair]

Published in *Tentamen Pteridographiae* 227, pl. 10, f. 5–6. 1836

Synonyms: *Elaphoglossum piloselloides* (L.) Keyserl.; *Lemmaphyllum piloselloides* (L.) Luerss.; *Notholaena piloselloides* (L.) Kaulf.; *Oetosis piloselloides* (L.) Kuntze; *Pteris piloselloides* L.; *Pteropsis piloselloides* (L.) Desv.; *Pyrrosia piloselloides* (L.) M.G. Price; *Taenitis piloselloides* (L.) Mett.

Local name: *sisik naga* (Kedayan)

Common name: dragon scale

Habitat: roadsides, parks, forests

Geographical distribution: from Bangladesh to Papua New Guinea

Botanical description: this hemiparasitic fern grows up to a length of about 2 m. Rhizomes: thin, scaly, lignose. Fronds: strongly dimorphic. Stipes: tiny. Blades of sterile fronds: 1–2 cm × 1–7 cm, fleshy, somewhat coriaceous, ellipsoid to globose to somewhat tongue-shaped, smooth, dull light green, with a thin dark purplish midrib. Blades of fertile fronds: linear, dull light green, 3–1.5 cm × 4–16 cm. Sori: near to the margin. Spores: brownish (Figure 2.6).

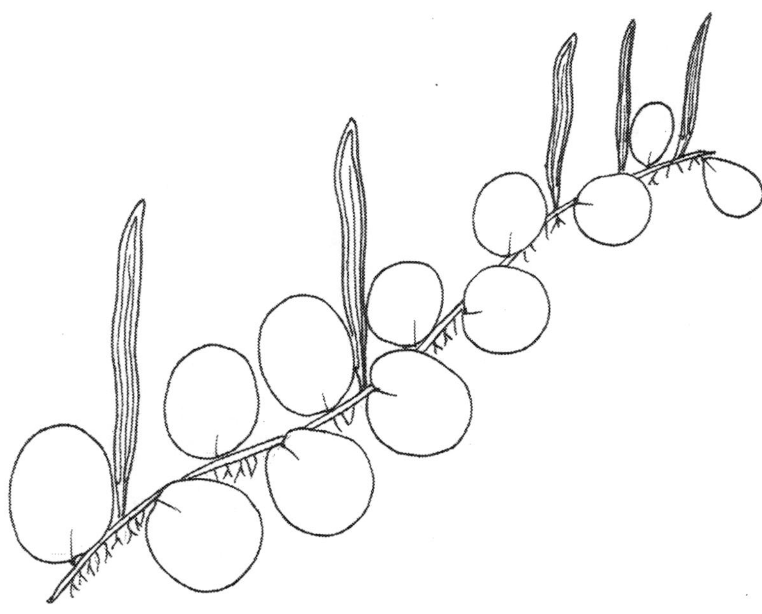

FIGURE 2.6 *Drymoglossum piloselloides* (L.) C. Presl.

Traditional therapeutic indications: diuretic, gallstones, hypertension (Kedayan)

Pharmacology and phytochemistry: antibacterial, antifungal (Somchit et al. 2011), toxic to cervical cancer cells (HeLa) (Sul'ain et al. 2019), antipyretic (Amoroso et al. 2014).

Toxicity, side effects, and drug interaction: not known

REFERENCES

Amoroso, V.B., Antesa, D.A., Buenavista, D.P. and Coritico, F.P. 2014. Antimicrobial, antipyretic, and anti-inflammatory activities of selected Philippine medicinal pteridophytes. *Asian Journal of Biodiversity*. 5(1): 18–40.

Somchit, M.N., Hassan, H., Zuraini, A., Chong, L.C., Mohamed, Z. and Zakaria, Z.A. 2011. *In vitro* anti-fungal and anti-bacterial activity of *Drymoglossum piloselloides* L. Presl. against numerous fungi responsible for Athlete's foot and common pathogenic bacteria. *African Journal of Microbiology Research*. 5(21): 3537–3541.

Sul'ain, M.D., Zakaria, F. and Johan, M.F. 2019. Anti-proliferative effects of methanol and water extracts of *Pyrrosia piloselloides* on the hela human cervical carcinoma cell line. *Asian Pacific Journal of Cancer Prevention: APJCP*. 20(1): 185.

***Drynaria sparsisora* (Desv.) T. Moore**

[From Greek *drys* = oak and Latin *sparsus* = scattered]

Published in *Index Filicum* 348. 1862

Synonyms: *Aglaomorpha sparsisora* (Desv.) Hovenkamp & S. Linds.; *Drynaria linnei* Bory ex Bedd.; *Polypodium sparsisorum* Desv.

Local name: *tapako* (Dusun, Kadazan)
Habitat: on trees or rocks, near seashores, swamps
Geographical distribution: Southeast Asia, Papua New Guinea, Australia, Pacific Islands
Botanical identification: this fern reaches a height of about 1 m. Rhizomes: somewhat snake-like, fleshy, and scaly. Stipes: 12–18 cm long, winged. Blades: deeply lobed, tapering at base, coriaceous, light dull green, 15–35 cm × 30–80 cm, oblong, the lobes up to the number of 20, somewhat lanceolate to oblong and somewhat asymmetrical. The basal leaves are sessile. Sori: discoid, about 2 mm across, on the lower surface of blades. Spores: spiny (Figure 2.7).
Traditional therapeutic indications: asthma, heart diseases (Dusun, Kadazan)

FIGURE 2.7 *Drynaria sparsisora* (Desv.) T. Moore.

Pharmacology and phytochemistry: phenolic compounds (Tan & Lim 2015).
Toxicity, side effects, and drug interaction: not known

Comment: Parts of this plant are sold in the wet markets of Sabah as medicine.

REFERENCE

Tan, J.B.L. and Lim, Y.Y. 2015. Antioxidant and tyrosinase inhibition activity of the fertile fronds and rhizomes of three different *Drynaria* species. *BMC Research Notes*. 8: 1–6.

***Pyrrosia lanceolata* (L.) Fawr.**

[From Greek *purros* = reddish and Latin *lanceolata* = lanceolate]

Published in *American Midland Naturalist* 12(8): 245. 1931

Synonyms: *Acrostichum lanceolatum* L.; *Acrostichum linnaeanum* Hook.; *Bolbitis linnaeana* (Hook.) C. Chr.; *Campium linnaeanum* (Hook.) Copel.; *Candollea lanceolata* (L.) Mirb.; *Chrysodium linnaeanum* (Hook.) Kuhn; *Cyclophorus lanceolatus* (L.) Alston; *Dendroglossa lanceolata* (L.) C. Presl; *Dendroglossa linnaeana* (Hook.) Fée; *Niphobolus lanceolatus* (L.) Trimen; *Gymnopteris lanceolata* (L.) T. Moore; *Gymnopteris linnaeana* (Hook.) Christ; *Leptochilus lanceolatus* (L.) Zoll.; *Leptochilus linnaeanus* Fée
Local name: *ubat alib* (Lundayeh)
Common name: lance leaf tongue fern
Habitat: on trees or rocks, mangroves, seashores
Geographical distribution: from India to Pacific Islands
Botanical description: this hemiparasitic fern grows up to a length of about 2 m. Rhizomes: creeping, scaly. Fronds: monomorphic. Stipes: stout, about 3 mm long. Blades: somewhat lanceolate, linear, spathulate, or oblong, fleshy, coriaceous, smooth, glossy, attenuate at base, obtuse at apex, and 5–8 mm × 0.5–1.3 cm. Sori: discoid, covering the apex of blade beneath (Figure 2.8).

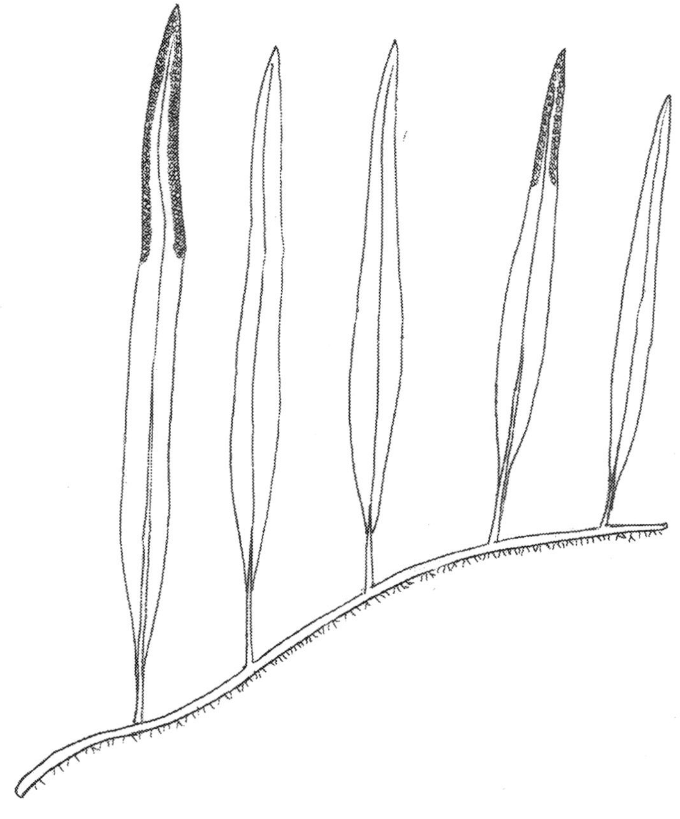

FIGURE 2.8 *Pyrrosia lanceolata* (L.) Fawr.

Traditional therapeutic indications: swollen pancreas (Lundayeh)
Pharmacology and phytochemistry: phenolic compounds (Kotrnon et al. 2007).
Toxicity, side effects, and drug interaction: not known

Comment: This fern is common in Kota Kinabalu.

REFERENCE

Kotrnon, K., Thammathaworn, A. and Chantaranothai, P. 2007. Comparative anatomy of the genus Pyrrosia Mirbel (Polypodiaceae) in Thailand. *Tropical Natural History*. 7(1): 75–85.

2.2.4 Order Schizaeales Schimp. (1869)

2.2.4.1 Family Lygodiaceae M. Roem. (1840), the Climbing Ferns Family

The family Lygodiaceae consists of the single genus: *Lygodium* Sw. (1801).

***Lygodium circinnatum* Sw.**

[From Greek *lugos* = flexible and Latin *circino* = I make round]

Published in *Synopsis Filicum* 153. 1806

Synonyms: *Hydroglossum circinnatum* (Burm.f.) Willd.; *Hydroglossum pedatum* (Burm.f.) Willd.; *Lygodium basilanicum* Christ; *Lygodium conforme* C. Chr.; *Lygodium dichotomum* (Cav.) Sw.; *Lygodium pedatum* (Burm.f.) Sw.; *Ophioglossum circinnatum* Burm.f.; *Ophioglossum pedatum* Burm.f.; *Ugena dichotoma* Cav.; *Ugena macrostachya* Cav.

Local names: *waratang* (Lundayeh); *taribu mianai* (Dusun)

Habitat: forests

Geographical distribution: from India to Pacific Islands

Botanical description: this climbing fern grows from a creeping, hairy rhizome. Stipes: thin, hairy at base. Blades: somewhat digitate, with up to 4 or 5 lobes, oblong, dull green, somewhat membranous, cartilaginous at margin, and up to 3.5 cm × 25 cm. Rachis of climbing fronds: thin and can reach a length of 10 m. Sori: at margin of lobes. Spores: verrucose (Figure 2.9).

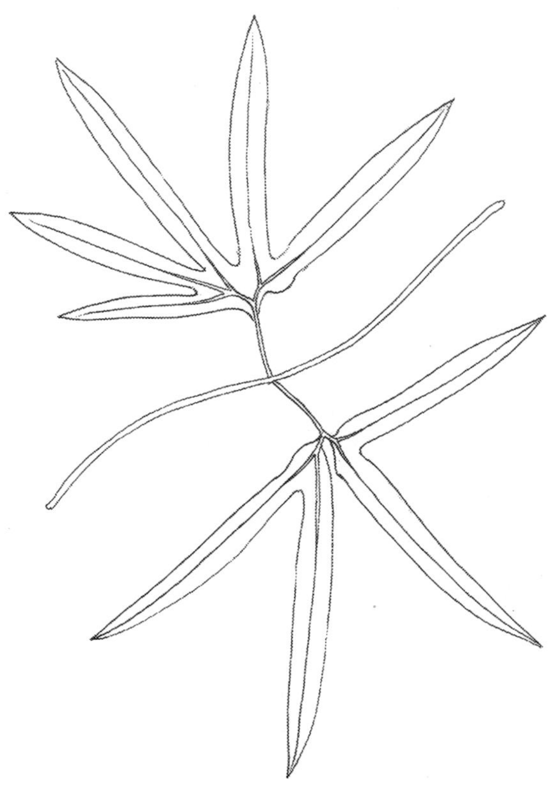

FIGURE 2.9 *Lygodium circinnatum* Sw.

Traditional therapeutic indications: venereal diseases (Lundayeh); womb diseases (Dusun)
Pharmacology and phytochemistry: phenolic compounds (Saman et al. 2017).
Toxicity, side effects, and drug interaction: not known

Comment: the rachis is used to make baskets, cordages, and fibers in the Philippines.

REFERENCE

Saman, R.B.A., Mokhtar, R.A.M. and Iqbal, M. 2017. Identification of bioactive compounds, quantitative measurement of phenolics and flavonoids content, and radical scavenging activity of Lygodium circinnatum. *Transactions on Science and Technology*. 4(3): 354–359.

Lygodium salicifolium C. Presl.

[From Greek *lugos* = flexible and Latin *salix* = willow, and *folium* = leaf]

Published in *Supplementum Tentaminis Pteridographiae.* 102. 1845

Synonyms: *Lygodium andamanicum* R.D. Dixit, J.B. Bhandari & R. Mukhop.; *Lygodium kingii* Copel.
Local name: *ubat amur* (Lundayeh)
Habitat: forests
Geographical distribution: from India to Papua New Guinea
Botanical description: this climbing fern grows from a creeping, hairy rhizome. Stipes: thin, hairy. Blades: bipinnate, with up to 4 or 6 lobes, oblong to narrowly lanceolate and enlarged at base, glossy, serrate, smooth, up to 0.5–2 cm × 4–15 cm, truncate to cordate at base. The terminal folioles are deeply bilobed. Rachis of climbing fronds: thin, can reach a length of 12 m. Sori: at margin of lobes. Spores: verrucose.
Traditional therapeutic indications: prevention of chicken pox and smallpox (Lundayeh)
Pharmacology and phytochemistry: not known
Toxicity, side effects, and drug interaction: not known

Comment: the rachis is used as weaving material in Thailand.

Lygodium flexuosum (L.) Sw.

[From Greek *lugos* = flexible and Latin *flexuosum* = flexuous]

Published in *Journal für die Botanik* 1800(2): 106. 1801

Synonyms: *Hydroglossum flexuosum* (L.) Willd.; *Hydroglossum scandens* (L.) Willd.; *Lygodium altum* (C.B. Clarke) Alderw.; *Lygodium pilosum* Desv.; *Lygodium semibipinnatum* R.Br.; *Odontopteris scandens* (L.) Bernh.; *Ophioglossum flexuosum* L.; *Ophioglossum scandens* L.; *Ramondia flexuosa* (L.) Mirb.; *Ramondia scandens* (L.) Mirb.
Local name: *ribu ribu besar* (Bajau)
Habitat: forests
Geographical distribution: from India to Australia
Botanical description: this climbing fern grows from a creeping, hairy rhizome. Stipes: about 50 cm long, thin, winged, hairy. Blades: bipinnate. Folioles: 4–6 per pinnae, oblong to narrowly lanceolate, and enlarged at base, or to trilobed at base, glossy, serrate, marked above with numerous and thin nervations, 1–2 cm × 6–25 cm, acute or rounded at apex. The terminal folioles are deeply bilobed. Rachis: winged. Sori: at margin of lobes. Spores: verrucose.
Traditional therapeutic indications: coughing up blood, coughs, fever (Bajau)
Pharmacology and phytochemistry: anti-inflammatory (Das et al. 2012), hepatoprotective (Wills & Asha 2006), toxic to human hepatoma cells (Wills & Asha 2009), antibacterial (Nayak et al. 2013). The plant produces a spirofuropyran-perhydrophenanthrene derivatives (lygodinolide) (Achari et al. 1990), triterpenes (dryocrassol), and an anthraquinone (tectoquinone) (Achari et al. 1986).
Toxicity, side effects, and drug interaction: aqueous extract administred orally at the dose of 5 g/kg/day for 30 days to rats was found to cause no signs of toxicity (Wills & Asha 2012).

Comments:

(i) The rachis is used to make baskets, cordages, and fibers in the Philippines.
(ii) *Lygodium microphyllum* (Cav.) R.Br. is used by the Dusun (local name: *taribu indu*) for womb diseases.

REFERENCES

Achari, B., Basu, K., Saha, C.R. and Pakrashi, S.C. 1986. A new triterpene ester, an anthraquinone and other constituents of the fern *Lygodium flexuosum*. *Planta Medica*. 52(4): 329–330.

Achari, B., Chaudhuri, C., Saha, C.R., Pakrashi, S.C., McPhail, D.R. and McPhail, A.T. 1990. X-ray crystal structure of lygodinolide: A novel spiro furopyran-perhydrophenanthrene derivative from *Lygodium flexuosum*. *The Journal of Organic Chemistry*. 55(16): 4977–4978.

Das, B., Talekar, Y.P., Apte, G.K. and Rajendra, C. 2012. A preliminary study of antiinflammatory activity and antioxidant property of *Lygodium flexuosum*, a climbing fern. *International Journal of Pharmacy and Pharmaceutical Sciences*. 4(4): 358–361.

Nayak, N., Rath, S., Mishra, M.P., Ghosh, G. and Padhy, R.N. 2013. Antibacterial activity of the terrestrial fern *Lygodium flexuosum* (L.) Sw. against multidrug resistant enteric-and uro-pathogenic bacteria. *Journal of Acute Disease*. 2(4): 270–276.

Wills, P.J. and Asha, V.V. 2006. Protective effect of *Lygodium flexuosum* (L.) Sw. (Lygodiaceae) against D-galactosamine induced liver injury in rats. *Journal of Ethnopharmacology*. 108(1): 116–123.

Wills, P.J. and Asha, V.V. 2009. Chemopreventive action of *Lygodium flexuosum* extract in human hepatoma PLC/PRF/5 and Hep 3B cells. *Journal of Ethnopharmacology*. 122(2): 294–303.

Wills, P.J. and Asha, V.V. 2012. Acute and subacute toxicity studies of *Lygodium flexuosum* extracts in rats. *Asian Pacific Journal of Tropical Biomedicine*. 2(1): S200–S202.

2.2.4.2 Family Thelypteridaceae Ching ex Pic. Serm. (1970), the Marsh Fern Family

The family Thelypteridaceae consists of about 20 genera and 1,000 species of ferns growing from scaly rhizomes. Fronds: monomorphic. Blades: pinnate. Sori: orbicular, oblong, or linear on veins. Sporangia: long stalked. Spores: bilateral.

***Pronephrium asperum* (C. Presl) Holttum**

[From Latin *aspera* = hard]

Published in *Blumea* 20(1): 112. 1972

Synonyms: *Abacopteris aspera* (C. Presl) Ching; *Abacopteris presliana* Ching; *Abacopteris urophylla* (Wall. ex Hook) Ching; *Aspidium asperum* (C. Presl) Mett.; *Aspidium urophyllum* (Wall. ex Hook) Christ; *Christella urophylla* (Wall. ex Hook) H. Lév.; *Cyclosorus asperus* (C. Presl) B.K. Nayar & S. Kaur; *Cyclosorus urophyllus* (Wall. ex Hook) Tardieu; *Dryopteris presliana* Ching; *Dryopteris urophylla* (Wall. ex Hook) C. Chr.; *Goniopteris aspera* C. Presl; *Meniscium urophyllum* (Wall. ex Hook) H. Itô; *Nephrodium urophyllum* (Wall. ex Hook) Keyserl.; *Phegopteris urophylla* Mett; *Polypodium asperum* C. Presl; *Polypodium urophyllum* Wall. ex Hook; *Thelypteris aspera* (C. Presl.) K. Iwats.; *Thelypteris urophylla* (Wall. ex Hook) K. Iwats.
Local name: *ingkakahas* (Murut)
Habitat: forests
Geographical distribution: from Thailand to Pacific Islands
Botanical description: this fern reaches a height of about 1.5 m. Rhizomes: creeping. Stipes: 50–75 cm long, channeled above, scaly at base. Rachis: hairy beneath, scaly. Blades: pinnate, 40–50 cm × 50–75 cm. Folioles: 4–8 pairs, narrowly oblong– lanceolate, 4–5.6 cm × 20–30 cm, round to truncate or attenuate at base, long acuminate at apex, wavy, serrate to crenate, coriaceous, hairy beneath, with 12–22 pairs of thin secondary nerves. Sori: indusiate, round, on veins. Sporangia: with a pair of setae. Spores: dark brown, monolete, winged (Figure 2.10).
Traditional therapeutic indication: medicinal (Murut)
Pharmacology and phytochemistry: not known
Toxicity, side effects, and drug interaction: not known

Comments:

(i) In the family Cytheaceae Kaulf. (1827), a plant of the genus *Cibotium* Kaulf. (1820) is used by Dusun and Kadazan (local name: *paku*) as an antidote for snakebites and for boils. A number of plants are used, especially by locals living in remote areas of primary forests for snake bites. This may seem incredible, but we cannot deny the fact that some of these plants have actually saved the lives of many people.
(ii) In the family Thelypteridaceae Ching ex Pic. Serm. (1970), a plant of the genus *Pneumatopteris* Nakai (1933) is used by the Dusun and Kadazan (local name: *menampun*) for postpartum and flatulence.
(iii) In the family Nephrolepidaceae Pic. Serm. (1975), a plant of the genus *Nephrolepis* Schott (1834) is used by the Dusun and Kadazan (local name: *monumpuru*) as a remedy for headaches.
(iv) *Nephrolepis discksonioides* Christ is used by the Dusun for itchiness and chills in children (local name: *ngkubuk*).
(v) In the family Schizaeaceae Kaulf. (1827), *Schizaea dichotoma* (L.) Sw. is used by the Dusun (local name: *pitagar payong*) as aphrodisiac.
(vi) In the family Pteridaceae E.D.M. Kirchn. (1831), a plant of the genus *Pteris* L. (1853) is used by the Dusun as aphrodisiac.

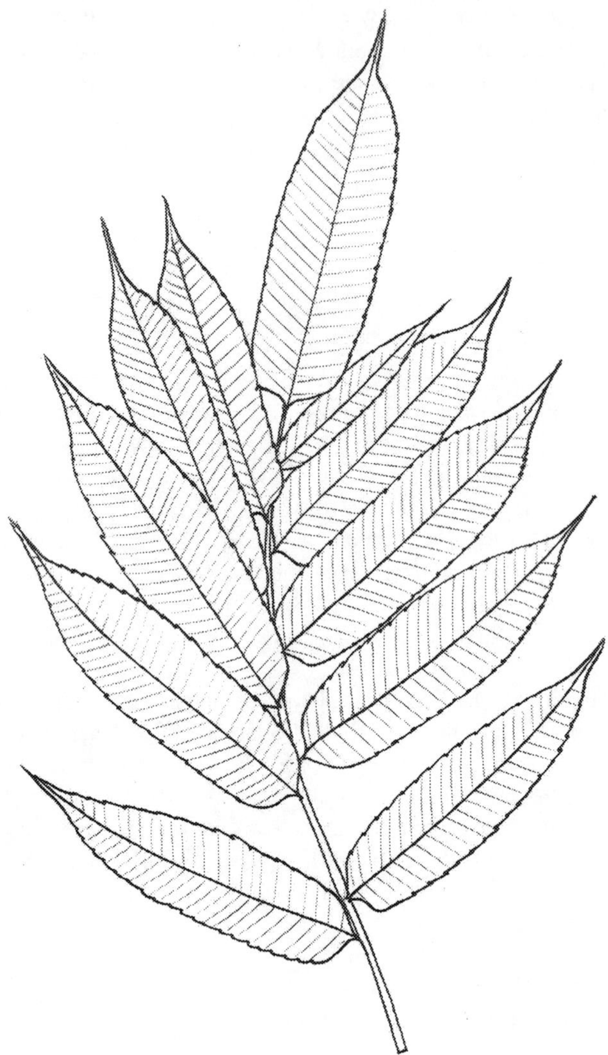

FIGURE 2.10 *Pronephrium asperum* (C. Presl) Holttum.

3 The Medicinal Plants of Sabah (North Borneo)

The Gymnosperms

3.1 SUBCLASS PINIDAE CRONQUIST, TAKHT. & W. ZIMM. (1966)

3.1.1 Order Auraucariales Gorozh. (1904)

3.1.1.1 Family Auraucariaceae Henkel & W. Hochst. (1865), the Araucaria Family

The family Auraucariaceae consists of 2 genera and 41 species of resinous pine trees. Leaves: coriaceous, spirally arranged or decussate, sessile. Inflorescences: cones. Seeds: flat, woody, winged or not.

Agathis borneensis **Warb.**

[From Greek *agatis* = ball of thread and Latin *borneensis* = from Borneo]

Published in *Monsunia* 1: 184.1900

Synonyms: *Abies sumatrana* Desf.; *Agathis beccarii* Warb.; *Agathis beckingii* Meijer Drees; *Agathis endertii* Meijer Drees; *Agathis latifolia* Meijer Drees; *Agathis macrostachys* Warb.; *Agathis rhomboidalis* Warb.; *Pinus sumatrana* Mirb.
Local name: *manggilan* (Dusun)
Common name: Borneo kauri
Habitat: forests
Geographical distribution: Malaysia, Indonesia
Botanical description: this massive timber tree, on the verge of extinction, can grow up to about 50 m tall. Trunk: up to 3.5 m wide. Bark: smooth, dimpled, grayish to light brown, yields an abundant amber-like resin once incised. Leaves: opposite, somewhat sessile. Blades: lanceolate, glossy, coriaceous, 2–4 cm × 6–14 cm, somewhat asymmetrical. Cones: about 7 cm long, broadly ovoid (Figure 3.1).
Traditional therapeutic indications: medicinal
Pharmacology and phytochemistry: essential oil (Alet et al. 2012; Adam et al. 2017). Antiplasmodial (Noor Rain et al. 2007).
Toxicity, side effects, and drug interaction: not known

Comments:

(i) If you drive through Sabah, you can pass for hours through vast expanses of oil palm plantations as far as the eye can see. Before the 1980s and 1990s, these plantations were non-existent and in their place shone a rainforest that was perhaps the richest and most magnificent on the planet. There is practically nothing left apart from a few islets of rainforests which, logically, will disappear. Some experts say it is possible to replant primary rainforests, but that doesn't make much sense because primary rainforests are unique biomes whose organisms (which interact in precise and multiple ways) have put millions of years to appear. How many global viral pandemics must we endure to

DOI:10.1201/9781003402886-3

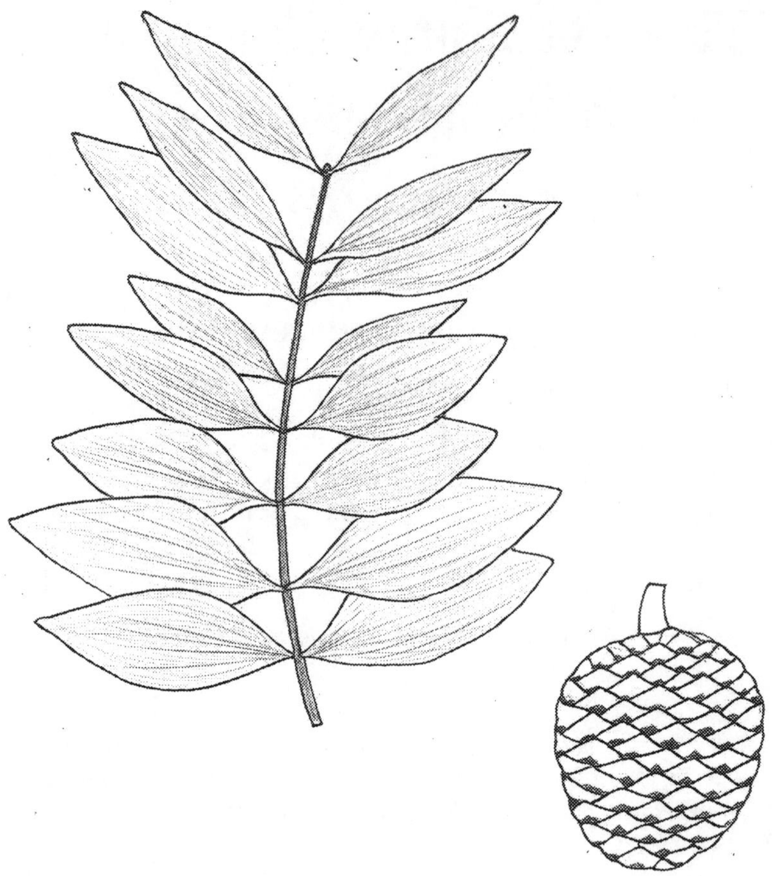

FIGURE 3.1 *Agathis borneensis* Warb.

finally understand that the destruction of the primary rainforests puts the world's population in mortal danger?

(ii) In the order Pinales Gorozh. (1904), family Pinaceae Spreng. & Rudolphi (1830), a plant of the genus *Pinus* L. (1753) is used by the Bajau for fever (local name: *nanas batu*).

REFERENCES

Adam, A.Z., Juiling, S., Lee, S.Y., Jumaat, S.R. and Mohamed, R. 2017. Phytochemical composition of Agathis borneensis (Araucariaceae) and their biological activities. *Malaysian Forester.* 80: 169–177.

Alet, F.B., Assim, Z.B., bin Jusoh, I. and Ahmad, F.B. 2012. Chemical constituents of essential oils from resin and bark of Agathis borneensis. *Borneo Journal of Resource Science and Technology*. 2(1): 28–32.

Noor Rain, A., Khozirah, S., Mohd Ridzuan, M.A., Ong, B.K., Rohaya, C., Rosilawati, M., Hamdino, I., Badrul, A. and Zakiah, I. 2007. Antiplasmodial properties of some Malaysian medicinal plants. *Trop Biomed*. 24(1): 29–35.

3.2 SUBCLASS GNETIDAE PAX (1894)

3.2.1 Order Gnetales Blume (1835)

3.2.1.1 Family Gnetaceae Blume (1833), the Gnetum Family

The family Gnetaceae consists of the single genus *Gnetum* L. (1767)

***Gnetum macrostachyum* Hook.f.**

[Probably from ancient Malay *ganemu* = a plant of the genus *Gnetum* and Greek *makros* = large and *stachys* = spike]

Published in *The Flora of British India* 5: 642. 1890

Synonym: *Thoa macrostachya* (Hook.f.) Doweld
Local name: *kokos* (Dusun, Kadazan)
Habitat: forests near streams
Geographical distribution: Cambodia, Laos, Vietnam, Thailand, Malaysia, Indonesia, Papua New Guinea
Botanical description: this ligneous climbing plant grows up to a length of 20 m. Stems: terete, smooth, articulate with swollen nodes. Leaves: simple, opposite. Petioles: about 1 cm long, stout. Blades: elliptic, lanceolate to lanceolate, coriaceous, attenuate at base, acuminate to cuspidate at apex, glossy, up to 4–8 cm × 10–20 cm, with about 6 pairs of secondary nerves. Spikes: axillary, cylindrical, stout, about 5–10 cm long, which can make one think of caterpillars or worms. In male flowers, the perianth is bifid and one stamen is present. In female flowers there is no perianth, and an ovule with a style-like tube is present. Drupes: about 1.2 cm × 2 cm, coriaceous, glossy, ovoid, apiculate at apex, on a basal hairy collar (Figure 3.2).

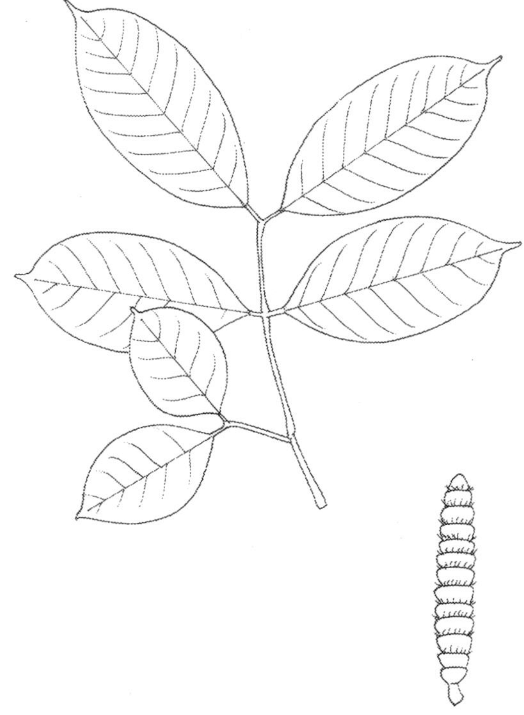

FIGURE 3.2 *Gnetum macrostachyum* Hook.f.

Traditional therapeutic indications: fatigue, postpartum (Dusun, Kadazan)

Pharmacology and phytochemistry: oligostilbenes toxic to cervical cancer cells (HeLa) (macrostachyol D; IC_{50}: 4.1 μM) (Sri-in et al. 2011). Stilbenes inhibiting platelet aggregation (resveratrol, *trans*-resveratrol, isorhapotigenin, gnetol) (Kloypan et al. 2012). Flavones (5,7,2′-trihydroxy-5′-methoxyflavone) (Saisin et al. 2009). Oligostilene inhibiting platelet aggregation (macrostachyol C) (Surapinit & Baisaeng 2021).

Toxicity, side effects, and drug interaction: not known

Comments:

(i) A plant of the genus *Gnetum* L. (1767) with the local name of "*lautan seribu*" is sold in the wet markets of Sabah as a medicine.

(ii) *Gnetum gnemon* L. is used by the Dusun for fatigue (local name: *bagu*).

REFERENCES

Kloypan, C., Jeenapongsa, R., Sri-in, P., Chanta, S., Dokpuang, D., Tip-pyang, S. and Surapinit, N. 2012. Stilbenoids from *Gnetum macrostachyum* attenuate human platelet aggregation and adhesion. *Phytotherapy Research.* 26(10): 1564–1568.

Saisin, S., Tip-pyang, S. and Phuwapraisirisan, P. 2009. A new antioxidant flavonoid from the lianas of *Gnetum macrostachyum. Natural Product Research.* 23(16): 1472–1477.

Sri-in, P., Sichaem, J., Siripong, P. and Tip-pyang, S. 2011. Macrostachyols A—D, new oligostilbenoids from the roots of *Gnetum macrostachyum. Fitoterapia.* 82(3): 460–465.

Surapinit, S. and Baisaeng, N. 2021. Macrostachyols AD, oligostilbenes from *Gnetum macrostachyum* inhibited *in vitro* human platelet aggregation. *Journal of Herbmed Pharmacology.* 10(3): 339–343.

4 The Medicinal Plants of Sabah (North Borneo)

The Angiosperms

4.1 SUBCLASS MAGNOLIIDAE NOVÁK EX TAKHT. (1967)

4.1.1 Superorder Austrobaileyanae Doweld ex M.W. Chase & Reveal (2009), the Protomagnoliids

4.1.1.1 Order Austrobaileyales Takht. ex Reveal (1992)

4.1.1.1.1 Family Schisandraceae Blume (1830), the Star Vine Family

The family Schisandraceae consists of about 2 genera and 40 species of aromatic ligneous climbing plants. Leaves: simple, alternate, exstipulate. Flowers: small, solitary, axillary, regular, hypogynous, with a somewhat elongated receptacle. Tepals: 5–24, spirally arranged. Stamens: up to 80, spirally arranged. Carpels: 12–300, free, each carpel contains 2–5 ovules. Fruits: an aggregate of fruiting carpels.

Kadsura borneensis **A.C. Sm**

[From the Japanese name of a plant of the genus *Kadzura* and Latin *borneensis* = from Borneo]

Published in *Sargentia* 7: 205 1947

Local name: *putu urat* (Lundayeh)
Habitat: forests
Geographical distribution: Borneo
Botanical description: this ligneous climbing plant grows up to a length of about 6 m. Leaves: simple, alternate, exstipulate. Petioles: 1.5–3.5 cm long. Blades: coriaceous, elliptic to ovate, somewhat wavy, irregular, 6.5–15 cm × 10.5–21.5 cm, with 7–8 pairs of secondary nerves, obtuse to truncate at base, acute to acuminate at apex. Flowers: solitary, axillary, male, or female. Tepals: 12–20, yellow, of irregular length (2 mm–1 cm long). Stamens: 18–28, pink, sessile. Carpels: about 35. Fruiting carpels: reddish-purple, globose, glossy, across, arranged in some kind of globose grappes with a diameter of 2 cm. Seeds: reniform (Figure 4.1).
Traditional therapeutic indications: muscular pains (Lundayeh)
Pharmacology and phytochemistry: not known
Toxicity, side effects, and drug interaction: not known

Comments:

(i) *Kadsura lanceolata* King is used by the Dusun for boils and swellings (local name: *topis*).
(ii) In the family Chloranthaceae R. Brown ex Sims (1820), the Dusun and Kadazan use plants of the genus *Chloranthus* Sw. (1787) as medicines: "*totol*" as styptic and "*kosup*" for skin diseases, body pains, and internal wounds.

DOI: 10.1201/9781003402886-4

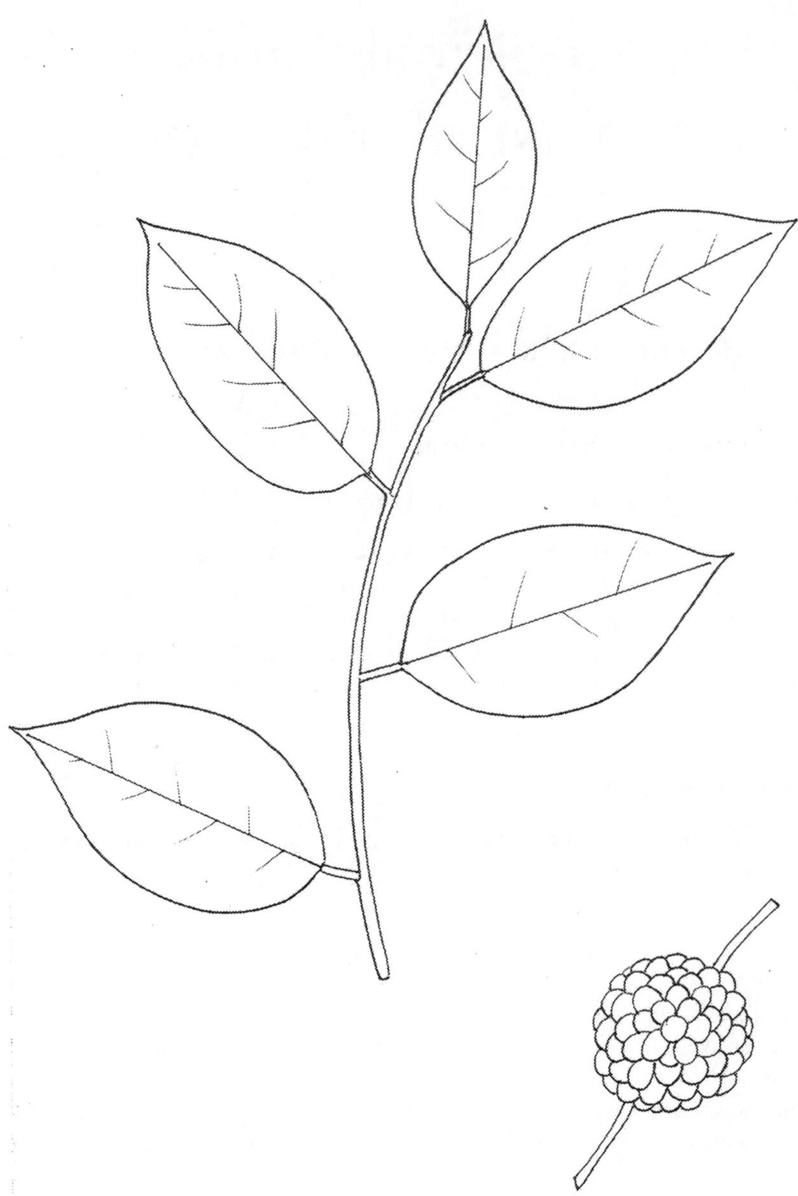

FIGURE 4.1 *Kadsura borneensis* A.C. Sm.

4.1.2 Superorder Magnolianae Takht. (1967), the Magnoliids

4.1.2.1 Order Piperales Bercht. & C. Presl. (1820)

4.1.2.1.1 Family Aristolochiaceae Juss. (1789), the Dutchman's Pipe Family

The family Aristolochiaceae consists of about 9 genera and 600 species of poisonous climbing plants, herbaceous plants, or shrubs. The cross sections of stems show characteristic patterns of bicycle-wheel-like medullary rays. Leaves: simple, alternate, exstipulate. Blades: with secondary nerves emerging from the base (as in the Dioscoreaceae). Flowers: solitary or cymose, terminal or axillary, pendulous, and with a grace and a multitude of colors and shapes that surpass understanding. Calyx: more or less trilobed, often pipe-shaped in a very unique manner to enormous and showy. Stamens: 6–12. Carpels: 4–6, forming a unilocular or 4–6 locular ovary, each locule with 50–many ovules. Styles: 1–6. Fruits: dehiscent capsules. Seeds: numerous.

***Aristolochia minutiflora* Ridl. ex Gamble**

[From Greek *aristo* = best, *locheia* = childbirth, and Latin *minutiflora* = with very small flower]

Published in *Bulletin of Miscellaneous Information, Royal Gardens, Kew*, 79. 1910

Local name: *lapad talang* (Lundayeh)
Habitat: forests
Geographical distribution: Malaysia, Borneo
Botanical description: this climbing shrub grows up to a length of about 10 m. Leaves: simple, alternate, exstipulate. Petioles: about 3–5 cm long. Blades: membranous, profoundly cordate at base, sagittate, 3–7 cm × 5–15 cm, with 5–7 pairs of secondary nerves, acuminate at apex. Racemes: axillary. Perianth: 1.5 cm long, tubular, swollen at base, the tube developing a bifid upper lobe and a linear inferior lobe. Stamens: 6. Ovary: 6-lobed. Capsules: oblong to ovoid, about 1.5–5 cm long, 6-lobed, striated. Seeds: pyriform, about 5 mm long, rugose.
Traditional therapeutic indications: diarrhea, vomiting (Lundayeh)
Pharmacology and phytochemistry: not known
Toxicity, side effects, and drug interaction: not known

***Aristolochia papillifolia* Ding Hou**

[From Greek *aristo* = best, *locheia* = childbirth, and Latin *papilla* = a nipple, and *folia* = leaf]

Published in *Blumea* 28: 346. 1983

Local name: *babas lontong* (Murut)
Habitat: forests
Geographical distribution: Borneo
Botanical description: climbing plant grows up to a length of about 9 m. Bark: dirty gray to dark brown and deeply fissured. Stems: terete, stout, swollen at nodes, up to 2.5 cm across. Leaves: simple, alternate, exstipulate. Petioles: thin, 2–6 cm long. Blades: broadly lanceolate, 9–14.5 cm × 13–19 cm, coriaceous, truncate at base, acuminate at apex, with 2 pairs of secondary nerves emerging from the base. Flowers: tiny. Inflorescences: cauliflorous fascicles on about 4 cm long peduncles. Perianth: straight, the tube about 2 cm long, the lip linear and about 5 cm long. Stamens: 6. Ovary: 6-lobed. Capsules: cylindrical, 6.5 cm × 1.2 cm, 6-lobed, obtuse on both ends, purplish green. Seeds: numerous, ovate, about 4 mm long (Figure 4.2).

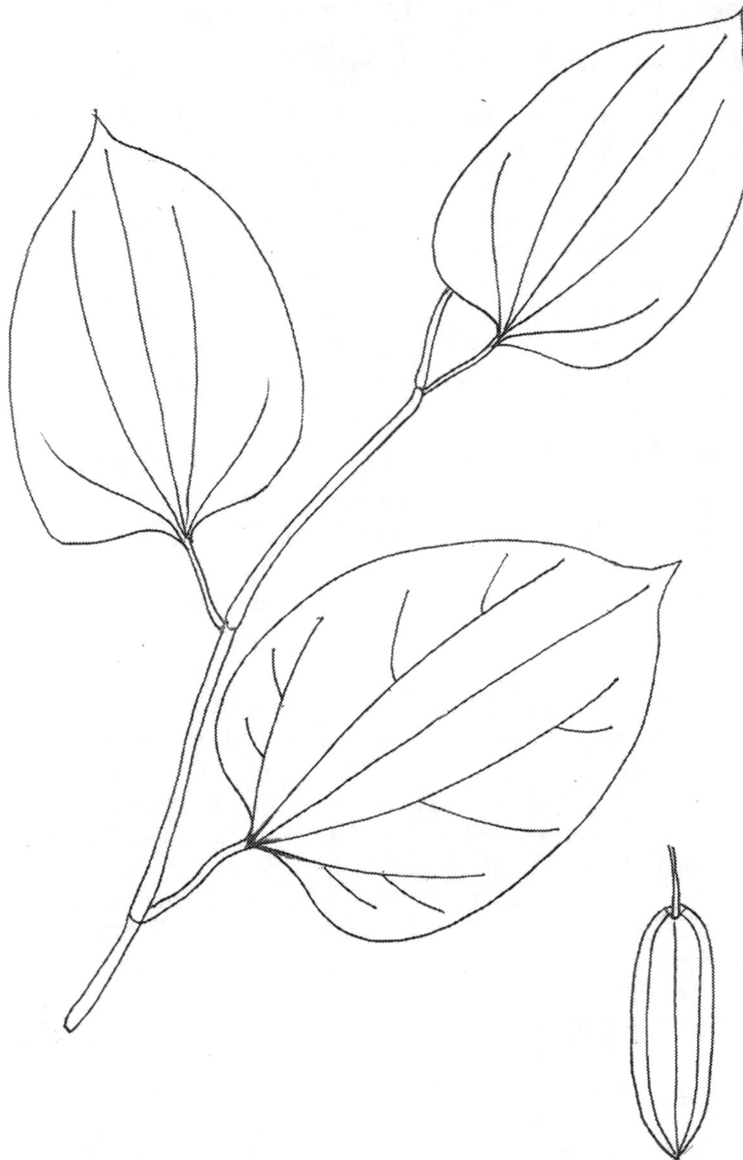

FIGURE 4.2 *Aristolochia papillifolia* Ding Hou.

Traditional therapeutic indications: poison antidote, jaundice, hepatitis, pancreatitis, diarrhea (Murut)

Pharmacology and phytochemistry: not known

Toxicity, side effects, and drug interaction: not known

4.1.2.1.2 Family Piperaceae Giseke (1792), the Pepper Family

The family Piperaceae comprises about 10 genera and about 3,000 species of herbaceous plants, climbing plants, or treelets. Stems: smooth, terete, articulate, often swollen at nodes. Leaves: simple, alternate, sometimes stipulate. Blades: often with longitudinal secondary nerves emerging from the base. Inflorescences: axillary spikes. Flowers: tiny. Perianth: none. Stamens: 1–10. Carpels: 2–4, forming a unilocular ovary with 1 ovule. Fruits: drupes, capsules.

***Peperomia pellucida* (L.) Kunth**

[From the Greek *peperi* = pepper and Latin *pellucida* = translucent]

Published in *Nova Genera et Species Plantarum (quarto ed.)* 1: 64. 1815

Synonyms: *Micropiper pellucidum* (L.) Miq.; *Peperomia concinna* (Haw.) A. Dietr.; *Peperomia ephemera* Ekman; *Peperomia knoblecheriana* Schott; *Peperomia praetenuis* Trel.; *Peperomia translucens* Trel.; *Piper concinnum* Haw.; *Piper pellucidum* L.; *Verhuellia knoblecheriana* (Schott) C. DC.

Local names: *ketumpangan air* (Kedayan); *limpananas, terang mata* (Bajau)

Common name: man to man

Habitat: in cracks in drainage walls and other stony, damp, and shady places in towns and villages

Geographical distribution: tropical

Botanical description: this herbaceous plant reaches a height of about 20 cm. Stems: thin, fleshy, terete, somewhat translucent, glabrous. Leaves: simple, spiral, exstipulate. Petioles: 1–2 cm long. Blades: deltoid, 1–3.5 cm × 1–3.5 cm, membranous, translucent, fleshy, cordate at base, acute or obtuse at apex, marked with 2–3 pairs of secondary nerves. Spikes: terminal or leaf-opposed, thin, up to about 6 cm long. Perianth absent. Stamens: 2. Ovary: ellipsoid. Drupes: tiny, black, globose.

Traditional therapeutic indications: abscesses, blisters, coughs, flu, pimples, rheumatisms (Kedayan); sore eyes, diabetes, blood diseases (Bajau); hypertension (Bajau, Kedayan)

Pharmacology and phytochemistry: antibacterial (Khan & Omoloso 2002), hypotensive (Saputri et al. 2021), analgesic, hypoglycemic (Sheikh et al. 2013). Phenylpropanoids (apiole) (Bayma 2000). Anti-inflammatory lignans (sesamin). Antiviral lignans (pellucidin, HIV, EC_{50} = 6.6 μM) (Fanhchaksai et al. 2016; Thongphichai et al. 2019). Secolignan (peperomin A) toxic to acute propromyelocytic leukemia cells (HL-60), breast cancer cells, and cervical cancer cells (HeLa) (Xu et al. 2006).

Toxicity, side effects, and drug interaction: the median lethal dose (LD_{50}) of an aqueous extract administered orally to mice was 5 g/kg (de Fátima Arrigoni-Blank et al. 2004).

Comment: *Aristolochia foveolata* Merr. is used by the Murut as antidote for poison blow-pipe dart (local name: *tabar kedayan*).

REFERENCES

Bayma, J.D.C., Arruda, M.S.P., Müller, A.H., Arruda, A.C. and Canto, W.C. 2000. A dimeric ArC2 compound from *Peperomia pellucida*. *Phytochemistry*. 55(7): 779–782.

de Fátima Arrigoni-Blank, M., Dmitrieva, E.G., Franzotti, E.M., Antoniolli, A.R., Andrade, M.R. and Marchioro, M. 2004. Anti-inflammatory and analgesic activity of *Peperomia pellucida* (L.) HBK (Piperaceae). *Journal of Ethnopharmacology*. 91(2–3): 215–218.

Fanhchaksai, K., Kodchakorn, K., Pothacharoen, P. and Kongtawelert, P. 2016. Effect of sesamin against cytokine production from influenza type A H1N1-induced peripheral blood mononuclear cells: Computational and experimental studies. *In vitro Cellular & Developmental Biology-Animal*. 52(1): 107–119.

Khan, M.R. and Omoloso, A.D. 2002. Antibacterial activity of *Hygrophila stricta* and *Peperomia pellucida*. *Fitoterapia*. 73(3): 251–254.

Saputri, F.C., Hutahaean, I. and Mun'im, A. 2021. *Peperomia pellucida* (L.) Kunth as an angiotensin-converting enzyme inhibitor in two-kidney, one-clip Goldblatt hypertensive rats. *Saudi Journal of Biological Sciences*. 28(11): 6191–6197.

Sheikh, H., Sikder, S., Paul, S.K., Hasan, A.R., Rahaman, M. and Kundu, S.P. 2013. Hypoglycemic, anti-inflammatory and analgesic activity of *Peperomia pellucida* (L.) HBK (piperaceae). *International Journal of Pharmaceutical Sciences and Research*. 4(1): 458–463.

Thongphichai, W., Tuchinda, P., Pohmakotr, M., Reutrakul, V., Akkarawongsapat, R., Napaswad, C., Limthongkul, J., Jenjittikul, T. and Saithong, S. 2019. Anti-HIV activities of constituents from the rhizomes of *Boesenbergia thorelii*. *Fitoterapia*. 139: 104388.

Xu, S., Li, N., Ning, M.M., Zhou, C.H., Yang, Q.R. and Wang, M.W. 2006. Bioactive compounds from *Peperomia pellucida*. *Journal of Natural Products*. 69(2): 247–250.

***Piper betle* L.**

[From the Greek *peperi* = pepper and possibly from the ancient Malayalam/Tamil name of betel vine = *vettila*]

Published in *Species Plantarum* 1: 28–29. 1753

Synonym: *Chavica betle* Miq.
Local names: *daun sireh* (Bajau, Kedayan); *daing* (Dusun, Kadazan); *sirih* (Bajau, Dusun); *molur malat* (Murut)
Common names: betel pepper, betel vine
Habitat: cultivated
Geographical distribution: India, Bangladesh, Myanmar, Cambodia, Laos, Vietnam, Thailand, Southern China, Malaysia, Indonesia, the Philippines
Botanical description: this climbing plant grows up to a length of about 3 m. Stems: articulate, swollen and rooting at nodes, ligneous. Leaves: simple, spiral, stipulate. Stipules: about 1.5 cm long, caducous. Petioles: 2–5 cm long, channeled, somewhat hairy. Blades: 5–11 cm × 7–5 cm, ovate to oblong, light green beneath, glossy above, with 3 pairs of secondary nerves coming from the base, cordate, often asymmetrical at base, acuminate at apex. Spikes: axillary, cylindrical whitish, about 5.5 cm long. Stamens: 2. Stigma: 4–5-lobed. Drupes: globose, about 3 mm diameter, greenish.
Traditional therapeutic indications: bone pains, coughs, cuts, scabies (Dusun); stomachache (Bajau; Dusun); hypertension (Bajau, Rungus); fever (Kedayan); toothaches (Bajau, Iranum); diabetes (Bajau, Dusun, Kadazan); bad breath, coughs, nosebleeds (Bajau, Dusun, Kadazan); cuts, boils, scabies, nose bleeds, flatulence, skin diseases, insect bites, foul body odor (Dusun, Kadazan); abscesses, asthma, gout, skin itchiness, skin diseases, ulcers (Bajau)
Pharmacology and phytochemistry: this plant has been the subject of numerous pharmacological studies which have highlighted, among other things, antibacterial (Limsuwan et al. 2009; Wannissorn et al. 2005), antiviral (dengue virus) (Dewi et al. 2021), antiarthritic (Hegde et al. 2018), analgesic anti-inflammatory (Alam et al. 2013; De et al. 2013; Reddy et al. 2016), anthelmintic (Ali & Mehta 1970), leishmanicidal (Sarkar et al. 2008), antinematodal (Singh et al. 2009), antiplasmodial (Al-Adhroey et al. 2011), anticoccidial (Leesombun et al. 2017), antiamoebic (Pecková et al. 2018; Sawangjaroen et al. 2006), cardioprotective (Arya et al. 2010), antidiabetic (Arambewela et al. 2005), and antidepressant activity (Gulhane et al. 2015). Piperine alkaloids mainly piperine (Bao et al. 2014; Tiwari et al. 2020), phenylpropanoids (chavicol, allylpyrocatechol) (Evans et al. 1984), lignans (-)-acuminatin, (-)-denudatin B, and puberulin D (Sun et al. 2016).
Toxicity, side effects, and drug interaction: the median lethal dose (LD_{50}) of an hydroalcoholic extract administered to rabbits was above 2 g/kg (Hegde et al. 2018).

Comments:

(i) The British anthropologist Ivor Hugh Norman Evans (1886–1957), who served as a colonial administrator in northern Borneo in the early 20th century, writes on page 115 of his book *Among the Primitive Peoples of Borneo* (1922):

> The ingredients of the quid, as made up in Borneo, are a sireh leaf – is a climbing pepper – a piece of nut from the betel palm, a piece of gambier, which is bought from the Chinese in small cubes, a little native-grown tobacco, and a smear of lime obtained by burning sea- or fresh-water shells or coral.

The climbing pepper in question is *Piper betle* L., betel palm stands for *Areca catechu* L., and gambier is an aqueous extract of *Uncaria gambier* (W. Hunter) Roxb. Reading through Evans' texts, one can feel a great affection and attachment for the Dusun and the land of North Borneo. This refined and intelligent scholar died in Labuan in 1957.

(ii) The Dusun use plants of the genus *Piper* L. (1753), such as "*tangkar*" for skin diseases, "*bohuton*" for caterpillar stings, and "*bonsodon*" for ant bites.

REFERENCES

Al-Adhroey, A.H., Nor, Z.M., Al-Mekhlafi, H.M., Amran, A.A. and Mahmud, R. 2011. Antimalarial activity of methanolic leaf extract of Piper betle L. *Molecules*. 16(1): 107–118.

Alam, B., Akter, F., Parvin, N., Sharmin Pia, R., Akter, S., Chowdhury, J., Sifath-E-Jahan, K. and Haque, E. 2013. Antioxidant, analgesic and anti-inflammatory activities of the methanolic extract of Piper betle leaves. *Avicenna Journal of Phytomedicine*. 3(2): 112.

Ali, S.M. and Mehta, R.K. 1970. Preliminary pharmacological and anthelmintic studies of the essential oil of Piper betle. *Indian Journal of Pharmacy*. 32: 132–133.

Arambewela, L.S.R., Arawwawala, L.D.A.M. and Ratnasooriya, W.D. 2005. Antidiabetic activities of aqueous and ethanolic extracts of Piper betle leaves in rats. *Journal of Ethnopharmacology*. 102(2): 239–245.

Arya, D.S., Arora, S., Malik, S., Nepal, S., Kumari, S. and Ojha, S. 2010. Effect of Piper betle on cardiac function, marker enzymes, and oxidative stress in isoproterenol-induced cardiotoxicity in rats. *Toxicology Mechanisms and Methods*. 20(9): 564–571.

Bao, N., Ochir, S., Sun, Z., Borjihan, G. and Yamagishi, T. 2014. Occurrence of piperidine alkaloids in Piper species collected in different areas. *Journal of Natural Medicines*. 68(1): 211–214.

De, S., Maroo, N., Saha, P., Hazra, S. and Chatterjee, M. 2013. Ethanolic extract of *Piper betle* Linn. leaves reduces nociception via modulation of arachidonic acid pathway. *Indian Journal of Pharmacology*. 45(5): 479.

Dewi, B.E., Tannardi, K. and Sudiro, T.M. 2021. The potency of *Piper betle* leave extract as antiviral drug to dengue virus *in vitro*. *Teikyo Medical Journal*. 44(2): 677–686.

Evans, P.H., Bowers, W.S. and Funk, E.J. 1984. Identification of fungicidal and nematocidal components in the leaves of *Piper betle* (Piperaceae). *Journal of Agricultural and Food Chemistry*. 32(6): 1254–1256.

Gulhane, H., Misra, A.K., Reddy, P., Pandey, D., Gulhane, R. and Varma, S.K. 2015. Effects of Piper betle leaves (paan) extract as anti-depressant and anti-anxiety in experimental animals. *Mintage Journal of Ph armaceutical and Medical Sciences*. 12–15.

Hegde, K., Emani, A., Shrijani, J.K. and Shabaraya, A.R. 2018. Anti arthritic potentials of *Piper betle* – A preclinical study. *Indian Journal of Pharmacy and Pharmacology*. 5(1): 21–28.

Leesombun, A., Boonmasawai, S. and Nishikawa, Y. 2017. Effects of Thai piperaceae plant extracts on *Neospora caninum* infection. *Parasitology International*. 66(3): 219–226.

Limsuwan, S., Subhadhirasakul, S. and Voravuthikunchai, S.P. 2009. Medicinal plants with significant activity against important pathogenic bacteria. *Pharmaceutical Biology*. 47(8): 683–689.

Pecková, R., Doležal, K., Sak, B., Květoňová, D., Kváč, M., Nurcahyo, W. and Foitová, I. 2018. Effect of *Piper betle* on *Giardia intestinalis* infection *in vivo*. *Experimental Parasitology*. 184: 39–45.

Reddy, P.S., Gupta, R.K. and Reddy, S.M. 2016. Analgesic and anti-inflammatory activity of hydroalcoholic extract of Piper betle leaves in experimental animals. *International Journal of Basic & Clinical Pharmacology*. 5: 979–985.

Sarkar, A., Sen, R., Saha, P., Ganguly, S., Mandal, G. and Chatterjee, M. 2008. An ethanolic extract of leaves of *Piper betle* (Paan) Linn mediates its leishmanicidal activity via apoptosis. *Parasitology Research*. 102(6): 1249.

Sawangjaroen, N., Phongpaichit, S., Subhadhirasakul, S., Visutthi, M., Srisuwan, N. and Thammapalerd, N. 2006. The anti-amoebic activity of some medicinal plants used by AIDS patients in southern Thailand. *Parasitology Research*. 98(6): 588–592.

Singh, M., Shakya, S., Soni, V.K., Dangi, A., Kumar, N. and Bhattacharya, S.M. 2009. The n-hexane and chloroform fractions of *Piper betle* L. trigger different arms of immune responses in BALB/c mice and exhibit antifilarial activity against human lymphatic filarid Brugia malayi. *International Immunopharmacology*. 9(6): 716–728.

Sun, Z.L., He, J.M., Wang, S.Y., Ma, R., Khondkar, P., Kaatz, G.W., Gibbons, S. and Mu, Q. 2016. Benzocyclohexane oxide derivatives and neolignans from Piper betle inhibit efflux-related resistance in Staphylococcus aureus. *RSC Advances*. 6(49): 43518–43525.

Tiwari, A., Mahadik, K.R. and Gabhe, S.Y. 2020. Piperine: A comprehensive review of methods of isolation, purification, and biological properties. *Medicine in Drug Discovery*. 7: 100027.

Wannissorn, B., Jarikasem, S., Siriwangchai, T. and Thubthimthed, S. 2005. Antibacterial properties of essential oils from Thai medicinal plants. *Fitoterapia*. 76(2): 233–236.

Piper caducibracteum **C. DC**

[From the Greek *peperi* = pepper and Latin *caducus* = falling, and *bracteatum* = bearing bracts]

Published in *An Interpretation of Rumphius's Herbarium Amboinense* 183. 1917

Local name: *sirih hutan* (Kedayan)
Habitat: forests
Geographical distribution: Borneo, Maluku Islands, Papua New Guinea
Botanical description: this climbing plant grows up to a length of about 4 m. Stems: articulate and swollen at nodes. Leaves: simple, spiral, graceful, exstipulate. Petioles: about 2 cm long, forming a sheath at base. Blades: about 8 cm × 18 cm, broadly lanceolate, with 2–3 pairs of longitudinal secondary nerves coming from the base, membranous, rounded, and asymmetrical at base, acuminate at apex, somewhat wavy. Spikes: thin, up to about 8 cm long. Stamens: 2. Stigma: 4-5-lobed. Drupes: obovoid, tiny diameter, black (Figure 4.3).

FIGURE 4.3 *Piper caducibracteum* C. DC.

Traditional therapeutic indications: earache, excessive menstruation, itchy skin (Kedayan)
Pharmacology and phytochemistry: inhibition of antiplatelet (Fakhrudin et al. 2021).
Toxicity, side effects, and drug interaction: not known

Comment:

In Lombok, Indonesia, the plant is called "*lekoq gawah*" and is used as a dewormer.

REFERENCE

Fakhrudin, N., Mufinnah, F.F., Husni, M.F., Wardana, A.E., Wulandari, E.I., Putra, A.R., Santosa, D., Nurrochmad, A. and Wahyuono, S. 2021. Screening of selected Indonesian plants for antiplatelet activity. *Biodiversitas Journal of Biological Diversity*. 22(12): 5268–5273.

***Piper caninum* Bl.**

[From the Greek *peperi* = pepper and Latin *canis* = dog]

Published in *Verhandelingen van het Bataviaasch Genootschap van Kunsten en Wetenschappen* 11: 214, f. 26. 1826

Synonyms: *Cubeba canina* (Bl.) Miq.; *Piper lanatum* Roxb.
Local name: *kimput pilot* (Murut)
Habitat: near streams and waterfalls in forests
Geographical distribution: from India to the Solomon Islands
Botanical description: this climbing plant grows up to a length of 20 m. Stems: swollen and rooting at nodes, hairy at apex. Leaves: simple, spiral, stipulate. Stipules: lanceolate, tiny. Petioles: 0.7–2 cm long, hairy, curved. Blades: elliptic, oblong, ovate, or cordate, somewhat asymmetrical, 2.5–7 cm × 3–15.5 cm, glossy above, glaucous beneath, rounded, attenuate, or cordate at base, acuminate at apex, with 3–4 pairs of secondary nerves sunken above and coming from the base. Spikes: cylindrical, up to about 2.5 cm long, pure white. Stamens: 3. Ovary: elliptic. Stigma: 4-lobed. Drupes: dull light red, globose, pedicelled, about 4 mm across.
Traditional therapeutic indications: medicinal (Murut)
Pharmacology and phytochemistry: antibacterial (Salleh et al. 2011), antifungal (Suriani et al. 2015). Antibacterial stilbenes, flavones, and alkylamide alkaloids (Salleh et al. 2015). Antibacterial caffeic acid derivatives (+)-bornyl *p*-coumarate and bornyl caffeate) (Setzer et al. 1999). Amide alkaloids (*N*-cis-feruloyl tyramine, *N*-trans-feruloyltyramine) and pyrrolidine alkaloids (1-cinnamoylpyrrolidine) (Ma et al. 2004). Aporphine alkaloids (cepharadione A) (Ma et al. 2004a).
Toxicity, side effects, and drug interaction: the median lethal dose (LD_{50}) of an ethanol extract administered to mice was above 2 g/kg (Suarni et al. 2022).

Comment: In the Philippines, the plant is used by the Ati Negritos of Guimara Island (local name: *kanuyom*) for toothaches. They are the first inhabitants of the Philippines indicating that this plant might have been used in this part of the world since the dawn of time.

REFERENCES

Ma, J., Jones, S.H. and Hecht, S.M. 2004. Phenolic acid amides: A new kind of DNA strand scission agent from *Piper caninum*. *Bioorganic & Medicinal Chemistry*. 12(14): 3885–3889.

Ma, J., Jones, S.H., Marshall, R., Johnson, R.K. and Hecht, S.M. 2004a. A DNA-damaging oxoaporphine alkaloid from *Piper caninum*. *Journal of Natural Products*. 67(7): 1162–1164.

Salleh, W.M.N.H.W., Ahmad, F. and Yen, K.H. 2015. Chemical constituents from *Piper caninum* and antibacterial activity. *Journal of Applied Pharmaceutical Science*. 5(6): 20–25.

Salleh, W.M.N.H.W., Ahmad, F., Yen, K.H. and Sirat, H.M. 2011. Chemical compositions, antioxidant and antimicrobial activities of essential oils of *Piper caninum* Blume. *International Journal of Molecular Sciences*. 12(11): 7720–7731.

Setzer, W.N., Setzer, M.C., Bates, R.B., Nakkiew, P., Jackes, B.R., Chen, L., McFerrin, M.B. and Meehan, E.J. 1999. Antibacterial hydroxycinnamic esters from *Piper caninum* from Paluma, north Queensland, Australia. The crystal and molecular structure of (+)-bornyl coumarate. *Planta Medica*. 65(8): 747–749.

Suarni, N.M.R., Ermayanti, N.G.A.M., Suyasa, D.G.N.A., Sudatri, N.W. and Dewi, G.A.M.K. 2022. Quality of male mice (*Mus musculus* L.) spermatozoes which is given ethanol leaves extract of *Piper caninum*. *Eastern Journal of Agricultural and Biological Sciences*. 2(3): 23–30.

Suriani, N.L., Suprapta, D.N., Sudana, I.M., Temaja, I.R.M. and Indonesia, D.B. 2015. Antifungal activity of *Piper caninum* against *Pyricularia oryzae* Cav. The cause of rice blast disease on rice. *Methods*. 5(8): 72–78.

***Piper sarmentosum* Roxb.**

[From the Greek *peperi* = pepper and Latin *sarmentosus* = full of twigs]

Published in *Flora Indica; or descriptions of Indian Plants* 1: 162–163. 1820

Synonyms: *Chavica hainana* C. DC.; *Chavica sarmentosa* (Roxb.) Miq.; *Piper albispicum* C. DC.; *Piper baronii* C. DC.; *Piper brevicaule* C. DC.; *Piper lolot* C. DC.; *Piper pierrei* C. DC.

Local name: *kaduk* (Bajau)

Habitat: cultivated, villages

Geographical distribution: from India to Australia

Botanical description: this climbing plant grows up to a length of about 10 m. Stems: swollen, rooting at nodes. Leaves: simple, spiral, exstipulate. Petioles: 2–5 cm long, pubescent. Blades: ovate, somewhat glossy, 6–13 cm × 7–14 cm, cordate at base, acute at apex, with 3 longitudinal pairs of secondary nerves coming from the base as well as scalariform tertiary nerves sunken above. Spikes: up to about 5 cm long, white. Stamens: 2. Stigma: 4-lobed, hairy. Drupes: somewhat globose, dark green, globose, about 3 mm across.

Traditional therapeutic indications: coughs, flu, malaria, ringworms, toothaches (Kedayan)

Pharmacology and phytochemistry: hypoglycemic (Peungvicha et al. 1998). Antiplasmodial, antimycobacterial, antifungal (sarmetine, 1-piperettyl pyrrolidine) amide alkaloids (Shi et al. 2017). Lignans (sesamine) (Rukachaisirikul et al. 2004; Tuntiwachwuttikul et al. 2006). Antibacterial phenylpropanoids (1-allyl-2,6-dimethoxy-3,4-methylenedioxybenzene) (Masuda et al. 1991).

Toxicity, side effects, and drug interaction: the median lethal dose (LD_{50}) of an aqueous extract administered to rats was above 10 g/kg (Peungvicha et al. 1998).

REFERENCES

Masuda, T., Inazumi, A., Yamada, Y., Padolina, W.G., Kikuzaki, H. and Nakatani, N. 1991. Antimicrobial phenylpropanoids from *Piper sarmentosum*. *Phytochemistry*. 30(10): 3227–3228.

Peungvicha, P., Thirawarapan, S.S., Temsiririrkkul, R., Watanabe, H., Prasain, J.K. and Kadota, S. 1998. Hypoglycemic effect of the water extract of *Piper sarmentosum* in rats. *Journal of Ethnopharmacology*. 60(1): 27–32.

Rukachaisirikul, T., Siriwattanakit, P., Sukcharoenphol, K., Wongvein, C., Ruttanaweang, P., Wongwattanavuch, P. and Suksamrarn, A. 2004. Chemical constituents and bioactivity of *Piper sarmentosum*. *Journal of Ethnopharmacology*. 93(2–3): 173–176.

Shi, Y.N., Liu, F.F., Jacob, M.R., Li, X.C., Zhu, H.T., Wang, D., Cheng, R.R., Yang, C.R., Xu, M. and Zhang, Y.J. 2017. Antifungal amide alkaloids from the aerial parts of *Piper flaviflorum* and *Piper sarmentosum*. *Planta Medica*. 83(1/2): 143–150.

Tuntiwachwuttikul, P., Phansa, P., Pootaeng-On, Y. and Taylor, W.C. 2006. Chemical constituents of the roots of *Piper sarmentosum*. *Chemical and Pharmaceutical Bulletin*. 54(2): 149–151.

Piper umbellatum **L.**

[From the Greek *peperi* = pepper and Latin *umbellatum* = umbellate]

Published in *Species Plantarum* 1: 30. 1753

Synonyms: *Heckeria umbellata* (L.) Kunth; *Lepianthes umbellata* (L.) Raf. ex Ramamoorthy; *Pothomorphe umbellata* (L.) Miq.
Local name: *kuyoh* (Dusun)
Common name: cow-foot leaf
Habitat: forests, roadsides
Geographical distribution: tropical
Botanical description: this shrubby herbaceous plant reaches a height of about 4 m. Stems: striated. Leaves: simple, spiral, exstipulate. Petioles: thin, 15–30 cm long, channeled, forming a sheath at base. Blades: somewhat broadly cordate, up to about 30 cm across, pale green above, glaucous beneath, somewhat serrulate, with 11–13 pairs of secondary nerves coming from the base. Spikes: 5–10 cm long, whitish, linear, which can make one think of repulsive worms, arranged in some kind in axillary fascicles. Flowers: tiny. Stamens: 2. Stigma: trifid. Drupes: trigonal, tiny, brownish.
Traditional therapeutic indications: boils in the ears (Dusun)
Pharmacology and phytochemistry: antibacterial (da Silva et al. 2014), toxic to Ehrlich solid tumor cells (Iwamoto et al. 2015), antifilarial (Cho-Ngwa et al. 2016).
Aristolactam alkaloids (piperumbellactams A—D), amide alkaloids (*N*-p-coumaroyl tyramine, *N*-trans-feruloyl tyramine), phenolic compound (4-nerolidylcatechol) (Tabopda et al. 2008). 4-Nerolidylcatechol inhibited phospholipase A2 (Núñez et al. 2005).
Toxicity, side effects, and drug interaction: not known

Comments:

(i) The plant is sold as a vegetable in the wet markets of Sabah.
(ii) In Sarawak, the plant is used as food and affords a remedy for fatigue and malaria by the Ulu.

REFERENCES

Cho-Ngwa, F., Monya, E., Azantsa, B.K., Manfo, F.P.T., Babiaka, S.B., Mbah, J.A. and Samje, M. 2016. Filaricidal activities on *Onchocerca ochengi* and Loa loa, toxicity and phytochemical screening of extracts of *Tragia benthami* and *Piper umbellatum*. *BMC Complementary and Alternative Medicine*. 16: 1–9.

da Silva Jr, I.F., de Oliveira, R.G., Soares, I.M., da Costa Alvim, T., Ascêncio, S.D. and de Oliveira Martins, D.T. 2014. Evaluation of acute toxicity, antibacterial activity, and mode of action of the hydroethanolic extract of *Piper umbellatum* L. *Journal of Ethnopharmacology*. 151(1): 137–143.

Iwamoto, L.H., Vendramini-Costa, D.B., Monteiro, P.A., Ruiz, A.L.T.G., Sousa, I.M.D.O., Foglio, M.A., de Carvalho, J.E. and Rodrigues, R.A.F. 2015. Anticancer and anti-inflammatory activities of a standardized dichloromethane extract from *Piper umbellatum* L. leaves. *Evidence-Based Complementary and Alternative Medicine*. 2015: 1–8.

Núñez, V., Castro, V., Murillo, R., Ponce-Soto, L.A., Merfort, I. and Lomonte, B. 2005. Inhibitory effects of *Piper umbellatum* and *Piper peltatum* extracts towards myotoxic phospholipases A2 from Bothrops snake venoms: Isolation of 4-nerolidylcatechol as active principle. *Phytochemistry*. 66(9): 1017–1025.

Tabopda, T.K., Ngoupayo, J., Liu, J., Mitaine-Offer, A.C., Tanoli, S.A.K., Khan, S.N., Ali, M.S., Ngadjui, B.T., Tsamo, E., Lacaille-Dubois, M.A. and Luu, B. 2008. Bioactive aristolactams from *Piper umbellatum*. *Phytochemistry*. 69(8): 1726–1731.

4.1.2.2 Order Laurales Juss. ex Bercht. & J. Presl (1820)

4.1.2.2.1 Family Lauraceae Juss. (1789), the Laurel Family

The family Lauraceae consists of about 50 genera and 2,000 species of trees, shrubs, climbing plants, and herbaceous plants. Leaves: simple, opposite, spiral, whorled, or alternate, exstipulate. Inflorescences: racemes or clusters of tiny flowers. Tepals: 2 sets of 3. Stamens: 8–16, poricidal anthers. Carpel: 1, forming a unilocular ovary with 1 ovule. Fruits: drupes or berries characteristically on a persistent eggcup like perianth.

***Cinnamomum iners* Reinw. ex Blume**

[From Hebrew *ganeh* = a reed and Latin *iners* = inert]

Published in *Bijdragen tot de flora van Nederlandsch Indië* (11): 570. 1825

Local name: *medang teja* (Bajau)
Common name: wild cinnamon
Habitat: roadsides, parks, cultivated
Geographical distribution: Southeast Asia
Botanical description: this handsome tree reaches a height of about 10 m. Bark: grayish, thick, slightly aromatic. Leaves: somewhat opposite, simple, exstipulate. Petioles: about 5 mm–1 cm long, channeled, stout, somewhat curved. Blades: coriaceous, fragrant when crushed, light red when young, with 1 pair of secondary nerves coming from the base, oblong-elliptic to lanceolate, 4–7 cm × 8–12 cm, rounded to attenuate at base, acute at apex. Panicles: axillary, thin, up to 16 cm long, light yellowish. Tepals: 6, elliptic, about 2 mm long, somewhat hairy. Stamens: 9, tiny. Ovary: ovoid. Stigma: enlarged, thin. Berries: olive-shaped, having a vague resemblance to the fruits of the laurel tree, about 1 cm long, on a persistent eggcup-like perianth.
Traditional therapeutic indications: joints pain. coughs (Dusun, Kadazan)
Pharmacology and phytochemistry: antibacterial (Mustaffa et al. 2011). Toxic to human colorectal tumor cells (Ghalib et al. 2012). Essential oil (Son et al. 2013), flavanols (5,7-dimethoxy-3′,4′-methylenedioxyflavan-3-ol) (Espineli et al. 2013).
Toxicity, side effects, and drug interaction: not known

Comment: Dusun and Kadazan use plants of the genus *Cinnamomum* Schaeff (1760) as medicines: "*kusur*" for stomach aches, "*makalabau*" for beriberi, whilst the Bajau use "*kayu manis*" for colds and sinusitis.

REFERENCES

Espineli, D.L., Agoo, E.M.G., Shen, C.C. and Ragasa, C.Y. 2013. Chemical constituents of *Cinnamomum iners*. *Chemistry of Natural Compounds*. 49: 932–933.

Ghalib, R.M., Hashim, R., Sulaiman, O., Mehdi, S.H., Anis, Z., Rahman, S.Z., Ahamed, B.K. and Abdul Majid, A.M.S. 2012. Phytochemical analysis, cytotoxic activity and constituents—activity relationships of the leaves of *Cinnamomum iners* (Reinw. ex Blume-Lauraceae). *Natural Product Research*. 26(22): 2155–2158.

Mustaffa, F., Indurkar, J., Ismail, S., Shah, M. and Mansor, S.M. 2011. An antimicrobial compound isolated from *Cinnamomum iners* leaves with activity against methicillin-resistant *Staphylococcus aureus*. *Molecules*. 16(4): 3037–3047.

Son, L.C., Dai, D.N., Thai, T.H., Huyen, D.D., Thang, T.D. and Ogunwande, I.A. 2013. The leaf essential oils of four Vietnamese species of *Cinnamomum* (Lauraceae). *Journal of Essential Oil Research*. 25(4): 267–271.

***Lindera pipericarpa* Boerl.**

[After the 18th century Swedish botanist Johan Linder and Latin *pipericarpa* = pepper-like fruit]

Published in *Handleiding tot de Kennis der Flora van Nederlandsch Indië* 3: 147. 1900

Synonyms: *Aperula pipericarpa* Meisn.; *Benzoin pipericarpum* (Meisn.) Kuntze; *Litsea pipericarpa* (Miq.) Kosterm.; *Polyadenia pipericarpa* Miq.

Local names: *laindos* (Dusun; Murut); *tanom* (Lundayeh)

Habitat: forests

Geographical distribution: Malaysia, Indonesia

Botanical description: this tree reaches a height of about 10 m. Stems: hairy at apex. Leaves: simple, alternate, exstipulate. Petioles: about 1–2 cm long. Blades: membranous, lanceolate to oblong, glossy, somewhat hairy and glaucous beneath, 2–5 cm × 6–15 cm, acute to rounded at base, acuminate at apex, finely marked with 5–12 pairs of secondary nerves. Inflorescences: axillary racemes of umbellules with a length of about 1 cm, hairy. Tepals: 6. Flowers: tiny. Stamens: 9. Drupes: globose, with a delicate lemon fragrance, about 5 mm across, shortly beaked, black, on a persistent eggcup-like perianth.

Traditional therapeutic indications: snakebite antidote (Murut); flatulence (Dusun, Kadazan); colic (Lundayeh)

Pharmacology and phytochemistry: essential oil (Ali & Jantan 1993). Aporphine alkaloids (laurotetanine, lindcarpine) (Kiang et al. 1961, Kiang & Sim 1967; Kam 1999). Long chain phenolics (linderone, methyl linderone) (Kiang et al. 1962).

Toxicity, side effects, and drug interaction: the plant is poisonous (Kiang et al. 1962).

Comments:

(i) The plant is used medicinally by the Kenyah of East Kalimantan.
(ii) Murut use the fruits to poison chicken.

REFERENCES

Ali, N.A.M. and Jantan, I. 1993. The essential oils of *Lindera pipericarpa*. *Journal of Tropical Forest Science*. 124–130.

Kam, T.S. 1999. Alkaloids from Malaysian flora. In *Alkaloids: Chemical and Biological Perspectives* (14: 285–435). Pergamon.

Kiang, A.K., Douglas, B. and Morsingh, F. 1961. A phytochemical survey of Malaya: Part II. Alkaloids. *Journal of Pharmacy and Pharmacology*. 13(1): 98–104.

Kiang, A.K., Lee, H.H. and Sim, K.Y., 1962. 843. The structure of linderone and methyl-linderone. *Journal of the Chemical Society (Resumed)*, 4338–4345.

Kiang, A.K. and Sim, K.Y. 1967. Lindcarpine, an alkaloid from *Lindera pipericarpa* Boerl (Lauraceae). *Journal of the Chemical Society C: Organic*: 282–283.

***Litsea accedens* (Bl.) Boerl.**

[From Chinese *li* = small, *tse* = plum, and Latin *accedens* = close to]

Published in *Handleiding tot de Kennis der Flora van Nederlandsch Indië* 3: 145. 1900

Synonyms: *Cylicodaphne costata* Bl.; *Cylicodaphne ochracea* Bl.; *Litsea singaporensis* Gamble; *Litsea kunstleri* Gamble; *Litsea perakensis* Gamble; *Litsea patellaris* Gamble; *Litsea patellaris* Gamble; *Litsea pustulata* Gamble; *Litsea wrayi* Gamble; *Litsea oblanceolata* Gamble
Habitat: forests
Geographical distribution: Malaysia, Indonesia, the Philippines
Botanical description: this tree reaches a height of about 25 m. Bark: grayish brown to gray. Wood: fragrant. Stems: terete or angular, glabrous. Leaves: simple, alternate, exstipulate. Petioles: 1–3 cm long, somewhat swollen. Blades: coriaceous, elliptic, 3–12 cm × 7–35 cm, attenuate at base, obtuse to sharply acute at apex, somewhat asymmetrical, with 7–14 pairs of secondary nerves, dark green and glossy above, somewhat glaucous beneath. Inflorescences: clusters of few-flowered umbels, 2 cm long, cauliflorous, or axillary. Perianth: tiny, 6-lobed, yellowish to whitish, the lobes linear-oblong, about 5 mm long, hairy. Stamens: 9–16. Stigma: peltate. Drupes: globose, about 1.5 cm long, shortly beaked at apex, smooth, pure white with somewhat the appearance of ceramic marbles, turning red, glossy, fragrant, on woody *Lithocarpus*-like cupules.
Traditional therapeutic indications: medicinal
Pharmacology and phytochemistry: not known
Toxicity, side effects, and drug interaction: not known

***Litsea garciae* Vidal.**

[From Chinese *li* = small, *tse* = plum, *garciae* probably after a botanist of the Philippines]

Published in *Revision de Plantas Vasculares Filipinas* 228. 1886

Synonyms: *Cylicodaphne garciae* (Vidal) Nakai; *Lepidadenia kawakamii* (Hayata) Masamune; *Litsea aurea* Kosterm.; *Litsea griseola* Elmer; *Litsea kawakamii* Hayata; *Litsea macrophylla* Elmer; *Litsea sebifera* Pers.; *Tetranthera calophylla* Miq.; *Tetradenia kawakamii* (Hayata) Makino & Nemoto
Local names: *novolo, pengolaban* (Murut); *pengalaban* (Kedayan); *pengalaban burong* (Brunei); *talus dara* (Dusun)
Common name: Borneo avocado
Habitat: forests
Geographical distribution: Borneo, the Philippines, Taiwan
Botanical description: this tree reaches a height of about 35 m. Trunk: straight. Bark: light gray, smooth, lenticelled. Sapwood: white. Stems: stout. Leaves: simple, alternate, exstipulate. Petioles: stout, up to 5 mm across, channeled, about 2 cm long. Blades: stout, ovate to lanceolate, somewhat asymmetrical, 6–15 cm × 25–40 cm, attenuate at base, obtuse at apex, with 15–20 pairs of secondary nerves, somewhat fleshy, glossy above glaucous beneath. Inflorescences: axillary umbels. Tepals: 6. Flowers: tiny. Drupes: somewhat oblate, on leafless stems, up to about 4 cm across, light pink, somewhat jambu colored, on a persistent eggcup-like perianth. The fruits are eaten by hornbills (Figure 4.4).
Traditional therapeutic indications: joint dislocation, sprains (Murut)
Pharmacology and phytochemistry: antibacterial (Wulandari et al. 2018), antifungal (Johnny et al. 2010), cytotoxic (Kutoi et al. 2012).
Toxicity, side effects, and drug interaction: not known

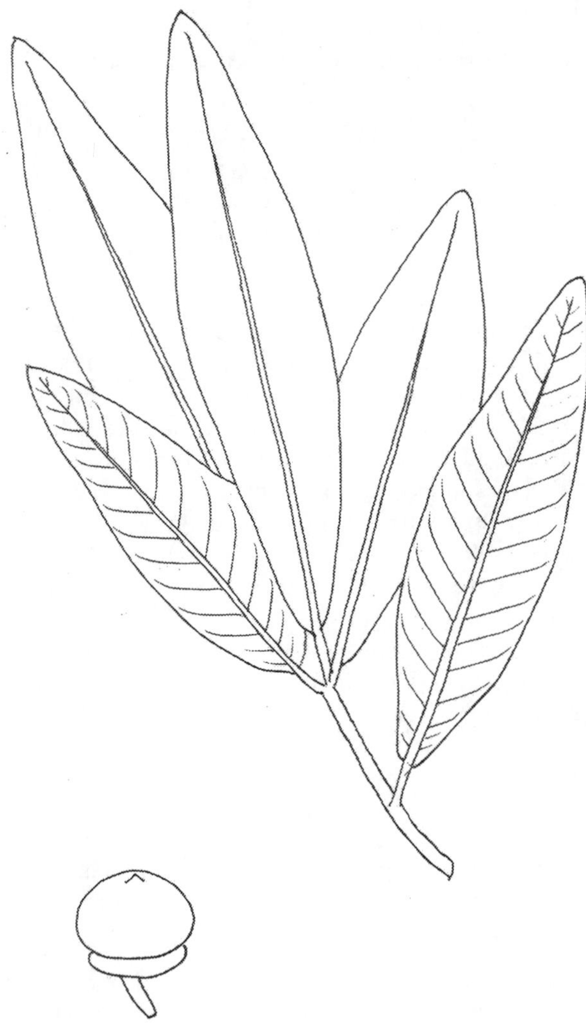

FIGURE 4.4 *Litsea garciae* Vidal.

Comments:

(i) In Kalimantan, the plant is called "*kangkala*" and the fruits are used as food.
(ii) In Sarawak, the plant is called "*engkala*" and is used for poisonous animal bites.

REFERENCES

Johnny, L., Yusuf, U.K. and Nulit, R. 2010. The effect of herbal plant extracts on the growth and sporulation of *Colletotrichum gloeosporioides*. *Journal of Applied Biosciences*. 34: 2218–2224.

Kutoi, C.J., Yen, K.H. and Seruji, N.M.U. 2012, April. Pharmacology evaluation of *Litsea garciae* (Lauraceae). In *2012 IEEE Business, Engineering & Industrial Applications Colloquium (BEIAC)* (pp. 31–33). IEEE.

Wulandari, I., Kusuma, I.W. and Kuspradini, H. 2018, April. Antioxidant and antibacterial activity of *Litsea garciae*. In *IOP Conference Series: Earth and Environmental Science* (144, No. 1: 012024). IOP Publishing.

***Litsea odorifera* Valeton**

[From Chinese *li* = small, *tse* = plum, and Latin *odorifera* = fragrant]

Published in *Icones Bogorienses* 3: t. 276. 1909

Synonyms: *Listea elliptica* Bl.; *Litsea petiolata* Hook.f.; *Litsea scortechinii* Gamble; *Malapoenna petiolata* (Hook.f.) Kuntze; *Tetranthera elliptica* var. *sanguinea* Hassk.
Local names: *lawang* (Murut); *tindas* (Bajau)
Common name: trawas oil tree
Habitat: forests
Geographical distribution: from Thailand to Papua New Guinea
Botanical description: this tree reaches a height of about 40 m. Trunk: straight, buttressed. Bark: smooth, with lenticels, grayish brown, fragrant. Leaves: simple, alternate, exstipulate. Petioles: about 1–3 cm long. Blades: elliptic, somewhat coriaceous, fragrant, elliptic to lanceolate, 2–7.8 cm × 6–20 cm, dark green, attenuate at base, acute to rounded, or acuminate at apex, with 4–10 pairs of secondary nerves sunken above. Inflorescences: umbels, about 1–3 cm long, cauliflorous, or axillary. Perianth: 6-lobed, the lobes obovate to lanceolate, somewhat hairy, and about 4 mm long. Stamens: 9–12. Ovary: globose, tiny. Stigma: peltate. Drupes: olive-shaped, up to about 1 cm long, glossy, and ripening reddish-purple to black, on a persistent eggcup-like perianth. Seeds: elliptic, pointed at apex.
Traditional therapeutic indications: diarrhea, skin diseases, fatigue, bone aches (Dusun, Kadazan); gastritis, stomachaches (Dusun, Kadazan, Murut)
Pharmacology and phytochemistry: essential oil (undecan-2-one) (Van Romburgh 1911). Toxic to non-small cell lung cancer cells (A549) (Goh et al. 2022). Aporphine alkaloids (boldine, actinodaphnine), benzylisoquinoline alkaloids (reticuline) (Goh 1999).
Toxicity, side effects, and drug interaction: the median lethal dose (LD_{50}) of essential oil administered to rats was 3.4 g/kg (Budin et al. 2012). At a dose of 500 mg/kg, the essential oil induced the formation of echinocytes in rats (Taib et al. 2009).

Comment: Pure essential oils should never be ingested or applied to mucosas.

REFERENCES

Budin, S.B., Siti Nor Ain, S.M., Omar, B., Taib, I.S. and Hidayatulfathi, O. 2012. Acute and subacute oral toxicity of *Litsea elliptica* Blume essential oil in rats. *Journal of Zhejiang University Science B*. 13(10): 783–790.

Goh, M.P.Y., Kamaluddin, A.F., Tan, T.J.L., Yasin, H., Taha, H., Jama, A. and Ahmad, N. 2022. An evaluation of the phytochemical composition, antioxidant and cytotoxicity of the leaves of *Litsea elliptica* Blume—An ethnomedicinal plant from Kedayan Darussalam. *Saudi Journal of Biological Sciences*. 29(1): 304–317.

Goh, S.H. 1999. Alkaloids from *Litsea ellipticea* (Lauraceae) *Malaysian Journal of Science*. 18(1): 63–66.

Taib, I.S., Budin, S.B., Siti Nor Ain, S.M., Mohamed, J., Louis, S.R., Das, S., Sallehudin, S., Rajab, N.F. and Hidayatulfathi, O. 2009. Toxic effects of *Litsea elliptica* Blume essential oil on red blood cells of Sprague-Dawley rats. *Journal of Zhejiang University Science B*. 10: 813–819.

Van Romburgh, P. 1911. The essential oil of *Litsea odorifera* Val (trawas oil). *Proceedings of the Koninklijke Nederlandse Akademie van Wetenschappen*. 14(Part 1): 325–327.

Litsea umbellata **(Lour.) Merr.**

[From Chinese *li* = small, *tse* = plum, and Latin *umbellata* = umbellate]

Published in *Philippine Journal of Science* 14(2): 242. 1919

Synonyms: *Litsea amara* Bl.; *Litsea gracilis* Gamble; *Litsea cylindrocarpa* Gamble; *Litsea cinerascens* Ridl; *Tetranthera amara* (Blume) Nees; *Tetranthera angusta* Nees; *Tetranthera firma* Bl.

Local name: *medang wangi hitam* (Bajau)

Common name: hairy medang

Habitat: forests

Description: from Myanmar to Papua New Guinea

Botanical description: this tree reaches a height of about 40 m. Trunk: straight, buttressed. Bark: brownish, fragrant. Stems: hairy at apex. Leaves: simple, alternate, exstipulate. Petioles: about 2 cm long, somewhat hairy. Blades: fragrant, somewhat coriaceous, elliptic to obovate or lanceolate, 1.2–7.5 cm × 3–20 cm, rounded to attenuate to oblique at base, acute to rounded to acuminate at apex, with 5–17 pairs of secondary nerves, dark green above, somewhat glaucous beneath. Inflorescences: clusters of umbels, about 2 cm long, axillary, or cauliflorous. Tepals: 5–6, hairy, triangular, about 3 mm long. Stamens: 8–14. Ovary: somewhat olive-shaped. Stigma: peltate. Drupes: somewhat olive-shaped, about 1 cm long, smooth, dark red to purple, turning black, glossy, on a persistent hairy eggcup-like perianth.

Pharmacology and phytochemistry: essential oil (Dai et al. 2020).

Toxicity, side effects, and drug interaction: not known

Comments:

(i) Plants of the genus *Litsea* Lam. (1792) are used by the Dusun as medicines: "*lamou lamou*" for cuts and scabies, "*sileu*" for flatulence, and "*sesulang kupes*" for boils. The Dusun and Kadazan use a plant of the genus *Litsea* Lam. (1792) "*lindos*" to assuage the pain caused by leeches' bites.

(ii) Dusun and Kadazan use a plant of the genus *Actinodaphne* Nees (1831) (local name: *pongulobon kusai*) for waist pain.

(iii) *Eusideroxylon zwageri* Teijsm. & Binn. (local name: *belian*) is used by the Murut to make the poison for their blowgun darts.

(iv) Avocado (*Persea americana* Mill.) is used by the Murut (local names: *buah lemak*, *buah susu*) for the treatment of viral fever.

REFERENCE

Dai, D.N., Hung, N.D., Chung, N.T., Huong, L.T., Hung, N.H. and Ogunwande, I.A. 2020. Chemical constituents of the essential oils from the leaves of *Litsea umbellata* and *Litsea iteodaphne* and their mosquito larvicidal activity. *Journal of Essential Oil Bearing Plants*. 23(6): 1334–1344.

4.1.2.2.2 Family Annonaceae Juss. (1789), the Custard Apple Family

The Family Annonaceae consists of about 130 genera and 2,400 species of tropical trees, shrubs, or woody climbing plants. The inner bark often has a slight delicate fragrance. Bark: often fibrous. Leaves: simple, alternate, exstipulate. Flowers: often terminal or axillary, solitary, and somewhat coriaceous, which can make one think of a monocotyledon. Blades: often more or less coriaceous. Sepals: 3. Petals: 3–6. Stamens: numerous, short, often packed in some kind of disc-like structures. Carpels: 10–100, free, each with 1 or 2 ovules. Fruits: free or fused ripe carpels, fleshy. When free, each ripe carpel consists of a peduncle and a monocarp. Seeds: 1 to many.

***Annona muricata* L.**

[From Latin *annona* = food and *muricata* = muricate]

Published in *Species Plantarum* 1: 536. 1753

Synonyms: *Annona bonplandiana* Kunth; *Annona cearensis* Barb. Rodr.; *Annona macrocarpa* Wercklé; *Guanabanus muricatus* (L.) M. Gómez

Local names: *durian makka, seri kaya Belandah* (Lundayeh); *durian salat* (Kedayan); *hampun kapal* (Dusun, Kadazan); *lampunjong, lampun tuan* (Murut); *nangka Belanda* (Bajau); *durian Belanda* (Bajau, Murut)

Common name: soursop

Habitat: gardens, in villages

Geographical distribution: tropical Asia

Botanical description: this tree reaches a height of about 10 m. Leaves: simple, alternate, exstipulate. Petioles: about 5 mm long. Blades: oblong, lanceolate to obovate, 5.5–18 cm × 2–7 cm, attenuate, acute, or rounded at base, coriaceous, acuminate, or acute at apex, with 6–13 pairs of secondary nerves, hairy beneath. Inflorescences: solitary, axillary. Sepals: 3, ovate-triangular, coriaceous, up to 5 mm long. Petals: 6, coriaceous, broadly lanceolate, greenish-yellow, up to 4 cm long. Stamens: numerous, 5 mm long. Carpels: numerous, 5 mm long, free. Infructescences: fleshy, ovoid, somewhat asymmetrical, muricate, green, heavy, 15–35 cm × 10–15 cm. Seeds: numerous, blackish, in a whitish edible flesh.

Traditional therapeutic indications: coughs (Murut, Dusun, Kadazan); stomachaches, hypertension, nausea, fever, diarrhea (Dusun, Kadazan), cancer (Bajau), asthma, laxative (Murut)

Pharmacology and phytochemistry: antibacterial (Silva et al. 2021), anti-inflammatory, analgesic (Ishola et al. 2014), hypotensive (Nwokocha et al. 2012), antiplasmodial (Nguyen-Pouplin et al. 2007; Boyom et al. 2011), anthelmintic (Ferreira et al. 2013). Cytotoxic acetogenins (solamin) (Myint et al. 1991; Wu et al. 1995). Leishmanicidal acetogenins (annonacinone and corossolone) (Vila-Nova et al. 2011). Isoquinoline alkaloids (Fofana et al. 2012; Castañeda-Ramírez et al. 2020).

Toxicity, side effects, and drug interaction: risk of developing Parkinson's disease (Matsushige et al. 2012).

REFERENCES

Boyom, F.F., Fokou, P.V.T., Yamthe, L.R.T., Mfopa, A.N., Kemgne, E.M., Mbacham, W.F., Tsamo, E., Zollo, P.H.A., Gut, J. and Rosenthal, P.J. 2011. Potent antiplasmodial extracts from Cameroonian Annonaceae. *Journal of Ethnopharmacology*. 134(3): 717–724.

Castañeda-Ramírez, G.S., Torres-Acosta, J.F.J., Mendoza-de-Gives, P., Tun-Garrido, J., Rosado-Aguilar, J.A., Chan-Pérez, J.I., Hernández-Bolio, G.I., Ventura-Cordero, J., Acosta-Viana, K.Y. and Jímenez-Coello, M. 2020. Effects of different extracts of three *Annona* species on egg-hatching processes of *Haemonchus contortus*. *Journal of Helminthology*. 94: 1–8.

Ferreira, L.E., Castro, P.M.N., Chagas, A.C.S., França, S.C. and Beleboni, R.O. 2013. *In vitro* anthelmintic activity of aqueous leaf extract of *Annona muricata* L. (Annonaceae) against *Haemonchus contortus* from sheep. *Experimental Parasitology*. 134(3): 327–332.

Fofana, S., Keita, A., Balde, S., Ziyaev, R. and Aripova, S.F. 2012. Alkaloids from leaves of *Annona muricata*. *Chemistry of Natural Compounds*. 48(4): 714–714.

Ishola, I.O., Awodele, O., Olusayero, A.M. and Ochieng, C.O. 2014. Mechanisms of analgesic and anti-inflammatory properties of *Annona muricata* Linn. (Annonaceae) fruit extract in rodents. *Journal of Medicinal Food*. 17(12): 1375–1382.

Matsushige, A., Kotake, Y., Matsunami, K., Otsuka, H., Ohta, S. and Takeda, Y. 2012. Annonamine, a new aporphine alkaloid from the leaves of *Annona muricata*. *Chemical and Pharmaceutical Bulletin*. 60(2): 257–259.

Myint, S.H., Cortes, D., Laurens, A., Hocquemiller, R., Lebȩuf, M., Cavé, A., Cotte, J. and Quéro, A.M. 1991. Solamin, a cytotoxic mono-tetrahydrofuranic γ-lactone acetogenin from *Annona muricata* seeds. *Phytochemistry*. 30(10): 3335–3338.

Nguyen-Pouplin, J., Tran, H., Tran, H., Phan, T.A., Dolecek, C., Farrar, J., Tran, T.H., Caron, P., Bodo, B. and Grellier, P. 2007. Antimalarial and cytotoxic activities of ethnopharmacologically selected medicinal plants from South Vietnam. *Journal of Ethnopharmacology*. 109(3): 417–427.

Nwokocha, C.R., Owu, D.U., Gordon, A., Thaxter, K., McCalla, G., Ozolua, R.I. and Young, L. 2012. Possible mechanisms of action of the hypotensive effect of *Annona muricata* (soursop) in normotensive Sprague—Dawley rats. *Pharmaceutical Biology*. 50(11): 1436–1441.

Silva, R.M.D., Silva, I.D.M.M.D., Estevinho, M.M. and Estevinho, L.M. 2021. Anti-bacterial activity of *Annona muricata* Linnaeus extracts: A systematic review. *Food Science and Technology*. 42: 1–10.

Vila-Nova, N.S., Morais, S.M.D., Falcão, M.J.C., Machado, L.K.A., Beviláqua, C.M.L., Costa, I.R.S., Brasil, N.V.G.P.D. and Andrade Júnior, H.F.D. 2011. Leishmanicidal activity and cytotoxicity of compounds from two Annonacea species cultivated in Northeastern Brazil. *Revista da Sociedade Brasileira de Medicina Tropical*. 44(5): 567–571.

Wu, F.E., Gu, Z.M., Zeng, L.U., Zhao, G.X., Zhang, Y., McLaughlin, J.L. and Sastrodihardjo, S. 1995. Two new cytotoxic monotetrahydrofuran Annonaceous acetogenins, annomuricins A and B, from the leaves of *Annona muricata*. *Journal of Natural Products*. 58(6): 830–836.

***Artabotrys roseus* Boerl.**

[From Greek *artao* = hung up and *botrys* = cluster and Latin *roseus* = pink]

Published in *Icon. Bogor.* 1: 122, t. 53.1899

Synonym: *Artabotrys pleianthus* Diels
Local name: *gangon* (Dusun, Kadazan)
Habitat: forests
Geographical distribution: Borneo
Botanical description: this climbing plant grows up to a length of about 10 m. Stems: rough, with revolute woody hooks, hairy at apex. Leaves: simple, alternate, exstipulate. Petioles: about 1 cm long. Blades: somewhat coriaceous, ovate, elliptic to obovate, 1.2–4.5 cm × 2.5–12 cm, obtuse to rounded at base, apiculate to acuminate at apex, with 9–10 pairs of secondary nerves. Flowers: born at the apex of hooks on 0.8–2 cm thin peduncles. Sepals: 3, broadly ovate, 4 mm long, acute at apex. Outer petals: 3, triangular, about 1.5 cm long, hairy. Inner petals: 3, clawed, linear, hairy. Stamens: numerous. Carpels: about 15. Ripe carpels: 8. Monocarps: somewhat ovoid, about 3.5 × 2 cm. Seeds: 2.
Traditional therapeutic indications: medicinal (Dusun, Kadazan)
Pharmacology and phytochemistry: antibacterial (Chung et al. 2004), serotoninergic (Chung et al. 2005).
Toxicity, side effects, and drug interaction: not known

Comment: Parts of the plant are sold in the wet markets of Sabah as medicine.

REFERENCES

Chung, L.Y., Goh, S.H. and Imiyabir, Z. 2005. Central nervous system receptor activities of some Malaysian plant species. *Pharmaceutical Biology*. 43(3): 280–288.
Chung, P.Y., Chung, L.Y., Ngeow, Y.F., Goh, S.H. and Imiyabir, Z. 2004. Antimicrobial activities of Malaysian plant species. *Pharmaceutical Biology*. 42(4–5): 292–300.

***Desmos teysmannii* (Boerl.) Merr.**

[From the Greek *desma* = chains and after the 19th century Dutch botanist Johannes Elias Teijsmann]

Published in *Philipp. J. Sci.*, C 10: 235. 1915

Synonyms: *Desmos acutus* (Teijsm. & Binn.) I.M.Turner; *Unona teysmannii* Boerl.; *Uvaria acuta* Teijsm. & Binn.
Local name: *molisun rumungkut* (Murut)
Habitat: forests
Geographical distribution: Malaysia, Indonesia
Botanical description: this woody climbing plant grows up to a length of 15 m. Stems: hairy at apex. Leaves: simple, alternate, exstipulate. Petioles: 5–7 mm long. Blades: membranous, glaucous beneath, ovate or oblong-elliptic to oblong-lanceolate, 5–13 cm × 2–6 cm, rounded to acute at base, acute at apex, with 11–13 pairs of secondary nerves. Inflorescences: solitary, opposite to the leaves on stout 3–6 cm long peduncles. Sepals: 3, valvate, triangular, up to about 1 cm long. Petals: 2 sets of 3, greenish-yellow, lanceolate, 4–7 cm × 1–1.8 cm. Stamens: numerous, tiny. Carpels: numerous. Ripe carpels: about 5 cm long, moniliform. Seeds: about 7 mm long, smooth.
Traditional therapeutic indications: headaches (Murut)
Pharmacology and phytochemistry: not known
Toxicity, side effects, and drug interaction: not known

***Fissistigma fulgens* Merr.**

[From Latin *fissus* = split, the Greek *stigma* = mark of a pointed instrument, and the Latin *fulgens* = shining]

Published in *Philipp. J. Sci.*, C 10: 235.1915

Synonyms: *Melodorum fulgens* Hook.f. & Thomson; *Melodorum parviflorum* var. *angustifolium* Boerl.
Habitat: forests
Geographical distribution: Malaysia, Indonesia, the Philippines
Botanical description: this woody climbing plant grows up to a length of about 6 m. Stems: hairy at apex. Leaves: simple, alternate, exstipulate. Petioles: 6 mm–1.1 cm long. Blades: lanceolate, rounded at base, acuminate at apex, 1–5 cm × 4.5–18 cm, with 7–11 pairs of secondary nerves, glaucous beneath. Inflorescences: fragrant fascicles opposite the leaves on about 1 cm long peduncles. Sepals: 3, valvate, triangular, about 3 mm long, hairy. Outer petals: 3, valvate, coriaceous, ovate-lanceolate, light orangish, about 1 cm long, hairy. Inner petals: 3, valvate, ovate-lanceolate, 7–8 mm long. Stamens: red, numerous, tiny. Carpels: numerous. Ripe carpels: hairy. Monocarps: 2.5 cm long. Seeds: dark brown, smooth, glossy, about 1 cm long.
Traditional therapeutic indications: fatigue, headaches, toothaches (Dusun, Kadazan)
Pharmacology and phytochemistry: isoquinoline alkaloids (Hadi 2000).
Toxicity, side effects, and drug interaction: not known

REFERENCE

Hadi, A.H.A. 2000. Alkaloids from *Fissistigma fulgens* Merr. Annonaceae. *Malaysian Journal of Science*. 19(1): 41–44.

***Fissistigma latifolium* (Dunal) Merr.**

[From Latin *fissus* = split, the Greek *stigma* = mark of a pointed instrument, and the Latin *lati* = broad and *folium* = leaf]

Published in *Philipp. J. Sci.* 15: 132. 1919

Synonyms: *Annona rufa* C. Presl; *Melodorum borneense* Miq.; *Melodorum clementis* Merr.; *Melodorum latifolium* (Dunal) Hook.f. & Thomson; *Melodorum molissimum* Miq.; *Uvaria longifolia* Bl.; *Unona latifolia* Dunal
Local name: *sumbun* (Dusun, Kadazan)
Habitat: forests, swamps
Geographical distribution: Cambodia, Laos, Vietnam, Malaysia, Indonesia, the Philippines
Botanical description: this woody climbing plant grows up to a length of about 10 m. Stems: hairy at apex. Leaves: simple, alternate, exstipulate. Petioles: 1–1.5 cm long, channeled, hairy. Blades: dull green, broadly elliptic to obovate, cordate, rounded, or obtuse at base, notched to obtuse at apex, 8–35 cm × 3–15 cm, with about 10–15 pairs of secondary nerves sunken above, hairy beneath. Sepals: 3, valvate, ovate, about 4 mm long, hairy. Petals: 2 sets of 3. Outer petals: valvate, coriaceous, reddish to orangish, broadly lanceolate, about 2 cm long, somewhat hairy beneath. Inner petals: lanceolate, about 1 cm long. Stamens: red, numerous, tiny. Carpels: numerous, oblong, hairy. Ripe carpels: globose, hairy. Monocarps: about 2 cm across on about 3 cm long peduncles.
Traditional therapeutic indications: colic (Dusun, Kadazan)
Pharmacology and phytochemistry: aporphine alkaloids (lysicamine, anonaine, asimilobine) (Alias et al. 2010), flavans (Lan et al. 2012), chalcones (2-hydroxy-4,5,6-trimethoxydihydrochalcone) (Lan et al. 2011; Geny et al. 2017; Teng et al. 2022).
Toxicity, side effects, and drug interaction: not known

REFERENCES

Alias, A., Hazni, H., Mohd Jaafar, F., Awang, K. and Ismail, N.H. 2010. Alkaloids from *Fissistigma latifolium* (Dunal) Merr. *Molecules*. 15(7): 4583–4588.

Geny, C., Abou Samra, A., Retailleau, P., Iorga, B.I., Nedev, H., Awang, K., Roussi, F., Litaudon, M. and Dumontet, V. 2017. (+)-and (−)-ecarlottones, uncommon chalconoids from *Fissistigma latifolium* with pro-apoptotic activity. *Journal of Natural Products*. 80(12): 3179–3185.

Lan, Y.H., Leu, Y.L., Peng, Y.T., Thang, T.D., Lin, C.C. and Bao, B.Y. 2011. The first bis-retrochalcone from Fissistigma latifolium. *Planta Medica*. 77(18): 2019–2022.

Lan, Y.H., Peng, Y.T., Thang, T.D., Hwang, T.L., Dai, D.N., Leu, Y.L., Lai, W.C. and Wu, Y.C. 2012. New flavan and benzil isolated from *Fissistigma latifolium*. *Chemical and Pharmaceutical Bulletin*. 60(2): 280–282.

Teng, Y.N., Hung, C.C., Kao, P.H., Chang, Y.T. and Lan, Y.H. 2022. Reversal of multidrug resistance by *Fissistigma latifolium*—derived chalconoid 2-hydroxy-4, 5, 6-trimethoxydihydrochalcone in cancer cell lines overexpressing human P-glycoprotein. *Biomedicine & Pharmacotherapy*. 156: 113832.

***Fissistigma manubriatum* Merr.**

[From Latin *fissus* = split and *manubriatum* = handle]

Published in *Philipp. J. Sci.* 15: 124. 1919

Synonyms: *Melodorum manubriatum* Hook.f. & Thomson; *Melodorum korthalsii* Miq.; *Melodorum bancanum* Scheff.; *Uvaria manubriata* Wall.

Local name: *gagon* (Kadazan)

Habitat: forests

Geographical distribution: Malaysia, Indonesia

Botanical description: this woody climbing plant grows up to a length of about 10 m. Stems: hairy at apex. Leaves: simple, alternate, exstipulate. Petioles: about 1 cm long, hairy, channeled. Blades: lanceolate, 5–20 cm × 1.5–6.5 cm, rounded at base, hairy beneath, acuminate at apex, with about 15 pairs of secondary nerves. Inflorescences: few flowered clusters opposite the leaves on about 1.5 cm long, hairy peduncles. Sepals: 3, broadly lanceolate, about 7 mm long, hairy. Petals: 2 sets of 3. Outer petals: coriaceous, about 1.5 cm long, oblong, pointed at apex, hairy. Inner petals: narrowly lanceolate, about 1 cm long. Stamens: numerous, tiny. Ripe carpels: 20. Monocarps: globose, about 2.5 cm across, on about 7 cm long peduncles.

Traditional therapeutic indication: fatigue (Dusun, Kadazan)

Pharmacology and phytochemistry: oxoaporphine alkaloids (lanuginosine, lysicamine and liriodenine) (Kam 1999; Rozana et al. 1999).

Toxicity, side effects, and drug interaction: not known

REFERENCES

Kam, T.S. 1999. Alkaloids from Malaysian flora. In *Alkaloids: Chemical and Biological Perspectives* (14: 285–435). Pergamon.

Rozana, O., Mazdida, S., Shasya, U., Awang, K., Hamid, H.H. 1999. Chemical constituents of three Malaysian Ancnonaceae. *Malaysian Journal of Science*. 18: 35–40.

***Goniothalamus roseus* Stapf**

[From Greek *gonia* = angle, *thalamos* = inner chamber, and Latin *roseus* = pink]

Published in *Trans. Linn. Soc. London*, Bot. 4: 130.1894

Synonym: *Goniothalamus elmeri* var. *longistalklatus* Bân
Local names: *limpanas* (Bajau); *kayu panas* (Kedayan)
Habitat: forests
Geographical distribution: Borneo
Botanical description: this treelet reaches a height of about 10 m. Leaves: simple, alternate, exstipulate. Petioles: about 1 cm long. Blades: oblong, acute to somewhat attenuate at base, acuminate to caudate at apex, about 5–6.5 cm × 20–26 cm, the secondary nerves unclearly visible. Inflorescences: showy, cauliflorous, 5 cm long, somewhat obconic, stout, glossy, and reddish peduncles. Sepals: 3, merged at base, triangular, 6–8 mm long. Petals: 2 sets of 6, pinkish, lanceolate, about 2–5 cm long. Outer petals: coriaceous, lanceolate (which can make one think of strange wings or cicadas). Inner petals: shorty clawed, merged into a cap at apex. Stamens: linear, tiny, numerous. Carpels: numerous, hairy (Figure 4.5).

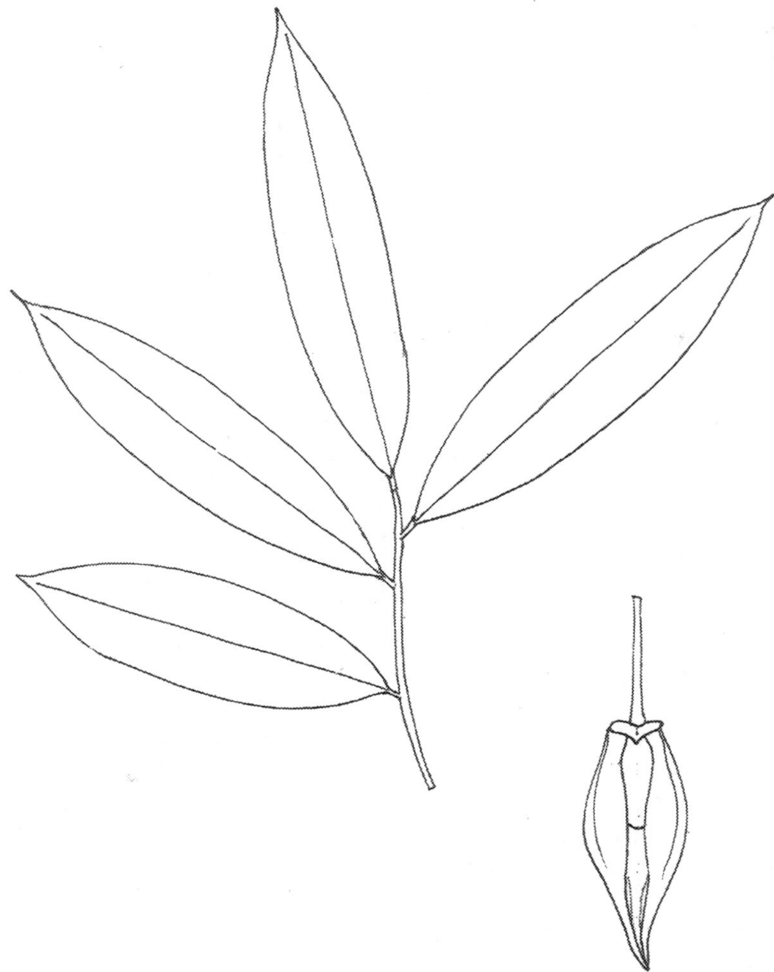

FIGURE 4.5 *Goniothalamus roseus* Stapf.

Traditional therapeutic indications: fever, fatigue (Bajau)
Pharmacology and phytochemistry: not known
Toxicity, side effects, and drug interaction: not known

Comments:

(i) The stems are sold in the wet markets of Sabah as medicine and for protection against evil spirits.
(ii) The Dusun and Kadazan use a plant of the genus *Goniothalamus* (Bl.) Hook.f. & Thomson (1875) (local name: *tuub*) to treat stomachaches whilst the Murut use *Goniothalamus woodii* Merr. (local name: *tampaliu*) to repel ghosts.
(iii) Scholars who favor rationalism and live in our Huxleyan and spiritually empty developed countries may deride local Sabah beliefs regarding ghosts or evil spirits. But rationalism does not explain everything. Cases of diseases cured by strange and non-rational practices are common in Southeast Asia, especially in poor societies living in remote areas, far from our post-industrial way of life, and, oh so much closer to our true human nature and identity. We lack humility, we have lost our roots, and we are disconnected from Mother Nature. We have much to learn from the ancient wisdom of these people.
(iv) *Goniothalamus velutinus* Airy Shaw is used by the Dusun for magical rituals (local name: *kalampanas*) as well as *Goniothalamus umbrosus* J. Sinclair (local name: *limpanas purak*). The Dusun believe that illnesses come from evil spirits. Typically, a patient will consult a high animist priestess Dusun called "*bobolian*" who will come into contact with the spirits who will give her the description of the plant to use and where and when to harvest it. As incredible as it may seem, these plants are often effective.

***Phaeanthus ophthalmicus* (Roxb. ex G. Don) J. Sinclair**

[From Greek *phae* = fairy, *anthos* = flower, and *ophthalmos*= eye]

Published in *Gard. Bull. Singapore* 14: 374.1955

Synonyms: *Monoon macropodum* Miq; *Phaeanthus cumingii* Miq; *Phaeanthus ebracteolatus* (Pers) Merr; *Phaeanthus nigrescens* Elmer; *Phaeanthus nitidus* Merr.; *Phaeanthus pubescens* Merr.; *Phaeanthus schefferi* Boerl. ex Koord *Polyalthia macropoda* (Miq.) F. Muell; *Uvarua ebracteolata* C. Presl.; *Uvaria tripelata* Roxb.
Local name: *korokos* (Murut)
Habitat: forests
Geographical distribution: Borneo, Indonesia, the Philippines, Papua New Guinea
Botanical description: this tree reaches a height of about 20 m. Bark: dark gray. Leaves: simple, alternate, exstipulate. Petioles: 3–7 mm long. Blades: 7–25 cm × 2–10 cm, hairy, with 9–14 pairs of secondary nerves, acute at base, acuminate at apex. Inflorescences: fascicles, cauliflorous, on 1.2–3 cm long peduncles. Sepal: 3, tiny, hairy. Petals: 3, lanceolate, fleshy, somewhat triangular, light brown to yellow, about 2 cm long. Stamens: numerous, tiny. Carpels: numerous, tiny, surrounded by a sheath, giving the whole structure the appearance of some kind of a hideous eye. Ripe carpels: ovoid, glossy, green turning yellow. Monocarps: about 1 cm long, pointed at apex, on about 3–4 cm long peduncles. Seeds: 1.
Traditional therapeutic indications: sore eyes (Murut)
Pharmacology and phytochemistry: antibacterial and cyclooxygenase-2 inhibitory tetrahydrobisbenzylisoquinoline alkaloids (tetrandrine, limacusine) (Magpantay et al. 2021; Malaluan et al. 2022).
Toxicity, side effects, and drug interaction: not known

Comment: In the northern Philippines (Pampanga Province), the plant is used externally to treat burns (local name: *amyung*).

REFERENCES

Magpantay, H.D., Malaluan, I.N., Manzano, J.A.H., Quimque, M.T., Pueblos, K.R., Moor, N., Budde, S., Bangcaya, P.S., Lim-Valle, D., Dahse, H.M. and Khan, A. 2021. Antibacterial and COX-2 inhibitory tetrahydrobisbenzylisoquinoline alkaloids from the Philippine medicinal plant *Phaeanthus ophthalmicus*. *Plants*. 10(3): 462.

Malaluan, I.N., Manzano, J.A.H., Muñoz, J.E.R., Bautista, T.J.L., Dahse, H.M. and Quimque, M.T.J. 2022. Antituberculosis and antiproliferative activities of the extracts and tetrahydrobisbenzylisoquinoline alkaloids from *phaeanthus ophthalmicus*: *In vitro* and *in silico* investigations. *Philippine Journal of Science*. 151: 371–381.

Polyalthia bullata **King**

[From Greek *poly* = many, *altheo* = cure, and Latin *bullata* = bullate]

Published in *Journal of the Asiatic Society of Bengal. Part 2. Natural History* 61: 64. 1893

Local names: *serat* (Murut); *lapad ruai* (Lundayeh)
Habitat: forests
Geographical distribution: Thailand, Malaysia, Borneo
Botanical description: this shrub reaches a height of about 3 m. Stems: hairy at apex. Leaves: simple, alternate, exstipulate. Petioles: 3–5 mm long, stout. Blades: thinly coriaceous, bullate, lanceolate, or oblong-lanceolate to oblanceolate, acute at apex, cordate to auricled at base, 3–12 cm × 28–37 cm, dull green, somewhat hairy beneath, with 25–40 pairs of secondary nerves. Inflorescences: solitary, cauliflorous, on 2–5 cm long peduncles. Sepals: 3, lanceolate, 5–7 mm long. Petals: 6, 2–5 cm long, dull yellow, narrowly triangular. Stamens: numerous, short, forming a button-like androecium. Carpels: few, free, somewhat hairy. Monocarps: somewhat globose, hairy, 1 cm across, on 5 mm long peduncles. Seeds: 2.
Traditional therapeutic indications: epileptic seizures (Lundayeh)
Pharmacology and phytochemistry: cytotoxic flavones (Nantapap et al. 2017), aporphine alkaloids (Connolly et al. 1996).
Toxicity, side effects, and drug interaction: not known

Comment: The plant has a reputation of being an aphrodisiac in Peninsular Malaysia.

REFERENCES

Connolly, J.D., Haque, M.E. and Kadir, A.A. 1996. Two 7, 7′-bisdehydroaporphine alkaloids from Polyalthia bullata. *Phytochemistry*. 43(1): 295–297.

Nantapap, S., Punyanitya, S., Nuntasaen, N., Pompimon, W. and Meepowpan, P. 2017. Flavones from aerial parts of *Polyalthia bullata* and cytotoxicity against cancer cell lines. *Chemistry of Natural Compounds*. 53(4): 762–763.

***Polyalthia insignis* (Hook.f.) Airy Shaw**

[From Greek *poly* = many, *altheo* = cure, and Latin *insignis* = striking]

Published in *Bull. Misc. Inform.* Kew 1939: 279. 1939

Synonyms: *Polyalthia dolichophylla* Merr.; *Sphaerothalamus insignis* Hook.f.; *Unona miniata* Elmer
Habitat: forests
Geographical distribution: Borneo, the Philippines
Botanical description: this tree reaches a height of about 6 m. Leaves: simple, sessile, alternate, exstipulate. Petioles: about 5 mm long, stout. Blades: coriaceous, cordate, and somewhat asymmetrical at base, obovate to oblong, about 7.5 cm × 22 cm, acuminate at apex, dark green, and glossy above, with about 24 pairs of secondary nerves arching at margin. Inflorescences: solitary, axillary, on about 2 cm long peduncles. Sepals: 3, ovate, brownish, flat, membranous, 1–5 cm long. Petals: 6, bright plain orange, oblong, fleshy, with a few longitudinal nerves, rounded at apex, 2.3–8 cm long. Stamens: about 4 mm, forming button-like androecium. Carpels: tiny, free, hairy.
Traditional therapeutic indications: coughs, thrush (Dusun, Kadazan)
Pharmacology and phytochemistry: antiplasmodial (Khozirah et al. 2011).
Benzyltetrahydroisoquinoline alkaloids (Lee et al. 1997)
Toxicity, side effects, and drug interaction: not known

REFERENCES

Khozirah, S., Noor Rain, A., Siti Najila, M.J., Imiyabir, Z., Madani, L. and Rodaya, C. 2011. *In vitro* antiplasmodial properties of selected plants of Sabah. *Journal of Science & Technology*. 19: 11–17.

Lee, K.H., Chuah, C.H. and Goh, S.H. 1997. seco-Benzyltetrahydroisoquinolines from Polyalthia insignis (annonaceae). *Tetrahedron Letters*. 38(7): 1253–1256.

***Polyalthia sumatrana* (Miq.) Kurz**

[From Greek *poly* = many, *altheo* = cure, and Latin *sumatrana* = from Sumatra]

Journal of the Asiatic Society of Bengal. Part 2. Natural History 43: 53. 1874.

Synonyms: *Guatteria sumatrana* Miq.; *Monoon sumatranum* (Miq.) Miq.
Habitat: forests
Geographical distribution: Sumatra, Borneo
Botanical description: this tree reaches a height of about 20 m. Stems: glabrous, striated. Leaves: simple, alternate, exstipulate. Petioles: 5–7 mm long, stout. Blades: dark green above and glossy, glaucous beneath, oblong to lanceolate, acute, or acuminate at apex, acute at base, finely marked with numerous pairs of secondary nerves, 3–6.5 cm × 12–20 cm. Inflorescences: cauliforous fascicles of up to 3 flowers on 1 cm long peduncles. Sepals: 3, tiny, somewhat orbicular, and obtuse at apex. Petals: 6, greenish, linear, 3.5–4 cm long. Stamens: tiny and form a button-like androecium. Carpels: tiny. Ripe carpels: few, the monocarps red, ovoid, about 3 cm long on 1–5 cm long peduncles. Seeds: 1.
Traditional therapeutic indications: medicinal
Pharmacology and phytochemistry: essential oil (Shakri et al. 2020).
Toxicity, side effects, and drug interaction: not known

Comments:

(i) A plant of the genus *Polyalthia* Bl. (1830) is used by the Dusun (local name: *dolipanas*) for black skin and for a disease locally termed "*menyusuk nyusuk*."
(ii) A plant of the genus *Polyalthia* Bl. (1830) is used by the Murut medicinally (local name: *ubat puru*).
(iii) *Polyalthia tenuipes* Merr. is used by the Dusun (local name: *kabangking*) for sick children.

REFERENCE

Shakri, N.M., Salleh, W.M.N.H.W., Khamis, S., Ali, N.A.M. and Shaharudin, S.M. 2020. Chemical composition of the essential oils of four Polyalthia species from Malaysia. *Zeitschrift für Naturforschung C.* 75(11–12): 473–478.

Uvaria cuneifolia (Hook.f. & Thomson) L.L. Zhou, Y.C.F. Su, & R.M.K. Saunders

[From Latin *uvar* = grapevine, *cuneus* = wedge-shaped, and *folia* = leaf]

Published in *Syst. Biodivers.* 7: 255.2009

Synonyms: *Ellipeia cuneifolia* Hook.f. & Thomson; *Ellipeia gilva* Miq.; *Uvaria gilva* (Miq.) L.L. Zhou, Y.C.F. Su & R.M.K. Saunders
Local name: *kayu bibiris* (Bajau)
Habitat: forests
Geographical distribution: the Andaman Islands, Malaysia, Borneo
Botanical description: this large woody climbing plant grows up to a length of about 10 m. Stems: hairy at apex. Leaves: simple, alternate, exstipulate. Petioles: stout and about 5 mm long. Blades: hairy, somewhat papery, glossy above, ovate to obovate to oblanceolate, 2–10 cm × 5–25 cm, rounded at base, acuminate at apex, with about 15–20 pairs of secondary nerves sunken above. Inflorescences: solitary, terminal, on about 1.5 cm long hairy peduncles. Sepals: 3, broadly triangular to ovate, about 4 mm long. Petals: 2 sets of 3, creamy yellow. Outer petals: ovate, about 1.5 cm long. Inner petals: triangular, about 4 mm long. Stamens: numerous, tiny, giving the androecium the shape of a button. Carpels: numerous, tiny. Ripe carpels: about 20, the peduncles about 1.5–2 cm long. Monocarps: ovoid, bumpy, about 1 cm long. Seeds: 1.
Traditional therapeutic indications: fever (Bajau)
Pharmacology and phytochemistry: not known
Toxicity, side effects, and drug interaction: not known

Uvaria grandiflora Roxb. ex Hornem

[From Latin *uvar* = grapevine and *grandis*= large, and *flora* = flower]

Synonyms: *Unona grandiflora* Lesch. ex DC.; *Uvaria platypetala* Champ. ex Benth.; *Uvaria purpurea* Blume; *Uvaria rhodantha* Hance; *Uvaria rubra* C.B. Rob.
Local names: *nolilitan* (Murut); *potudung* (Dusun, Kadazan)
Habitat: forests
Geographical distribution: from India to Papua New Guinea
Botanical description: this woody climbing plant grows up to a length of about 10 m. Stems: hairy at apex. Leaves: simple, alternate, exstipulate. Petioles: stout, hairy, curved, about 5 mm long. Blades: hairy, somewhat coriaceous, glossy above, obovate to oblong, 9–30 cm × 3–15 cm, rounded to cordate at base, acuminate at apex, with about 7–15 pairs of secondary nerves sunken above. Inflorescences: solitary, axillary, cauliflorous on about 2.5 cm long peduncles (can make one think of flowers of *Helenium sp.*). Sepals: 3, orbicular, pale yellow, rounded, about 2 cm across. Petals: 5–6, ovate to lanceolate, coriaceous, about 1.5 cm × 3 cm, dull burgundy red. Stamens: numerous, tiny, giving the androecium the shape of a button. Carpels: numerous, tiny. Ripe carpels: about 30 or more, somewhat oblong, yellow, about 5 × 1.5 cm. Seeds: about 20, moon-shaped, about 1 cm long. Infructescence: can make one think of a small bunch of ripe bananas (hence the Malay name: *pisang-pisang*).
Traditional therapeutic indications: colds and shivers in children, intestinal worms, jaundice, fatigue (Dusun, Kadazan); stomachaches (Murut, Dusun, Kadazan); waist pain (Murut)
Pharmacology and phytochemistry: antibacterial alkaloids (Kongkum et al. 2021), anti-inflammatory and cytotoxic cyclohexene derivatives (zeylenol) (Seangphakdee et al. 2013; Ho et al. 2015).
Toxicity, side effects, and drug interaction: not known

***Uvaria sorzogonensis* C. Presl.**

[From Latin *uvar* = grapewine and *sorsogon*= a province in the Philippines]

Published in *Reliq. Haenk*. 2: 76.1835

Synonyms: *Guatteria cordata* Dunal; *Guatteria rufa* Lindl.; *Unona camphorata* Blanco; *Unona littoralis* Bl.; *Uvaria cordata* (Dunal) Kuntze; *Uvaria gamopetala* Zoll.; *Uvaria littoralis* (Blume) Bl.; *Uvaria macrophylla* Roxb. ex Wall.; *Uvaria ovalifolia* Blume; *Uvaria rufescens* A.DC.; *Uvaria synsepala* (Miq.) Kuntze

Local name: *sogombong* (Kadazan)

Habitat: forests

Geographical distribution: from India to the Solomon Islands

Botanical description: this large woody climbing plant grows up to a length of about 10 m. Stems: hairy at apex. Leaves: simple, alternate, exstipulate. Petioles: stout, somewhat swollen, hairy, 1.3–3 cm long. Blades: thin, ovate to lanceolate to obovate, 7–13 cm × 13–31 cm, rounded to truncate to cordate at base, acuminate at apex, with about 10–20 pairs of secondary nerves. Inflorescences: solitary, cauliflorous, terminal on 2.5 cm long peduncles. Sepals: 3, hairy, broadly triangular, about 5 mm long. Petals: 6, obovate, hooded at apex, red to somewhat dark purplish brown, about 1 cm long. Stamens: numerous, tiny, giving the androecium the shape of a button. Carpels: numerous, tiny. Ripe carpels: about 15 or more, orange, the peduncles up to about 3 cm long. Monocarps: about 1.5 cm long, somewhat oblong, finely beaked. Seeds: light brown, glossy (Figure 4.6).

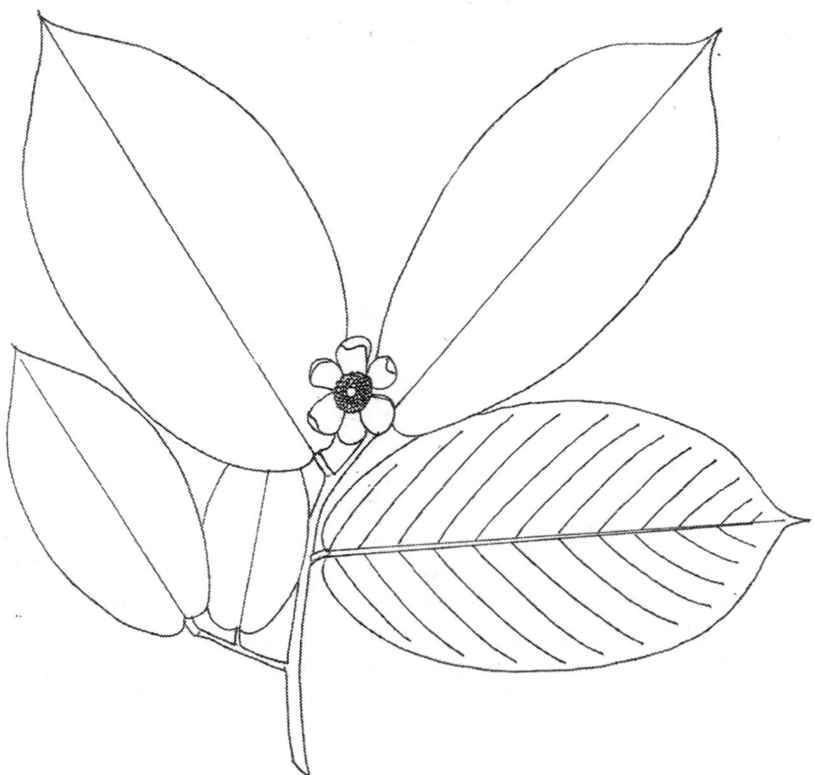

FIGURE 4.6 *Uvaria sorzogonensis* C. Presl.

Traditional therapeutic indications: fatigue (Dusun, Kadazan)
Pharmacology and phytochemistry: inhibition of cholinesterase (Rahman et al. 2017).
Toxicity, side effects, and drug interaction: not known

Comment: Plants of the genus *Uvaria* L. (1753) are used as medicines by the Dusun and Kadazan: "*bab*" for swellings and "*langad langad*" for rheumatisms.

REFERENCE

Ho, D.V., Kodama, T., Le, H.T.B., Van Phan, K., Do, T.T., Bui, T.H., Le, A.T., Win, N.N., Imagawa, H., Ito, T. and Morita, H., 2015. A new polyoxygenated cyclohexene and a new megastigmane glycoside from Uvaria grandiflora. *Bioorganic & Medicinal Chemistry Letters*, 25(16): 3246–3250.

Kongkum, N., Thanasansurapong, S., Surapanich, N., Kumutanat, W., Sukkum, C., Kuhakarn, C., Jaipetch, T., Reutrakul, V. and Hongthong, S., 2021. Antioxidant and antimicrobial alkaloids isolated from twigs of Uvaria grandiflora Roxb. *ScienceAsia*, 47.

Rahman, A., Haque, A., Uddin, M.S., Mian, M.M., Sufian, M.A. and Rahman, M.M. 2017. *In vitro* screening for antioxidant and anticholinesterase effects of *Uvaria littoralis* Blume. *A nootropic phytotherapeutic remedy. Journal of Intellectual Disability – Diagnosis and Treatment*. 5(2): 50–60.

Seangphakdee, P., Pompimon, W., Meepowpan, P., Panthong, A., Chiranthanut, N., Banjerdpongchai, R., Wudtiwai, B., Nuntasaen, N. and Pitchuanchom, S., 2013. Anti-inflammatory and anticancer activities of (−)-zeylenol from stems of Uvaria grandiflora. *ScienceAsia*, 39(6): 610–614.

***Xylopia dehiscens* (Blanco) Merr.**

[From the Greek *xylon* = wood, *pikron* = bitter, and Latin *dehiscere* = to split open]

Published in *Bull. Bur. Forest. Philipp. Islands* 1: 20.1903

Synonyms: *Unona dehiscens* Blanco; *Xylopia blancoi* S. Vidal
Local name: *mizas pizas* (Rungus)
Habitat: forests, near the sea
Geographical distribution: Borneo, the Philippines
Botanical description: this tree reaches a height of about 40 m. Trunk: straight with large buttresses. Bark: brown to reddish-brown somewhat scaly, yields a yellow to cream yellow sap once incised. Leaves: simple, alternate, exstipulate. Petioles: 4–9 mm long, channeled. Blades: 6.5–10.9 cm × 2.4–4.8 cm, somewhat coriaceous, elliptic to oblong, attenuate at base, blunt at apex, with 7–12 pairs of secondary nerves. Inflorescences: fascicles, cauliflorous, 1–5 flowered. Sepals: 3, valvate, about 2 mm long, triangular. Petals: 6, yellowish-green, linear, 1–3 cm long. Stamens: numerous, tiny. Carpels: 3–4. Ripe carpels: brownish drying black, dehiscent. Monocarps: 3–6 cm × 2.5–3.5 cm, pod-shaped, with a 3–5 mm long stipes. Seeds: 8–12 in a red mass.
Traditional therapeutic indications: medicinal (Rungus)
Pharmacology and phytochemistry: not known
Toxicity, side effects, and drug interaction: not known

Comments:

(i) The Dusun and Kadazan use a plant of the genus *Xylopia* L. (1759) for swellings of the armpit (local name: *linsou linsou*).
(ii) Parts of a plant of the genus *Enicosanthum* Becc. (1871) are sold in the wet markets of Sabah as medicine.

4.1.3 Superorder Lilianae Takht. (1967), the Monocots

4.1.3.1 Order Acorales Link (1835)

4.1.3.1.1 Family Acoraceae Martinov (1820), the Sweet Flag Family

The family Acoraceae consists of the single genus *Acorus* L. (1753).

***Acorus calamus* L.**

[From Greek *akoron* = sweet flag and Latin *calamus* = a reed]

Published in *Species Plantarum* 1: 324. 1753

Synonyms: *Acorus americanus* (Raf.) Raf.; *Acorus angustatus* Raf.; *Acorus angustifolius* Schott; *Acorus asiaticus* Nakai; Engl.; *Acorus cochinchinensis* (Lour.) Schott; *Acorus griffithii* Schott; *Acorus spurius* Schott; *Acorus triqueter* Turcz. ex Schott; *Acorus verus* (L.) Houtt.; *Calamus aromaticus* Garsault; *Orontium cochinchinense* Lour.
Local names: *kusul* (Murut); *kamburongoh, komburongoh, komburungoh* (Dusun; Kadazan); *jerangau putih* (Bajau; Kedayan); *guo sanit* (Rungus).
Common name: sweet flag
Habitat: ponds, lakes, slow rivers, wetlands, roadsides, marshes
Botanical description: this herbaceous plant reaches a height of about 2 m. Rhizomes: strongly aromatic, up to 20 cm long. Leaves: erect, ensiform, light green, coriaceous, 1–2.5 cm × 70–100 cm. Spathes: 15–75 cm long, ensiform, upright. Spadices: green, sessile, cylindrical, densely flowered, 5–10 cm long. Tepals: 6, orbicular. Stamens: 6. Ovary: lobed. Berries: tiny.

Traditional therapeutic indications: skin diseases (Dusun); fever, stomachaches (Murut); poison antidote, insect stings, fatigue, diarrhea, gastritis (Dusun, Kadazan)
Pharmacology and phytochemistry: this plant has been the subject of numerous pharmacological studies which have highlighted, among other things, analgesic (Jayaraman et al. 2010), sedative, hypotensive, antipyretic (Agarwal et al. 1956), antihelminthic (McGaw et al. 2000), and antibacterial properties (Rani et al. 2003). Essential oil (β-asarone) (Keller & Stahl 1983; Balakumbahan et al. 2010). β-Asarone is antifungal (Lee et al. 2004; Rajput & Karuppayil 2013; Jacob et al. 2015), toxic to glioma cells (Wang et al. 2018), and neuroprotective (Zhang et al. 2016). Antiviral (dengue) phenolics (tatanan A) (Yao et al. 2018).
Toxicity, side effects, and drug interaction: β-Asarone is carcinogenic (Cartus et al. 2015).

Comment: The plant is used for protection against evil spirits by the Dusun whilst the Rungus use it as insect repellent. We might tend to smile at and ignore the beliefs and legends of the Dusun, Rungus, Murut, and other native ethnic groups of Sabah. However, these beliefs are very valuable for ethnopharmacologists because they can contribute to the discovery of molecules of therapeutic interest. In Pakistan for instance, pucks of roots of a plant in the family Valerianaceae are hung around the neck of epileptics for protection against evil forces, and studies have demonstrated the presence of a volatile substance in this plant with the ability to mitigate epileptic seizures via inhibition of GABA receptors (Iqbal Choudhary, personal communication). It is also thanks to phytochemical and pharmacological studies of our belladonna or digitalis. which were used by medieval witches, that atropine and digitalis were discovered. The examples are numerous and include opium used for magic rituals by the Canaanites from which morphine was identified.

REFERENCES

Agarwal, S.L., Dandiya, P.C., Singh, K.P. and Arora, R.B. 1956. A note on the preliminary studies of certain pharmacological actions of *Acorus calamus* L. *Journal of the American Pharmaceutical Association*. 45(9): 655–656.

Balakumbahan, R., Rajamani, K. and Kumanan, K. 2010. *Acorus calamus*: An overview. *Journal of Medicinal Plants Research*. 4(25): 2740–2745.

Cartus, A.T., Stegmüller, S., Simson, N., Wahl, A., Neef, S., Kelm, H. and Schrenk, D. 2015. Hepatic metabolism of carcinogenic β-asarone. *Chemical Research in Toxicology*. 28(9): 1760–1773.

Jacob, J., Aravind, S.R., Nishanth Kumar, S., Sreelekha, T.T. and Kumar, D. 2015. Asarones from *Acorus calamus* in combination with azoles and amphotericin B: A novel synergistic combination to compete against human pathogenic *Candida* species *in vitro*. *Biotechnology and Applied Biochemistry*. 175(8): 3683–3695.

Jayaraman, R., Anitha, T. and Joshi, V.D. 2010. Analgesic and anticonvulsant effects of *Acorus calamus* roots in mice. *International Journal of PharmTech Research*. 2: 552–555.

Keller, K. and Stahl, E. 1983. Composition of the essential oil from beta-Asarone free calamus. *Planta Medica*. 47(2): 71–74.

Lee, J.Y., Lee, J.Y., Yun, B.S. and Hwang, B.K. 2004. Antifungal activity of β-asarone from rhizomes of *Acorus gramineus*. *Journal of Agricultural and Food Chemistry*. 52(4): 776–780.

McGaw, L.J., Jäger, A.K. and Van Staden, J. 2000. Antibacterial, anthelmintic and anti-amoebic activity in South African medicinal plants. *Journal of Ethnopharmacology*. 72(1–2): 247–263.

Rajput, S.B. and Karuppayil, S.M. 2013. β-Asarone, an active principle of *Acorus calamus* rhizome, inhibits morphogenesis, biofilm formation and ergosterol biosynthesis in *Candida albicans*. *Phytomedicine*. 20(2): 139–142.

Rani, A.S., Satyakala, M., Devi, V.S. and Murty, U.S. 2003. Evaluation of antibacterial activity from rhizome extract of *Acorus calamus* Linn. *AIP Conference Proceedings*. 2155: 020054.

Wang, N., Han, Y., Luo, L., Zhang, Q., Ning, B. and Fang, Y. 2018. β-asarone induces cell apoptosis, inhibits cell proliferation and decreases migration and invasion of glioma cells. *Biomedicine & Pharmacotherapy*. 106: 655–664.

Yao, X., Ling, Y., Guo, S., Wu, W., He, S., Zhang, Q., Zou, M., Nandakumar, K.S., Chen, X. and Liu, S. 2018. Tatanan A from the *Acorus calamus* L. root inhibited dengue virus proliferation and infections. *Phytomedicine*. 42: 258–267.

Zhang, Q.S., Wang, Z.H., Zhang, J.L., Duan, Y.L., Li, G.F. and Zheng, D.L. 2016. Beta-asarone protects against MPTO-induced Parkinson's disease via regulating long non-coding RNA MALAT1 and inhibiting α-synuclein protein expression. *Biomedicine & Pharmacotherapy*. 83: 153–159.

4.1.3.1.2 Family Araceae Juss. (1789), the Arum Family

The family Araceae consists of about 110 genera and 1,800 species of poisonous herbaceous plants growing from creeping or tuberous rhizomes or corms. Stems: often fleshy to sappy. Leaves: simple, alternate, the petiole often forming a sheath at base. Blades: simple, entire to variously cleft to perforate or compound, fleshy. Inflorescences: spadices more or less enclosed in spathes. Flowers: tiny, fragrant. Perianth: 4-6-lobed. Stamens: 2–8. Carpels: often 3, forming a unilocular or bilocular ovary, each locule with 1–15 ovules. Fruits: berries.

***Aglaonema oblongifolium* (Roxb.) Kunth**

[From Greek *aglaos* = beautiful, *nema* = threat, and Latin *oblongifolium* = with oblong leaves]

Published in *Enumeratio Plantarum Omnium Hucusque Cognitarum* 3: 54. 1841

Synonyms: *Aglaonema marantifolium* BI; *Calla oblongifolia* Roxb.
Local name: *pilonos* (Murut)
Habitat: forests, cultivated
Description: Borneo, Indonesia, Papua New Guinea
Botanical description: this erect herbaceous plant reaches a height of about 1 m. Leaves: simple, basal. Petioles: channeled, 8–12 cm long. Blades: 40–60 cm × 6–9.5 cm, fleshy, glossy, with yellow blotches, marked with 6–10 pairs of secondary nerves. Spadices: oblong, glossy, on about 7 cm long peduncles. Spathes: globose. Berries: bright orange, 1.8 cm × 5 mm, marked at apex by a tiny disc.
Traditional therapeutic indications: boils (Murut)
Pharmacology and phytochemistry: antibacterial (Silaban et al. 2022).
Toxicity, side effects, and drug interaction: calcium oxalate crystals make this plant poisonous (Genua & Hillson 1985).

Comments:

(i) In Indonesia, this plant used for swollen joints and in the Philippines for skin diseases.
(ii) Antibacterial principles in this plant need to be identified.

REFERENCES

Genua, J.M. and Hillson, C.J. 1985. The occurrence, type and location of calcium oxalate crystals in the leaves of fourteen species of Araceae. *Annals of Botany*. 56(3): 351–361.

Silaban, S., Nainggolan, B., Simorangkir, M., Zega, T.S., Pakpahan, P.M. and Gurning, K. 2022. Antibacterial activities test and brine shrimp lethality test of Simargaolgaol (Aglaonema modestum Schott ex Engl.) leaves from North Sumatera, Indonesia. *Rasayan Journal of Chemistry*. 15(2): 745–750.

Alocasia macrorrhizos **(L.) G.Don**

[From Greek *a* = without, *lokasia* = lotus root, *makros* = large, and *rhiza* = root]

Published in *Hortus Britannicus*, ed. 3 631. 1839

Synonyms: *Alocasia cordifolia* (Bory) Cordem.; *Alocasia grandis* N.E.Br.; *Alocasia indica* (Lour.) Spach; *Alocasia marginata* N.E.Br.; *Alocasia metallica* Schott; *Alocasia pallida* K. Koch & C.D. Bouché; *Alocasia plumbea* Van Houtte; *Alocasia variegata* K. Koch & C.D. Bouché; *Arum cordifolium* Bory; *Arum indicum* Lour.; *Arum macrorrhizon* L.; *Arum mucronatum* Lam.; *Arum peregrinum* L.; *Caladium indica* (Lour.) K. Koch; *Caladium macrorrhizon* (L.) R.Br.; *Caladium metallicum* (Schott) Engl.; *Caladium odoratum* Lodd.; *Caladium plumbeum* K. Koch; *Calla badian* Blanco; *Calla maxima* Blanco; *Colocasia boryi* Kunth; *Colocasia indica* (Lour.) Kunth; *Colocasia mucronata* (Lam.) Kunth; *Colocasia peregrina* (L.) Raf.; *Colocasia rapiformis* Kunth; *Philodendron peregrinum* (L.) Kunth; *Philodendron punctatum* (Desf.) Kunth

Local name: *buntui* (Murut)

Common names: giant taro, elephant's ears

Habitat: roadsides, villages, jungle paths, cultivated

Geographical distributions: tropical

Botanical description: this giant herbaceous plant reaches a height of about 4 m. Leaves: simple, spiral, basal, or on top of approximately 50 cm tall pseudostems. Blades: fleshy, more or less glossy, somewhat broadly sagittate, up to 50 cm × 120 cm, with about 20 pairs of stout secondary nerves. Spathes: 13–35 cm long, pale yellow, broadly oblong to lanceolate. Spadices: 10–30 cm long, whitish. Ovary: obovoid. Stigma: flat. Berries: about 1 cm long, red, ellipsoid.

Traditional therapeutic indications: itchiness (Murut)

Pharmacology and phytochemistry: sphingolipid (1-*O*-*β*-D-glucopyranosyl-(2*S*, 3*R*, 4*E*, 8*Z*)-2-[(2(*R*)-hydroctadecanoyl) amido]-4,8-octadecadiene-1,3-diol), indole alkaloid (hyrtiosin B) toxic to larynx cancer cells (Elsbaey et al. 2017).

Toxicity, side effects, and drug interaction: calcium oxalate crystals make this plant poisonous (Chan et al. 1995; Karnawat et al. 2015).

Comments:

(i) According to Burkhill (1935), the plant is used as a counterirritant in Peninsular Malaysia, Java, and the Philippines, whilst the leaves are used for magic rituals.

(ii) Plants of the genus *Alocasia* (Schott) G.Don (1839) are used as medicines by the Dusun and Kadazan: "*sisial*" to treat itchiness induced by caterpillar stings and "*tanom*" for measles.

REFERENCES

Chan, T.Y.K., Chan, L.Y., Tam, L.S. and Critchley, J.A.J.H. 1995. Neurotoxicity following the ingestion of a Chinese medicinal plant, *Alocasia macrorrhiza. Human & Experimental Toxicology*. 14(9): 727–728.

Elsbaey, M., Ahmed, K.F., Elsebai, M.F., Zaghloul, A., Amer, M.M. and Lahloub, M.F.I. 2017. Cytotoxic constituents of Alocasia macrorrhiza. *Zeitschrift für Naturforschung C*. 72(1–2): 21–25.

Karnawat, B.S., Joshi, A., Narayan, J.P. and Sharma, V. 2015. Alocasia macrorrhiza: A decorative but dangerous plant. *International Journal of Scientific Study*. 3(1): 221–223.

Amydrium medium **Nicolson**

[From Greek *amydros* = indistinct and Latin *medium* = middle]

Published in *Blumea* 16: 124. 1968

Synonyms: *Epipremnum truncatum* Engl. & K. Krause; *Rhaphidophora huegelii* Schott; *Scindapsus medius* Zoll. & A. Mortizi
Local names: *lapad bara* (Lundayeh); *kulimpiau* (Dusun, Kadazan)
Habitat: forests
Geographic distribution: Southeast Asia
Botanical description: this climbing plant reaches a length of about 10 m. Stems: with adhesive roots. Leaves: simple, spiral. Petioles: 15–35 cm long, forming a sheath at base. Blades: cordate to ovate, deeply incised to hollowed, 9–25 cm × 12–45 cm, glossy. Spadices: oblong, 4–6 × 1 cm, white to cream. Spathes: conchiform to ovate, apiculate, 9 cm × 7 cm, whitish, fleshy. Stamens: 4. Ovary: conical. Berries: somewhat globose, 1 cm across, white.
Traditional therapeutic indications: flu (Dusun, Kadazan); swollen legs (Lundayeh)
Pharmacology and phytochemistry: not known
Toxicity, side effects, and drug interaction: probably poisonous

Homalomena propinqua **Schott**

[From Greek *homalos* = smooth and Latin *propinquus* = near]

Published in *Annales Musei Botanici Lugduno-Batavi* 1: 280. 1864

Synonyms: *Calla humulis* Jack; *Homalomena humilis* (Jack) Hook.f.
Local name: *nyato* (Murut)
Habitat: forests, cultivated
Geographical distribution: Southeast Asia
Botanical description: this stoloniferous herbaceous plant grows up to a length of about 1 m. Leaves: simple, basal. Petioles: about 3–15 cm long, forming a sheath at base, fleshy, somewhat pinkish. Blades: membranous, about 2.5–6 cm × 4–15 cm, somewhat serrate, wavy, brownish, elliptic to oblong, asymmetrical, rounded, or attenuate at base, acute at apex, with 4–8 pairs of secondary nerves. Spadices: about 2 cm long. Spathes: elliptic cuspidate or acuminate at apex, and membranous. Berries: globose, red, about 1 cm across (Figure 4.7).
Traditional therapeutic indications: feverish colds (Murut)
Pharmacology and phytochemistry: not known
Toxicity, side effects, and drug interaction: probably poisonous

Comments:

(i) A plant of the genus *Homalomena* Schott (1832) is used by the Dusun and Kadazan for insect bites whilst the Murut use it for fever.
(ii) A plant of the genus *Homalomena* Schott (1832) is used by the Dusun as vegetable (local name: *buntui*).
(iii) A plant of the genus *Homalomena* Schott (1832) is used by the Dusun for magic rituals (local name: *latu*).
(iv) *Scindapsus longistipitatus* Merr. is used for skin diseases (local name: *timbalung lanut*) whilst a plant of the genus *Scindapsus* Schott (1832) is used to reduce swellings by the Dusun and Kadazan (local name: *timbolung lolu*).

FIGURE 4.7 *Homalomena propinqua* Schott.

(v) The Murut use a plant of the genus *Schismatoglottis* Zoll. & Moritzi (1846) (local name: *pongongondog*) as a medicine.

(vi) It should be noted that the number of unidentified medicinal plants in Sabah remains high and unstudied. Most of the work remains to be done but, due to a lack of local research funds and enormous local and international administrative barriers, these plants may never been studied. We need medicine and these plants could be the solution.

***Rhaphidophora korthalsii* Schott**

[From Greek *rhaphidos* = needle, and after the 19th century Dutch botanist Pieter Willem Korthals]

Published in *An. Mus. Lugduno-Batavi* 1: 129. 1863

Synonyms: *Epipremnum multicephalum* Elmer; *Pothos bifarius* Wall. ex Hook.f.; *Pothos celatocaulis* N.E.Br.; *Rhaphidophora celatocaulis* (N.E.Br.) Alderw.; *Rhaphidophora copelandii* Engl.; *Rhaphidophora grandifolia* K. Krause; *Rhaphidophora grandis* Ridl.; *Rhaphidophora latifolia* Alderw.; *Rhaphidophora maxima* Engl.; *Rhaphidophora palawanensis* Merr.; *Rhaphidophora ridleyi* Merr.; *Rhaphidophora trinervia* Elmer; *Rhaphidophora trukensis* Hosok.; *Scindapsus anomalus* Carrière

Local name: *ubat ugut* (Lundayeh)

Habitat: forests

Geographical distribution: from Bangladesh to the Pacific Islands

Botanical description: this stout climbing plant grows up to a length of about 20 m. Stems: smooth, with adhesive roots. Leaves: simple, regularly alternate (distichous). Petioles: of variable length, channeled. Blades: lanceolate, 3.5–45 cm × 5–90 cm, cordate to truncate at base, entire or deeply incised, acute to acuminate at apex. Spathes: narrowly canoe-shaped, stoutly beaked, 10-30 cm × 3–5 cm, coriaceous, greenish to dull yellow. Spadices: cylindrical 1.5–2 cm × 9–26 cm, dull green to dirty white. Infructescences: 14–27 cm × 3 cm.

Traditional therapeutic indications: gout (Lundayeh)

Pharmacology and phytochemistry: indole alkaloid (5, 6-dihydroxyindole) (Wong et al. 1996). Immunomodulatory (Yeap et al. 2012), wound healing (Dandannavar et al. 2019), cytotoxic (Wong & Tan 1996).

Toxicity, side effects, and drug interaction: probably poisonous

REFERENCES

Dandannavar, V.S., Balakrishna, J.P., Dhawale, L., Joseph, J.P., Gollapalli, S.S.R., Pillai, A.A. and Veeraraghavan, V.P. 2019. Wound-healing potential of methanolic extract of *Rhaphidophora korthalsii* leaves possibly mediated by collagen and fibronectin expression in L929 cell line. *National Journal of Physiology, Pharmacy and Pharmacology*. 9(8): 813–813.

Wong, K.T. and Tan, B.K.H. 1996. *In vitro* cytotoxicity and immunomodulating property of *Rhaphidophora korthalsii*. *Journal of Ethnopharmacology*. 52(1): 53–57.

Wong, K.T., Tan, B.K.H., Sim, K.Y. and Goh, S.H. 1996. A cytotoxic melanin precursor, 5, 6-dihydroxyindole, from the folkloric anti-cancer plant *Rhaphidophora korthalsii*. *Natural Product Letters*. 9(2): 137–140.

Yeap, S.K., Omar, A.R., Ali, A.M., Ho, W.Y., Beh, B.K. and Alitheen, N.B. 2012. Immunomodulatory effect of *Rhaphidophora korthalsii* on natural killer cell cytotoxicity. *Evidence-Based Complementary and Alternative Medicine*. 1–7.

***Scindapsus perakensis* Hook.f.**

[Probably from the Greek *skindapsos* = an instrument with four strings and Latin *perakensis* = from Perak]

Published in *The Flora of British India* 6: 542. 1893

Synonym: *Scindapsus longipetiolatus* Ridl.
Local name: *pagawangan* (Murut)
Habitat: forests, ornamental
Geographical distribution: Bangladesh, Myanmar, Cambodia, Laos, Thailand, Malaysia, Indonesia
Botanical description: this stout climbing plant grows up to a length of about 12 m. Stems: terete, smooth. Leaves: simple, alternate. Petioles: 10–15 cm long, broadly winged. Blades: broadly oblong to elliptic, coriaceous, 6–12 cm × 21–30 cm, acute or rounded at base, acuminate at apex, glossy, with numerous thin secondary nerves. Spathes: creamy white, ovate, cuspidate, fleshy, about 14 cm long. Spadices: 9–10 cm long, green, oblong. Stamens: 4. Styles: Berries: slate blue.
Traditional therapeutic indications: medicinal (Murut)
Pharmacology and phytochemistry: not known
Toxicity, side effects, and drug interaction: not known

4.1.3.2 Order Asparagales Link (1829)

4.1.3.2.1 Family Asparagaceae Juss. (1789), the Asparagus Family

The family Asparagaceae consists of about 100 genera and 2,500 species of shrubs, climbing plants, or herbaceous plants. Leaves: basal or apical. Blades: with longitudinal secondary nerves. Inflorescences: racemes, panicles, cymes, spikes. Tepals: merged in a 6-lobed tube or free. Stamens: 6. Carpels: 3, forming a trilocular ovary, each locule with 2–12 ovules. Fruits: capsules, berries.

***Cordyline fruticosa* (L.) A. Chev.**

[From Greek *kordyle* = a club and Latin *fruticosus* = bushy]

Published in *Catalogue des plantes du Jardin botanique de Saigon* 66. 1919

Synonyms: *Aletris chinensis* Lam.; *Asparagus terminalis* L.; *Convallaria fruticosa* L.; *Cordyline terminalis* (L.) Kunth; *Dracaena terminalis* L.; *Taetsia fruticosa* (L.) Merr.; *Taetsia terminalis* (L.) W. Wight ex Saff.
Local name: *pipisokalaganan* (Murut)
Common name: ti plant
Habitat: gardens
Geographical distribution: tropical
Botanical description: this ornamental herbaceous plant reaches a height of about 3 m. Stems: ligneous, terete in full-grown plant. Leaves: simple, spiral. Petioles: about 10–30 cm long, channeled, and forming a sheath at base. Blades: green or of a heavenly purple or pink, oblong-lanceolate, elliptic-lanceolate, or narrowly oblong, 25–50 cm × 5–10 cm, acute at base, aristate at apex, somewhat coriaceous, smooth, wavy. Panicles: 30–60 cm long, lax, purplish, coriaceous. Tepals: merged into 6-lobed and 5 mm long tube, the lobes revolute. Stamens: 6, coming out of the corolla. Berries: red, glossy, about 5 mm across.
Traditional therapeutic indications: flatulence (Murut); postpartum (Dusun, Kadazan)

Pharmacology and phytochemistry: steroidal saponins (antibacterial, cytotoxic) (Fouedjou et al. 2014; Nguyen et al. 2022; Ponou et al. 2019). Anti-atherosclerosis (Bogoriani et al. 2015).

Toxicity, side effects, and drug interaction: not known

Comments:

(i) A plant of the genus *Cordyline* Comm. ex R. Br. (1810) is medicinal for the Dusun and Kadazan (local name: *rolok*).

(ii) Plants of the Asparagaceae family Juss. (1798) produce steroidal saponins, which are generally responsible for their pharmacological activities.

REFERENCES

Bogoriani, N.W., Manuaba, I.B.P., Suastika, K. and Wita, I.W. 2015. *Cordyline terminalis* Kunth leaves's saponin lowered plasma cholesterol and bile acids levels by increased the excretion of fecal total bile acids and cholesterol in male wistar rats. *European Journal of Biomedical and Pharmaceutical Sciences*. 2(5): 122–134.

Fouedjou, R.T., Teponno, R.B., Quassinti, L., Bramucci, M., Petrelli, D., Vitali, L.A., Fiorini, D., Tapondjou, L.A. and Barboni, L. 2014. Steroidal saponins from the leaves of *Cordyline fruticosa* (L.) A. Chev. and their cytotoxic and antimicrobial activity. *Phytochemistry Letters*. 7: 62–68.

Nguyen, T.H.Y., Chu, H.M. and Nguyen, D.H. 2022. Two new steroidal saponins from the roots of Cordyline fruticosa (L.) A. Chev. *Natural Product Research*. 1–6.

Ponou, B.K., Teponno, R.B., Tapondjou, A.L., Lacaille-Dubois, M.A., Quassinti, L., Bramucci, M. and Barboni, L. 2019. Steroidal saponins from the aerial parts of *Cordyline fruticosa* L. var. strawberries. *Fitoterapia*. 134: 454–458.

Dracaena elliptica **Thunb.**

[From Greek *drakon* = dragon and Latin *elliptica* = elliptical]

Published in *Dracaena* 6. 1808

Synonyms: *Draco elliptica* (Thunb. & Dalm.) Kuntze; *Pleomele elliptica* (Thunb.) N.E.Br.
Local name: *sipak* (Murut)
Habitat: forests
Geographical distribution: Southeast Asia
Botanical description: this shrub reaches a height of about 3 m. Stems: terete, somewhat fleshy, with leaf scars. Leaves: simple, spiral. Petioles: about 1 cm long, forming a sheath at base. Blades: narrowly lanceolate to narrowly elliptic, fleshy, glossy, 2–3 cm × 10–35 cm, somewhat asymmetrical, tapering at base, long acuminate at apex, with numerous longitudinal secondary nerves. Panicles: lax, terminal, or axillary, up to 10–25 cm long, many-flowered on about 1 cm long peduncles. Perianth: cylindrical, deeply 6-lobed, thin, whitish green, about 1.8–2.5 cm long. Stamens: 6, as long as the perianth. Styles: 1, thin, protuding. Berries: globose, 1.5 cm across, dull orangish. Seeds: 3 (Figure 4.8).
Traditional therapeutic indications: fatigue (Murut)
Pharmacology and phytochemistry: not known
Toxicity, side effects, and drug interaction: not known

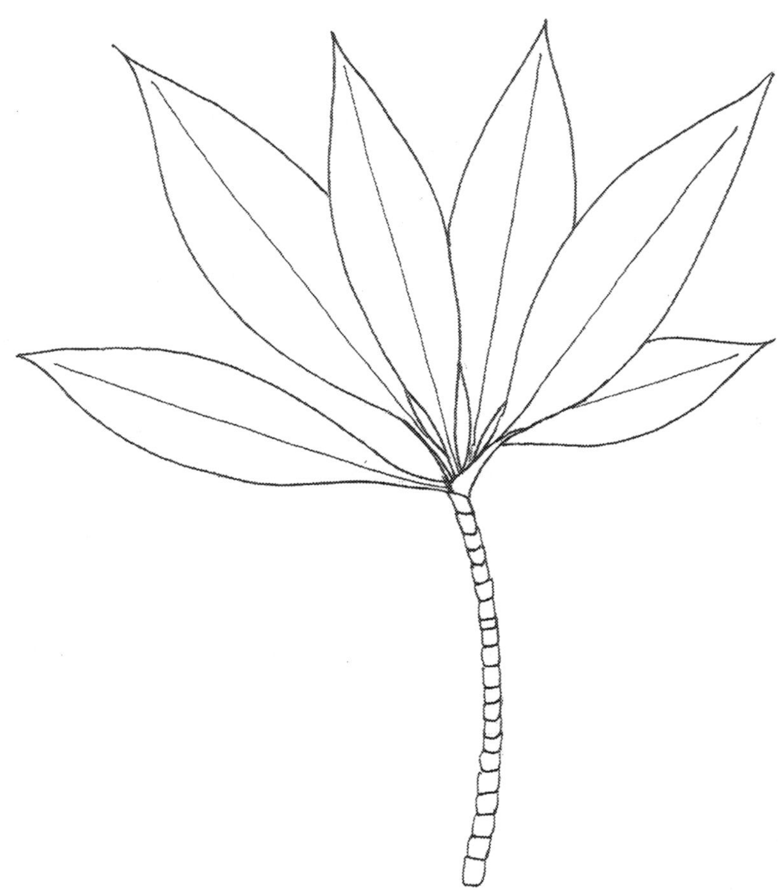

FIGURE 4.8 *Dracaena elliptica* Thunb.

***Dracaena umbratica* Ridl.**

[From Greek *drakon* = dragon and Latin *umbra* = shadow]

Published in *The Flora of the Malay Peninsula* 4: 334. 1924

Local name: *dolol apui* (Murut)
Habitat: forests
Geographical description: Thailand, Malaysia, Borneo
Botanical description: this shrub reaches a height of about 1 m. Stems: terete. Leaves: simple, spiral. Petioles: about 2.5–5 cm long, channeled, sheathing at base. Blades: elliptic to elliptic-lanceolate, 1–4 cm × 5–10 cm, somewhat asymmetrical, wavy, tapering at base, caudate at apex, with numerous longitudinal secondary nerves. Racemes: lax, terminal, up to about 20 cm in length, with numerous pendulous flowers on about 3 mm long peduncles. Perianth: about 2.5 cm long, white, split halfway, 6-lobed, the 6 lobes linear. Stamens: 6. Styles: 1, thin. Berries: globose to vaguely lobed, 6 mm across, orange. Seeds: 3.
Traditional therapeutic indications: medicinal (Murut)
Pharmacology and phytochemistry: steroidal saponins toxic of murine leukemia cells (P-388) (Raudhatul Jannah et al. 2013).
Toxicity, side effects, and drug interaction: not known

Comments:

(i) Plants of the genus *Dracaena* Vand. (1767) are used by the Dusun: "*patidong*" for magic rituals and "*sambangun*" for fatigue.
(ii) It should be noted that the Dusun, Kadazan, and other ethnic groups living in areas less accessible to the so-called "modern world" tend to feel guilty and intimidated when it comes to their medicinal traditions.

REFERENCE

Raudhatul Jannah, M.Z., Zurina, M., Zaini, Y., Shaida, F.S., Juliawaty, L.D. and Nordin, L. 2013, Cytotoxicity and antioxidant activity of a steroidal saponin isolated from *Dracaena umbratica*. *The Open Conference Proceedings Journal*. 44: 100.

***Sansevieria trifasciata* Prain**

[After a 18th century Italian aristocrat and Latin *trifasciata* = with 3 bundles]

Published in *Bengal Plants* 2: 1054. 1903

Local name: *lidah jin* (Bajau, Kedayan, Murut)
Common names: mother-in-law's tongue, snake plant
Habitat: gardens, indoor plant
Distribution: tropical
Botanical description: this rhizomatous herbaceous plant reaches a height of about 80 cm. Leaves: broadly linear, 3–7 cm × 30–100 cm long, coriaceous, fibrous, marbled with whitish to yellow streaks, somewhat wavy, sappy, subulate at apex Racemes: 30–75 cm long. Tepals: 6, up to about 2 cm long, whitish, revolute. Stamens: 6, as long as the tepals, with white and straight filaments. Styles: 1., thin, as long as the stamens. Drupes: rare, orange or reddish, globose, about 8 mm across.
Traditional therapeutic indications: earache, itchy skin, toothaches (Kedayan); antidote for poisons, wound, blood circulation (Bajau), kidney stones (Bajau, Dusun)
Pharmacology and phytochemistry: antibacterial (Dewatisari et al. 2022), wound healing (Yuniarsih et al. 2023), analgesic (Anbu et al. 2009), anti-inflammatory, anti-anaphylactic (Andhare et al. 2012), antiulcer (Ighodaro et al. 2017). Homoisoflavonoid (trifasciatine A), steroidal saponins (Mimaki et al. 1996; Tchegnitegni et al. 2015) of which trifasciatoside B toxic to cervical cancer cells (HeLa) (Teponno et al. 2016).
Toxicity, side effects, and drug interaction: allergy (Valencia 2022). Ethanol extract administered to rats up to a dose of 18 g/kg was found to cause no signs of toxicity (Ighodaro et al. 2017).

Comment: In the family Amaryllidaceae J.St.-Hil. (1805), shallot (*Allium ascalonicum* L.) is used by the Lundayeh (local name: *bawang Siam*) for skin cancer and wounds. The Rungus use shallot for stomach aches, headaches, and joints pain. It is used for fever by locals of Javanese descent and for flatulence by the Dusun and Kadazan. Onion (*Allium cepa* L.) is used by the Bugis (local name: *lasuna cellak*) for cancer, fever, and skin diseases while garlic (*Allium sativum* L.) is used (local name: *lasuna puteh*) for fever.

REFERENCES

Anbu, J.S.J., Jayaraj, P., Varatharajan, R., Thomas, J., Jisha, J. and Muthappan, M. 2009. Analgesic and antipyretic effects of *Sansevieria trifasciata* leaves. *African Journal of Traditional, Complementary, and Alternative Medicines*. 6(4): 529.

Andhare, R.N., Raut, M.K. and Naik, S.R. 2012. Evaluation of antiallergic and anti-anaphylactic activity of ethanolic extract of *Sanseveiria trifasciata* leaves (EEST) in rodents. *Journal of Ethnopharmacology*. 142(3): 627–633.

Dewatisari, W., Nugroho, L.H. and Retnaningrum, E. 2022. Antibacterial and anti-biofilm-forming activity of secondary metabolites from Sansevieria trifasciata-leaves against pseudomonas aeruginosa. *Indonesian Journal of Pharmacy*. 100–109.

Ighodaro, O.M., Adeosun, A.M., Ojiko, B.F., Akorede, A.T. and Fuyi-Williams, O. 2017. Toxicity status and antiulcerative potential of *Sansevieria trifasciata* leaf extract in Wistar rats. *Journal of Intercultural Ethnopharmacology*. 6(2): 234.

Mimaki, Y., Inoue, T., Kuroda, M. and Sashida, Y. 1996. Steroidal saponins from *Sansevieria trifasciata*. *Phytochemistry*. 43(6): 1325–1331.

Tchegnitegni, B.T., Teponno, R.B., Tanaka, C., Gabriel, A.F., Tapondjou, L.A. and Miyamoto, T. 2015. Sappanin-type homoisoflavonoids from *Sansevieria trifasciata Prain*. *Phytochemistry Letters*. 12: 262–266.

Teponno, R.B., Tanaka, C., Jie, B., Tapondjou, L.A. and Miyamoto, T. 2016. Trifasciatosides A—J, steroidal saponins from *Sansevieria trifasciata*. *Chemical and Pharmaceutical Bulletin*. 64(9): 1347–1355.

Valencia, A.N. 2022. Allergic Rhinitis due to the ornamental plant *Sansevieria trifasciata*. *The Journal of Investigational Allergology and Clinical Immunology*. 32: 1–7.

Yuniarsih, N., Hidayah, H., Gunarti, N.S., Kusumawati, A.H., Farhamzah, F., Sadino, A. and Alkandahri, M.Y. 2023. Evaluation of wound-healing activity of hydrogel extract of *Sansevieria trifasciata* Leaves (Asparagaceae). *Advances in Pharmacological and Pharmaceutical Sciences*. 2023: 1–10.

4.1.3.2.2 Family Hypoxidaceae R.Br. (1814), the Star-Grass Family

The family Hypoxidaceae consists of about 9 genera and 120 species of herbaceous plants growing from rhizomes or corms. Leaves: simple, spiral, basal. Inflorescences: solitary or racemes, spikes, umbels. Tepals: 6. Stamens: 6. Carpels: 3 forming a 3-locular ovary, each locule containing numerous ovules. Styles: 1.Capsules: fleshy. Seeds: numerous.

Curculigo latifolia **Dryand. ex W.T. Aiton**

[From Latin *curculio* = weevil, *latius* = broad, and *folia* = leaf]

Published in *A Voyage to Terra Australis* 2: 576. 1814

Synonyms: *Aurota latifolia* (Dryand. ex W.T. Aiton) Raf.; *Molineria latifolia* (Dryand. ex W.T. Aiton) Kurz
Local name: *tambaka* (Dusun, Lundayeh, Murut)
Common name: weevil lily
Habitat: forests
Geographical distribution: from India to Borneo
Botanical description: this orchid-like plant reaches a height of about 40 cm. Roots: fibrous. Leaves: simple, basal, turn blackish in herbarium samples. Petioles: 7–13 cm long. Blades: coriaceous, plicate, elliptic, tapering at base, narrowly acuminate at apex, 6.5–10.5 cm × 20–35 cm, somewhat hairy beneath. Racemes: hairy, thin, up to 10 cm long. Tepals: 2 sets of 3, oblong, 6 mm–1 cm long, ciliate, yellow. Stamens: 6, about 5 mm long, plain buttercup yellow. Ovary: hairy, oblong, about 7 mm long. Styles 1, thin, yellow. Stigma: trifid, yellow. Berries: somewhat bottle-shaped to fusiform or weevil-like, about 1.5 cm long, with a repulsive aura, glossy, light gray, marked at apex by a beak (Figure 4.9).

FIGURE 4.9 *Curculigo latifolia* Dryand. ex W.T. Aiton.

Traditional therapeutic indications: asthma, bone aches, skin diseases (Dusun, Kadazan); coughs (Dusun); fever, cuts, wounds (Murut); abdominal pain (Lundayeh)
Pharmacology and phytochemistry: antidiabetic (Ishak et al. 2013), cytotoxic (Sheh-Hong & Darah 2013), antibacterial (Hong & Ibrahim 2012). Phenolic compounds (Ooi et al. 2016).
Toxicity, side effects, and drug interaction: not known

Comment: The Murut use the fruits as food.

REFERENCES

Hong, L., S., Ibrahim, D. 2012. Studies on antibiotic compounds of methanol extract of *Curculigo Latifolia* Dryand. In *2nd Syiah Kuala University Annual International Conference 2012, Banda Aceh, Indonesia, November 2012*. Syiah Kuala University.

Ishak, N.A., Ismail, M., Hamid, M., Ahmad, Z. and Abd Ghafar, S.A. 2013. Antidiabetic and hypolipidemic activities of *Curculigo latifolia* fruit: Root extract in high fat fed diet and low dose STZ induced diabetic rats. *Evidence-based Complementary and Alternative Medicine*. 2013: 1–13.

Ooi, D.J., Chan, K.W., Sarega, N., Alitheen, N.B., Ithnin, H. and Ismail, M. 2016. Bioprospecting the curculigoside-cinnamic acid-rich fraction from *Molineria latifolia* rhizome as a potential antioxidant therapeutic agent. *Molecules*. 21(6): 682.

Sheh-Hong, L. and Darah, I. 2013. Assessment of anticandidal activity and cytotoxicity of root extract from *curculigo latifolia* on pathogenic *Candida albicans*. *The Journal of Medical Sciences*. 13(3): 193–200.

4.1.3.2.3 Family Aspholedaceae Juss. (1789), the Asphodel Family

The family Aspholedaceae consists of about 40 genera and 900 species of herbaceous plants. Leaves: simple, basal, spiral, coriaceous or fleshy, sappy, exuding a yellow, red, or brown resin once incised. Inflorescences: racemes, panicles. Tepals: 6. Stamens: 6. Carpels: 3, forming a 3-locular ovary, each locule with 2-many ovules. Styles: 1 Fruits: capsules. Seeds: arillate.

Aloe vera **(L.) Burm.f.**

[Possible from Arabic *alloeh* = shining bitter matter and Latin *vera* = true]

Published in *Flora Indica. . . nec non Prodromus Florae Capensis* 83. 1768

Synonyms: *Aloe barbadensis* Mill.; *Aloe chinensis* (Haw.) Baker

Local names: *lidah buaya* (Bajau, Kedayan, Murut); *lilla buaja* (Bugis); *dihabuazo* (Dusun, Kadazan)

Common names: Barbado aloe, true aloe

Habitat: cultivated

Geographical distribution: tropical

Botanical description: this herbaceous plant reaches a height of about 1 m. Leaves: basal, fleshy, linear-lanceolate, slightly stiff, almost terete, 4–5 cm × 15–50 cm, loosely spiny at margin, yielding an orangish latex once incised, and, within, filled with a translucent gel. Spikes: erect, 60–90 cm tall. Perianth: 6-lobed, pale yellow, about 3 cm long. Stamens: 6, about 5 mm long. Styles: 1., filiform.

Traditional therapeutic indications: kidney stones, postpartum (Bajau); sore throats, fever, asthma (Kedayan); skin diseases (Bugis); burns, cuts (Kedayan, Dusun); pimples (Bajau, Rungus); wounds (Bajau, Bugis, Dusun, Kadazan, Rungus); stings (Iranun); hair loss, acne (Bajau); itchiness, cuts, burns, stomach aches, shampoo (Dusun, Kadazan)

Pharmacology and phytochemistry: this plant has been the subject of numerous pharmacological studies (Sahu et al. 2013) which have highlighted analgesic, anti-inflammatory (Egesie et al. 2011), and antiulcer properties (Langmead et al. 2004; Sahu et al. 2013; Iftikhar et al. 2015). The yellow resin contains anthraquinones (aloe-emodin, chysophanol, and aloin B) (Sahu et al. 2013).

Toxicity, side effects, and drug interaction: the yellow latex is toxic (Goodyear-Smith. 2011). The gel given at the volume of 100 ml twice daily for 4 weeks did not produce side effects in volunteers (Langmead et al. 2004).

REFERENCES

Egesie, U.G., Chima, K.E. and Galam, N.Z. 2011. Anti-inflammatory and analgesic effects of aqueous extract of Aloe Vera (*Aloe barbadensis*) in rats. *African Journal of Biomedical Research*. 14(3): 209–212.

Goodyear-Smith, F. 2011. Potion or poison? *Aloe vera. Journal of Primary Health Care*. 3(4): 322–322.

Iftikhar, A., Hasan, I.J., Sarfraz, M., Jafri, L. and Ashraf, M.A. 2015. Nephroprotective effect of the leaves of *Aloe barbadensis* (Aloe Vera) against toxicity induced by diclofenac sodium in albino rabbits. *The West Indian Medical Journal*. 64(5): 462.

Langmead, L., Feakins, R.M., Goldthorpe, S., Holt, H., Tsironi, E., De Silva, A., Jewell, D.P. and Rampton, D.S. 2004. Randomized, double-blind, placebo-controlled trial of oral aloe vera gel for active ulcerative colitis. *Alimentary Pharmacology & Therapeutics*. 19(7): 739–747.

Sahu, P.K., Giri, D.D., Singh, R., Pandey, P., Gupta, S., Shrivastava, A.K., Kumar, A. and Pandey, K.D. 2013. Therapeutic and medicinal uses of *Aloe vera*: A review. *Pharmacology & Pharmacy*. 4(8): 599.

Dianella ensifolia **(L.) Redouté**

[From Latin *Diana*, the Roman deity of hunting and *ensifolia* = sword-shaped leaves]

First pulished in *Les Liliacées. . . a Paris* 1(1): pl. 1. (1802)

Synonyms: *Anthericum adenanthera* G. Forst.; *Dianella javanica* Kunth; *Dianella mauritiana* Bl.; *Dianella montana* Bl.; *Dianella nemorosa* Lam.; *Dianella odorata* Bl.; *Dianella sandwicensis* Hook. & Arn.; *Dracaena ensata* Thunb. & Dallm.; *Dracaena ensifolia* L.; *Rhuacophila javanica* Bl.; *Walleria paniculata* Fritsch
Local names: *lepi lepi*, *tagari* (Dusun)
Common name: umbrella dracaena
Habitat: parks, gardens
Geographical distribution: from India to Pacific Islands
Botanical description: this herbaceous plant reaches a height of about 1 m. Rhizomes: present. Leaves: ensiform, up to 80 cm × 2.5 cm, serrate, obtuse at apex. Panicles: thin on about 50 cm long peduncles. Tepals: 6, white, yellowish, or purplish, linear-lanceolate, up to 7 mm long. Stamens: 6, about the same length as the tepals, with oblong anthers. Styles: 1, about 6 mm long. Berries: of a strange sort of purplish blue, glossy, with some kind of sinister aura, somewhat globose, about 5 mm across, glossy.
Traditional therapeutic indications: headaches, aphrodisiac (Dusun)
Pharmacology and phytochemistry: antiviral (Liu et al. 2018). Essential oil (antibacterial) (He et al. 2019). Antibacterial anthraquinone (chrysophanol) (Hatano et al. 1999. (2*S*)-2′,4′-Dihydroxy-7-methoxy-8-methylflavan toxic to cancer cells (Tang et al. 2017; Le Thi Hong Nhung et al. 2019).
Toxicity, side effects, and drug interaction: poisonous (Le Thi Hong Nhung et al. 2019).

Comment: The plant is used for magic rituals in Indonesia.

REFERENCES

Hatano, T., Uebayashi, H., Ito, H., Shiota, S., Tsuchiya, T. and Yoshida, T. 1999. Phenolic constituents of *Cassia* seed and antibacterial effect of some naphthalenes and antraquinones on Methicillin-Resistant Staphylococcus aureus. *Chemical Pharmacology Bulletin*. 47: 1121–1127.

He, Z.Q., Shen, X.Y., Cheng, Z.Y., Wang, R.L., Lai, P.X. and Xing, X. 2019. Chemical composition, antibacterial, antioxidant and cytotoxic activities of the essential oil of dianella ensifolia. *Records of Natural Products*. 14(2): 160–165.

Le Thi Hong Nhung, N.T., Linh, T., Cham, B.T., Thuy, T.T., Tam, N.T., Thien, D.D., Huong, P.T.M., Tan, V.M., Tai, B.H. and Anh, N.T.H. 2019. New phenolics from Dianella ensifolia. *Natural Products Communication*. 1–8.

Liu, J., Zu, M., Chen, K., Gao, L., Min, H., Zhuo, W., Chen, W. and Liu, A. 2018. Screening of neuraminidase inhibitory activities of some medicinal plants traditionally used in Lingnan Chinese medicines. *BMC Complementary and Alternative Medicine*. 18: 102.

Tang, B.Q., Huang, S.S., Liang, Y.E., Sun, J.B., Ma, Y., Zeng, B., Lee, S.M.Y. and Lu, J.L. 2017. Two new flavans from the roots of Dianella ensifolia (L.) DC. *Natural Product Research*. 31(13): 1561–1565.

4.1.3.2.4 Family Orchidaceae Juss. (1789), the Orchid Family

The family Orchidaceae is a vast taxon consisting of about 800 genera and 25,000 species of terrestrial or epiphytic herbaceous plants growing from rhizomes, corms, or tubers. The base of stems in epiphytic species forms a pseudobulb. Leaves: simple, alternate, often fleshy. Inflorescences: racemes, spikes, panicles. Flowers: often with bird or insect shapes, and a grace that surpasses understanding. Tepals: 6, including a labellum. Stamens and style merged in a column (gynostemium). Carpels: 3, forming a unilocular ovary with 30–many ovules. Fruits: capsules. Seeds: numerous, tiny.

Dendrobium umbellatum **Rchb. f.**

[From Greek *dendron* = tree and Latin *umbellata* = umbellate]

Published in *W. G. Walpers, Ann. Bot. Syst.* 6: 303. 1861

Synonyms: *Cadetia homochroma* (J.J.Sm.) Schltr.; *Cadetia pseudoumbellata* (J.J.Sm.) Schltr.; *Cadetia opacifolia* (J.J.Sm.) Schltr.; *Cadetia umbellata* Gaudich.; *Dendrobium homochromum* J.J.Sm.; *Dendrobium opacifolium* J.J.Sm.; *Dendrobium pseudoumbellatum* J.J.Sm.

Local name: *tingasu* (Murut)

Common name: umbeled cadetia

Habitat: forests

Geographical distribution: Indonesia, Borneo, Papua New Guinea, Australia

Botanical description: this epiphyte orchid reaches a height of about 20 cm. Leaves: simple, basal, vaguely forming some kinds of umbels. Petioles: terete, about 5–8 cm long. Blades: narrowly oblong, somewhat curved, dark green, fleshy, wedge-shaped at base, rounded at apex, about 1 cm × 4–7 cm. Flowers: tiny, arise from the leaf axis. Perianth: pure white, with a pair of horizontal broad tepals, light purple at the throat. Ovary: spiny (Figure 4.10).

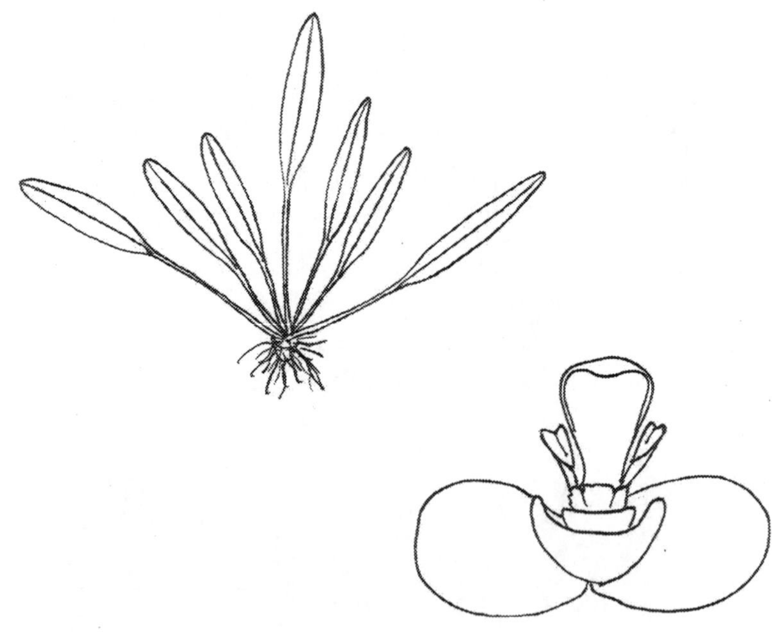

FIGURE 4.10 *Dendrobium umbellatum* Rchb. f.

Traditional therapeutic indications: medicinal (Murut)
Pharmacology and phytochemistry: not known
Toxicity, side effects, and drug interaction: not known

Comments:

(i) A plant of the genus *Flickingeria* A.D. Hawkes (1961) is used by the Dusun and Kadazan for the treatment of snakebites.
(ii) A plant of the genus *Epigeneium* Gagnep. (1932) is used by the Dusun and Kadazan (local name: *tapako*) for snakebites and feverish colds.
(iii) *Bromheadia finlaysionana* (Lindl.) Miq. is used by the Dusun as an aphrodisiac.

4.1.3.3 Order Liliales Pelerb (1826)

4.1.3.3.1 Family Smilacaceae Vent. (1799), the Catbrier Family

The family Smilacaceae consists of about 3 genera and 375 species of rhizomatous climbing plants. Stems: often spiny, ligneous. Leaves: simple, alternate, or opposite, sometimes stipulate. Petioles: often curved. Blades: often coriaceous, glossy, elliptic, with 3 clearly visible longitudinal nerves. Inflorescences: axillary umbels. Tepals: 6. Stamens: 6. Carpels: 3, forming a unilocular or trilocular ovary, each locule with up to 2 ovules. Style: 1. Styles: 1. Stigma: trifid. Berries: globose, red to black.

***Smilax odoratissima* Bl.**

[From the Greek nymph *Smilax* and Latin *odoratissima* = most fragrant]

Published in *Enumeratio Plantarum Javae* 19. 1827

Local name: *lapad makar* (Lundayeh)
Habitat: forests, near streams
Geographical distribution: from Bangladesh to Indonesia
Botanical description: this climbing plant grows up to a length of about 20 m. Stems: finely spiny, woody, sometimes purplish. Tendrils: linear. Leaves: simple, alternate, stipulate. Stipules: lanceolate, about 1 cm long. Petioles: 1–2 cm long. Blades: oblong to ovate, 4–8 cm × 6–13 cm, obtuse-acute at base, acuminate at apex, with 3 longitudinal nerves. Inflorescences: terminal or axillary racemes of umbels on 5 cm long peduncles. Tepals: 6, linear-oblong, tiny. Stamens: 6. Berries: red to purplish black, 5–6 mm across. Seeds: globose to ovoid-angular (Figure 4.11).

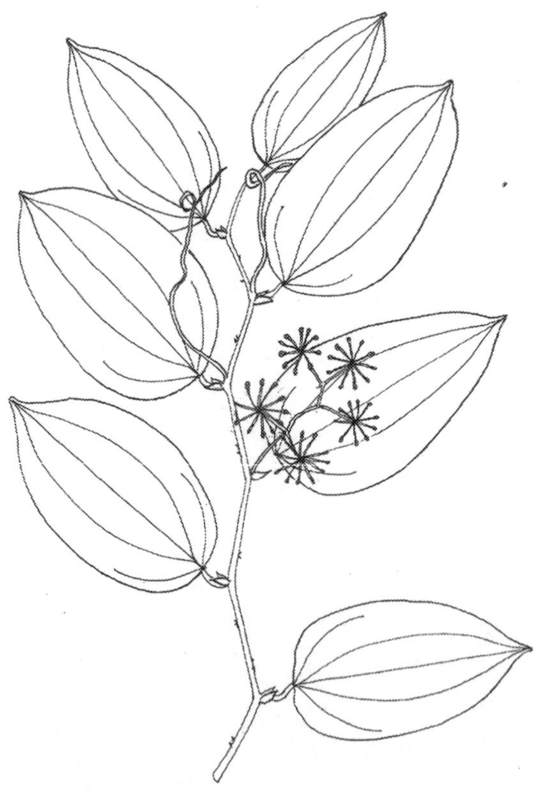

FIGURE 4.11 *Smilax odoratissima* Bl.

Traditional therapeutic indications: sore throats (Lundayeh)
Pharmacology and phytochemistry: not known
Toxicity, side effects, and drug interaction: not known

Comments:

(i) Plants of the genus *Smilax* L. (1753) are used as medicines by the Dusun and Kadazan: "*tunda*" for back and waist pain and "*tongkung kowilan*" for food poisoning.
(ii) In the order Dioscoreales R.Br. (1835), Family Dioscoreaceae R.Br. (1810), a plant of the genus *Dioscorea* R.Br. (1810) is used by the Dusun and Kadazan (local name: *kolonton aiso due*) for fatigue.

4.1.3.4 Order Pandanales R.Br. ex Bercht. & J. Presl (1820)

4.1.3.4.1 Family Pandanaceae R.Br. (1810), the Screwpine Family

The family Pandanaceae consists of about 3 genera and 800 species of shrubs or small trees. Aerial roots: present. Leaves: simple, terminal, spiral, spinous at the margin. Inflorescences: racemes, panicles. Tepals: none or 3–4. Stamens: 10–100. Carpels: numerous, forming a plurilocular ovary, each locule with 1–5 ovules. Fruits: woody drupes packed into a massive fruiting head.

***Pandanus amaryllifolius* Roxb.**

[From the Malay name of the plant *pandan* and Latin *amaryllifolius* = with leaves of *Amaryllis*]

Published in *Flora indica; or, descriptions of Indian Plants* 3: 743. 1832

Synonyms: *Pandanus latifolius* Hassk.; *Pandanus odorus* Ridl.
Local names: *pandan wangi* (Bajau); *pamdan* (Murut)
Common name: pandan
Habitat: cultivated
Geographical distribution: Southeast Asia
Botanical description: this shrub reaches a height of about 4 m. Stems: stout, terete. Leaves: terminal, spinous serrulate, ensiform, 2–5 cm × 25–75 cm, somewhat fleshy, fragrant.
Traditional therapeutic indication: high cholesterol (Bajau)
Pharmacology and phytochemistry: antibacterial, toxic to neck squamous cell carcinoma cells (Suwannakul et al. 2018), antidiabetic (Saenthaweesuk et al. 2016), hepatoprotective (Shameenii et al. 2021).
Pyrrolidine alkaloids (pandamarine B, pandalizines C and D) (Hu et al. 2015; Tsai et al. 2015; Takayama et al. 2000).
Toxicity, side effects, and drug interaction: not known

Comment: The leaves are used to give flavor to dishes by locals.

REFERENCES

Hu, H.C., Cheng, Y.B. and Chang, F.R. 2015. Studies on the chemical constituent and bioactivities of *Pandanus amaryllifolius*. *Planta Medica*. 81(16): 129.

Saenthaweesuk, S., Naowaboot, J. and Somparn, N. 2016. *Pandanus amaryllifolius* leaf extract increases insulin sensitivity in high-fat diet-induced obese mice. *Asian Pacific Journal of Tropical Biomedicine*. 6(10): 866–871.

Shameenii, A., Thanebal, P.P., Vun-Sang, S. and Iqbal, M. 2021. Hepatoprotective effects of *Pandanus amaryllifolius* against carbon tetrachloride (CCl4) induced toxicity: A biochemical and histopathological study. *Arabian Journal of Chemistry*. 14(10): 103390.

Suwannakul, S., Chaibenjawong, P. and Suwannakul, S. 2018. Antioxidant anti-cancer and antimicrobial activities of ethanol Pandanus amaryllifolius Roxb. leaf extract (*in vitro*) – A potential medical application. *Journal of International Dental and Medical Research*. 11(2): 383–389.

Takayama, H., Ichikawa, T., Kuwajima, T., Kitajima, M., Seki, H., Aimi, N. and Nonato, M.G. 2000. Structure characterization, biomimetic total synthesis, and optical purity of two new pyrrolidine alkaloids, pandamarilactonine-A and-B, isolated from *Pandanus amaryllifolius* Roxb. *Journal of the American Chemical Society*. 122: 8635–8639.

Tsai, Y.C., Yu, M.L., El-Shazly, M., Beerhues, L., Cheng, Y.B., Chen, L.C., Hwang, T.L., Chen, H.F., Chung, Y.M., Hou, M.F. and Wu, Y.C. 2015. Alkaloids from *Pandanus amaryllifolius*: Isolation and their plausible biosynthetic formation. *Journal of Natural Products*. 78(10): 2346–2354.

4.1.4 Superorder Lilianae Takht. (1967), the Commelinids

4.1.4.1 Order Arecales Bromhead (1840)

4.1.4.1.1 Family Arecaceae Bercht. & J. Presl (1820), the Palm Family

The family Arecaceae consists of about 180 genera and 2,500 species of palm trees, shrubs, climbing plants, or herbaceous plants. Bark: often smooth, graysish, and regularly marked with scars of fallen leaves. Leaves: palmate, coriaceous, or fleshy, made of a sheath, petiole, and blade. Inflorescences: axillary fascicles of tiny flowers. Sepals: 3. Petals: 3. Stamens: 6–24. Carpels: 3 forming trilocular ovary, each locule with 1 ovule. Fruits: nuts encapsulated in a strongly fibrous husk, drupes.

***Areca catechu* L.**

[From Malayalam *aṭaykka* = *Areca catechu* L. and probably from the Malay *kacu* = *Uncaria gambier* (W. Hunter) Roxb.]

Published in *Species Plantarum* 2: 1189. 1753

Synonyms: *Areca faufel* Gaertn.; *Areca himalayana* Griff. ex H. Wendl.; *Areca hortensis* Lour.; *Areca nigra* Giseke ex H. Wendl.; *Sublimia areca* Comm. ex Mart.
Local names: *logus, lugus* (Dusun, Kadazan); *pinang* (Bajau, Murut); *kusauh, kusob* (Murut)
Common names: betel palm, areca palm, betel nut palm
Habitat: cultivated, roadsides, parks
Geographical distribution: tropical Asia
Botanical description: this palm tree reaches a height of about 20 m. Trunk: straight, thin, 10–20 cm across, grayish, with clearly visible and regularly spaced leaf scars. Leaves: spiral, pinnate. Petioles: up to 5 cm long. Rachis up to 2 m long with 20–30 pinnae. Pinnae: 30–60 × 3–7 cm. Inflorescences: up to 25 cm long. Sepals: 3. Petals: 3. Stamens: 6. Female flowers: larger than male flowers. Nuts: yellow, orange, or plain dull red, ovoid, hard, about 7 cm × 4 cm with persistent calyx. Transversal section of kernels presents brown veinations.
Traditional therapeutic indications: poison antidote (Dusun, Kadazan); hypertension (Bajau); fatigue, cuts, scabies (Kadazan Dusun, Rungus); anemia, stomach aches, bloating, gastritis (Bajau); toothaches (Bajau, Dusun, Kadazan, Rungus)
Pharmacology and phytochemistry: nematicidal (Mubarokah et al. 2019), amoebicidal (Sawangjaroen & Sawangjaroen 2005), antifungal (Anthikat et al. 2014). Piperidine alkaloid (arecoline) with muscarinic agonistic activity (Papke et al. 2015; Wu et al. 2020). Antiviral (Japanese encephalitis virus) anthraquinones (aloe-emodin) (Lin et al. 2008). Antifungal triterpene (fernenol) (Yenjit et al. 2010). Antiviral (transmissible gastroenteritis coronavirus) flavan (catechin) (Liang et al. 2015).
Toxicity, side effects, and drug interaction: long-term use of "*sireh*" has been linked to the development of oral cancers (Deng et al. 2001; Lin et al. 2009).

Comment: The spathes of *Nipa fruticans* (Wurmb) Thunb. are occasionally used to make cigarette paper.

REFERENCES

Anthikat, R.R.N., Michael, A., Kinsalin, V.A. and Ignacimuthu, S. 2014. Antifungal activity of Areca catechu L. *International Journal of Pharmacology and Clinical Sciences*. 4(1): 1–3.
Deng, J.F., Ger, J., Tsai, W.J., Kao, W.F. and Yang, C.C. 2001. Acute toxicities of betel nut: Rare but probably overlooked events. *Journal of Toxicology: Clinical Toxicology*. 39(4): 355–360.
Liang, W., He, L., Ning, P., Lin, J., Li, H., Lin, Z., Kang, K. and Zhang, Y. 2015. (+)-Catechin inhibition of transmissible gastroenteritis coronavirus in swine testicular cells is involved its antioxidation. *Research in Veterinary Science*. 103: 28–33.

Lin, C.W., Wu, C.F., Hsiao, N.W., Chang, C.Y., Li, S.W., Wan, L., Lin, Y.J. and Lin, W.Y. 2008. Aloe-emodin is an interferon-inducing agent with antiviral activity against Japanese encephalitis virus and enterovirus 71. *International Journal of Antimicrobial Agents*. 32(4): 355–359.

Lin, T.J., Nelson, L.S., Tsai, J.L., Hung, D.Z., Hu, S.C., Chan, H.M. and Deng, J.F. 2009. Common toxidromes of plant poisonings in Taiwan. *Clinical Toxicology*. 47(2): 161–168.

Mubarokah, W.W., Nurcahyo, W., Prastowo, J. and Kurniasih, K. 2019. *In vitro* and *in vivo Areca catechu* crude aqueous extract as an anthelmintic against *Ascaridia galli* infection in chickens. *Veterinary World*. 12(6): 877.

Papke, R.L., Horenstein, N.A. and Stokes, C. 2015. Nicotinic activity of arecoline, the psychoactive element of "Betel Nuts", suggests a basis for habitual use and anti-inflammatory activity. *PLoS ONE*. 10(10): e0140907.

Sawangjaroen, N. and Sawangjaroen, K. 2005. The effects of extracts from anti-diarrheic Thai medicinal plants on the *in vitro* growth of the intestinal protozoa parasite: *Blastocystis hominis*. *Journal of Ethnopharmacology*. 98(1–2): 67–72.

Wu, J., Chen, L., Deng, M., Ye, X., Jiang, X., Fu, J., Zhou, P. and Zhang, C. 2020. Arecoline induces dual modulation of blood pressure in rat, including an initial downregulation and a subsequent upregulation. *Tropical Journal of Pharmaceutical Research*. 19(8): 1637–1641.

Yenjit, P., Issarakraisila, M., Intana, W. and Chantrapromma, K. 2010. Fungicidal activity of compounds extracted from the pericarp of Areca catechu against Colletotrichum gloeosporioides *in vitro* and in mango fruit. *Postharvest Biology and Technology*. 55(2): 129–132.

***Caryota mitis* Lour.**

[From the Greek *karyon* = a nut and Latin *mitis* = gentle]

Published in *Flora Cochinchinensis* 2: 569–570. 1790

Synonyms: *Caryota furfuracea* Bl.; *Caryota griffithii* Becc.; *Caryota minor* Wall.; *Caryota nana* Linden; *Caryota propinqua* Blume; *Caryota speciosa* Linden; *Drymophloeus zippellii* Hassk; *Thuessinkia speciosa* Korth.
Local name: *botu* (Dusun)
Common name: fishtail palm
Habitat: forests, cultivated
Geographical distribution: from India to the Philippines
Botanical description: this palm tree reaches a height of about 10 m and emits a somewhat dismal aura. Trunk: light gray, vaguely bamboo-like. Leaves: spiral, pinnate. Petioles: 80 cm–1 m long. Blades: up to 3 m long. Pinnae: 9–23, somewhat glossy, wavy. Spikes: pendulous, numerous, form a hideous mass of up to 1 m long. Sepals: 3, about 3 mm long. Petals: 3, light brown, about 5 mm–1 cm long, hooded. Stamens: 12–24, elongated anthers. Nuts: dull pinkish red turning blackish purple, extremely irritating, globose, smooth, apiculate, up to 3 cm long, eaten by the Asian palm civets (*Paradoxus hermaphoditus*).
Traditional therapeutic indications: to promote lactation (Dusun, Kadazan)
Pharmacology and phytochemistry: antibacterial (Abdelhakim et al. 2017). Flavone glycosides (isoquercitrin, rutin) (Tran 2014). Alkaloids (Zaher et al. 2018).
Toxicity, side effects, and drug interaction: the plant is toxic because its organs are saturated with crystals of calcium oxalate (Snyder et al. 1979).

Comments:

(i) The Dusun and Kadazan use a plant of the genus *Caryota* L. (1753) to promote lactation.
(ii) A plant of the genus *Calamus* L. (1753) is used by the Dusun (local name: *lambah*) for fever, while the Kadazan use "*sarae*" and "*tuai bondig*" for beriberi and impotence.

REFERENCES

Abdelhakim, I.A., El-Mokhtar, M.A., El-Baky, A.M.A. and Bishay, D.W. 2017. Chemical constituents and antimicrobial activity of the leaves of *Caryota mitis* Lour. (Arecaceae). *Journal of Medicinal Plants*. 5(5): 250–255.

Snyder, D.S., Hatfield, G.M. and Lampe, K.F. 1979. Examination of the itch response from the raphides of the fishtail palm Caryota mitis. *Toxicology and Applied Pharmacology*. 48(2): 287–292.

Tran, C.N.B. 2014. Isolation of active compounds from skin fruits of *Caryota Mitis* L (Doctoral dissertation, International University HCMC).

Zaher, A.M., Abdel-Hakim, I.A., Abdel-Baky, A.M. and Bishay, D.W. 2018. Rapid detection of eight volatile alkaloids from *Caryota mitis* Lour. by LC-MS/MS and antimicrobial effects of their extracts. *Medicinal & Aromatic Plants* (Los Angeles). 7: 314.

***Licuala spinosa* Wurmb.**

[From the Macassar name of the plant = *leko wala* and Latin *spinosa* = spiny]

Published in *erhandelingen van het Bataviaasch Genootschap van Kunsten en Wetenschappen* 2: 474. 1780

Synonyms: *Corypha pilearia* Lour.; *Licuala horrida* Bl.; *Licuala pilearia* (Lour.) Bl.; *Licuala ramosa* Bl.
Local name: *palma* (Suluk)
Common names: mangrove fan palm, spiny licuala palm
Habitat: swampy forests, cultivated
Geographical distribution: Bangladesh to the Philippines
Botanical description: this palm reaches a height of about 4 m. Stems: with leaf scars. Leaves; palmate, spiral. Leaf sheaths: up to 40 cm long. Ligules: up to 10 cm long. Petioles: up to about 2 m long, spiny at the margin. Blades: somewhat disc-shaped, coriaceous, 45–65 cm × 80–150 cm, Pinnae: 15–19, narrowly triangular, costulate, 4–9 cm × 41–51 cm. Inflorescences: panicle-like, lax, 0.9–3 m long, on 20–50 cm long peduncles. Calyx: cupular, trilobed, hairy, 3 mm long. Corolla: hairy, up to 4 mm long, whitish, trilobed, lobes acute. Stamens: 6. Ovary: tiny, amphora-shaped. Styles: 1, thin. Nuts: globose, about 8 mm across, globose smooth, orange to red. Seeds: globose, smooth, about 5 mm across.
Traditional therapeutic indications: dehydration (Suluk)
Pharmacology and phytochemistry: antibacterial (Othman 2014). Steroidal saponins, flavonol glycosides (Asami et al. 1991).
Toxicity, side effects, and drug interaction: not known

Comments:

(i) The leaves of this plant and others plants of the genus *Licuala* Wurm. (1790) including *Licualia bidentata* Becc. are used for magic rituals by the Dusun (local name: *silad*). It is interesting to note that about the same practice is found in the Philippines (Palawan).
(ii) *Arenga brevipes* Becc. and *Daemonorops didymophylla* Becc. (local name: *lomu lomu*) are medicinal for the Dusun. They also use *Korthalsia jala* J. Dransf. to expedite childbirth (local name: *rukatan*) whilst *Plectocomia mulleri* Bl. is used to make men and women barren (local name: *mangkawaian*).

REFERENCES

Asami, A., Hirai, Y. and Shoji, J. 1991. Studies on the constituents of palmae plants. VI. Steroid saponins and flavonoids of leaves of *Phoenix canariensis* hort. ex CHABAUD, P. humilis ROYLE var. hanceana BECC., P. dactylifera L., and Licuala spinosa WURMB. *Chemical and Pharmaceutical Bulletin.* 39(8): 2053–2056.

Othman, N.A. 2014. Evaluation of antibacterial activity and toxicity of *Licuala spinosa* stems. Final Year Project Report Submitted in Partial Fulfillment of the Requirements for the Degree of Bachelor of Science (Hons.) Biology in the Faculty of Applied Sciences Universiti Teknologi MARA. Available at https://ir.uitm.edu.my/id/eprint/24873/

Metroxylon sagu **Rottb.**

[From Greek *metra* = uterus, *xylon* = wood, and Austronesian *sagu* = sago]

Published in *Nye Samling af det Kongelige Danske Videnskabers Selskabs Skrifter* 2: 527. 1783

Synonyms: *Metroxylon hermaphroditum* Hassk.; *Metroxylon inerme* (Roxb.) Mart. *Metroxylon laeve* (Giseke) Mart.; *Metroxylon longispinum* (Giseke) Mart.; *Metroxylon micracanthum* Mart.; *Metroxylon oxybracteatum* Warb. ex K.Schum. & Lauterb.; *Metroxylon rumphii* (Willd.) Mart.; *Metroxylon sago* K.D. Koenig; *Metroxylon squarrosum* Becc.; *Metroxylon sylvestre* (Giseke) Mart.; *Sagus americana* Poir.; *Sagus genuina* Giseke; *Sagus inermis* Roxb.; *Sagus koenigii* Griff.; *Sagus laevis* Jack; *Sagus longispina* (Giseke) Blume; *Sagus micracantha* (Mart.) Blume; *Sagus rumphii* Willd.; *Sagus sagu* (Rottb.) H. Karst.; *Sagus spinosa* Roxb.; *Sagus sylvestris* (Giseke) Blume

Local name: *rumbio* (Dusun, Kadazan)

Common name: sago palm

Habitat: swamps, on the banks of rivers, ponds

Geographical distribution: Southeast Asia, Papua New Guinea, Pacific Islands

Botanical description: this palm tree reaches a height of about 30 m. Roots: spongy. Pneumatophores present. Wood: starchy. Leaves: pinnate, spiral. Leaf sheath 5–8 m long. Petioles: stout, clasping the stem at base armed with about 20 cm long needlelike spines. Pinnae: numerous, coriaceous, spiny, 3–6 cm × 50–160 cm, with filiform appendages at apex. Panicles: terminal, up to 5 m long, many-flowered, the flowers tiny. Stamens: 6. Stigma: trifid. Drupes: globose, hard, scaly in a pangolin manner, about 3–5 cm across. Seed: subglobose, about 3 cm across, immersed in a whitish mass.

Traditional therapeutic indications: fatigue (Dusun, Kadazan)

Pharmacology and phytochemistry: antibacterial (Nurlila et al. 2021), immunomodulatory (Pulla et al. 2014).

Toxicity, side effects, and drug interaction: not known

Comment: In Papua New Guinea, the trunk is used as food.

REFERENCES

Nurlila, R.U., Sudiana, S. and La Fua, J. 2021. Efek antibakteri daun sagu (*Metroxylon sagu* Rottb.) terhadap pertumbuhan *Staphylococcus aureus* dan *Escherichia coli*. *Jurnal Mandala Pharmacon Indonesia*. 7(2): 285–322.

Pulla, S., Sannithi, N. and Challa, S.R. 2014. Immunomodulatory effect of water soluble polysaccharides isolated from *Metroxylon sagu* in animal models of immunosuppression. *Pharmacognosy Journal*. 6(5): 55.

***Plectocomiopsis geminiflora* (Griff.) Becc.**

[From the Greek *plektos* = twisted and *poma* = a lid, and Latin *geminiflora* = with twin flowers]

Published in *The Flora of British India* 6: 479. 1893

Synonyms: *Calamus geminiflorus* Griff; *Calamus turbinatus* Ridl.
Local name: *ambarua* (Murut)
Habitat: forests
Geographical distribution: Cambodia, Laos, Vietnam, Malaysia, Borneo
Botanical description: this climbing plant reaches a length of about 50 m. Leaves: pinnate, spiral. Leaf sheaths covered with thin woody spines up to about 3 cm long. Petioles: about 12 cm long. Rachis: about 70 cm–1.4 m long, regularly covered with verticils of hooks with a length of about 1 cm long. Cirrus: 1 m long. Pinnae: 15–40 pairs, up to about 1.5–4 cm × 25–40 cm, attenuate at base, lanceolate, glossy, sessile, long acuminate at apex. Racemes: up to 40 cm long. Flowers: 5 mm long. Nuts: globose to somewhat oblate, to obovate, scaly, about 3 cm across, on persistent perianth.
Traditional therapeutic indications: medicinal (Murut)
Pharmacology and phytochemistry: not known
Toxicity, side effects, and drug interaction: not known

Comment: in Sarawak, the fruits used as food.

***Salacca zalacca* (Gaertn.) Voss**

[From the Malay name of the plant = *salak*]

Published in *Vilmorins Blumengärtnerei. . . Dritte neubearbeite Auflage* 1: 1152. 1895

Synonyms: *Calamus zalacca* Gaertn.; *Calamus salakka* Willd. ex Steud.; *Salacca blumeana* Mart.; *Salacca edulis* Reinw.; *Salacca rumphii* Wall.
Local name: *salak* (Bajau)
Common names: salak palm, snake fruit
Habitat: forests
Geographical distribution: Vietnam, Malaysia, Indonesia
Botanical description: this palm tree reaches a height of about 4 m. Leaves: pinnate, spiral. Leaf sheaths spiny. Petioles: spiny. Blades: 3–7 m long. Pinnae: spiny, 2–7.5 cm × 20–70 cm. Racemes: up to 1 m long. Perianth: tubular, yellow outside, red inside, tiny. Stamens: 6. Styles: 1, trifid. Berries: somewhat pear-shaped, 5 cm × 5–7 cm, tapering toward the base, scaly, rounded at apex. Seeds: 3, about 5 mm long, somewhat globose, black, glossy, in a garlic-colored flesh.
Traditional therapeutic indications: body pains, gastritis (Bajau)
Pharmacology and phytochemistry: antidiabetic, antihypuricemic, cytotoxic (Saleh et al. 2018).
Toxicity, side effects, and drug interaction: not known

Comments:

(i) The fruits are used as food by locals.
(ii) In the family Arecaeae, coconut locally called "*kelapa*" (*Cocos nucifera* L.) is used for smallpox (Sungai, Bajau, Dusun, Kadazan, Rungus, Kedayan, Iranun), as poison antidote, for mumps, headaches, itchiness, cancer (Bugis, local name: *kaluku*), diarrhea, headaches (Dusun, Kadazan, local name: *piasau*), fever (Dusun, Bajau),

high cholesterol (Bajau), hair care, kidney stones (Kedayan), **blood detoxification** at postpartum (Murut), scalds (Dusun, Kadazan, Kedayan), and hypertension (Bajau, Kedayan).

(iii) *Oncosperma horridum* (Griff.) Scheff. is used by the Dusun as food (local name: *nibung*) as well as *Daemonorops periacantha* Miq. (local name: *uwai lambat*).

REFERENCE

Saleh, M.S., Siddiqui, M.J., Mediani, A., Ismail, N.H., Ahmed, Q.U., So'ad, S.Z.M. and Saidi-Besbes, S. 2018. Salacca zalacca: A short review of the palm botany, pharmacological uses and phytochemistry. *Asian Pacific Journal of Tropical Medicine*. 11(12): 645–652.

4.1.4.2 Order Commelinales Mirb. ex Bercht. & J. Presl (1820)

4.1.4.2.1 Family Commelinaceae Mirbel (1804), the Spiderwort Family

The family Commelinaceae consists of about 40 genera and 650 species of herbaceous plants. Leaves: simple, alternate, or spiral. Petioles: sometimes present. Inflorescences: cincinni in panicles or solitary, heads, fascicles. Sepals: 3, free or merged at base often boat-shaped or carinate. Corolla: 3 petals or tubular. Stamens: 2–6, free. Staminodes: 1–3. Carpels: 3, forming a trilocular ovary, each locule with 1–50 ovules. Fruits: capsules or berries.

***Commelina communis* L.**

[After the 17th century Dutch botanist Jan Commelin and Latin *communis* = common]

Published in *Species Plantarum* 1: 40–41. 1753

Synonyms: *Commelina coreana* H. Lév.; *Commelina ludens* Miq.
Local name: *obat batu* (Lundayeh)
Common name: Asiatic dayflower
Habitat: humid and shady places, gardens
Geographical distribution: Cambodia, Laos, Thailand, Vietnam, Malaysia, China, Japan, Korea, Russia
Botanical description: this herbaceous plant grows up to a length of about 80 cm. Stems: creeping, rooting at nodes. Leaves: simple, spiral. Leaf sheaths: 1–2 cm long, glabrous. Blades: narrowly lanceolate to ovate-lanceolate, 1.5–2 cm × 3–9 cm. Inflorescences: cincinni in a folded and cordate involucral bract with a length of 1.5–3 cm long. Sepals: 3, 5 mm long, membranous. Petals: 3, dark blue, 5 mm–1 cm long. Capsules: ellipsoid, about 5 mm long, 2-valved. Seeds: 2 per valve, brown-yellow, tiny, truncate at one end.
Traditional therapeutic indication: bladder stones (Lundayeh)
Pharmacology and phytochemistry: antiviral (influenza) (Bing et al. 2009), antibacterial (Liu et al. 2023). Piperidine alkaloids (1-deoxynojirimycin) (Kim et al. 1999), flavonols (Zhang et al. 2018).
Toxicity, side effects, and drug interaction: not known

REFERENCES

Bing, F.H., Liu, J., Li, Z., Zhang, G.B., Liao, Y.F., Li, J. and Dong, C.Y. 2009. Anti-influenza-virus activity of total alkaloids from *Commelina communis* L. *Archives of Virology*. 154: 1837–1840.

Kim, H.S., Kim, Y.H., Hong, Y.S., Paek, N.S., Lee, H.S., Kim, T.M., Kim, K.W. and Lee, J.J. 1999. α-Glucosidase inhibitors from *Commelina communis*. *Planta Medica*. 65(5): 437–439.

Liu, Y., Tang, Y., Ren, S. and Chen, L. 2023. Antibacterial components and modes of the methanol-phase extract from *Commelina communis* Linn. *Plants*. 12(4): 890.

Zhang, X., Liang, C., Li, C., Bu, M., Bu, L., Xiao, Y., Sun, H. and Zhang, L. 2018. Simultaneous qualitative and quantitative study of main compounds in *Commelina communis* Linn. by UHPLC—Q-TOF-MS-MS and HPLC—ESI-MS-MS. *Journal of Chromatographic Science*. 56(7): 582–594.

***Commelina nudiflora* L.**

[After the 17th century Dutch botanist Jan Commelin and Latin *nudus* = naked and *flos* = flower]

Published in *Species Plantarum* 1: 41–42. 1753

Synonym: *Murdannia nudiflora* (L.) Brenan
Local name: *soriau ngadau* (Dusun)
Common name: doveweed
Habitat: wet soils, on the banks of rivers and streams
Geographical distribution: from Pakistan to Pacific Islands
Botanical description: this herbaceous plant grows up to a length of up to about 30 cm. Stems: creeping, diffuse. Leaves: simple, spiral, extipulate. Leaf sheaths: about 5 cm long, hairy. Blades: linear to lanceolate, 0.5–1 cm × 2.5–10 cm, obtuse to acuminate at apex. Cymes: terminal, lax. Sepals: 3, 3 mm long, ovate to elliptic. Petals: 3, of a peculiar kind of blue, obovate to spathulate, membranous, about 4 mm long. Stamens: 2, blue anthers. Staminodes: white, sagittate. Capsules: ovoid to globose, about 4 mm long, trigonous. Seeds: 2 per valve, irregular, brown-yellow, tiny.
Traditional therapeutic indication: fever (Dusun)
Pharmacology and phytochemistry: antibacterial (Kuppusamy et al. 2015), hepatoprotective (Shah et al. 2017).
Toxicity, side effects, and drug interaction: not known

REFERENCES

Kuppusamy, P., Yusoff, M.M., Parine, N.R. and Govindan, N. 2015. Evaluation of in-vitro antioxidant and antibacterial properties of *Commelina nudiflora* L. extracts prepared by different polar solvents. *Saudi Journal of Biological Sciences*. 22(3): 293–301.

Shah, M.D., D'souza, U.J. and Iqbal, M. 2017. The potential protective effect of *Commelina nudiflora* L. against carbon tetrachloride (CCl4)-induced hepatotoxicity in rats, mediated by suppression of oxidative stress and inflammation. *Environmental Health and Preventive Medicine*. 22(1): 1–19.

***Forrestia griffithii* C.B. Clarke**

[After the 18th century British navigator Thomas Forrest and after the 19th century British botanist William Griffith, civil surgeon in Malacca]

Published in *Monographiae Phanerogamarum* 3: 236. 1881
Synonym: *Amischotolype griffithii* (C.B. Clarke) I.M. Turner
Local name: *tatapis da aputulan* (Murut)
Habitat: forests near streams
Geographical distribution: Malaysia, Indonesia
Botanical description: this herbaceous plant reaches a height of about 1 m. Rhizomes: present. Stems: erect, stout, smooth, terete, diffuse. Leaves: simple, spiral. Leaf sheaths: about 3–4 cm long, hairy, somewhat cylindrical. Petioles: about 1 cm long. Blades: elliptic, 4.5–7.5 cm × 25–35 cm, tapering at base, long acuminate to obtuse at apex, with about 4–5 pairs of longitudinal nerves, hairy beneath. Inflorescences: axillary heads, about 3 cm across. Sepals: 3, about 5 mm long, purplish. Petals: 3, white, oblanceolate, membranous, caducous, about 4 mm long. Stamens: 6. Styles: 1. Capsules: pyriform, about 1 cm long, fleshy, glossy, white to purple, trigonous. Seeds: 2 per valve, trilobed.
Traditional therapeutic indications: medicinal (Murut)
Pharmacology and phytochemistry: phenolic compounds (Mohd Noor et al. 2020).
Toxicity, side effects, and drug interaction: not known

Comment: *Amischotolype sphagnorrhiza* Cowley is used by the Dusun to stop the vomiting of blood.

REFERENCE

Mohd Noor, H.S., Abu Bakar, M.F., Abu Bakar, F.I., Ismail, N.A., Sanusi, S.B. and Mohamed, M. 2020. Phytochemical content and antioxidant activity of selected wild ulam/vegetables consumed by indigenous Jakun community in Taman Negara Johor Endau Rompin (TNJER), Malaysia. *Food Research.* 4(1): 28–33.

4.1.4.2.2 Family Hanguanaceae Airy Shaw (1965)

The family Hanguanaceae consists of the single genus *Hanguana* Bl. (1827).

***Hanguana malayana* (Jack.) Merr.**

[From the Indonesian name of *Hanguana kassintu* Bl. = *hanguana* and Latin *malayana* = from Malaya]

Published in *Enumeratio Plantarum Javae* 14. 1827

Synonyms: *Susum anthelminticum* Blume ex Schult. & Schult. f.; *Susum malayanum* (Jack) Planch. ex Hook.f.; *Veratronia malayana* (Jack) Miq.; *Veratrum malayanum* Jack
Local names: *bunga* (Lundayeh); *tambaka* (Dusun, Kadazan); *nalu kapar* (Kadazan); *tatapis da umbir* (Murut)
Habitat: swamps, slow rivers, lakes
Geographical distribution: Cambodia, Malaysia, Borneo, Indonesia, Papua New Guinea
Botanical description: this herbaceous plant herbaceous plant reaches a height of about 2 m. Leaves: simple, spiral, basal. Petioles: none to about 20–30 cm long, forming a sheath at base. Blades: coriaceous, lanceolate, tapering at base, acuminate at apex, 1.5–15 cm × 20–120 cm, with numerous pairs of longitudinal nerves, dark green, glossy above, somewhat lighter green beneath. Panicles: 20–120 cm long, lax. Tepals: 2 sets of 3, yellowish, and punctuated with red or oval blotches, about 3 mm long. Stamens: 6. Staminodes: 6. Ovary: ovoid, globose. Stigma: sessile, trifid. Berries: oblong, deep red turning dark purple, glossy, about 2 cm long, with persistent style. Seeds: 1–3.
Traditional therapeutic indications: fungal infection (Lundayeh); poison antidote, sprains, fatigue (Dusun, Kadazan); gastritis (Murut); muscle cramps (Dusun, Kadazan)
Pharmacology and phytochemistry: antibacterial (Borela et al. 2020).
Toxicity, side effects, and drug interaction: not known

Comment: A plant of the genus *Hanguana* Bl. (1827) is used by the Dusun and Kadazan as styptic (local name: *tambaka*).

REFERENCE

Borela, V.T., Urbano, J.A.B., Tayag, A.J.A. and Teresa, M.G. 2020. Antibacterial property of *Hanguana malayana* (Bakong) crude leaf ethanolic extract against *Staphylococcus aureus*. *Journal La Lifesci*. 1(5): 1–11.

4.1.4.2.3 Family Pontederiaceae Kunth (1816), the Pickerel-Weed Family

The family Pontederiaceae consists of about 6 genera and 40 species of aquatic herbaceous plants. Leaves: in basal rosettes or spiral, fleshy. Petioles: form a sheath at base. Inflorescences: terminal racemes, spikes, or umbels, or solitary. Tepals: 6. Stamens: 6. Carpels: 3, forming a trilocular ovary, each locule containing 1 to numerous ovules. Styles: 1. Stigma: capitate. Fruits: 3-valved capsules. Seeds: tiny.

***Eichhornia crassipes* (Mart.) Solms**

[After the 19th century Polish statesman Johann Albrecht Freidrich Eichhorn and Latin *crassus* = thick]

Published in *Monographiae Phanerogamarum* 4: 527. 1883

Synonyms: *Eichhornia cordifolia* Gand.; *Eichhornia crassicaulis* Schltdl.; *Eichhornia speciosa* Kunth; *Piaropus crassipes* (Mart.) Raf.; *Piaropus mesomelas* Raf.; *Pondeteria crassipes* Mart.; *Pontederia elongata* Balf.

Local name: *keladi agas* (Kedayan)

Common names: common water hyacinth, water lily

Habitat: ponds, lakes, slow rivers

Distribution: tropical

Botanical description: this floating herbaceous plant reaches a height of about 80 cm. Roots: numerous and somewhat plumose. Leaves: spongy, simple, in basal rosettes. Petioles: 10–40 cm long, spongy, bottle-shaped, smooth, glossy. Blades: broadly ovate to rhomboidal, glossy, about 5 cm × 5 cm, fleshy, cordate, rounded, broadly attenuate at base, somewhat rounded at apex. Panicles: upright, terminal, 7–15-flowered (the flowers somewhat crocus-like), on 35–45 cm long peduncles. Tepals: 6, membranous, somewhat light purplish, or of a heavenly blue, ovate, the upper lobe with a yellow blotch. Stamens: 6, the filaments curved. Styles: 1, thin. Stigma: glandular, hairy. Capsules: ovoid. Seeds: numerous, tiny, winged longitudinally.

Traditional therapeutic indications: asthma, coughs, stomach aches, toothaches (Kedayan)

Pharmacology and phytochemistry: antibacterial (Zhou et al. 2009; Joshi et al. 2013), antifungal (Haggag et al. 2017), anti-inflammatory (Jayanthi et al. 2013).

Phenalenones (Della Greca et al. 1992; Lalitha et al. 2012). Steroids (Della Greca et al. 1991). Anticandidal benzoindenones (2,5-dimethoxyl-4-phenyl-benzoindenone) (Della Greca et al. 1991a). Benzyl amide alkaloid (N^1-acetyl-N^2-formyl-5-methoxykynuramine) (Hardeland et al. 2009).

Toxicity, side effects, and drug interaction: the median lethal dose (LD_{50}) of aqueous extract administered to mice was about 2 g/kg (Lalitha et al. 2012a).

Comment: In Indonesia, this plant is used as food.

REFERENCES

Della Greca, M., Monaco, P. and Previtera, L. 1991. New oxygenated sterols from the weed Eichhornia crassipes solms. *Tetrahedron*. 47(34): 7129–7134.

Della Greca, M., Lanzetta, R., Molinaro, A., Monaco, P. and Previtera, L. 1992. Phenalene metabolites from Eichhornia crassipes. *Bioorganic & Medicinal Chemistry Letters*. 2(4): 311–314.

Della Greca, M., Lanzetta, R., Mangoni, L., Monaco, P. and Previtera, L. 1991a. A bioactive benzoindenone from Eichhornia crassipes solms. *Bioorganic & Medicinal Chemistry Letters*. 1(11): 599–600.

Haggag, M.W., Abou El Ella, S.M. and Abouziena, H.F. 2017. Phytochemical analysis, antifungal, antimicrobial activities and application of *Eichhornia crassipes* against some plant pathogens. *Planta Daninha*. 35: e017159560.

Hardeland, R., Tan, D.X. and Reiter, R.J. 2009. Kynuramines, metabolites of melatonin and other indoles: The resurrection of an almost forgotten class of biogenic amines. *Journal of Pineal Research*. 47(2): 109–126.

Jayanthi, P., Lalitha, P., Sujitha, R. and Thamaraiselvi, A. 2013. Anti-inflammatory activity of the various solvent extracts of *Eichhornia crassipes* (Mart.) Solms. *International Journal of PharmTech Research*. 5(2): 641–645.

Joshi, M. and Kaur, S. 2013. *In vitro* evaluation of antimicrobial activity and phytochemical analysis of *Calotropis procera*, *Eichhornia crassipes* and *Datura innoxia* leaves. *Asian Journal of Pharmaceutical and Clinical Research*. 6(5): 25–28.

Lalitha, P., Sripathi, S.K. and Jayanthi, P. 2012. Secondary metabolites of *Eichhornia crassipes* (waterhyacinth): A review (1949 to 2011). *Natural Product Communications*. 7(9): 1249–1256.

Lalitha, P., Sripathi, S.K. and Jayanthi, P. 2012a. Acute toxicity study of extracts of *Eichhornia crassipes* (Mart.) Solms. *Asian Journal of Pharmaceutical and Clinical Research*. 5(4): 59–61.

Zhou, B., Peng, J., Guo, J. and Tang, S. 2009. Research on the antibacterial activities of extract from *Eichhornia crassipes*. *Jiangsu Journal of Agricultural Sciences*. 25(3): 547–550.

4.1.4.3 Order Poales Small (1903)

4.1.4.3.1 Family Bromeliaceae Juss. (1789), the Pineapple Family

The family Bromeliaceae includes about 50 genera and 2,500 species of epiphytic or terrestrial herbaceous plants or shrubs. Leaves: simple, spiral, sessile, serrate, coriaceous. Inflorescences: terminal or axillary racemes, spikes, or heads. Sepals: 3. Petals: 3. Stamens: 6. Carpels: 3, forming a trilocular ovary, each locule with 5–50 ovules. Fruits: berries, capsules. Seeds: winged or hairy.

Ananas comosus **(L.) Merr.**

[Probably from the ancient Central American Indian name of pineapple = *nana* and Latin *comosus* = comose]

Published in *An Interpretation of Rumphius's Herbarium Amboinense* 133. 1917

Synonyms: *Ananas ananas* (L.) Voss; *Ananas domestica* Rumph.; *Ananas parguazensis* L.A. Camargo & L.B. Sm.; *Ananas sativa* Lindl.; *Ananas sativus* Schult. & Schult. f.; *Ananassa sativa* Lindl.; *Bromelia ananas* L.; *Bromelia comosa* L.

Local names: *nanas*, *tingkauran* (Murut)

Common name: pineapple

Habitat: cultivated

Geographical distribution: tropical

Botanical description: this coriaceous herbaceous plant reaches a height of about 1.5 m. Leaves: narrowly lanceolate, in basal rosettes, coriaceous, spiny at margin. Scapes: short. Spikes: about 20 cm long. Sepals: 3. Corolla: purplish. Stamens: 6. Ovary: globose. Styles: 1. Synarp: comose, ovoid, up to 30 cm long, fleshy, fragrant, heavy, palatable.

Traditional therapeutic indications: dandruff (Murut)

Pharmacology and phytochemistry: this plant has been the subject of numerous pharmacological studies (da Paixão et al. 2021) which have highlighted, among other things, antifungal activity (Octaviani et al. 2020). Proteolytic enzyme (bromelain) (Taussig et al. 1988).

REFERENCES

da Paixão, J.A., de Araújo Neto, J.F., do Nascimento, B.O., da Costa, D.M., Brandão, H.N., Souza, F.V.D., de Souza, E.H., Alves, Q.L., Erling, S.B.L. and de Lima David, J.P. 2021. Pharmacological actions of *Ananas comosus* L. Merril: Revision of the works published from 1966 to 2020. *Pharmacognosy Reviews*. 15(29): 57.

Octaviani, M., Fikrani, D. and Susanti, E. 2020. Aktivitas antijamur ekstrak etanol kulit buah *Ananas comosus* (L) Merr. terhadap Trichophyton mentagrophytes dan *Malassezia furfur. Jurnal Farmasi Indonesia*. 12(2): 159–165.

Taussig, S.J. and Batkin, S. 1988. Bromelain, the enzyme complex of pineapple (*Ananas comosus*) and its clinical application. An update. *Journal of Ethnopharmacology*. 22(2): 191–203.

4.1.4.3.2 Family Cyperaceae Juss. (1789), the Sedge Family

The family Cyperaceae consists of about 70 genera and 4,000 species of sedges growing from rhizomes and bulbs, and developing bulbils and stolons. Culms: often scapose, somewhat stiff, trigonous. Leaf sheaths: present. Ligules: present or none. Leaves: simple, spiral. Inflorescences: spikes, multiple spikes, anthelia, or anthelodia. Glumes: spirally or distichously arranged. Perianth: bristle-like or none. Stamens: 1–3. Carpels: 2–3, forming a unilocular ovary with 1 ovule. Stigma: bifid or trifid, plumose. Achenes: trigonous or biconvex, winged or not. Seed: 1.

***Cyperus rotundus* L.**

[From Greek *kuperos* = sedge and Latin *rotare* = rotate]

Published in *Species Plantarum* 1: 44. 1753

Synonyms: *Chlorocyperus rotundus* (L.) Palla; *Pycreus rotundus* (L.) Hayek
Local name: *rumput halia hitam* (Bajau)
Habitat: grassland, on the banks of streams, ditches, wetlands
Geographical description: tropical
Botanical description: this sedge reaches a height of about 90 cm. Stolons: thin, with ellipsoidal tubers. Culms: solitary, thin, trigonous, smooth. Leaf sheaths: brown. Blades: narrow. Inflorescences: simple or compound anthela. Rays: 3–10, unequal in length, spreading. Spikes: obdeltoid, with 3–10 slightly loosely arranged spikelets. Spikelets: obliquely spreading, linear, 1–3 cm long, 8–28-flowered. Rachilla wings: white, slightly broad, hyaline. Glumes: red to purplish brown on both surfaces, 3 mm long, 5–7-veined. Stamens: 3, anthers linear. Stigma: trifid, protuding. Achenes: obovoid-oblong, tiny, trigonous, puncticulate.
Traditional therapeutic indications: coughing up blood, hematuria nosebleed (Dusun, Kadazan)
Pharmacology and phytochemistry: this plant has been the subject of numerous pharmacological studies (Peerzada et al. 2015). Essential oil (antibacterial) (Kilani et al. 2005). Antiplatelet sesquiterpene: (nootkatone) (Seo et al. 2011). Anti-inflammatory sesquiterpene (caryophyllene α-oxide) (Jin et al. 2011). Antiplasmodial sesquiterpene (10, 12-peroxycalamenene) (Thebtaranonth et al. 1995). Sesquiterpene alkaloids (Jeong et al. 2000).
Toxicity, side effects, and drug interaction: ethanol extract administered orally at the dose of 1 g/kg/day for 14 days to rats did not cause toxic effects (Thanabhorn et al. 2005).

Comment: The Dusun use a plant of the genus *Cyperus* L. (1753) (local name: *wallang*) to make a bath for jaundiced babies.

REFERENCES

Jeong, S.J., Miyamoto, T., Inagaki, M., Kim, Y.C. and Higuchi, R. 2000. Rotundines A-C, three novel sesquiterpene alkaloids from *Cyperus rotundus*. *Journal of Natural Products*. 63(5): 673–675.

Jin, J.H., Lee, D.U., Kim, Y.S. and Kim, H.P. 2011. Anti-allergic activity of sesquiterpenes from the rhizomes of *Cyperus rotundus*. *Archives of Pharmacal Research*. 34: 223–228.

Kilani, S., Abdelwahed, A., Ammar, R.B., Hayder, N., Ghedira, K., Chraief, I., Hammami, M. and Chekir-Ghedira, L. 2005. Chemical composition, antibacterial and antimutagenic activities of essential oil from (Tunisian) *Cyperus rotundus*. *Journal of Essential Oil Research*. 17(6): 695–700.

Peerzada, A.M., Ali, H.H., Naeem, M., Latif, M., Bukhari, A.H. and Tanveer, A. 2015. *Cyperus rotundus* L.: Traditional uses, phytochemistry, and pharmacological activities. *Journal of Ethnopharmacology*. 174: 540–560.

Seo, E.J., Lee, D.U., Kwak, J.H., Lee, S.M., Kim, Y.S. and Jung, Y.S. 2011. Antiplatelet effects of Cyperus rotundus and its component (+)-nootkatone. *Journal of Ethnopharmacology*. 135(1): 48–54.

Thanabhorn, S., Jaijoy, K., Thamaree, S., Ingkaninan, K. and Panthong, A. 2005. Acute and subacute toxicities of the ethanol extract from the rhizomes of *Cyperus rotundus* Linn. *Mahidol University Journal of Pharmaceutical Sciences*. 32: 15–22.

Thebtaranonth, C., Thebtaranonth, Y., Wanauppathamkul, S. and Yuthavong, Y. 1995. Antimalarial sesquiterpenes from tubers of *Cyperus rotundus*: Structure of 10, 12-peroxycalamenene, a sesquiterpene endoperoxide. *Phytochemistry*. 40(1): 125–128.

***Hypolytrum nemorum* (Vahl) Spreng.**

[From Greek *hypo* = beneath, *elytron* = sheath, and Latin *nemorum* = from the woods]

Published in *Systema Vegetabilium, editio decima sexta* 1: 233. 1825

Synonyms: *Hypolytrum formosanum* Ohwi; *Hypolytrum latifolium* Rich. ex Pers.; *Hypolytrum proliferum* Boeckeler; *Schoenus nemorum* Vahl; *Tunga diandra* Roxb.
Local name: *balasan sungei* (Murut)
Habitat: forests near rivers, streams, pools
Geographical distribution: from India to Pacific Islands
Botanical description: this sedge reaches a height of about 1 m. Rhizomes: short, somewhat stout, covered with reddish scales. Culms: 3–4 mm across, trigonous, stiff. Leaf sheaths: brownish, 5–15 cm long, membranous at margin. Leaves: simple, spiral, basal, or cauline. Blades: linear, 0.8–2.6 cm × 35–120 cm, dark green and glossy above, somewhat glaucous beneath, somewhat folded and finely serrate, ensiform. Involucral bracts: 3–5, leaflike, much longer than inflorescences. Inflorescences: paniculate, somewhat globose, up to 7 cm across, with many spikes. Spikes: somewhat oblong, 3–7 mm long. Glumes: brown, cymbiform, tiny, membranous, keeled. Stamens: 2. Stigmas 2–3, white, thin. Achenes: olive green, somewhat pyriform to ovoid, pointed at apex, about 2 mm, compressed or not, longitudinally rugose, apex conically rostrate.
Traditional therapeutic indications: medicinal (Murut)
Pharmacology and phytochemistry: not known
Toxicity, side effects, and drug interaction: not known

Comment: The pharmacological properties of most Sabah medicinal plants have not been studied.

4.1.4.3.3 Family Flagellariaceae Dumort. (1829), the Bushcane Family

The family Flagellariaceae consists of the genus *Flagellaria* L. (1753)

***Flagellaria indica* L.**

[From Latin *flagellum* = a whip and *indica* = from India]

Published in *Species Plantarum* 1: 333. 1753

Synonyms: *Flagellaria angustifolia* Wall: *Flagellaria minor* Bl. ex Schultes & Schultes
Local names: *sogoto tumolong* (Rungus); *waau* (Murut)
Common names: false rattan, whip vine
Habitat: Grassy and swampy roadsides, seashores, mangroves
Botanical description: this climbing plant grows up to a length of about 15 m. Stems: flexuous, terete, somewhat glossy, vaguely bamboo-like. Leaves: simple, alternate. Leaf sheaths: about 2–7 cm long. Blades: narrowly lanceolate, 0.5–2.6 cm × 7–25 cm, rounded at base, somewhat glossy, wavy, characteristically extending into a curly tendril. Panicles: terminal, about 15 cm long. Perianth: creamy white, tiny, tubular, develops 6 oblong and revolute lobes. Stamens: 6, about 7 mm long. Stigma: trifid. Drupes: light pink to red, globose, finely mucronate at apex, about 6 mm across. Seeds: 1, globose, black.
Traditional therapeutic indications: paralysis, stroke (Murut); vomiting, flu, coughs (Rungus); jaundice (Dusun, Kadazan)
Pharmacology and phytochemistry: antiviral (dengue) (Klawikkan et al. 2011), hepatoprotective (Gnanaraj et al. 2016).
Toxicity, side effects, and drug interaction: not known

REFERENCES

Gnanaraj, C., Shah, M.D., Makki, J.S. and Iqbal, M. 2016. Hepatoprotective effects of *Flagellaria indica* are mediated through the suppression of pro-inflammatory cytokines and oxidative stress markers in rats. *Pharmaceutical Biology*. 54(8): 1420–1433.

Klawikkan, N., Nukoolkarn, V., Jirakanjanakit, N., Yoksan, S. and Wiwat, C. 2011. Effect of Thai medicinal plant extracts against Dengue virus *in vitro*. *Mahidol University Journal of Pharmaceutical Science*. 38(1–2): 13–18.

***Scleria bancana* Miq.**

[From Greek *skleros* = hard and from Sundanese *bancana* = deception]

Published in *Flora van Nederlandsch Indië, Eerste Bijvoegsel* (Supplement) (3): 602. 1860

Synonyms: *Scleria ciliaris* Nees; *Scleria chinensis* Kunth; *Scleria malaccensis* Boeckeler; *Scleria scrobiculata* Moritzi
Local name: *onininsil* (Murut)
Common name: winged scleria
Habitat: at the edges of forests
Geographical distribution: Bangladesh to the Solomon Islands
Botanical description: this sedge reaches a height of about 1.5 m. Rhizomes: present. Culms: 2.5–6 mm across, trigonous. Leaves: simple, cauline, spiral. Blades: linear, up to about 1.5 cm × 40 cm, flat. Leaf sheaths: 4–8 cm long, winged. Involucral bracts: leaflike, up to 20 cm long. Panicles: conical, up to 60 cm long. Spikelets: lanceolate, about 4 mm long. Glumes: ovate to lanceolate, about 4 mm long, acuminate at apex, membranous at margin, and reddish-brown. Stamens: 3. Ovary: ovoid. Stigma: trifid. Achenes: globose, about 2 mm across, glossy, reticulate.
Traditional therapeutic indications: medicinal (Murut)
Pharmacology and phytochemistry: not known
Toxicity, side effects, and drug interaction: not known

4.1.4.3.4 Family Poaceae Barnhart (1895), the Grass Family

The family Poaceae is a vast taxon consisting of about 700 genera and 12,000 species of grasses. Culms: often soft. Leaves: simple, alternate. Ligules: present. Leaf sheaths: present. Inflorescences: spikes, spikelets. Perianth: none or bristles. Stamens: 1–6. Carpels: 2, forming a unilocular ovary with 1 ovule. Stigma: plumose. Fruit: caryospsis.

***Bambusa vulgaris* Schrad. ex J.C. Wendl.**

[From the ancient Malay name of the plant = *mambu* and Latin *vulgaris* = common]

Published in *Collectio Plantarum* 2: 26, pl. 47. 1808

Synonyms: *Arundarbor blancoi* (Steud.) Kuntze; *Arundarbor fera* (Oken) Kuntze; *Arundarbor mitis* (Lour.) Kuntze; *Arundarbor monogyna* (Blanco) Kuntze; *Arundarbor striata* (Lodd. ex Lindl.) Kuntze; *Arundo fera* Oken; *Arundo mitis* Lour.; *Bambusa auriculata* Kurz; *Bambusa blancoi* Steud.; *Bambusa fera* (Oken) Miq.; *Bambusa humilis* Rchb. ex Rupr.; *Bambusa latiflora* (Balansa) T.Q. Nguyen *Bambusa madagascariensis* Rivière & C.Rivière; *Bambusa mitis* Blanco; *Bambusa monogyna* Blanco; *Bambusa nguyenii* Ohrnb.; *Bambusa sieberi* Griseb.; *Bambusa striata* Lodd. ex Lindl.; *Bambusa surinamensis* Rupr.; *Bambusa thouarsii* (Raspail) Kunth; *Bambusa wamin* E.G. Camus; *Gigantochloa auriculata* (Kurz) Kurz; *Leleba vulgaris* (Schrad. ex J.C. Wendl.) Nakai; *Nastus thouarsii* Raspail; *Nastus viviparus* Raspail; *Phyllostachys mitis* (Lour.) Rivière & C.Rivière
Local names: *tamahang* (Dusun, Kadazan); *tamalang, tambalang* (Dusun, Kadazan, Murut); *tamalang silau* (Dusun)
Common name: golden bamboo
Habitat: around villages, at the edges of forests, along rivers
Geographical distribution: tropical

Botanical description: this bamboo reaches a height of about 20 m. Culms: cylindrical, smooth, olive green at first then yellow, hollowed, up to 9 cm across, with 20–35 cm long internodes, the nodes visible. Culm sheaths: deciduous, striated, hairy. Auricles: oblong to reniform, about 1 cm long. Leaves: simple, alternate. Ligules: about 4 mm long, serrate. Blades: lanceolate, coriaceous, rounded at base, 1.3–4 cm × 10–25 cm, acute at apex. Inflorescences: spikelets. Glumes: 1–2, hairy near apex, the apex apiculate. Stamens: 6, 6 mm long. Stigmas: 3. Caryopsis: terete.

Traditional therapeutic indication: poison antidote (Dusun, Kadazan)

Pharmacology and phytochemistry: anti-inflammatory (Carey et al. 2009), analgesic (Khairani et al. 2021), wound healing (Lodhi et al. 2016), hypoglycemic (Fernando et al. 1990), anticonvulsant, antiamnesic, anxiolytic (Adebayo et al. 2020), antimalarial (Anigboro 2018), anti-aging (Elekofehinti et al. 2023). Cyanogenic glycosides (taxiphyllin) (Padilla-González et al. 2021).

Toxicity, side effects, and drug interaction: abortifacient (Yakubu & Bukoye 2009). Contains cyanide (Caasi-Lit et al. 2010) and crystals of calcium oxalate (Alade et al. 2015). Ethanol extract given daily for 14 days to rats at the dose of 500 mg/kg has caused an increase in white blood cell count (Alade et al. 2015).

Comments:

(i) The young shoots are used as food in Sabah.
(ii) Bamboos were previously part of the family Bambusaceae Nakai (1943), which seems to be much more logical from a morphological point of view.
(iii) A plant of the genus *Bambusa* Schreb. (1789) is used by the Bajau for postpartum (local name: *buluh*).
(iv) *Saccharum officinarum* L. is used by the Dusun for flu (local name: *tebu*).

REFERENCES

Adebayo, M.A., Akinpelu, L.A., Okwuofu, E.O., Ibia, D.E., Lawson-Jack, A.F. and Igbe, I. 2020. Anticonvulsant, antiamnesic and anxiolytic activities of methanol leaf extract of *Bambusa vulgaris* (Poaceae) in mice. *Journal of African Association of Physiological Sciences*. 8(2): 149–157.

Alade, G.O., Ajibesin, K.K. and Omobuwajo, O.R. 2015. Pharmacognostic evaluation of *Bambusa vulgaris* var. *vulgaris* leaf and its toxicity studies in male wistar rats. *Nigerian Journal of Natural Products and Medicine*. 19: 106–114.

Anigboro, A.A. 2018. Antimalarial efficacy and chemopreventive capacity of bamboo leaf (*Bambusa vulgaris*) in malaria parasitized mice. *Journal of Applied Sciences and Environmental Management*. 22(7): 1141–1145.

Caasi-Lit, M.T., Mabesa, L.B. and Candelaria, R.B. 2010. Bamboo shoot resources of the Philippines: II. Proximate analysis, cyanide content, shoot characteristics and sensory evaluation of local bamboo species. *Philippine Journal of Crop Science*. 35(3): 31–40.

Carey, W.M., Dasi, J.M.B., Rao, N.V. and Gottumukkala, K.M. 2009. Anti-inflammatory activity of methanolic extract of *Bambusa vulgaris* leaves. *International Journal of Green Pharmacy (IJGP)*. 3(3).

Elekofehinti, O.O., Aladenika, Y.V., Iwaloye, O., Okon, E.I.A. and Adanlawo, I.G. 2023. *Bambusa vulgaris* leaves reverse mitochondria dysfunction in diabetic rats through modulation of mitochondria biogenic genes. *Hormone Molecular Biology and Clinical Investigation*. 1–22.

Fernando, M.R., Thabrew, M.I. and Karunanayake, E.H. 1990. Hypoglycaemic activity of some medicinal plants in Sri-Lanka. *General Pharmacology: The Vascular System*. 21(5): 779–782.

Khairani, S., Rahayu, L., Sandhiutami, N.M., Dewi, R.S. and Rahmawati, I. 2021. Uji efek anti-inflamasi dan analgesik dari rebusan daun bambu kuning (*Bambusa vulgaris* Schard). *Jurnal Ilmu Kefarmasian Indonesia*. 19(2): 266–271.

Lodhi, S., Jain, A.P., Rai, G. and Yadav, A.K. 2016. Preliminary investigation for wound healing and anti-inflammatory effects of *Bambusa vulgaris* leaves in rats. *Journal of Ayurveda and Integrative Medicine*. 7(1): 14–22.

Padilla-González, G.F., Sadgrove, N.J., Rosselli, A., Langat, M.K., Fang, R. and Simmonds, M.S. 2021. Cyanogenic derivatives as chemical markers for the authentication of commercial products of bamboo shoots. *Journal of Agricultural and Food Chemistry*. 69(34): 9915–9923.

***Coix lacryma-jobi* L.**

[From the Greek name of an Egyptian palm tree = *koix*, Latin *lacryma* = tear, and *jobi* = Job]

Published in *Species Plantarum* 2: 972. 1753

Synonyms: *Coix agrestis* Lour.; *Coix arundinacea* Lam.; *Coix exaltata* J. Jacq.; *Coix lacryma* L.; *Coix ovata* Stokes; *Coix pendula* Salisb.; *Lithagrostis lacryma-jobi* (L.) Gaertn.; *Sphaerium lacryma* (L.) Kuntze

Local names: *dalai, tigiu* (Dusun, Kadazan)

Common name: Job's tears

Habitat: on the banks of rivers, watery soils, forests, cultivated

Geographical distribution: tropical Asia, Pacific Islands

Botanical description: this stout herbaceous plant reaches a height of about 3 m. Leaves: simple, alternate, cauline. Ligules: tiny. Blades: linear-lanceolate, 10–45 cm × 1.5–5 cm, somewhat rounded at base, acute at apex. Racemes: up to 4 cm long with a few spikelets arranged in pairs or by 3, oblong, and up to 8 mm long. Stamens: 3, 5 mm long. Stigmas: 2. Nuts: magnificent (used to make prayer beads, rosaries, and necklaces) somewhat bluish, pyriform, pear-shaped, glossy, up to 1.5 cm long, containing an edible caryopsis.

Traditional therapeutic indications: intestinal worms, coughs, fever, flu, headaches (Dusun, Kadazan)

Pharmacology and phytochemistry: toxic to cervical cancer cells (HeLa) (Son et al. 2019), antibacterial, antifungal (Ishiguro et al. 1993), antivenom (Rajesh et al. 2017), hypouricemic, nephroprotective (Taejarernwiriyakul et al. 2015), hypolipidemic (Gu et al. 2021), angiotensin-converting enzyme 2 inhibition (Diningrat et al. 2021). Antifungal, antiviral (COVID), anti-inflammatory benzoxazolinones (coixol) (Otsuka et al. 1988; Wang et al. 2020). Anti-inflammatory, analgesic polysaccharides (Sui & Xu 2022).

Toxicity, side effects, and drug interaction: not known

Comment: Quinones are often nematicidal (Esteves et al. 2017) and one could examine the nematicidal effects of coixol.

REFERENCES

Diningrat, D.S., Sari, A.N. and Harahap, N.S. 2021. Potential of Hanjeli (Coix lacryma-jobi) essential oil in preventing SARS-CoV-2 infection via blocking the angiotensin converting enzyme 2 (ACE2) receptor. *Journal of Plant Biotechnology*. 48(4): 289–303.

Esteves, I., Maleita, C., Fonseca, L., Braga, M.E., Abrantes, I. and De Sousa, H.C. 2017. *In vitro* nematicidal activity of naphthoquinones against the root-lesion nematode *Pratylenchus thornei*. *Phytopathologia Mediterranea*. 127–132.

Gu, L., Zhang, Y., Zhang, S., Zhao, H., Wang, Y., Kan, D., Zhang, Y., Guo, L., Lv, J., Hao, Q. and Tian, X. 2021. *Coix lacryma-jobi* seed oil reduces fat accumulation in nonalcoholic fatty liver disease by inhibiting the activation of the p-AMPK/SePP1/ApoER2 pathway. *Journal of Oleo Science*. 70(5): 685–696.

Ishiguro, Y., Okamoto, K. and Sonoda, Y. 1993. Antimicrobial activity in etiolated seedlings of adlay. *Nippon Shokuhin Kogyo Gakkaishi*. 40(5): 353–356.

Otsuka, H., Hirai, Y., Nagao, T. and Yamasaki, K. 1988. Anti-inflammatory activity of benzoxazinoids from roots of *Coix lachryma-jobi* var. *ma-yuen*. *Journal of Natural Products*. 51(1): 74–79.

Rajesh, K.S., Bharath, B.R., Rao, C.V., Bhat, K.I., Bhat, K.C. and Bhat, P. 2017. Neutralization of *Naja naja* venom induced lethality, edema and myonecrosis by ethanolic root extract of Coix lacryma-jobi. *Toxicology Reports*. 4: 637–645.

Son, E.S., Kim, S.H., Kim, Y.O., Lee, Y.E., Kyung, S.Y., Jeong, S.H., Kim, Y.J. and Park, J.W. 2019. *Coix lacryma-jobi* var. *ma-yuen* Stapf sprout extract induces cell cycle arrest and apoptosis in human cervical carcinoma cells. *BMC Complementary and Alternative Medicine*. 19(1): 1–9.

Sui, Y. and Xu, D. 2022. Isolation and identification of anti-inflammatory and analgesic polysaccharides from Coix seed (*Coix lacryma-jobi* L. var. a-yuen (Roman.) Stapf). *Natural Product Research*. 1–10.

Taejarernwiriyakul, O., Anzai, N., Jutabha, P., Kruanamkam, W. and Chanluang, S. 2015. Hypouricemia and nephroprotection of *Coix lacryma-jobi* L. seed extract. *Songklanakarin Journal of Science & Technology*. 37(4): 441–447.

Wang, S.X., Wang, Y., Lu, Y.B., Li, J.Y., Song, Y.J., Nyamgerelt, M. and Wang, X.X. 2020. Diagnosis and treatment of novel coronavirus pneumonia based on the theory of traditional Chinese medicine. *Journal of Integrative Medicine*. 18(4): 275–283.

***Cymbopogon citratus* (DC.) Stapf.**

[From the Greek *kymbe* = boat, *pogon* = beard, and Latin *citratus* = lemonlike]

Published in *Species Plantarum* 2: 972. 1753

Synonyms: *Andropogon cerifer* Hack.; *Andropogon citriodorum* hort. ex Desf.; *Andropogon roxburghii* Nees ex Wight & Arn.; *Andropogon schoenanthus* L.; *Cymbopogon nardus* (L.) Rendle

Local names: *sagumau, segumau* (Dusun); *serai makan, serai mandi, serai wangi* (Kedayan, Bajau); *sohumo* (Murut); *serai* (Bajau); *capai-pimping, sohumau* (Murut)

Common names: citronella grass, fever grass, lemon grass

Habitat: cultivated

Geographical distribution: tropical Asia

Botanical description: this tufted herbaceous plant reaches a height of about 2 m. Rhizomes: present. Leaves: at base form a sort of elongated bulb, which is light yellow to somewhat pinkish and with a slight lemony odor when crushed. Blades: linear-lanceolate, dull light green, nodding, 1–2 cm × 15–60 cm. Spikelets: about 2 cm long.

Traditional therapeutic indications: coughs (Murut); headaches, stomach aches (Kedayan); fever, jaundice (Murut); colds, gastritis, vomiting, breathlessness, itchiness (Bajau); fatigue (Dusun, Murut); flatulence (Bajau, Kedayan, Dusun); fever (Dusun, Murut)

Pharmacology and phytochemistry: this plant has been the subject of numerous pharmacological studies which have highlighted, among other things, antibacterial, antifungal (Onawunmi et al. 1984; Silva et al. 2008; Irkin & Korukluoglu 2009; Miller et al. 2015), antiplasmodial, antitrypanosomal (Kpoviessi et al. 2014; Chukwuocha et al. 2016), giardicidal (Méabed et al. 2018), cytotoxic (Koba et al. 2009), antiviral (adenovirus, dengue virus) (Chiamenti et al. 2019; Rosmalena et al. 2019), and cholesterol-lowering activity (Agbafor & Akubugwo 2007).

Antibacterial monoterpenes (myrcene, citral, citronellal, citronellol, geraniol) (Onawunmi et al. 1984). Antitrypanosomal, antimalarial, anticandidal monoterpenes (citral, limonene) (Moura et al. 2001; Abe et al. 2003; Cardoso & Soares 2010). Anti-inflammatory and antipyretic monoterpene (citral) to rats (Emílio-Silva et al. 2017). Antiviral (respiratory syncytial virus) flavonol *C*-glycoside (isoorientin) (Zhu et al. 2015). Antiviral (coronavirus) hydroxycinnamic acid derivatives (chlorogenic acid, caffeic acid) (Weng et al. 2019; Kayinda 2020).

Toxicity, side effects, and drug interaction: aqueous extract administered orally at the dose of 4 g/kg in mice did not cause signs of toxicity (Méabed et al. 2018).

Comments:

(i) The plant is a common sight in the wet markets of Sabah and gives a slight lemony taste to local soups.
(ii) In the Philippines (Cebu) the plant is used for hypertension (local name: *tanglad*).

REFERENCES

Abe, S., Sato, Y., Inoue, S., Ishibashi, H., Maruyama, N., Takizawa, T., Oshima, H. and Yamaguchi, H. 2003. Anti-*Candida albicans* activity of essential oils including lemongrass (*Cymbopogon citratus*) oil and its component, citral. *Nippon Ishinkin Gakkai Zasshi*. 44(4): 285–291.

Agbafor, K.N. and Akubugwo, E.I. 2007. Hypocholesterolaemic effect of ethanolic extract of fresh leaves of *Cymbopogon citratus* (lemongrass). *African Journal of Biotechnology*. 6(5): 596.

Cardoso, J. and Soares, M.J. 2010. *In vitro* effects of citral on *Trypanosoma cruzi* metacyclogenesis. *Memorias do Instituto Oswaldo Cruz*. 105(8): 1026–1032.

Chiamenti, L., Silva, F.P.D., Schallemberger, K., Demoliner, M., Rigotto, C. and Fleck, J.D. 2019. Cytotoxicity and antiviral activity evaluation of *Cymbopogon* spp hydroethanolic extracts. *Brazilian Journal of Pharmaceutical Sciences*. 55: 1–9.

Chukwuocha, U.M., Fernández-Rivera, O. and Legorreta-Herrera, M. 2016. Exploring the antimalarial potential of whole *Cymbopogon citratus* plant therapy. *Journal of Ethnopharmacology*. 193: 517–523.

Emílio-Silva, M.T., Mota, C.M., Hiruma-Lima, C.A., Antunes-Rodrigues, J., Cárnio, E.C. and Branco, L.G. 2017. Antipyretic effects of citral and possible mechanisms of action. *Inflammation*. 40: 1735–1741.

Irkin, R. and Korukluoglu, M. 2009. Effectiveness of *Cymbopogon citratus* L. essential oil to inhibit the growth of some filamentous fungi and yeasts. *Journal of Medicinal Food*. 12(1): 193–197.

Kanyinda, J.N.M. 2020. Coronavirus (COVID-19): A protocol for prevention and treatment (Covalyse®). *European Journal of Medical and Health Sciences*. 2: 1–4.

Koba, K., Sanda, K., Guyon, C., Raynaud, C., Chaumont, J.P. and Nicod, L. 2009. *In vitro* cytotoxic activity of *Cymbopogon citratus* L. and *Cymbopogon nardus* L. essential oils from Togo. *Bangladesh Journal of Pharmacology*. 4(1): 29–34.

Kpoviessi, S., Bero, J., Agbani, P., Gbaguidi, F., Kpadonou-Kpoviessi, B., Sinsin, B., Accrombessi, G., Frédérich, M., Moudachirou, M. and Quetin-Leclercq, J. 2014. Chemical composition, cytotoxicity and *in vitro* antitrypanosomal and antiplasmodial activity of the essential oils of four *Cymbopogon* species from Benin. *Journal of Ethnopharmacology*. 151(1): 652–659.

Méabed, E.M., Abou-Sreea, A.I. and Roby, M.H. 2018. Chemical analysis and giardicidal effectiveness of the aqueous extract of *Cymbopogon citratus* Stapf. *Parasitology Research*. 117: 1745–1755.

Miller, A.B., Cates, R.G., Lawrence, M., Soria, J.A.F., Espinoza, L.V., Martinez, J.V. and Arbizú, D.A. 2015. The antibacterial and antifungal activity of essential oils extracted from Guatemalan medicinal plants. *Pharmaceutical Biology*. 53(4): 548–554.

Moura, I.C., Wunderlich, G., Uhrig, M.L., Couto, A.S., Peres, V.J., Katzin, A.M. and Kimura, E.A. 2001. Limonene arrests parasite development and inhibits isoprenylation of proteins in *Plasmodium falciparum*. *Antimicrobial Agents and Chemotherapy*. 45(9): 2553–2558.

Onawunmi, G.O., Yisak, W.A. and Ogunlana, E.O. 1984. Antibacterial constituents in the essential oil of *Cymbopogon citratus* (DC.) Stapf. *Journal of Ethnopharmacology*. 12(3): 279–286.

Rosmalena, R., Elya, B., Dewi, B.E., Fithriyah, F., Desti, H., Angelina, M., Hanafi, M., Lotulung, P.D., Prasasty, V.D. and Seto, D. 2019. The antiviral effect of indonesian medicinal plant extracts against dengue virus *in vitro* and in silico. *Pathogens*. 8(2): 85.

Silva, C.D.B.D., Guterres, S.S., Weisheimer, V. and Schapoval, E.E. 2008. Antifungal activity of the lemongrass oil and citral against Candida spp. *Brazilian Journal of Infectious Diseases*. 12(1): 63–66.

Weng, J.R., Lin, C.S., Lai, H.C., Lin, Y.P., Wang, C.Y., Tsai, Y.C., Wu, K.C., Huang, S.H. and Lin, C.W. 2019. Antiviral activity of *Sambucus formosana* Nakai ethanol extract and related phenolic acid constituents against human coronavirus NL63. *Virus Research*. 273: 197767.

Zhu, X.Z., Shen, W.W., Gong, C.Y., Wang, Y., Ye, W.C., Li, Y.L. and Li, M.M. 2015. Antiviral activity of isoorientin against respiratory syncytial virus *in vitro* and *in vivo*. *Journal of Sun Yat-sen University (Medical Sciences)*. 3: 5.

Dinochloa scandens **(Blume) Kuntze**

[From Greek *deinos* = mighty, *chloa* = grass, and Latin *scandens* = climbing]

Published in *Revisio Generum Plantarum* 2: 773. 1891

Synonyms: *Bambusa scandens* Blume; *Chusquea amplopaniculata* Steud.; *Dinochloa macrocarpa* Elmer; *Dinochloa tjankorreh* (Schult.) Buse; *Nastus tjankorreh* Schult.; *Schizostachyum parviflorum* Munro

Habitat: at the edges of forests

Geographical distribution: Thailand, Malaysia, Indonesia

Botanical description: this climbing bamboo grows up to a length of about 10 m. Rhizomes: short, stout. Culms: caespitose, zigzag-shaped, about 1.5–2.5 cm across, ligneous. Internodes: 17–24 cm long, smooth. Culm sheaths: 8–14 cm long, somewhat hairy. Leaves: simple, alternate. Blades: dark green, glossy, lanceolate, 1–3.5 cm × 12–25 cm, attenuate at base, long acuminate at apex. Inflorescences: clustered at the nodes, up to 50 cm long. Spikelets: elliptic, about 4 mm long. Glumes: shorter than spikelets. Stamens: 6, 4 mm long. Ovary: umbonate, glabrous. Stigma: trifid. Caryopsis: orbicular, about 7 mm long, smooth, glabrous, black (Figure 4.12).

Traditional therapeutic indications medicinal (Dusun)

Pharmacology and phytochemistry: not known

Toxicity, side effects, and drug interaction: not known

FIGURE 4.12 *Dinochloa scandens* (Blume) Kuntze

***Dinochloa scabrida* S. Dransf.**

[From Greek *deinos* = mighty, *chloa* = grass, and Latin *scaber* = rough]

Published in *Kew Bulletin* 36(3): 628, f. 7. 1981

Local names: *baran* (Murut); *wadan* (Dusun)
Habitat: at the edges of forests
Geographical distribution: Borneo
Botanical description: this climbing bamboo grows up to a length of about 10 m. Rhizomes: short, stout. Culms: caespitose, zigzag-shaped, about 2 cm across, ligneous. Internodes: 20–25 cm long, smooth. Culm sheaths: glabrous, with white wax. Leaves: simple, alternate. Blades: dark green, glossy, lanceolate, 1.5–2.5 cm × 12–25 cm, hairy, pubescent at the margin, broadly attenuate at base, long acuminate at apex. Inflorescences: clustered at the nodes, up to 3 m long. Spikelets: oblong, greenish-yellow, laterally compressed, about 2 mm long. Glumes: 2, persistent, shorter than spikelets. Stamens: 6. Stigma: trifid. Ovary: umbonate, glabrous. Caryopsis: orbicular, about 5 mm long, smooth, glabrous, black.
Traditional therapeutic indication: medicinal (Murut)
Pharmacology and phytochemistry: not known
Toxicity, side effects, and drug interaction: not known

***Dinochloa sublaevigata* S. Dransf.**

[From Greek *deinos* = mighty, *chloa* = grass, *laevigata* = smooth]

Published in *Kew Bulletin* 36(3): 626, f. 1A—D, 6. 1981.

Local names: *bambu badan, bambu wadan, wadan* (Dusun)
Habitat: forests
Geographical distribution: Borneo
Botanical description: this climbing bamboo grows up to a length of about 10 m. Culms: caespitose, zigzag-shaped, hairy at first, about 2 cm across. Internodes: 30–35 cm long. Culm sheaths: smooth, glabrous, purplish with white wax. Blades: dark green, glossy, narrowly lanceolate, attenuate at base, long acuminate at apex, 2–6.5 cm × 15–40 cm, smooth, glabrous. Inflorescences: clustered at the nodes, up to 5 m long. Spikelets: 4 mm long. Glumes: 2 mm long. Caryopsis: globose, rugulose, purplish to black, about 1 cm long.
Traditional therapeutic indications: sore eyes, flatulence, sore throats, (Dusun); bleeding from cut (Dusun, Kadazan)
Pharmacology and phytochemistry: toxic to *Artemia nauplii* (Benjamin et al. 2022).
Toxicity, side effects, and drug interaction: not known

Comment: A plant of the genus *Dinochloa* Buse (1854) is medicinal for the Murut (local name: *baran*). Kadazan use another plant in this genus as styptic.

***Dinochloa trichogona* S. Dransf.**

[From Greek *deinos* = mighty, *chloa* = grass, *thrix* = hair, and *gonia* = angle]

Published in *Kew Bulletin* 36(3): 624, f. 1E—F, 5. 1981

Local names: *buluh badan, bambu wadan* (Dusun)
Habitat: at the edges of forests

Geographical distribution: Borneo

Botanical description: this climbing bamboo grows up to a length of about 10 m. Rhizomes: short, stout. Culms: caespitose, zigzag-shaped, about 2–3 cm across, ligneous. Internodes: 20–25 cm long, hairy. Culm sheaths: hairy at base, purplish with white wax. Leaves: simple, alternate. Blades: dark green, glossy, lanceolate, 3–7 cm × 25–30 cm, rounded at base, glabrous, long acuminate at apex. Inflorescences: clustered at the nodes, up to 3 m long. Spikelets: oblong, greenish-yellow, laterally compressed, about 2 mm long. Glumes: 2, persistent, shorter than spikelets. Stamens: 6. Syles: 3. Ovary: umbonate, glabrous. Caryopsis: orbicular, about 1 cm long, smooth, glabrous, black.

Traditional therapeutic indication: sore eyes (Dusun)

Pharmacology and phytochemistry: not known

Toxicity, side effects, and drug interaction: not known

Comment: The discovery of this plant species is quite recent and it is safe to say that many other species are waiting to be discovered in Sabah, but they are on the verge of extinction owed to palm oil and other logging activities.

REFERENCE

Benjamin, M.A.Z., Ng, S.Y., Saikim, F.H. and Rusdi, N.A. 2022. The effects of drying techniques on phytochemical contents and biological activities on selected bamboo leaves. *Molecules*. 27(19): 6458.

Eleusine indica **(L.) Gaertn.**

[From Greek *eleusis* = an ancient town in Attica and Latin *indica* = from India]

Published in *De Fructibus et Seminibus Plantarum*. 1: 8. 1788

Synonyms: *Agropyron geminatum* (Spreng.) Schult.; *Cynodon indicus* (L.) Raspail; *Cynosurus indicus* L.; *Cynosurus pectinatus* Lam.; *Eleusine glabra* Schumach.; *Eleusine gonantha* Schrank; *Eleusine gouini* E.Fourn.; *Eleusine inaequalis* E.Fourn.; *Eleusine macrosperma* Stokes; *Eleusine polydactyla* Steud.; *Eleusine scabra* E.Fourn.; *Triticum geminatum* Spreng.
Local names: *liagon* (Murut); *solinatad* (Dusun, Kadazan)
Common name: Indian goose grass
Habitat: lawns, roadsides
Geographical distribution: tropical Asia
Botanical description: this common grass reaches a height of about 90 cm. Culms: flattened, glabrous. Leaves: simple, alternate. Blades: dull light green, 3–8 mm × 10–15 cm. Spikes: somewhat of a dog-tail appearance, arranged in umbels, up to 10 cm long.
Traditional therapeutic indications: asthma, flu (Dusun); bone aches, diarrhea, food poisoning (Murut); cut, flu, piles, postpartum, wounds (Dusun, Kadazan)
Pharmacology and phytochemistry: this plant has been the subject of numerous pharmacological studies (Zakri et al. 2021) which have highlighted, among other things, hepatoprotective (Iqbal & Gnanaraj 2012), antitryprotozoal (Ogtrunk et al. 2018), antiplasmodial (Ettebong et al. 2012), anticandidal (Wiart et al. 2004), antiviral (herpes simplex virus) (Hamidi et al. 1996), and antibacterial activity (Alaekwe et al. 2015). Angiotensin-converting enzyme inhibition (Tutor & Chichioco-Hernandez 2018). Anti-inflammatory *C*-glycosyl flavones (schaftoside, vitexin) (De Melo et al. 2005).
Toxicity, side effects, and drug interaction: not known

REFERENCES

Alaekwe, I.O., Ajiwe, V.I.E., Ajiwe, A.C. and Aningo, G.N. 2015. Phytochemical and anti-microbial screening of the aerial parts of *Eleusine indica*. *International Journal of Pure & Applied Bioscience*. 3(1): 257–264.

De Melo, G.O., Muzitano, M.F., Legora-Machado, A., Almeida, T.A., De Oliveira, D.B., Kaiser, C.R., Koatz, V.L.G. and Costa, S.S. 2005. C-glycosylflavones from the aerial parts of *Eleusine indica* inhibit LPS-induced mouse lung inflammation. *Planta Medica*. 71(4): 362–363.

Ettebong, E.O., Nwafor, P.A. and Okokon, J.E. 2012. *In vivo* antiplasmodial activities of ethanolic exract and fractions of *Eleucine indica*. *Asian Pacific Journal of Tropical Medicine*. 5(9): 673–676.

Hamidi, J.A., Ismaili, N.H., Ahmadi, F.B. and Lajis, N.H. 1996. Antiviral and cytotoxic activities of some plants used in Malaysian indigenous medicine. *Pertanika Journal of Tropical Agricultural Science*. 19(2/3): 129–136.

Iqbal, M. and Gnanaraj, C. 2012. *Eleusine indica* L. possesses antioxidant activity and precludes carbon tetrachloride (CCl4)-mediated oxidative hepatic damage in rats. *Environmental Health and Preventive Medicine*. 17(4): 307–315.

Ogtrunk, O.O., Segun, P.A. and Fasinu, P.S. 2018. Antimicrobial and antiprotozoal activities of twenty-four Nigerian medicinal plant extracts. *South African Journal of Botany*. 117: 240–246.

Tutor, J.T. and Chichioco-Hernandez, C.L. 2018. Angiotensin-converting enzyme inhibition of fractions from *Eleusine indica* leaf extracts. *Pharmacognosy Journal*. 10(1): 25–28.

Wiart, C., Mogana, S., Khalifah, S., Mahan, M., Ismail, S., Buckle, M., Narayana, A.K. and Sulaiman, M. 2004. Antimicrobial screening of plants used for traditional medicine in the state of Perak, Peninsular Malaysia. *Fitoterapia*. 75(1): 68–73.

Zakri, Z.H.M., Suleiman, M., Ng, S.Y., Ngaini, Z., Maili, S. and Salim, F. 2021. *Eleusine indica* for food and medicine. *Journal of Agrobiotechnology*. 12(2): 68–87.

***Garnotia acutigluma* (Steud.) Ohwi**

[After the 19th century French naturalist Prosper Garnot and from the Latin *acuo* = sharp, and *gluma* = husk]

Published in *Botanical Magazine*, Tokyo 55: 393. 1941

Synonyms: *Garnotia caespitosa* Santos; *Garnotia erecta* Santos *Garnotia flexuosa* Santos; *Garnotia griffithii* Munro ex Hook.f.; *Garnotia himalayensis* Santos; *Garnotia kengii* S.L.Chen; *Garnotia khasiana* Santos; *Garnotia longiaristata* Santos; *Garnotia sandwicensis* Hillebr.; *Garnotia tenuis* Keng f. ex S.L.Chen; *Garnotia triseta* Hitchc.; *Streptachne indica* Buse
Local name: *udu bulu* (Lundayeh)
Habitat: shady and moist places, on the banks of streams, forests
Geographical distribution: from India to Pacific Islands
Botanical description: this tufted grass reaches a height of about 80 cm. Nodes: pubescent. Leaves: simple, alternate. Leaf sheaths: hairy. Ligules: tiny. Blades: linear, 1.5–7 mm × 3.5–20 cm, somewhat hairy beneath, narrowly attenuate at base. Panicles: 5–25 cm long. Spikelets: about 3 mm long. Glumes: about 3 mm long, acute at apex. Stamens: 1, tiny. Stigma: 2, linear.
Traditional therapeutic indication: venereal diseases (Lundayeh)
Pharmacology and phytochemistry: not known
Toxicity, side effects, and drug interaction: not known

***Imperata cylindrica* (L.) Raeusch.**

[After the 17th century Italian apothecary Ferrante Imperato and Latin *cylindrica* = cylindrical]

Published in *Nomenclator Botanicus* [ed. 3] 3: 10. 1797

Synonyms: *Imperata allang* Jungh.; *Imperata arundinacea* Cirillo; *Lagurus cylindricus* L.; *Saccharum cylindricum* (L.) Lam.; *Saccharum koenigii* Retz.; *Saccharum thunbergii* Retz.
Local names: *paka* (Dusun); *lalang* (Kedayan, Murut)
Common name: cogon grass
Habitat: grassy areas, roadsides
Geographical distribution: Asia, Pacific Islands
Botanical description: this herbaceous plant reaches a height of about 1.2 m. Rhizomes: present. Leaves: simple, spiral. Leaf sheaths: somewhat hairy at margin and throat. Ligules: tiny. Blades: linear, up to 1 m × 2 cm, long acuminate at apex. Spikes: cylindrical, up to 20 cm long, whitish, of a heavenly grace under the sunlight at dawn. Glumes: 5–9-veined, hairy. Stamens: 2, yellow, about 3 mm long. Stigma: 2, dark purplish, plumose.
Traditional therapeutic indications: fever (Dusun, Kadazan, Murut, Sungai); thrush (Dusun); boils, rheumatisms (Dusun, Kadazan), cuts, wounds (Murut), smallpox (Dusun, Murut, Rungus), childbirth pain (Kedayan); chicken pox, hepatitis, measles, urinary diseases (Dusun); stomach aches (Bajau); flatulence (Sungai)
Pharmacology and phytochemistry: this plant has been the subject of numerous pharmacological studies which have highlighted, among other things, analgesic, anti-inflammatory (Razafindrakoto et al. 2021), immunostimulant (Rosnizar et al. 2022), antibacterial, antifungal (Okey-Nzekwe et al. 2019), antiviral (foot-and-mouth disease) (Chungsamarnyart et al. 2007), and hypoglycemic activity (Suraya et al. 2012).

Flavonol glycosides and chomone glycosides (Abdullah et al. 2016). Vasorelaxant lignans (graminones A and B) (Matsunaga et al. 1994). Hepatoprotective *C*-methylated phenylpropanoid glycosides, neolignans (Ma et al. 2018). Megastigmanes (tabanone) (Cerdeira et al. 2012). Chromones, flavone glycosides (Xuan et al. 2013).

Toxicity, side effects, and drug interaction: not known

Comments:

(i) In the Philippines, the plant is used for urinary tract infections by the Kalanguya of Luzon (local name: *golon*).
(ii) In Sulawesi, the plant is used for stomach aches (local name: *alang alang*).
(iii) In Sumatra, the plant is used for cancer (local name: *rih*).

REFERENCES

Abdullah, F.O., Hussaina, F.H.S., Sardarb, A.S., Vita-Finzic, C. and Vidaric, G. 2016. NPC natural product communications. *Natural Product Communications*. 11: 12.

Cerdeira, A.L., Cantrell, C.L., Dayan, F.E., Byrd, J.D. and Duke, S.O. 2012. Tabanone, a new phytotoxic constituent of cogongrass (Imperata cylindrica). *Weed Science*. 60(2): 212–218.

Chungsamarnyart, N., Sirinarumitr, T., Chumsing, W. and Wajjawalku, W. 2007. *In vitro* study of antiviral activity of plant crude-extracts against the foot and mouth disease virus. *Agriculture and Natural Resources*. 41(5): 97–103.

Ma, J., Sun, H., Liu, H., Shi, G.N., Zang, Y.D., Li, C.J., Yang, J.Z., Chen, F.Y., Huang, J.W., Zhang, D. and Zhang, D.M. 2018. Hepatoprotective glycosides from the rhizomes of Imperata cylindrical. *Journal of Asian Natural Products Research*. 20(5): 451–459.

Matsunaga, K., Shibuya, M. and Ohizumi, Y. 1994. Graminone B, a novel lignan with vasodilative activity from *Imperata cylindrica*. *Journal of Natural Products*. 57(12): 1734–1736.

Okey-Nzekwe, C.M., Ekwonu, A.M., Egwuatu, C.I. and Umennadi, P.U. 2019. Pharmacological activities of compounds of leaves and roots of *Imperata cylindrica* with its antimicrobial and structural elucidation. *American Academic & Scholarly Research Journal*. 11: 20–35.

Razafindrakoto, Z.R., Tombozara, N., Donno, D., Gamba, G., Nalimanana, N.R., Rakotondramanana, D.A., Andrianjara, C., Beccaro, G.L. and Ramanitrahasimbola, D. 2021. Antioxidant, analgesic, anti-inflammatory and antipyretic properties, and toxicity studies of the aerial parts of *Imperata cylindrica* (L.) Beauv. *South African Journal of Botany*. 142: 222–229.

Rosnizar, R., Muliani, F., Ramli, I.M. and Eriani, K. 2022, May. The immunostimulant effects of alang-alang (*Imperata cylindrica*) roots extract on BALB/c male mice (*Mus musculus*). In *7th International Conference on Biological Science (ICBS 2021)* (pp. 487–493). Atlantis Press.

Suraya, A.A., Bariah, S.S., Fattepur, S. and Halijah, H. 2012. Antidiabetic activity of ethanolic extract of Imperata cylindrical (lalang) leaves in alloxan induced diabetic rats. *Archives of Pharmacy Practice*. 3(1): 46.

Xuan, Liu, Zhang, B.F., Li, Yang., Gui-Xin, Chou and Zheng-Tao, Wang. 2013. Two new chromones and a new flavone glycoside from *Imperata cylindrica*. *Chinese Journal of Natural Medicines*. 11(1): 77–80.

***Lophatherum gracile* Brongn.**

[From Greek *lophos* = a crest, ather = *barb*, and Latin *gracilis* = thin]

Published in *Voyage Autour du Monde* 2(2): 50, pl. 8. 1829

Synonyms: *Acroelytrum japonicum* Steud.; *Acroelytrum urvillei* (Steud.) Steud. ex Miq.; *Allelotheca urvillei* Steud.; *Lophatherum annulatum* Franch. & Sav.; *Lophatherum dubium* Steud.; *Lophatherum elatum* Zoll. & Moritzi ex Steud.; *Lophatherum humile* Miq.; *Lophatherum japonicum* (Steud.) Steud.; *Lophatherum lehmannii* Nees ex Steud.; *Lophatherum multiflorum* Steud.; *Lophatherum pilosulum* Steud.; *Lophatherum zeylanicum* Hook.f.

Local name: *udu bulu* (Lundayeh)

Common names: bamboo grass, bamboo leaf

Habitat: shady and moist places

Geographical distribution: from India to Pacific Islands

Botanical description: this beautiful grass reaches a height of about 1.5 m. Tubers: present. Leaves: simple, alternate. Leaf sheaths: hairy or glabrous. Ligules: tiny, brown. Blades: lanceolate, somewhat glossy, base rounded with an approximately 1 cm long pseudopetiole, 2–5 cm × 5–30 cm, somewhat hairy, long acuminate at apex. Racemes: 5–10 cm long. Spikelets: about 1 cm long, narrowly lanceolate. Glumes: about 4 mm long, ovate, with 5 veinations. Stamens: 2–3.

Traditional therapeutic indications: pancreatitis (Lundayeh); postpartum (Dusun)

Pharmacology and phytochemistry: antiviral (Chen et al. 2019), anti-inflammatory (Lai et al. 2021; Ma et al. 2023), toxic to breast cancer cells (Istiqomah et al. 2015).

Antiviral *C*-glycosyl flavones (Zhang et al. 2009; Wang et al. 2011).

Toxicity, side effects, and drug interaction: deadly poisonous for goats and cows (Pratama et al. 2022).

Comment: We can observe a preponderance of medicinal plants used by the Dusun and the Kadazan to treat diseases linked to alcoholism such as pancreatitis and jaundice.

REFERENCES

Chen, L.F., Zhong, Y.L., Luo, D., Liu, Z., Tang, W., Cheng, W., Xiong, S., Li, Y.L. and Li, M.M. 2019. Antiviral activity of ethanol extract of *Lophatherum gracile* against respiratory syncytial virus infection. *Journal of Ethnopharmacology*. 242: 111575.

Istiqomah, A., Muti'ah, R. and Hayati, E.K. 2015. Anticancer activity against breast cancer cells T47D and identification of its compound from extracts and fractions of leaves bamboo grass (*Lophaterum gracile* B.). *ALCHEMY: Journal of Chemistry*. 4(1): 6–16.

Lai, K.H., Chen, P.J., Chen, C.C., Yang, S.H., El-Shazly, M., Chang, Y.C., Wu, Y.H., Wu, Y.H., Wang, Y.H., Hsieh, H.L. and Hwang, T.L. 2021. Lophatherum gracile Brongn. attenuates neutrophilic inflammation through inhibition of JNK and calcium. *Journal of Ethnopharmacology*. 264: 113224.

Ma, Y.L., Wu, Z.F., Li, Z., Wang, Y., Shang, Y.F., Thakur, K. and Wei, Z.J. 2023. *In vitro* digestibility and hepato-protective potential of Lophatherum gracile Brongn. leave extract. *Food Chemistry*. 137336.

Pratama, A.M., Herawati, O., Nuranisa, N.R., Hanifah, N., Wijayanti, A.D., Rahmatullah, S., Nuraini, E. and Budiyanto, A. 2022. Identification of poisonous plants and their solutions for traditional livestock in Bojonegoro District, East Java, Indonesia. *Biodiversitas Journal of Biological Diversity*. 23(1).

Wang, Y., Chen, M., Zhang, J., Zhang, X.L., Huang, X.J., Wu, X., Zhang, Q.W., Li, Y.L. and Ye, W.C. 2011. Flavone C-glycosides from the leaves of *Lophatherum gracile* and their *in vitro* antiviral activity. *Planta Medica*: 46–51.

Zhang, J., Wang, Y., Zhang, X.Q., Zhang, Q.W. and Ye, W.C. 2009. Chemical constituents from the leaves of *Lophatherum gracile*. *Chinese Journal of Natural Medicines*. 7(6): 428–431.

Miscanthus floridulus (Labill.) Warb. ex K. Schum. & Lauterb.

[From Greek *miskos* = stem, *anthos* = flower, and Latin *floridus* = profusely flowering]

Published in *Die Flora der Deutschen Schutzgebiete in der Südsee* 166. 1901

Synonyms: *Eulalia japonica* Trin.; *Miscanthus japonicus* (Trin.) Andersson; *Saccharum floridulum* Labill
Local name: *bidau* (Dusun)
Common name: Pacific Island silvergrass
Habitat: grassy lands
Geographical distribution: from Bangladesh to Pacific Islands
Botanical description: this grass reaches a height of about 4 m. Culms: stiff, up to 1.5 cm across. Leaf sheaths: hairy at the throat. Ligules: hairy, up to 3 mm long. Blades: linear, somewhat coriaceous, glaucous beneath, rounded at base, long acuminate at apex, up to 0.5–4 cm × 20–85 cm. Panicles: massive, with an almost supernatural silvery tone at sunset, somewhat oblong, up to 50 cm long. Glumes: membranous, golden brown, about 5 mm long. Stamens: 2. Stigmas: 2, plumose. Caryopsis: oblong, tiny.
Traditional therapeutic indications: cooling, flatulence (Dusun)
Pharmacology and phytochemistry: not known
Toxicity, side effects, and drug interaction: not known

Panicum palmifolium J. Koenig

[From Latin *panis* = bread and *palmifolium* = palmlike leaf]

Synonyms: *Agrostis plicata* Lour.; *Chaetochloa effusa* (E.Fourn.) Hitchc.; *Chaetochloa sulcata* (Aubl.) Hitchc.; *Chamaeraphis effusa* (E.Fourn.) Kuntze; *Chamaeraphis nepalensis* (Spreng.) Kuntze; *Chamaeraphis neurodes* (Schult.) Kuntze; *Chamaeraphis paniculifera* (Steud.) Kuntze; *Chamaeraphis sulcata* (Aubl.) Kuntze; *Panicum amplissimum* Steud.; *Panicum kleinianum* Nees ex Andersson; *Panicum lene* Steud.; *Panicum mexicanum* Scribn. & Merr.; *Panicum nepalense* Spreng.; *Panicum nervosum* Roxb.; *Panicum neurodes* Schult.; *Panicum palmatum* R.Schleich.; *Panicum paniculiferum* Steud.; *Panicum plicatum* Willd.; *Panicum plicatum* Roxb.; *Panicum sulcatum* Aubl.; *Setaria effusa* E.Fourn.; *Setaria lenis* (Steud.) Miq.; *Setaria paniculifera* (Steud.) E.Fourn. ex Hemsl.; *Setaria palmifolia* (J. Koenig) Stapf; *Setaria sulcata* (Aubl.) Desv.
Local name: *tandaki* (Dusun, Kadazan)
Common name: palm grass
Habitat: edges of forests, forest paths, bamboo thickets
Geographical distribution: from India to Australia
Botanical description: this grass reaches a height of about 1.7 m. Rhizome: short, knotty. Culms: about 5 mm across. Leaves: simple, alternate. Leaf sheaths: hairy, 10–25 cm long. Ligules: 2–3 mm long, membranous. Blades: lanceolate, graceful, finely plicate, somewhat glossy above, 2–7 cm × 20–60 cm, nodding, somewhat vaguely waffled beneath, rounded at base, somewhat truncate at base, acuminate at apex, ciliate. Panicles: up to 70 cm long. Spikelets: broadly lanceolate, 3–4 mm long, acute at apex. Glumes: triangular-ovate, tiny. Stigma: bifid, plumose.
Traditional therapeutic indications: malaria (Dusun, Kadazan)
Pharmacology and phytochemistry: not known
Toxicity, side effects, and drug interaction: not known

Comments:

(i) One could search for antiplasmodial principles in this plant. These could be terpenes.
(ii) The plant is used as of food by Dusun. It is a source of cereals in the Philippines (local name: *liyahan*) and vegetable in Papua New Guinea (local name: *mai*). It is important for developing countries to preserve their agricultural heritage. Local and ancient crops, such as palm grass, could offer an alternative to rice shortage.

***Paspalum conjugatum* P. J. Bergius**

[From Greek *paspalos* = some kind of millet and Latin *conjugatum* = connected]

Published in *Acta Helvetica, Physico-Mathematico-Anatomico-Botanico-Medica* 7: 129, pl. 8. 1772

Synonyms: *Digitaria conjugata* (P.J.Bergius) Schult.; *Panicum conjugatum* (P.J.Bergius) Roxb.; *Paspalum africanum* Poir.; *Paspalum bicrurum* Salzm. ex Döll; *Paspalum ciliatum* Lam.; *Paspalum dolichopus* Trin. ex Steud.; *Paspalum longissimum* Hochst. ex Steud.; *Paspalum renggeri* Steud.; *Paspalum sieberianum* Steud.; *Paspalum tenue* Gaertn.

Local names: *talinting* (Dusun); *belandak* (Murut)

Common names: carabao grass, hilo grass

Habitat: wet soils, paddy fields

Geographical distribution: tropical

Botanical description: this grass grows up to a length of about 40 cm. Culms: erect or creeping, rooting at nodes, reddish-purple. Leaves simple, alternate. Leaf sheaths: papery, 3–7 cm long, flat, keeled, hairy at the margin. Ligure: tiny, membranous. Blades: lanceolate, 1.2–1.6 cm × 10–15 cm, rounded at base and apex. Racemes: somewhat moniliform, straight, about 2.3–4 cm long. Spikelets: ovate, hairy, about 2 mm long, hairy, acute at apex. Glumes: membranous, tiny. Stamens: 3, anthers yellow. Stigma: bifid, plumose.

Traditional therapeutic indication: bone aches (Dusun)

Pharmacology and phytochemistry: antibacterial (Garduque et al. 2019), anti-angiogenic (Datorin et al. 2020).

Toxicity, side effects, and drug interaction: not known

REFERENCES

Datorin, X.V.T., Liboon, J.D., Barazan, P.N.T. and Guanzon, L.A.T. 2020. Anti-angiogenic activity of carabao grass (Paspalum conjugatum) extracts in cam assay. A Research Study Presented to the Faculty of the Integrated School University of Negros Occidental-Recoletos. Available at https://www.researchgate.net/profile/Xynnel-Datorin/publication/363467735_ANTI-ANGIOGENIC_ACTIVITY_OF_CARABAO_GRASS_Paspalum_conjugatum_EXTRACTS_IN_CAM_ASSAY/links/63c0100ea99551743e6204fb/ANTI-ANGIOGENIC-ACTIVITY-OF-CARABAO-GRASS-Paspalum-conjugatum-EXTRACTS-IN-CAM-ASSAY.pdf

Garduque, D.A.P., Mateo, K.R.G. and Oyinloye, S.M.A. 2019. Antimicrobial efficacy of carabao grass (Paspalum conjugatum) leaves on Staphylococcus aureus. *Abstract Proceedings International Scholars Conference*. 7(1): 384–397.

***Thysanolaena latifolia* (Roxb. ex Hornem.) Honda**

[From the Greek *thysanos* = fringe, *chlaena* = cloak, and Latin *latifolia* = broad leaf]

Published in *Journal of the Faculty of Science: University of Tokyo, Section 3, Botany* 3(1): 312–313. 1930

Synonyms: *Agrostis maxima* Roxb.; *Melica latifolia* Roxb. ex Hornem.; *Thysanolaena maxima* (Roxb.) Kuntze
Local name: *togiung* (Dusun)
Common names: bamboo grass, broom grass, tiger grass
Habitat: roadsides, grasslands, hillsides
Geographical distribution: from India to Pacific Islands
Botanical identification: this bamboo-like grass reaches a height of about 3 m. Culms: stiff. Leaf sheaths: smooth, hairy at the margin. Ligules: membranous, truncate, about 2 mm long. Blades: broadly lanceolate to oblong, somewhat coriaceous, glaucous beneath, rounded at base, long acuminate at apex, up to 3–7.5 cm × 30–65 cm. Panicles: massive, purplish brown, up to 1 m long. Glumes: tiny, ovate. Stamens: 2. Caryopsis: oblong, tiny, reddish.
Traditional therapeutic indications: flu (Dusun); headaches, fever (Dusun, Kadazan)
Pharmacology and phytochemistry: analeptic, analgesic (Hoque et al. 2021), hepatoprotective (Gnanaraj et al. 2012), antibacterial (Hoque et al. 2016). Flavonol glycosides (6″- *O* -acetylorientin-2″- *O*-α- L -rhamnopyranoside) (Shrestha et al. 2016).
Toxicity, side effects, and drug interaction: not known

Comments:

(i) Corn (*Zea mays* L.) is used for stomach aches by the Murut (local name: *halai*). Corn rice is burned to make a drink to substitute coffee by impoverished farmers of Mindanao in the Philippines.
(ii) A bamboo of the genus *Gigantochloa* Kurz ex Munro (1868) is used for pancreatitis by the Dusun and Kadazan (local name: *poring*).
(iii) *Schizostachyum latifolium* Gamble is used by the Dusun for bloody stools (local name: *bulu gana*).
(iv) In the family *Eriocaulaceae* Martinov (1820), *Eriocaulon longifolium* Nees ex Kunth is used by the Dusun for canker sores (local name: *kumpau sambangau*).
(v) Compared with other medicinal practices in Southeast Asia, the traditional system of medicine in Sabah seems to employ more Monocots. This could be due because Dusun and Kadazan tend to live on the slopes of Mount Kinabalu. It should be noted that the Dusun villages are charming, clean, quiet, and welcoming. The Dusun (a word that could be translated as farmers or people of the orchards) have green fingers and their gardens are among the most beautiful from Southeast Asia.

REFERENCES

Gnanaraj, C., Haque, A.T.M.E. and Iqbal, M. 2012. The chemopreventive effects of *Thysanolaena latifolia* against carbon tetrachloride (CCl4)-induced oxidative stress in rats. *Journal of Experimental and Integrative Medicine*. 2: 345–355.

Hoque, N., Fatemee, N., Hossain, M.J., Shanta, M.A. and Asaduzzaman, M. 2021. CNS depressant and analgesic activities of *Thysanolaena maxima* Roxb. available in Bangladesh. *Dhaka University Journal of Pharmaceutical Sciences*. 20(2): 227–233.

Hoque, N., Sohrab, M.H., Debnath, T. and Rana, M.S. 2016. Antioxidant, antibacterial and cytotoxic activities of various extracts of *Thysanolaena maxima* (Roxb) Kuntze available in Chittagong hill tracts of Bangladesh. *International Journal of Pharmacy and Pharmaceutical Sciences*. 8(7): 168–172.
Shrestha, S., Park, J.H., Cho, J.G., Lee, D.Y., Jeong, R.H., Song, M.C., Cho, S.K., Lee, D.S. and Baek, N.I. 2016. Phytochemical constituents from the florets of tiger grass *Thysanolaena latifolia* from Nepal. *Journal of Asian Natural Products Research*. 18(2): 206–213.

4.1.4.4 Order Zingiberales Griseb. (1854)

4.1.4.4.1 Family Cannaceae Juss. (1789), the Canna Family

The family Cannaceae consists of the single genus *Canna* L. (1753)

***Canna indica* L.**

[From Latin *canna* = reed and *indica* = from India]

Published in *Species Plantarum* 1: 1. 1753

Synonyms: *Canna elegans* Raf.; *Canna ellipticifolia* Stokes; *Canna orientalis* L.; *Canna variabilis* Willd.; *Cannacorus indicus* (L.) Medik.; *Cannacorus ovatus* Moench
Local name: *bunga canna* (Kedayan)
Common name: Indian shot
Habitat: cultivated
Geographical distribution: tropical
Botanical description: this graceful herbaceous plant reaches a height of about 2 m. Rhizomes: branched. Culms: stout. Leaves: simple, alternate, exstipulate. Leaf sheaths: green or purple. Petioles: tiny. Blades: ovate-oblong to lanceolate, smooth, somewhat fleshy and zingiberaceous, 10–20 cm × 30–60 cm, with numerous thin pairs of secondary nerves. Inflorescences: racemes of cincinni. Sepals: 3, pale purplish green, lanceolate, about 1.5 cm long. Petals: 3–6, heavenly pink to yellow, or bright red, lancelolate, about 4–6 cm long. Staminodes: 2–3, erect, spathulate, about 5 cm long, emarginate at apex. Stamens: 1, filament about 4 cm long, anthers about 1 cm long. Ovary: green, globose, about 6 mm across, spiny. Styles: red to apricot yellow, thin, about 4–5 cm long. Capsules: broadly ovoid, trilobed, spiny, up to about 3 cm long, marked at apex with vestiges of the perianth. Seeds: black, glossy, somewhat globose, with a vague resemblance to lead shot ammunition.
Traditional therapeutic indications: metrorrhagia (Kedayan)
Pharmacology and phytochemistry: hepatoprotective (Longo et al. 2015), anti-inflammatory (Chen et al. 2013), analgesic, anthelmintic (Nirmal et al. 2007), HIV reverse transcriptase inhibition (Woradulayapinij et al. 2005), neuroprotective (Ojha et al. 2022), antibacterial (George 2014), nephroprotective (Singh et al. 2020).
Antithrombotic hydroxycinammic acid derivatives (isorinic acid) (Nguyen et al. 2023).
Toxicity, side effects, and drug interaction: not known

Comment: The bright red color of the petals may have encouraged the use of this plant for the treatment of metrorrhagia. This is called the theory of signature. It may sound unscientific, but we cannot deny the obvious fact that walnuts are good for the brain. We do not know everything in the field of medicinal plants, actually, we are ignorant. I have observed, in Sabah, Dusun and Murut people in their 60s fit, without wrinkles and white hair, and I am often told that their secret to longevity is the use of medicinal plants on the advice of shamans and local herbalists

These herbalists and shamans are becoming rare and combined with Islamization and the adoption of our Western way of life, local knowledge of medicinal plants is decreasing day by day. A host of local administrative requirements further aggravate this situation and it is almost certain that we will miss the opportunity to find the medicinal plants from Sabah that could have been made into medicines. The fact is that Southeast Asia is home to medicinal plants that could be used as a source of drugs but almost everything is done locally and internationally (Nagoya Protocol) to make their impossible. Do we need another viral pandemic to understand that we need to find new drugs more than ever?

REFERENCES

Chen, H.J., Chen, C.N., Sung, M.L., Wu, Y.C., Ko, P.L. and Tso, T.K. 2013. *Canna indica* L. attenuates high-glucose-and lipopolysaccharide-induced inflammatory mediators in monocyte/macrophage. *Journal of Ethnopharmacology*. 148(1): 317–321.

George, J. 2014. Screening and antimicrobial activity of *Canna indica* against clinical pathogens bioactive. *International Journal of Life Sciences Education Research*. 2(3): 85–88.

Longo, F., Teuwa, A., Fogue, S.K., Spiteller, M. and Ngoa, L.E. 2015. Hepatoprotective effects of *Canna indica* L. rhizome against acetaminophen (paracetamol). *The World Journal of Pharmaceutical Sciences*. 4: 1609–1624.

Nguyen, T.V.A., Nguyen, T.M.H., Le, H.L. and Bui, D.H. 2023. Potential antithrombotic effect of two new phenylpropanoid sucrose esters and other secondary metabolites of *Canna indica* L. rhizome. *Natural Product Research*.: 1–9.

Nirmal, S.A., Shelke, S.M., Gagare, P.B., Jadhav, P.R. and Dethe, P.M. 2007. Antinociceptive and anthelmintic activity of *Canna indica*. *Natural Product Research*. 21(12): 1042–1047.

Ojha, P.S., Biradar, P.R., Tubachi, S. and Patil, V.S. 2022. Evaluation of neuroprotective effects of Canna indica L against aluminium chloride induced memory impairment in rats. *Advances in Traditional Medicine*: 1–18.

Singh, P., Sahoo, H.B. and Jain, A.P. 2020. Investigations on rhizomes for nephroprotective activity with *Canna indica* antioxidant effects. *Advance Pharmaceutical Journal*. 5(5): 172–179.

Woradulayapinij, W., Soonthornchareonnon, N. and Wiwat, C. 2005. *In vitro* HIV type 1 reverse transcriptase inhibitory activities of Thai medicinal plants and *Canna indica* L. rhizomes. *Journal of Ethnopharmacology*. 101(1–3): 84–89.

4.1.4.4.2 Family Costaceae Nakai (1941), the Costus Family

The family Costaceae consists of about 4 genera and 120 species of herbaceous plants growing from rhizomes. Culms: often fleshy, spiral. Leaves: simple, spiral, exstipulate. Inflorescences: in terminal spikes. Calyx: tubular, 2-3-lobed Corolla: tubular with irregular lobes. Stamens: 1. Carpels: 3, forming a 3 locular ovary, each locule with 15–50 ovules. Fruits: capsules. Seeds: numerous, black, arillate.

***Costus speciosus* (J. Koenig ex Retz.) Sm.**

[From Sanskrit *canda* = *Costus speciosus* (J. Koenig ex Retz.) Sm. and Latin *speciosus* = beautiful]

Published in *Transactions of the Linnean Society of London* 1: 249–250. 1791

Synonyms: *Banksea speciosa* J. Koenig ex Retz.; *Cheilocostus speciosus* (J. Koenig ex Retz.) C.D. Specht; *Costus formosanus* Nakai

Local names: *busu busu* (Rungus); *insasabu, linsabu* (Murut); *silok* (Lundayeh); *sidbu sidbu tongkur ongkur* (Dusun, Kadazan); *setawar halia* (Bajau)

Common name: spiral ginger

Habitat: gardens, villages, roadsides, cultivated

Geographical distribution: from India to Pacific Islands

Botanical description: this magnificent herbaceous plant reaches a height of about 3 m. Culms: reddish, sappy, spiral. Leaves: simple, spiral. Petioles: 5–7 mm long. Blades: lanceolate to oblong, 6–10 cm × 15–20 cm, hairy beneath, somewhat rounded at base, acuminate to caudate at apex. Spikes: ovate, reddish, scaly, up to about 15 cm long. Calyx: about 2 cm long, trilobed, red, hairy. Corolla: about 6 cm long, of a heavenly or somewhat supernatural pure white (early morning), membranous, caducous, yellow at the throat, showy, somewhat incised. Stamens: 1, about 4 cm long, hairy, orange. Capsules: red, globose, about 1.5 cm across. Seeds: numerous, glossy, smooth, beautiful.

Traditional therapeutic indications: coughs (Rungus); postpartum (Dusun); asthma, respiratory problems (Dusun, Kadazan, Lundayeh, Murut); asthma, flu, headaches (Dusun, Kadazan, Murut); chest pain, stomach aches, swellings (Murut); fever, smallpox (Kedayan); fever, bone aches (Dusun, Kadazan)

Pharmacology and phytochemistry: antiviral (herpes simplex virus, influenza virus) (Senevirathne et al. 2023), antipyretic, anti-inflammatory (Kassuya et al. 2009).

Antiparasitic steroids (diosgenin) and steroidal saponins (Dasgupta & Pandey 1970; Zheng et al. 2015). Antiparasitic, antifungal, cytotoxic, antiviral (hepatitis B virus), antipyretic, anti-inflammatory sesquiterpene lactone (costunolide) (Lee et al. 1971; Sánchez et al. 2007; Eliza et al. 2009; Castaño & Giraldo 2019). Antidiabetic and hypolipidemic sesquiterpene lactone (eremanthin) (Eliza et al. 2009).

Toxicity, side effects, and drug interaction: not known

Comments:

(i) The Malays of Perak (Peninsular Malaysia) hold the flowers of this plant in great repute in their protective magical rituals. They place the petals in a bucket of water that is used to shower (with mantras) a person touched by a magic spell.

(ii) A plant in the genus *Costus* L. (1753) is used by Dusun and Kadazan for sprains (local name: *subor subor*).

(iii) The Dusun use *Costus paradoxus* K.Schum. (local name: *badui*) as a nasal medicine.

REFERENCES

Castaño Osorio, J.C. and Giraldo García, A.M. 2019. Antiparasitic phytotherapy perspectives, scope and current development. *Infection*. 23(2): 189–204.

Dasgupta, B. and Pandey, V.B. 1970. A new Indian source of diosgenin (*Costus speciosus*). *Cellular and Molecular Life Sciences*. 26(5): 475–476.

Eliza, J., Daisy, P., Ignacimuthu, S. and Duraipandiyan, V. 2009. Antidiabetic and antilipidemic effect of eremanthin from *Costus speciosus* (Koen.) Sm., in STZ-induced diabetic rats. *Chemico-Biological Interactions*. 182(1): 67–72.

Kassuya, C.A.L., Cremoneze, A., Barros, L.F.L., Simas, A.S., da Rocha Lapa, F., Mello-Silva, R., Stefanello, M.É.A. and Zampronio, A.R. 2009. Antipyretic and anti-inflammatory properties of the ethanolic extract, dichloromethane fraction and costunolide from Magnolia ovata (Magnoliaceae). *Journal of Ethnopharmacology*. 124(3): 369–376.

Lee, K.H., Huang, E.S., Piantadosi, C., Pagano, J.S. and Geissman, T.A. 1971. Cytotoxicity of sesquiterpene lactones. *Cancer Research*. 31(11): 1649–1654.

Sánchez, L.A., Capitan, Z., Romero, L.I., Ortega-Barría, E., Gerwick, W.H. and Cubilla-Rios, L. 2007. Bioassay guided isolation of germacranes with anti-protozoan activity from *Magnolia sororum*. *Natural Product Communications*. 2(11): 1065–1069.

Senevirathne, A., Jayathilaka, E.T., Haluwana, D.K., Chathuranga, K., Senevirathne, M., Jeong, J.S., Kim, T.W., Lee, J.S. and De Zoysa, M. 2023. The aqueous leaf extract of the medicinal herb *Costus speciosus* suppresses influenza A H1N1 viral activity under *in vitro* and *in vivo* conditions. *Viruses*. 15(6): 1375.

Zheng, W., Yan, C.M., Zhang, Y.B., Li, Z.H., Li, Z., Li, X.Y., Wang, Z.W., Wang, X., Chen, W.Q. and Yu, X.H. 2015. Antiparasitic efficacy of gracillin and zingibernsis newsaponin from *Costus speciosus* (Koen ex. Retz) Sm. against *Ichthyophthirius multifiliis*. *Parasitology*. 142(3): 473–479.

4.1.4.4.3 Family Marantaceae R.Br. (1814), the Arrowroot Family

The family Marantaceae consists of about 30 genera and 530 species of herbaceous plants which can make one think of gingers. Leaves: simple, alternate. Petioles: forming a sheath at base. Inflorescences: terminal or axillary. Sepals: 3. Corolla: trilobed. Stamens: 1. Staminodes: present. Carpels: 3, forming a 3 locular ovary, each locule with 1 ovule. Styles: 1, cylindrical. Fruits: berries, capsules.

***Donax canniformis* (G. Forst.) K.Schum.**

[From Latin *donax* = reed and *canniformis* = cane-shaped]

Published in *Botanische Jahrbücher für Systematik, Pflanzengeschichte und Pflanzengeographie* 15(4): 440, in obs. 1893

Synonyms: *Actoplanes canniformis* (G. Forst.) K.Schum.; *Donax arundastrum* Lour.; *Thalia canniformis* G. Forst.
Local names: *lias* (Dusun, Murut); *babalit* (Lundayeh)
Habitat: cultivated
Geographical distribution: from India to the Philippines
Botanical description: this herbaceous plant reaches a height of about 4 m. Rhizomes: present. Culms: somewhat vaguely bamboo-like. Leaves: simple, alternate, exstipulate. Petioles: forming a sheath at base, about 10–20 cm long. Blades: broadly ovate to elliptic, 10–25 cm × 10–45 cm, almost horizontal, rounded to obtuse at base, acuminate at apex, with innumerable and discrete nervations. Panicles: terminal, emerging at base of leaves, up to 20 cm long. Sepals: 3, about 3 mm long, triangular-ovate, white. Corolla: tube about 1 cm long, pure white, with membranous and white lobes (which can make one think of an orchid). Staminodes: present. Stamens: 1, about 1 cm long. Ovary: hairy. Berries: globose, pale cream, almost glossy, about 1 cm, beaked at apex (Figure 4.13).

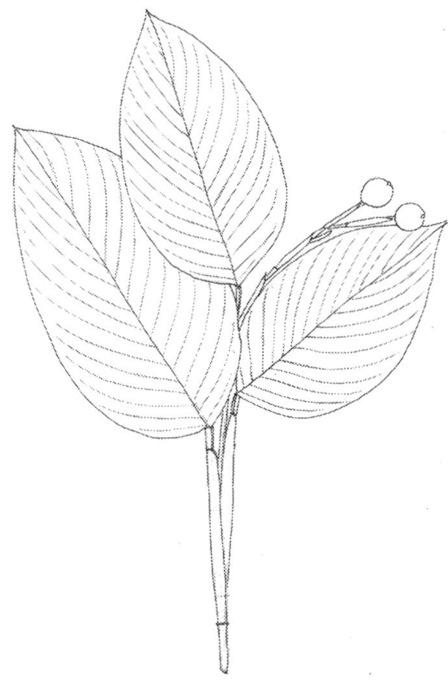

FIGURE 4.13 *Donax canniformis* (G. Forst.) K.Schum.

Traditional therapeutic indications: coughs (Dusun)
Pharmacology and phytochemistry: anti-inflammatory (Paramita et al. 2017).
Toxicity, side effects, and drug interaction: not known

Comments:

(i) The Lundayeh use the fruits as food whilst the Dusun use the culms for the manufacture of mats, baskets, and fish traps.
(ii) In the Philippines, this plant is called "*banban*" and is used externally for mental disorders.

REFERENCE

Paramita, S., Kosala, K., Dzulkifli, D., Saputri, D.I. and Wijayanti, E. 2017. Anti-inflammatory activities of ethnomedicinal plants from Dayak Abai in North Kalimantan, Indonesia. *Biodiversitas Journal of Biological Diversity*. 18(4): 1556–1561.

4.1.4.4.4 Family Musaceae Juss. (1789), the Banana Family

The family Musaceae consists of about 3 genera and 40 species of herbaceous plants growing from corms or rhizomes. Culms: composed of closely packed leaf sheaths. Leaves: simple, spiral, or alternate, exstipulate. Inflorescences: terminal or rarely axillary, cymose, often with colorful bracts. Tepals: 2 sets of 3, more or less connate. Stamens: 5–6, free. Carpels: 3, forming a trilocular ovary, each locule with 10–50 ovules. Styles: 1. Stigma: capitate. Fruits: berries, fleshy. Seeds: numerous, embedded in a pulp.

***Musa paradisiaca* L.**

[From Arabic *mauz* = banana and Latin *paradisiaca* = from paradise]

Published in *Species Plantarum* 2: 1043. 1753

Synonym: *Musa sapientum* L.
Local names: *daung loka* (Bugis); *pokok pisang* (Bajau)
Common name: plantain
Distribution: tropical
Botanical description: this treelike herbaceous plant reaches a height of about 9 m. Culms: massive, heavy, fibrous, sappy. Leaf sheaths: tubular, forming a thick trunk. Leaves: simple, spiral, exstipulate. Blades: up to about 3 m long, tearing with age, oblong, fleshy, smooth, a little cool to the touch, the midrib deeply sunken above and the secondary nerves almost at right angles to the midrib. Spikes: stout, about 1 m long, nodding. Bracts: 15–20 cm, ovate, concave, burgundy red to deep purple, somewhat fleshy. Tepals: 2 sets of 3, whitish cream, connate, about 2–2.5 cm long, pink. Stamens: 5, protuding. Berries: oblong, fleshy, 5–7 cm long. Seeds: black, glossy, in a whitish-yellow edible pulp. Flowers are visited by hummingbirds.
Traditional therapeutic indications: to clean the skin, fever, sore throats (Bugis)
Pharmacology and phytochemistry: this plant has been the subject of numerous pharmacological studies (Galani et al. 2019) which have highlighted, among other things, antipyretic (Maya 2015), hepatoprotective (Nirmala et al. 2012), antifungal (Karadi et al. 2011), anti-asthma (Patro et al. 2016), and antilithiatic activity (Prasad et al. 1993). Phenylphenalenones (irelonone, emenolone) (Luis et al. 1993).
Toxicity, side effects, and drug interaction: not known

Comments:

(i) A plant of the genus *Musa* L. (1753) is affords a remedy for diarrhea for the Dusun and Kadazan (local name: *togutui*).
(ii) In Cebu, the trunk is used to make fibers.

REFERENCES

Galani, V. 2019. Musa paradisiaca Linn.-A comprehensive review. *Scholars International Journal of Traditional and Complementary Medicine*: 45–56.

Karadi, R.V., Shah, A., Parekh, P. and Azmi, P. 2011. Antimicrobial activities of *Musa paradisiaca* and *Cocos nucifera. International Journal of Research in Pharmaceutical and Biomedical Sciences*. 2(1): 264–267.

Luis, J.G., Echeverri, F., Quinones, W., Brito, I., Lopez, M., Torres, F., Cardona, G., Aguiar, Z., Pelaez, C. and Rojas, M. 1993. Irenolone and emenolone: Two new types of phytoalexin from *Musa paradisiaca. The Journal of Organic Chemistry*. 58(16): 4306–4308.

Maya, S.W. 2015. Phytochemical screening and antipyretic effect of stem juice from kepok banana (*Musa paradisiaca* L) on white male rats stain wistar (Rattus norvegicus) induced with DTO-Hb. *Pharmacon*. 4(1): 1–11.

Nirmala, M., Girija, K., Lakshman, K. and Divya, T. 2012. Hepatoprotective activity of *Musa paradisiaca* on experimental animal models. *Asian Pacific Journal of Tropical Biomedicine*. 2(1): 11–15.

Patro, G., Panda, M., Das, P., Bhaiji, A., Panda, A. and Sahoo, H.B. 2016. Pharmacological evaluation of *Musa paradisiaca* (Linn.) on bronchial asthma. *Egyptian Pharmaceutical Journal*. 15(1): 25–30.

Prasad, K.V.S.R.G., Bharathi, K. and Srinivasan, K.K. 1993. Evaluation of Musa (Paradisiaca Linn. cultivar)- "Puttubale" stem juice for antilithiatic activity in albino rats. *Indian Journal of Physiology and Pharmacology*. 37: 337–337.

4.1.4.4.5 Family Zingiberaceae Martinov (1820), the Ginger Family

The family Zingiberaceae consists of about 50 genera and 1,300 species of gingers. Rhizomes: often aromatic. Culms: composed of closely packed leaf sheaths. Leaves: simple, alternate, or spiral. Inflorescences: cymes or spikes. Sepals: 3. Corolla: tubular, trilobed. Stamens: 1. Carpels: 3, forming a unilocular or trilocular ovary, each locule with 4–many ovules. Styles: 1. Stigma: papillate. Fruits: capsules. Seeds: numerous, often aromatic, arillate.

Alpinia galanga **(L.) Sw.**

[After the 17th century Italian botanist Propero Alpini and from the Arabic name of the plant = *caluegia*]

Published in *Species Plantarum. Editio quarta* 1(1): 12. 1797

Synonyms: *Amomum galanga* (L) Lour.; *Languas galanga* (L.) Stuntz; *Languas vulgare* (L.) Stuntz; *Maranta galanga* L.
Local name: *lengkuas* (Bajau)
Common name: greater galanga
Habitat: cultivated
Geographical distribution: India, Indonesia, Malaysia, Myanmar, Thailand, Vietnam, China, Taiwan
Botanical description: this ginger reaches a height of about 2 m. Rhizomes: aromatic, pungent, coriaceous. Leaves: simple, alternate. Ligules: suborbicular, 5 mm long. Petioles: 6 mm long. Blades: oblong, dull green, wavy, 6–10 cm × 25–35 cm, attenuate at base, acute at apex. Panicles: up to about 30 cm long. Flowers: green-white and fragrant. Calyx: tubular, 6 mm–1 cm long. Corolla: tubular. Labellum: white with red lines, obovate-spathulate, 2 cm long, penciled with blood vessel-like lines and bifid at apex. Staminodes: present. Stamens: 1, about 1.7 cm long. Capsules: dull red, oblong, slightly contracted in the middle, 1 cm–1.5 cm long. Seeds: numerous.
Traditional therapeutic indications: indigestion, skin diseases (Dusun, Kadazan); colds, fever (Bajau)
Pharmacology and phytochemistry: this plant has been the subject of numerous pharmacological studies (Khairullah et al. 2020) which have highlighted, among other things, antiamoebic activity (Sawangjaroen et al. 2006). Essential oil (1,8 cineole, myrcene, terpinen-4-ol) (Janssen & Scheffer 1985; Charles et al. 1992; Jirovetz et al. 2003). Cytotoxic flavanone (pinocembrin) (Rasul et al. 2013). Cytotoxic, antimycobacterial, and antifungal hydroxycinnamic acid derivative (acetoxychavicol acetate) (Janssen & Scheffer 1985; Awang et al. 2010; Baradwaj et al. 2017; Warit et al. 2017).
Toxicity, side effects, and drug interaction: none when consumed at normal food quantity.

Comments:

(i) Plants of the genus *Alpinia* L. (1753) are medicinal in Sabah. The Dusun and Kadazan use a plant they call "*tolidus*" for skin diseases. The Tidung use a plant thy call "*limpuyang*" for odory menstruations. The Dusun use "*sagang*" as food.
(ii) The rhizomes of *Alpinia galanga* (L.) Sw. are a common sight in the wet markets of Sabah.

REFERENCES

Awang, K., Azmi, M.N., Aun, L.I.L., Aziz, A.N., Ibrahim, H. and Nagoor, N.H. 2010. The apoptotic effect of 1'S-1'-acetoxychavicol acetate from *Alpinia conchigera* on human cancer cells. *Molecules*. 15(11): 8048–8059.

Baradwaj, R.G., Rao, M.V. and Kumar, T.S. 2017. Novel purification of 1'S-1'-Acetoxychavicol acetate from *Alpinia galanga* and its cytotoxic plus antiproliferative activity in colorectal adenocarcinoma cell line SW480. *Biomedicine & Pharmacotherapy*: 485–493.

Charles, D.J., Simon, J.E. and Singh, N.K. 1992. The essential oil of *Alpinia galanga* Willd. *Journal of Essential Oil Research*. 4(1): 81–82.

Janssen, A.M. and Scheffer, J.J.C. 1985. Acetoxychavicol acetate, an antifungal component of *Alpinia galanga*. *Planta Medica*. 51(6): 507–511.

Jirovetz, L., Buchbauer, G., Shafi, M.P. and Leela, N.K. 2003. Analysis of the essential oils of the leaves, stems, rhizomes and roots of the medicinal plant *Alpinia galanga* from southern India. *Acta Pharmaceutica-Zagreb*. 53(2): 73–82.

Khairullah, A.R., Solikhah, T.I., Ansori, A.N.M., Fadholly, A., Ram, S.C., Ansharieta, R., Widodo, A., Riwu, K.H.P., Putri, N., Proboningrat, A. and Kusala, M.K.J. 2020. A review of an important medicinal plant: *Alpinia galanga* (L.) willd. *Systematic Reviews in Pharmacy*. 11(10): 387–395.

Rasul, A., Millimouno, F.M., Ali Eltayb, W., Ali, M., Li, J. and Li, X. 2013. Pinocembrin: A novel natural compound with versatile pharmacological and biological activities. *BioMed Research International*. 2013: 1–9.

Sawangjaroen, N., Phongpaichit, S., Subhadhirasakul, S., Visutthi, M., Srisuwan, N. and Thammapalerd, N. 2006. The anti-amoebic activity of some medicinal plants used by AIDS patients in southern Thailand. *Parasitology Research*. 98(6): 588–592.

Warit, S., Rukseree, K., Prammananan, T., Hongmanee, P., Billamas, P., Jaitrong, S., Chaiprasert, A., Jaki, B.U., Pauli, G.F., Franzblau, S.G. and Palittapongarnpim, P. 2017. *In vitro* activities of enantiopure and racemic 1′-acetoxychavicol acetate against clinical isolates of *Mycobacterium tuberculosis*. *Scientia Pharmaceutica*. 85(3): 32.

Boesenbergia pulchella (Ridl.) Merr.

[After the 19th century Scottish botanist Francis Buchanan Hamilton and Latin *pulcher* = beautiful]

Published in *Journal of the Straits Branch of the Royal Asiatic Society* Special Number: 122. 1921

Synonym: *Gastrochilus pulchellus* Ridl.
Local name: *lipat* (Dusun, Kadazan)
Habitat: forests
Geographical distribution: Borneo
Botanical description: this orchid-like herbaceous plant reaches a height of about 50 cm. Rhizomes: elongated. Leaves: simple, basal. Petioles: about 10–30 cm long, thin. Ligules: bifid. Blades: broadly lanceolate, with numerous thin secondary nerves, dark green and glossy above, somewhat asymmetrical at base, 3–8 cm × 10–20 cm, acuminate at apex. Flowers: solitary growing close to the ground, emerging from leaf sheaths, white with a reddish throat. The calyx and corolla are tubular. Staminodes: present. Labellum: oblong, longer than the corolla lobes. Stamens: 1. Ovary: trilobed. Fruits: capsules.
Traditional therapeutic indication: skin diseases (Dusun, Kadazan)
Pharmacology and phytochemistry: not known
Toxicity, side effects, and drug interaction: not known

Boesenbergia rotunda (L.) Mansf.

[After the 19th century Scottish botanist Francis Buchanan Hamilton and Latin *rotunda* = round]

Published in *Kulturpflanze* 6: 239. 1958

Synonyms: *Boesenbergia cochinchinensis* (Gagnep.) Loes.; *Boesenbergia pandurata* (Roxb.) Schltr.; *Curcuma rotunda* L.; *Gastrochilus panduratus* (Roxb.) Ridl.; *Gastrochilus rotundus* (L.) Alston; *Kaempferia cochinchinensis* Gagnep.; *Kaempferia ovata* Roscoe; *Kaempferia pandurata* Roxb.
Local name: *temu kuci* (Bajau)
Common name: fingerroot
Habitat: cultivated
Geographical distribution: Bangladesh to Borneo
Botanical description: this ginger reaches a height of about 50 cm. Rhizomes: strongly aromatic, yellowish, with about 7 cm long roots, the entire underground part looking like some sort of monstrous fingers. Leaves: simple, alternate, sometimes basal. Ligules: bifid, about 5 mm long. Petioles: 7–16 cm long, channeled. Blades: ovate, oblong, elliptic to lanceolate, somewhat wavy and asymmetrical, 7–12 cm × 25–50 cm, base rounded to attenuate, apex apiculate. Cymes: terminal, appearing from within apical leaf sheaths, 3–8 cm long. Bracts: lanceolate, 4–5 cm long. Flowers: aromatic. Calyx: tubular, 1.5–2 cm long, apex dentate. Corolla: pink, beautiful, 4.5–5.5 cm long, bilobed, the lobes oblong. Staminodes: light pink, obovate, about 1.5 cm long. Labellum: white or pink, fiddle-shaped, 2.5–3.5 cm long, wavy. Stamens: 1, about 8 mm long. Ovary: cylindrical, about 3 mm long. Styles: 1, filiform, about 5 cm long. Stigma: flabellate. Capsules: cylindrical, white, glabrous. Seeds: brown.
Traditional therapeutic indications: hypertension, pain at postpartum, stomach aches, sprains (Kedayan)
Pharmacology and phytochemistry: anti-SARS-CoV-2 (Kanjanasirirat et al. 2020), vasorelaxant (Adhikari et al. 2020), analgesic (Wang et al. 2022), hepatoprotective (Salama et al. 2012), toxic to breast cancer cells (T47D) (Widyananda et al. 2022), anti-osteoporosis (Saah et al. 2021).

Anti-inflammatory chalcones such as boesenbergin A (Isa et al. 2012) and cardamonin (Voon et al. 2017). Antiviral (SARS-CoV-2) chalcone (panduratin) (Kanjanasirirat et al. 2020). Anti-adipogenic flavanone (pinostrobin) (San et al. 2022).

Toxicity, side effects, and drug interaction: the median lethal dose (LD_{50}) of ethanol extract administered to rats was about 4 g/kg (Rosdianto et al. 2020).

REFERENCES

Adhikari, D., Gong, D.S., Oh, S.H., Sung, E.H., Lee, S.O., Kim, D.W., Oak, M.H. and Kim, H.J. 2020. Vasorelaxant effect of *Boesenbergia rotunda* and its active ingredients on an isolated coronary artery. *Plants*. 9(12): 1688.

Isa, N.M., Abdelwahab, S.I., Mohan, S., Abdul, A.B., Sukari, M.A., Taha, M.M.E., Syam, S., Narrima, P., Cheah, S.C., Ahmad, S.J.B.J.O.M. and Mustafa, M.R. 2012. *In vitro* anti-inflammatory, cytotoxic and antioxidant activities of boesenbergin A, a chalcone isolated from *Boesenbergia rotunda* (L.) (finger-root). *Brazilian Journal of Medical and Biological Research*. 45: 524–530.

Kanjanasirirat, P., Suksatu, A., Manopwisedjaroen, S., Munyoo, B., Tuchinda, P., Jearawuttanakul, K., Seemakhan, S., Charoensutthivarakul, S., Wongtrakoongate, P., Rangkasenee, N., et al. 2020. High-content screening of Thai medicinal plants reveals *Boesenbergia rotunda* extract and its component panduratin A as anti-SARS-CoV-2 agents. *Scientific Reports*. 10: 19963.

Rosdianto, A.M., Puspitasari, I.M., Lesmana, R. and Levita, J. 2020. Inhibitory activity of *Boesenbergia rotunda* (L.) Mansf. rhizome towards the expression of Akt and NF-KappaB p65 in acetic acid-induced Wistar rats. *Evidence-Based Complementary and Alternative Medicine*. 2020: 1–13.

Saah, S., Siriwan, D. and Trisonthi, P. 2021. Biological activities of *Boesenbergia rotunda* parts and extracting solvents in promoting osteogenic differentiation of pre-osteoblasts. *Food Bioscience*. 41: 101011.

Salama, S.M., Bilgen, M., Al Rashdi, A.S. and Abdulla, M.A. 2012. Efficacy of *Boesenbergia rotunda* treatment against thioacetamide-induced liver cirrhosis in a rat model. *Evidence-based Complementary and Alternative Medicine*. 2012: 1–12.

San, H.T., Khine, H.E.E., Sritularak, B., Prompetchara, E., Chaotham, C., Che, C.T. and Likhitwitayawuid, K. 2022. Pinostrobin: An adipogenic suppressor from Fingerroot (*Boesenbergia rotunda*) and its possible mechanisms. *Foods*. 11(19): 3024.

Voon, F.L., Sulaiman, M.R., Akhtar, M.N., Idris, M.F., Akira, A., Perimal, E.K., Israf, D.A. and Ming-Tatt, L. 2017. Cardamonin (2′, 4′-dihydroxy-6′-methoxychalcone) isolated from *Boesenbergia rotunda* (L.) Mansf. inhibits CFA-induced rheumatoid arthritis in rats. *European Journal of Pharmacology*. 794: 127–134.

Wang, P., Wen, C. and Olatunji, O.J. 2022. Anti-inflammatory and antinociceptive effects of *Boesenbergia rotunda* polyphenol extract in diabetic peripheral neuropathic rats. *Journal of Pain Research*: 779–788.

Widyananda, M.H., Wicaksono, S.T., Rahmawati, K., Puspitarini, S., Ulfa, S.M., Jatmiko, Y.D., Masruri, M. and Widodo, N. 2022. A potential anticancer mechanism of finger root (*Boesenbergia rotunda*) extracts against a breast cancer cell line. *Scientifica*. 2022: 1–17.

Boesenbergia stenophylla **R.M. Sm.**

[After the 19th century Scottish botanist Francis Buchanan Hamilton, Greek *stenos* = narrow, and *phyllon* = leaf]

Published in *Notes from the Royal Botanic Garden, Edinburgh* 44: 230. 1987

Local names: *kaburo apad* (Lundayeh); *akar bumi, jerangau merah* (Bajau)
Synonym: *Gastrochilus pulchellus* Ridl.
Habitat: forests
Geographical distribution: Borneo
Botanical description: this herbaceous plant reaches a height of about 40 cm. Rhizomes: elongated, yellowish, aromatic. Leaves: simple, alternate, basal. Petioles: thin, 7–25 cm long. Ligules: bifid. Blades: oblong, somewhat asymmetrical, 1.5–3 cm × 12–20 cm, smooth, dark green and glossy above, without apparent secondary nerves, the midrib showy and sunken above, tapering at base, acute at apex, reddish beneath when young. Inflorescences: solitary, emerging from leaf sheaths. The calyx and corolla are tubular. Staminodes: present. Labellum: oblong, longer than the corolla lobes. Stamens: 1. Ovary: trilobed. Fruits: capsules (Figure 4.14).

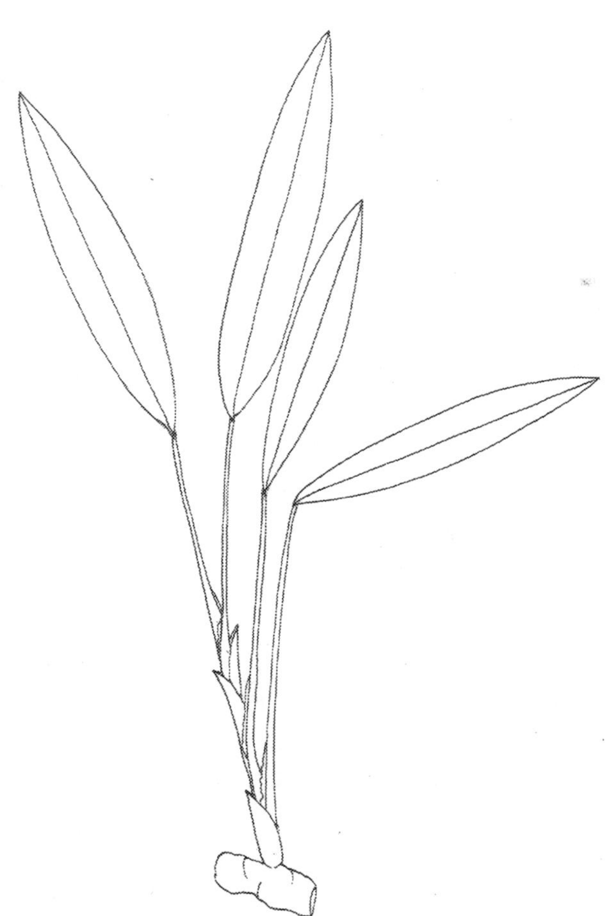

FIGURE 4.14 *Boesenbergia stenophylla* R.M. Sm.

Traditional therapeutic indications: skin diseases (Dusun, Kadazan); breathlessness, diabetes, hypertension, heart diseases, kidney stones, coughs, diarrhea, fever, malaria, asthma, fatigue, bleeding (Bajau); stomach aches, food poisoning (Bajau, Lundayeh)

Pharmacology and phytochemistry: essential oil (antibacterial) (Atiekah 2018). Chalcones toxic to neuroblastoma cells (Neuro-2a) (stenophyllol A) and breast cancer cells (stenophyllol B) (Lee et al. 2023). Arylheptanoids, flavones (Primus et al. 2022, 2022a). Hydroxycinnamic acid derivatives (methyl (*E*)-cinnamate), sesquiterpenes (δ-elemene, β-elemene, α-santalene, α-humulene) (Ahmad & Jantan 2003).

Toxicity, side effects, and drug interaction: not known

Comments:

(i) In Sarawak, the Kelabit chew the rhizome before intake of alcohol to avoid being drunk. It is also used for diarrhea and stomach aches (local name: *kaburoh apad*).

(ii) A plant of the genus *Boesenbergia* Kuntze (1891) is used by the Dusun for rheumatisms (local name: *layo tutumolong*).

REFERENCES

Ahmad, F.B. and Jantan, I.B. 2003. The essential oils of *Boesenbergia stenophylla* RM Sm. as natural sources of methyl (E)-cinnamate. *Flavour and Fragrance Journal*. 18(6): 485–486.

Atiekah, N. 2018. *Volatile Constituents and Antibacterial Activity of Essential Oils of Selected Boesenbergia Species* (Doctoral dissertation, University of Malaya).

Lee, M.Y., Shiau, J.P., Tang, J.Y., Hou, M.F., Primus, P.S., Kao, C.L., Choo, Y.M. and Chang, H.W. 2023. *Boesenbergia stenophylla*-derived Stenophyllol B exerts antiproliferative and oxidative stress responses in triple-negative breast cancer cells with few side effects in normal cells. *International Journal of Molecular Sciences*. 24(9): 7751.

Primus, P.S., Ismail, M.H., Adnan, N.E., Wu, C.H.Y., Kao, C.L. and Choo, Y.M. 2022a. Chemical constituents and anti-neuroblastoma activity from *Boesenbergia stenophylla*. *JSM*. 51(4): 1075–1084.

Primus, P.S., Ismail, M.H., Adnan, N.E., Wu, C.H.Y., Kao, C.L. and Choo, Y.M. 2022b. Stenophyllols AC, new compounds from *Boesenbergia stenophylla*. *Journal of Asian Natural Products Research*. 24(2): 146–152.

Curcuma caesia **Roxb.**

[From the Arabic *kurkum* = turmeric and Latin *caesia* = bluish gray]

Published in *Asiatic Researches* 11: 335. 1810

Local name: *kunyit hitam* (Bajau, Kedayan)
Synonym: *Curcuma kuchoor* Royle
Common name: black turmeric
Habitat: cultivated
Geographical distribution: from India to Borneo
Botanical description: this ginger reaches a height of about 1.2 m. Rhizomes: ovate, light brown, aromatic, about 10 cm long, of a strange intense blue in cross-section. Leaves: simple, alternate, up to 1 m long. Petioles: as long as blades. Blades: emerge from culms, broadly lanceolate to elliptic, dull green with a purple to brownish longitudinal blotch along the midrib, about 10–12 cm × 30–40 cm, acute at base, with numerous pairs of secondary nerves, long acuminate to acute at apex. Spikes: about 10–20 cm tall, somewhat obconical to cylindrical, emerging from the ground on 8–25 cm long peduncles. Bracts: numerous, up to about 7 cm long, oblong, of a heavenly purple or light pink. Calyx: about 1 cm long, trilobed, truncate, split on one side. Corolla: tubular, yellowish, about 3 cm long. Labellum: emarginate, yellow at apex. Staminodes: present. Stamens: 1, 7 mm long. Styles: 1, thin. Stigma: bifid.
Traditional therapeutic indications: coughs (Bajau)
Pharmacology and phytochemistry: this plant has been the subject of numerous pharmacological studies (Ibrahim et al. 2023) which have highlighted, among other things, anti-inflammatory (Sawant et al. 2014), antiasthmatic (Pathan et al. 2016), and vasorelaxant activity (Arulmozhi et al. 2006), as well as toxic effects against liver cancer cells (HepG2) (Mukunthan et al. 2017). Essential oil (antimicrobial, anti-inflammatory) (Borah et al. 2019). Sesquiterpenes (Mukunthan et al. 2014).
Toxicity, side effects, and drug interaction: not known

REFERENCES

Arulmozhi, D.K., Sridhar, N., Veeranjaneyulu, A. and Arora, S.K. 2006. Preliminary mechanistic studies on the smooth muscle relaxant effect of hydroalcoholic extract of *Curcuma caesia*. *Journal of Herbal Pharmacotherapy*. 6(3–4): 117–124.

Borah, A., Paw, M., Gogoi, R., Loying, R., Sarma, N., Munda, S., Pandey, S.K. and Lal, M. 2019. Chemical composition, antioxidant, anti-inflammatory, anti-microbial and in-vitro cytotoxic efficacy of essential oil of *Curcuma caesia* Roxb. leaves: An endangered medicinal plant of North East India. *Industrial Crops and Products*. 129: 448–454.

Ibrahim, N.N.A., Wan Mustapha, W.A., Sofian-Seng, N.S., Lim, S.J., Mohd Razali, N.S., Teh, A.H., Rahman, H.A. and Mediani, A. 2023. A Comprehensive review with future prospects on the medicinal properties and biological activities of *Curcuma caesia* Roxb. *Evidence-Based Complementary and Alternative Medicine*. 2023: 1–17.

Mukunthan, K.S., Anil Kumar, N.V., Balaji, S. and Trupti, N.P. 2014. Analysis of essential oil constituents in rhizome of *Curcuma caesia* Roxb. from South India. *Journal of Essential Oil Bearing Plants*. 17(4): 647–651.

Mukunthan, K.S., Satyan, R.S. and Patel, T.N. 2017. Pharmacological evaluation of phytochemicals from South Indian Black Turmeric (*Curcuma caesia* Roxb.) to target cancer apoptosis. *Journal of Ethnopharmacology*. 209: 82–90.

Pathan, A.R., Vadnere, G. and Sabu, M. 2016. *Curcuma caesia* rhizomes: Evaluation of antiasthmatic effect by using clonidine induced mast cell degranulation. *Neuropharmacology Journal*. 1: 7–12.

Sawant, S.B., Bihani, G., Mohod, S. and Bodhankar, S. 2014. Evaluation of analgesic and anti-inflammatory activity of methanolic extract of *Curcuma caesia* Roxb. rhizomes in laboratory animals. *International Journal of Pharmacy and Pharmaceutical Sciences*. 6(2): 243–247.

***Curcuma longa* L.**

[From the Arabic *kurkum* = turmeric and Latin *longa* = long]

Published in *Species Plantarum* 1: 2. 1753

Synonym: *Curcuma domestica* Valeton
Local names: *kunyit* (Bajau, Dusun, Kadazan, Murut); *kunyit biasa* (Kedayan); *onyik* (Bugis)
Common name: turmeric
Habitat: cultivated
Geographical distribution: tropical Asia
Botanical description: this ginger reaches a height of about 1 m. Rhizomes: orange or bright yellow, somewhat cylindrical, aromatic, with a strange sour taste. Leaves: simple, alternate. Petioles: 20–45 cm long. Blades: emerge from culms, broadly lanceolate to oblong, dull green, 15–18 cm × 30–45 cm, attenuate at base, with numerous pairs of secondary nerves, acuminate at apex. Spikes: about 12–18 cm tall, somewhat obconical to cylindrical, emerge from the ground on 12–20 cm long peduncles. Bracts: numerous, about 3–5 cm in length, whitish green, to pure white, and more or less tinged at apex with a heavenly light pink. Calyx: white, 0.8–1.2 cm long, puberulent, split on one side, trilobed at apex. Corolla: tubular, yellowish, up to 3 cm long, the lobes deltoid and 1–1.5 cm long. Labellum: yellowish, obovate, bifid, 1.2–2 cm long. Staminodes: present. Stamens: 1. Ovary: hairy.
Traditional therapeutic indications: cuts (Kedayan); fungal infection (Murut); coughs, fever (Bugis); stomach aches, insect bites, sprains (Dusun); itchiness (Kedayan, Dusun); fractures, skin diseases; flatulence (Sungai, Bajau, Iranum, Kedayan); flu (Rungus); pimples (Kedayan); poison antidote (Dusun, Kadazan); menstruation, fever (Bajau); jaundice (Bajau, Iranum, Kedayan Dusun, Sungai); postpartum (Bajau, locals of Javanese descent, Kedayan, Iranun)
Pharmacology and phytochemistry: the plant has been subject to numerous pharmacological studies (Jyotirmayee & Mahalik 2022). It produces diarylheptanoids such as curcumin which is among other things, anti-inflammatory (Sikora et al. 2010), hepatoprotective (Khan et al. 2019), and antiviral (hepatitis B virus) (Hesari et al. 2018).
Toxicity, side effects, and drug interaction: curcumin can induce bleeding if taken in excess (Gronich et al. 2022).

REFERENCES

Gronich, N., Hurani, H., Weisz, I. and Halabi, S. 2022. Spontaneous bleeding and curcumin: Case report. *Journal of Clinical Images and Medical Case Reports*. 3(11): 2141.

Hesari, A., Ghasemi, F., Salarinia, R., Biglari, H., Tabar Molla Hassan, A., Abdoli, V. and Mirzaei, H. 2018. Effects of curcumin on NF-κB, AP-1, and Wnt/β-catenin signaling pathway in hepatitis B virus infection. *Journal of Cellular Biochemistry*. 119(10): 7898–7904.

Jyotirmayee, B. and Mahalik, G. 2022. A review on selected pharmacological activities of *Curcuma longa* L. *International Journal of Food Properties*. 25(1): 1377–1398.

Khan, H., Ullah, H. and Nabavi, S.M. 2019. Mechanistic insights of hepatoprotective effects of curcumin: Therapeutic updates and future prospects. *Food and Chemical Toxicology*. 124: 182–191.

Sikora, E., Scapagnini, G. and Barbagallo, M. 2010. Curcumin, inflammation, ageing and age-related diseases. *Immunity & Ageing*. 7(1): 1–4.

Curcuma xanthorrhiza **Roxb.**

[From the Arabic *kurkum* = turmeric and Latin *xanthorrhiza* = with yellow roots]

Published in *Synopsis Plantarum* 1: 19. 1839

Synonym: *Curcuma zanthorrhiza* Roxb.
Local names: *temu lawak* (Bajau, Kedayan, Dusun, Kadazan)
Common names: false turmeric, Javanese ginger
Habitat: cultivated
Geographical distribution: India, Bangladesh, Southeast Asia
Botanical description: this ginger reaches a height of about 2 m. Rhizomes: deep bright orange within, somewhat cylindrical, aromatic. Leaves: simple, alternate. Petioles: 5–20 cm long. Ligules: bilobed, about 3 mm long. Blades: emerge from culms, oblong to lanceolate, dark green with red blotches along the midrib, 10–30 cm × 30–100 cm, attenuate at base, with numerous pairs of secondary nerves, long acuminate at apex. Spikes: about 15–25 cm tall, somewhat obconical to cylindrical, emerge from the ground on 10–30 cm long peduncles. Bracts: numerous, about 3–5 cm in length, whitish green, to pure white, and more or less tinged at apex with a heavenly light pink. Calyx: white, 1.5 cm long, hairy, trifid at apex. Corolla: pale purple, about 1.7 cm long. Labellum: yellowish, square, about 2 cm across. Stamens: 1, about 4 mm long.
Traditional therapeutic indications: toothaches (Sungai); coughs, postpartum (Bajau)
Pharmacology and phytochemistry: antiparasitic (Murnigsih et al. 2005), heptatoprotective (Sutha et al. 2010), analgesic (Devaraj et al. 2010).
Arylheptanoids (Masuda et al. 1992). Antiplasmodial and antibabesial sesquiterpenes (12*R*)-, and (12*S*)-12,13-dihydro-12,13-dihydroxyxanthorrhizols), and arylheptanoids (3′-demethoxycyclocurcumin) (Matsuura et al. 2007; Yamada et al. 2009). Antibacterial sesquiterpenes (xanthorrhizol) (Hwang et al. 2000). Anti-inflammatory sesquiterpenes (germacrone) (Ozaki 1990).
Toxicity, side effects, and drug interaction: ethanol extract administered orally to rats up to a dose of 5 g/kg was found to cause no signs of toxicity (Devaraj et al. 2010).

Comment: In Indonesia, the plant is used to treat liver ailments.

REFERENCES

Devaraj, S., Esfahani, A.S., Ismail, S., Ramanathan, S. and Yam, M.F. 2010. Evaluation of the antinociceptive activity and acute oral toxicity of standardized ethanolic extract of the rhizome of *Curcuma xanthorrhiza* Roxb. *Molecules*. 15(4): 2925–2934.

Hwang, J.K., Shim, J.S. and Pyun, Y.R. 2000. Antibacterial activity of xanthorrhizol from *Curcuma xanthorrhiza* against oral pathogens. *Fitoterapia*. 71(3): 321–323.

Masuda, T., Isobe, J., Jitoe, A. and Nakatani, N. 1992. Antioxidative curcuminoids from rhizomes of *Curcuma xanthorrhiza*. *Phytochemistry*. 31(10): 3645–3647.

Matsuura, H., Nomura, S., Subeki, Yamada, K., Yamasaki, M., Yamato, O., Maede, Y., Katakura, K., Trimurningsih, C. and Yoshihara, T. 2007. Anti-babesial compounds *from Curcuma xanthorrhiza*. *Natural Product Research*. 21(4): 328–333.

Murnigsih, T., Matsuura, H., Takahashi, K., Yamasaki, M., Yamato, O., Maede, Y., Katakura, K., Suzuki, M., Kobayashi, S. and Yoshihara, T. 2005. Evaluation of the inhibitory activities of the extracts of Indonesian traditional medicinal plants against *Plasmodium falciparum* and *Babesia gibsoni*. *Journal of Veterinary Medical Science*. 67(8): 829–831.

Ozaki, Y. 1990. Antiinflammatory effect of *Curcuma xanthorrhiza* Roxb. and its active principles. *Chemical and Pharmaceutical Bulletin*. 38(4): 1045–1048.

Sutha, D., Sabariah, I., Surash, R., Santhini, M. and Yam, M.F. 2010. Evaluation of the hepatoprotective activity of standardized ethanolic extract of *Curcuma xanthorrhiza* Roxb. *Journal of Medicinal Plants Research*. 4(23): 2512–2517.

Yamada, K., Nabeta, K., Yamasaki, M., Katakura, K. and Matsuura, H. 2009. Isolation of antibabesial compounds from *Brucea javanica, Curcuma xanthorrhiza*, and *Excoecaria cochinchinensis*. *Bioscience, Biotechnology, and Biochemistry*. 0902051308.

***Etlingera brevilabrum* (Valeton) R.M. Sm.**

[After the 18th century German botanist Andreas Ernest Etlinger and from the Latin *brevis* = short, and *labrum* = lip]

Published in *Notes from the Royal Botanic Garden*, Edinburgh 43: 243. 1986

Synonyms: *Achasma brevilabrum* Valeton; *Etlingera brevilabris* (Valeton) R.M. Sm. *Geanthus brevilabris* (Valeton) Valeton; *Hornstedtia brevilabris* (Valeton) Merr.
Local name: *sibu* (Dusun)
Habitat: forests
Geographical distribution: Borneo
Botanical description: this herbaceous plant reaches a height of about 2 m. Rhizomes: elongated, aromatic. Leaves: simple, alternate. Ligules: about 1.2 cm long, ovate. Blades: emerge from culms, oblong to elliptic, dark green, glossy, wavy, with a visible midrib, about 12 cm × 55 cm, asymmetrical at base, shortly accuminate at apex. Flowers: few, on the ground, dark red, about 12 cm long. Calyx: tubular, reddish, about 5.5 cm long, membranous, split on one side, trilobed at apex. Corolla: dull red, up to 5 cm long, tubular, trilobed. Labellum: elliptic, 3.5 cm long l. Stamens: 1, about 1.5 cm long. Stigma: white. Capsules: fleshy, smooth, about 1 cm long.
Traditional therapeutic indications: cuts (Dusun)
Pharmacology and phytochemistry: antibacterial essential oil (eucalyptol, β-pinene, caryophyllene oxide, and α-thujene) (Mahdavi et al. 2013, 2016).
Toxicity, side effects, and drug interaction: not known

Comment: The bright red color of the petals may have encouraged the use of this plant for the treatment of cuts.

REFERENCES

Mahdavi, B., Yaacob, W.A., Din, L.B., Heng, L.Y. and Ibrahim, N. 2016. Chemical composition, antioxidant, and antibacterial activities of essential oils from *Etlingera brevilabrum* valeton. *Records of Natural Products*. 10(1): 22.

Mahdavi, B., Yaacob, W.A., Din, L.B. and Siti Aisha, M.A. 2013. Essential oil composition of three air-dried parts of *Etlingera brevilabrum*. *Journal of Essential Oil Bearing Plants*. 16(1): 17–22.

***Etlingera coccinea* (Blume) S. Sakai & Nagam.**

[After the 18th century German botanist Andreas Ernest Etlinger and from the Latin *coccinea* = bright red]

Published in *Edinburgh Journal of Botany* 60: 190. 2003

Synonyms: *Achasma coccineum* (Blume) Valeton; *Alpinia coccinea* (Blume) D.Dietr.; *Amomum coccineum* (Blume) K.Schum.; *Cardamomum coccineum* (Blume) Kuntze; *Geanthus coccineus* Reinw.

Local name: *tuhau* (Dusun, Kadazan)

Habitat: forests, cultivated, roadsides, plantations

Geographical distribution: Southeast Asia

Botanical description: this massive ginger reaches a height of about 5 m. Rhizomes: present. Leaves: simple, alternate. Ligules: about 1.5 cm long, brown, triangular, hairy. Petioles: none. Blades: emerge from culms, oblong, dark green, glossy, wavy, with a visible midrib, about 15 cm × 70 cm, rounded at apex. Spikes: about 10 cm tall, on the ground, with vaguely the appearance of burning embers. Calyx: tubular, reddish, about 5 cm long, membranous, split on one side, trilobed. Corolla: dull red, up to 8 cm long, tubular, trilobed. Labellum: elongated, bilobed to capitate. Stamens: 1, 1 cm long. Ovary: oblong. Styles: 1, thin. Capsules: fleshy, smooth, about 1 cm long.

Traditional therapeutic indications: hypertension, blood detoxification, bloody stools, laxative (Dusun, Kadazan)

Pharmacology and phytochemistry: antibacterial (Daniel-Jambun et al. 2017). Essential oil (borneol) with antibacterial and toxic to cervical cancer cells (HeLa) (Jems et al. 2021; Vairappan et al. 2012).

Toxicity, side effects, and drug interaction: not known

Comments:

(i) The culms are used as food by the Dusun and Kadazan, and are commonly sold by the roadsides and markets.

(ii) The bright red color of the petals may have encouraged the use of this plant for the treatment of diseases related to the blood.

REFERENCES

Daniel-Jambun, D., Dwiyanto, J., Lim, Y.Y., Tan, J.B.L., Muhamad, A., Yap, S.W. and Lee, S.M. 2017. Investigation on the antimicrobial activities of gingers (*Etlingera coccinea* (Blume) S. Sakai & Nagam and *Etlingera sessilanthera* RM Sm.) endemic to Borneo. *Journal of Applied Microbiology*. 123(4): 810–818.

Jems, N.S., Rusdi, N.A., Mus, A.A. and Godoong, E. 2021. Chemical composition of essential oil from *Etlingera coccinea* (Blume) S. Sakai & Nagam in Kadamaian, Kota Belud, Sabah. *Journal of Tropical Biology & Conservation (JTBC)*, 18: 91–105.

Vairappan, C.S., Nagappan, T. and Palaniveloo, K. 2012. Essential oil composition, cytotoxic and antibacterial activities of five *Etlingera species* from Borneo. *Natural Product Communications*. 7(2): 1934578X1200700233.

***Etlingera elatior* (Jack) R.M. Sm.**

[After the 18th century German botanist Andreas Ernest Etlinger and from the Latin *elatus* = tall]

Published in *Notes from the Royal Botanic Garden, Edinburgh* 43(2): 244–245. 1986

Synonyms: *Alpinia elatior* Jack; *Alpinia speciosa* (Blume) D.Dietr.; *Elettaria speciosa* Blume; *Nicolaia speciosa* (Blume) Horan.; *Phaeomeria imperialis* (Roscoe) Lindl.; *Phaeomeria magnifica* (Roscoe) K.Schum.; *Phaeomeria speciosa* (Blume) Koord.

Local names: *bunga kantan* (Dusun); *topu* (Dusun, Kadazan); *kantan* (Murut)

Common names: kantan flower, Philippine waxflower, torch ginger

Habitat: cultivated

Distribution: Southeast Asia

Botanical description: this massive ginger reaches a height of about 5 m. Rhizome: thick Culms: coriaceous. Leaves: simple, alternate. Ligules: about 1.5 cm long, bilobed. Petioles: 1.5–4 cm long. Blades: emerge from culms, dark green, glossy, wavy, lanceolate, elliptical, or oblong, with a visible midrib, 10–20 cm × 20–90 cm, rounded to cordate at base, acute at apex. Spikes: magnificent, which can make one think of some kind of solitary flower, about 10 cm long on approximately 50 cm–1 m tall peduncles. Bracts: reddish-pink, about 2–18 cm long. Calyx: tubular, reddish, about 3–3.5 cm long, split on one side, trilobed. Corolla: pink to red, up to 4 cm long, tubular. Labellum: spathulate, about 4 cm long. Stamens: 1, 2.5 cm long with red anther and a white and hairy filament. Styles: 1, thin. Capsules: globose, about 2.5 cm across. Seeds: numerous, black.

Traditional therapeutic indications: fever, flatulence (Dusun, Kadazan)

Pharmacology and phytochemistry: antibacterial (Juwita et al. 2018), antidiabetic, hypotensive (Nor et al. 2020; Widyarini et al. 2022), anti-inflammatory (Kusriani 2023), tyrosinase inhibition (Chan et al. 2008). Essential oil (β-pinene, 1-dodecene, *E* -farnesene, *E* -caryophyllene) (Chan et al. 2011). Flavonol glycosides, hydroxycinnamic acid derivatives (3-*O*-caffeoylquinic acid, 5-*O*-caffeoylquinic acid, 5-*O*-caffeoylquinic acid methyl ester) (Chan et al. 2011).

Toxicity, side effects, and drug interaction: the median lethal dose (LD_{50}) of ethanol extract administered orally to mice was above 2 g/kg (Kusriani 2023).

Comments:

(i) Inflorescences and rhizomes are used as food.
(ii) In Aceh, the plant is called "*bak kala*" and is used for hypertension.
(iii) The Dusun use a plant of the genus *Etlingera Giseke* (1792) for piles and as emmenagogue (local names: *teriwad*, *tolivad*). Note here again a medicinal plant with red petals used to treat diseases linked to blood.

REFERENCES

Chan, E.W.C., Lim, Y.Y., Wong, L.F., Lianto, F.S., Wong, S.K., Lim, K.K., Joe, C.E. and Lim, T.Y. 2008. Antioxidant and tyrosinase inhibition properties of leaves and rhizomes of ginger species. *Food Chemistry*. 109(3): 477–483.

Chan, E.W.C., Lim, Y.Y. and Wong, S.K. 2011. Phytochemistry and pharmacological properties of *Etlingera elatior*: A review. *Pharmacognosy Journal*. 3(22): 6–10.

Juwita, T., Puspitasari, I.M. and Levita, J. 2018. Torch ginger (*Etlingera elatior*): A review on its botanical aspects, phytoconstituents and pharmacological activities. *Pakistan Journal of Biological Sciences*. 21(4): 151–165.

Kusriani, D.A.R.H. 2023. Antiinflamation and acute toxicity of etlingera elatior leaves extract. *International Journal of Biology, Pharmacy and Allied Sciences*. 10(9): 2906–2918.

Nor, N.A.M., Noordin, L., Bakar, N.H.A. and Ahmad, W.A.N.W. 2020. Evaluation of antidiabetic activities of *Etlingera elatior* flower aqueous extract *in vitro* and *in vivo*. *Journal of Applied Pharmaceutical Science*. 10(8): 43–51.

Widyarini, T., Indarto, D. and Purwanto, B. 2022. Modulation effects of *Etlingera elatior* ethanol extract as anti-inflammatory on chronic kidney disease in mice with hypertension and diabetes. *Journal of Population Therapeutics and Clinical Pharmacology*. 29(4): 140–149.

***Etlingera littoralis* (J. Koenig) Giseke**

[After the 18th century botanist German botanist Andreas Ernest Etlinger and from the Latin *littoralis* = from seashores]

Published in *Praelectiones in ordines naturales plantarum* 199, 209. 1792

Synonyms: *Achasma megalocheilos* Griff.; *Amomum littorale* J. Koenig; *Hornstedtia megalocheilos* (Griff.) Ridl.
Local name: *tepus* (Bajau)
Habitat: forests
Geographical distribution: from Bangladesh to Borneo
Botanical description: this massive ginger reaches a height of about 6 m. Rhizomes: thick. Culms: coriaceous. Leaves: simple, alternate. Ligules: 1.5–2 cm long, broadly rounded at apex. Petioles: 1.5–4.5 cm long. Blades: emerge from culms, oblong to lanceolate, asymmetrical, tapering at base, 9–15 cm × 50–70 cm, smooth, acuminate at apex. Flowering heads: emerge from the ground, about 10 cm across, made of 4–12 bright red flowers. Bracts: ovate to oblong, 3–5 cm long. Bracteoles: tubular, about 4 cm long. Calyx: tubular, pink, about 7–9 cm long, split on one side, trilobed. Corolla: red, hairy at the throat, up to 5 cm long, tubular. Labellum: obovate to oblong, about 5 cm long, red with yellow margin, emarginate at apex. Stamens: 1. Ovary: hairy. Styles: pink, about 6 cm long. Capsules: globose, smooth, about 3 cm across. Seeds: numerous.
Traditional therapeutic indications: fever, stomach aches (Bajau)
Pharmacology and phytochemistry: essential oil (methyl isoeugenol) (Wong et al. 2010). Hydroxycinnamic acid derivative (caffeoylquinic acid) (Chan et al. 2009). Antibacterial (Chan et al. 2007), tyrosinase inhibition (Chan et al. 2008).
Toxicity, side effects, and drug interaction: not known

REFERENCES

Chan, E.W.C., Lim, Y.Y., Ling, S.K., Tan, S.P., Lim, K.K. and Khoo, M.G. 2009. Caffeoylquinic acids from leaves of *Etlingera* species (Zingiberaceae). *LWT-Food Science and Technology*. 42(5): 1026–1030.

Chan, E.W.C., Lim, Y.Y. and Omar, M. 2007. Antioxidant and antibacterial activity of leaves of *Etlingera* species (Zingiberaceae) in Peninsular Malaysia. *Food Chemistry*. 104(4): 1586–1593.

Chan, E.W.C., Lim, Y.Y., Wong, L.F., Lianto, F.S., Wong, S.K., Lim, K.K., Joe, C.E. and Lim, T.Y. 2008. Antioxidant and tyrosinase inhibition properties of leaves and rhizomes of ginger species. *Food Chemistry*. 109(3): 477–483.

Wong, K.C., Sivasothy, Y., Boey, P.L., Osman, H. and Sulaiman, B. 2010. Essential oils of *Etlingera elatior* (Jack) RM Smith and *Etlingera littoralis* (Koenig) Giseke. *Journal of Essential Oil Research*. 22(5): 461–466.

***Etlingera punicea* (Roxb.) R.M. Sm.**

[After the 18th century botanist German botanist Andreas Ernest Etlinger and from the Latin *punicea* = crimson]

Published in *Notes from the Royal Botanic Garden, Edinburgh* 43: 249. 1986

Synomym: *Alpinia punicea* Roxb.
Local name: *tuhau* (Dusun, Kadazan)
Habitat: forests
Geographical distribution: Thailand, Malaysia, Indonesia
Botanical description: this massive ginger reaches a height of about 5 m. Rhizomes: thin. Culms: coriaceous. Leaves: simple, alternate. Ligules: 0.5–1 cm long, apex broadly rounded, hairy. Petioles: tiny. Blades: emerge from culms, tapering at base, 7–22 cm × 60–90 cm, smooth, acuminate at apex. Flowering heads: emerge from the ground, about 10 cm across, made of about 8 bright red flowers. Bracts: lanceolate, about 6 cm long. Bracteoles: about 4 cm long. Calyx: tubular, about 7 cm long, trilobed, split on one side. Corolla: crimson, tubular, up to 6 cm long. Labellum: somewhat spoon-shaped and wavy at margin, about 6 cm long, bilobed. Stamens: 1, about 1.3 cm long, thin. Ovary: 5 mm long, hairy. Styles: 1, thin, hairy. Stigma: pink. Capsules: obovoid, up to about 3 cm × 4.5 cm, smooth. Seeds: numerous, angular, about 4 mm across.
Traditional therapeutic indications: beriberi, fatigue (Dusun, Kadazan)
Pharmacology and phytochemistry: antibacterial and antifungal essential oil (methyl chavicol) (Tadtong et al. 2009).
Toxicity, side effects, and drug interaction: not known

REFERENCE

Tadtong, S., Wannakhot, P., Poolsawat, W., Athikomkulchai, S. and Ruangrungsi, N. 2009. Antimicrobial activities of essential oil from *Etlingera punicea* rhizome. *Journal of Health Research*. 23(2): 77–79.

Globba francisci Ridl.

[From ancient Indonesian *galoba* = ginger]

Published in *Journal of the Linnean Society, Botany* 42: 162. 1914

Synonym: *Globba argentiana* R.M. Sm.
Local name: *layo timbaan* (Dusun)
Habitat: on the banks of rivers
Geographical distribution: Borneo, the Philippines
Botanical description: this herbaceous plant reaches a height of about 1.5 m. Rhizomes: present. Culms: erect. Leaves: sessile. Ligules: bilobed, tiny. Blades: emerge from culms, oblong to lanceolate, 2–5 cm × 10–35 cm, somewhat asymmetrical at base, acuminate at apex. Racemes: terminal, conical, up to 30 cm long. Flowers: up to about 20, orange, about 3 cm long, somewhat swan-shaped. Calyx: trilobed, yellowish-green, about 3 mm long. Corolla: thin, about 1.5 cm long, the lobes elliptic. Labellum: triangular, with a thin filament, about 1 cm long. Staminodes: present. Stamens: 1. Capsules: dehiscent. Seeds: numerous, arillate. Bulbils: present.
Traditional therapeutic indications: cooling (Dusun)
Pharmacology and phytochemistry: not known
Toxicity, side effects, and drug interaction: not known

Comment: A plant of the genus *Globba* L. (1771) is used by the Dusun as an aphrodisiac.

Globba propinqua Ridl.

[From ancient Indonesian *galoba* = ginger and Latin *propinquus* = near]

Published in *Journal of the Straits Branch of the Royal Asiatic Society* 46: 230. 1906

Synonym: *Globba paucibractea* Valeton
Local name: *mazoloso* (Dusun, Kadazan)
Habitat: on the banks of rivers
Geographical distribution: Borneo, Sumatra
Botanical description: this herbaceous plant reaches a height of about 60 cm. Rhizomes: present. Culms: erect. Ligules: tiny, oblong, hairy. Blades: emerge from culms, oblong-lanceolate, about 3 cm × 10 cm, asymmetrical and attenuate at base, acuminate at apex. Racemes: nodding, thin and with few orange flowers which can make one think of flying birds. Calyx: trilobed, the lobes ovate and red. Corolla: thin, the lobes lanceolate. Labellum: linear-oblong with a red central spot. Staminodes: present. Stamens: 1. Capsules: dehiscent. Seeds: numerous, arillate.
Traditional therapeutic indications: menorrhagia (Dusun, Kadazan)
Pharmacology and phytochemistry: not known
Toxicity, side effects, and drug interaction: not known

Hedychium longicornutum Griff. ex Baker

[From Greek *hedys* = sweet, *chion* = snow, from Latin *longus* = long, and *cornutus* = having horns]

Published in *The Flora of British India* 6: 228. 1892

Synonym: *Hedychium crassifolium* Baker
Local name: *kunyit hantu* (Bajau)
Common name: hornbill's ginger

Habitat: forests, cultivated
Geographical distribution: from Thailand to Borneo
Botanical description: this epiphytic ginger grows on small trees. Roots: thick, fleshy, somewhat pseudobulb-like. Culms: up to 60 cm tall. Leaves: simple, spiral. Ligules: 6 cm long. Blades: emerge from culms. Spikes: terminal, somewhat globose, ressembling a hideous mass of worms, about 12 cm long. Calyx: spathaceous, oblique, pink, 3 cm long. Corolla: about 9 cm long, lobes narrow, linear, reddish-orange. Stamens: 1, curved, up to 15 cm long, orange. Capsules: oblong, hairy, brown. Seeds: about 15 in a crimson aril.
Traditional therapeutic indications: fever (Bajau)
Pharmacology and phytochemistry: not known
Toxicity, side effects, and drug interaction: not known

Comment: A plant of the genus *Hedychium* J. Koenig (1783) is used by the Dusun and Kadazan (local name: *sidbu*) for insect stings, as styptic, and for wounds.

***Kaempferia galanga* L.**

[After the 17th century German naturalist Engelbert Kaempfer and from Arabic *caluegia* = greater galangal]

Published in *Species Plantarum* 1: 2–3. 1753

Synonyms: *Alpinia sessilis* J. Koenig, *Kaempferia humilis* Salisb., *Kaempferia latifolia* Donn ex Hornem, *Kaempferia marginata* Carey ex Roscoe, *Kaempferia plantaginifolia* Salisb., *Kaempferia procumbens* Noronha
Local names: *cekur* (Bajau, Kedayan); *kusur* (Sungai); *kesur* (Bajau)
Common names: East Indian galangal, resurrection lily
Habitat: cultivated
Geographical distribution: from India to the Philippines
Botanical description: this ginger reaches a height of about 15 cm. Rhizomes: ovoid, aromatic, about 5 cm long. Leaves: on the ground, sessile, exstipulate. Ligules: triangular. Blades: pale green above, purplish beneath, broadly elliptic, wavy, somewhat fleshy, smooth, 5–10 cm × 6–15 cm, with longitudinal secondary nerves, cordate at base, acute at apex. Flowers: fragrant, of a peculiar kind of light purple, solitary, which can make one think of butterflies, basal. Calyx: 2–3 cm long, bifid. Corolla: tubular, the tube 2.5–5 cm long, the lobes linear and about 1.5–3 cm long. Labellum: obovate, about 3 cm long. Staminodes present. Stamens: 1, 1 cm long.
Traditional therapeutic indications: dandruff, food poisoning, sore throats (Kedayan); postpartum, colds, headaches, dysmenorrhea (Bajau); fatigue, stomach aches (Bajau, Dusun, Kadazan); coughs (Kedayan, Bajau)
Pharmacology and phytochemistry: anti-inflammatory, analgesic (Sulaiman et al. 2008; Vittalrao et al. 2011). Antibacterial essential oil (Munda et al. 2018).
Antiplasmodial diterpenes (Thongnest et al. 2005). Anti-inflammatory hydroxycinnamic acid derivatives (ethyl-*p*-methoxy-cinnamate) (Umar et al. 2012). Flavones (5-hydroxy-3,7-dimethoxyflavone) inhibiting viral proteases (Sookkongwaree et al. 2006).
Toxicity, side effects, and drug interaction: intake of ethanol extract at the dose of 200 mg/kg/day for 28 days induced in rats a decrease of lymphocytes count (Kanjanapothi et al. 2004).

Comment: This plant is used as food by locals.

REFERENCES

Kanjanapothi, D., Panthong, A., Lertorasertsuke, N., Taesotikul, T., Rujjanawate, C., Kaewpinit, D., Sudthayakorn, R., Choochote, W., Chaithong, U., Jitoakdi, A. and Pitasawat, B. 2004. Toxicity of crude rhizome extract of *Kaempferia galanga* L. (Proh Hom). *Journal of Ethnopharmacology*. 90(2–3): 359–365.

Munda, S., Saikia, P. and Lal, M. 2018. Chemical composition and biological activity of essential oil of Kaempferia galanga: A review. *Journal of Essential Oil Research*. 30(5): 303–308.

Sookkongwaree, K., Geitmann, M., Roengsumran, S., Petsom, A. and Danielson, U.H. 2006. Inhibition of viral proteases by Zingiberaceae extracts and flavones isolated from *Kaempferia parviflora*. *Die Pharmazie-An International Journal of Pharmaceutical Sciences*. 61(8): 717–721.

Sulaiman, M.R., Zakaria, Z.A., Daud, I.A., Ng, F.N., Ng, Y.C. and Hidayat, M.T. 2008. Antinociceptive and anti-inflammatory activities of the aqueous extract of *Kaempferia galanga* leaves in animal models. *Journal of Natural Medicines*. 62: 221–227.

Thongnest, S., Mahidol, C., Sutthivaiyakit, S. and Ruchirawat, S. 2005. Oxygenated pimarane diterpenes from *Kaempferia marginata*. *Journal of Natural Products*. 68(11): 1632–1636.

Umar, M.I., Asmawi, M.Z., Sadikun, A., Atangwho, I.J., Yam, M.F., Altaf, R. and Ahmed, A. 2012. Bioactivity-guided isolation of ethyl-p-methoxycinnamate, an anti-inflammatory constituent, from *Kaempferia galanga* L. extracts. *Molecules*. 17(7): 8720–8734.

Vittalrao, A.M., Shanbhag, T., Kumari, M., Bairy, K.L. and Shenoy, S. 2011. Evaluation of antiinflammatory and analgesic activities of alcoholic extract of Kaempferia galanga in rats. *Indian Journal of Physiology and Pharmacology*. 55(1): 13–24.

***Zingiber officinale* Roscoe**

[From the Greek *ziggiberis* = ancient spice known to Arabs and Latin *officinalis* = of medicinal value]

Published in *Transactions of the Linnean Society of London* 8: 348. 1807

Synonyms: *Amomum zingiber* L.; *Curcuma longifolia* Wall.; *Zingiber aromaticum* Noronha; *Zingiber majus* Rumph.; *Zingiber missionis* Wall.; *Zingiber sichuanense* Z.Y. Zhu, S.L. Zhang & S.X. Chen; *Zingiber zingiber* (L.) H. Karst.

Local names: *layo taporak* (Dusun); *hayo, hazo* (Dusun, Kadazan); *halia* (Bajau, Murut); *halia rajah* (Kedayan)

Common name: ginger

Habitat: cultivated

Geographical distribution: tropical Asia

Botanical description: this ginger reaches a height of about 1 m. Rhizomes: smooth, light brown, somewhat glossy, deliciously aromatic, with a pungent lemony taste, fibrous and slightly yellow inside. Leaves: simple, alternate. Ligules: bilobed, about 4 mm long. Petioles: 1.5 cm long. Blades: emerge from culms, narrowly elliptic, dull green, 2.5–15 cm × 15–30 cm. Spikes: bullet-shaped, 5 cm long, scaly, on about 25 cm long peduncles. Bracts: greenish-yellow, ovate, acute at apex. Calyx: 1 cm long. Corolla: yellowish, up to 2.5 cm long, and with a ovate and purplish lip. Stamens: 1, dark purple. Capsules: dehiscent. Seeds: black, arillate.

Traditional therapeutic indications: cooling, postpartum, sprains (Dusun, Kadazan); flatulence (Dusun, Murut); rheumatisms (Kedayan, Dusun, Kadazan); colds (Bajau)

Pharmacology and phytochemistry: this plant has been the subject of numerous pharmacological studies (Bitari et al. 2023) which have highlighted, among other things, antibacterial, antifungal (Akintobi et al. 2013), antiviral (chikungunya virus) (Kaushik et al. 2020), antiamoebic (Dyab et al. 2016; Luangboribun et al. 2014; Arbabi et al. 2016), nematicidal (Lin et al. 2010), and antidiabetics activity (Thomson et al. 2002; Akhani et al. 2004).

Alkylphenols ([6]-gingerol) (Wang et al. 2014; Amri et al. 2016; Mao et al. 2019).

Toxicity, side effects, and drug interaction: the median lethal dose (LD_{50}) of methanol extract administered to rats was 10.2 g/kg (Shalaby & Hamowieh 2010).

Comments:

(i) *Zingiber officinale var. rubrum* Theilade is used by the Bajau for blood circulation, swellings, fever, and to clean the skin of the face.

(ii) In the genus *Tamijia* S. Sakai & Nagam. (2000), the Dusun use "*kamlimigi arou*" for food poisoning whilst "*sisibu*" is used by the Dusun and Kadazan for stomach aches.

(iii) The Lundayeh use a plant of the genus *Hornstedia* Juss., (1816) for colds (local name: *baku tabu*).

(iv) A plant of the genus *Zingiber* Martinov (1820) is used by the Dusun and Kadazan for runny noses (local name: *layo*).

REFERENCES

Akhani, S.P., Vishwakarma, S.L. and Goyal, R.K. 2004. Anti-diabetic activity of *Zingiber officinale* in streptozotocin-induced type I diabetic rats. *Journal of Pharmacy and Pharmacology*. 56(1): 101–105.

Akintobi, O.A., Onoh, C.C., Ogele, J.O., Idowu, A.A., Ojo, O.V. and Okonko, I.O. 2013. Antimicrobial activity of *Zingiber officinale* (ginger) extract against some selected pathogenic bacteria. *Nature and Science*. 11(1): 7–15.

Amri, M. and Touil-Boukoffa, C. 2016. *In vitro* anti-hydatic and immunomodulatory effects of ginger and [6]-gingerol. *Asian Pacific Journal of Tropical Medicine*. 9(8): 749–756.

Arbabi, M., Delavari, M., Kashan, Z.F., Taghizadeh, M. and Hooshyar, H. 2016. Ginger (*Zingiber officinale*) induces apoptosis in *Trichomonas vaginalis in vitro*. *International Journal of Reproductive Biomedicine*. 14(11): 691.

Bitari, A., Oualdi, I., Touzani, R., Elachouri, M. and Legssyer, A. 2023. *Zingiber officinale* Roscoe: A comprehensive review of clinical properties. *Materials Today: Proceedings*. 72: 3757–3767.

Dyab, A.K., Yones, D.A., Ibraheim, Z.Z. and Hassan, T.M. 2016. Anti-giardial therapeutic potential of dichloromethane extracts of *Zingiber officinale* and Curcuma longa *in vitro* and *in vivo*. *Parasitology Research*. 115(7): 2637–2645.

Kaushik, S., Jangra, G., Kundu, V., Yadav, J.P. and Kaushik, S. 2020. Anti-viral activity of *Zingiber officinale* (Ginger) ingredients against the Chikungunya virus. *Virusdisease*. 31: 270–276.

Lin, R.J., Chen, C.Y., Chung, L.Y. and Yen, C.M. 2010. Larvicidal activities of ginger (*Zingiber officinale*) against *Angiostrongylus cantonensis*. *Acta Tropica*. 115(1–2): 69–76.

Luangboribun, T., Tiewcharoen, S., Rabablert, J., Auewarakul, P., Lumlerdkij, N., Junnu, V. and Roytrakul, S. 2014. Screening of anti-amoebic activity on *Naegleria fowleri* in extracts of Thai medicinal plants. *JITMM Proceedings*. 3: 16–22.

Mao, Q.Q., Xu, X.Y., Cao, S.Y., Gan, R.Y., Corke, H., Beta, T. and Li, H.B. 2019. Bioactive compounds and bioactivities of ginger (*Zingiber officinale* Roscoe). *Foods*. 8(6): 185.

Shalaby, M.A. and Hamowieh, A.R. 2010. Safety and efficacy of Zingiber officinale roots on fertility of male diabetic rats. *Food and Chemical Toxicology*. 48(10): 2920–2924.

Thomson, M., Al-Qattan, K.K., Al-Sawan, S.M., Alnaqeeb, M.A., Khan, I. and Ali, M. 2002. The use of ginger *(Zingiber officinale* Rosc.) as a potential anti-inflammatory and antithrombotic agent. *Prostaglandins, Leukotrienes and Essential Fatty Acids*. 67(6): 475–478.

Wang, S., Zhang, C., Yang, G. and Yang, Y. 2014. Biological properties of 6-gingerol: A brief review. *Natural Product Communications*. 9(7): 1027–1030.

***Zingiber purpureum* Roscoe**

[From the Greek *ziggiberis* = ancient spice known to Arabs and Latin *purpureum* = purplish]

Published in *Transactions of the Linnean Society of London* 8: 348. 1807

Synonyms: *Amomum cassumunar* (Roxb.) Don; *Zingiber cassumunar* Roxb.
Local names: *dangalai* (Rumanau); *bonglai* (Suluk)
Common names: Bengal ginger, cassumunar ginger, Javanese ginger
Habitat: cultivated
Geographical distribution: from India to the Philippines
Botanical description: this ginger reaches a height of about 2 m. Rhizomes: yellowish inside, aromatic. Leaves: simple, alternate. Ligules: tiny. Blades: emerge from culms, narrowly lanceolate, about 2–5 cm × 10–30 cm. Spikes: about 10 cm long, somewhat dark red to purplish, glossy, fleshy, pine cone-shaped to fusiform, on about 30 cm long peduncles Bracts: ovate, 4.5 cm long, reddish, hairy. Calyx: white, membranous, tubular, about 1.7 cm long, trilobed. Corolla: tubular, cream-colored, trilobed, membranous, about 5 cm long. Labellum: orbicular, about 2 cm across. Staminodes: present. Capsules: small and globose.
Traditional therapeutic indications: fatigue (Rumanau)
Pharmacology and phytochemistry: antibacterial (Tandirogang et al. 2022), anti-hyperlipidemic (Hasimun et al. 2019). Essential oil (Wang et al. 2015). Neurotrophic phenylbutenoids (Matsui et al. 2012; Kubo et al. 2015). Analgesic and anti-inflammatory phenylbutenoid (*E*)-1-(3, 4-dimethoxyphenyl)but-1-ene) (Ozaki et al. 1991).
The plant produces the sesquiterpene zerumbone, which exered anti-inflammatory, analgesic (Chien et al. 2016), antifungal (Kishore & Dwivedi 1992; Shin et al. 2019), antibacterial (Kumar et al. 2013; da Silva et al. 2018), antiviral (HIV) (Dai et al. 1997), immunomodulatory (Keong et al. 2010), analgesic, and anti-infl ammatory activity (Somchit et al. 2012). Zerumbone was toxic for various cancer cells, including liver and pancreatic cancers (Huang et al. 2005; Sharifah Sakinah et al. 2007; Shamoto et al. 2014).
Toxicity, side effects, and drug interaction: in healthy volunteers, intake of 200 mg of an extract for 12 weeks induced somnolence (Chongmelaxme et al. 2017).

REFERENCES

Chien, T.Y., Huang, S.K.H., Lee, C.J., Tsai, P.W. and Wang, C.C. 2016. Antinociceptive and anti-inflammatory effects of zerumbone against mono-iodoacetate-induced arthritis. *International Journal of Molecular Sciences*. 17(2): 249.

Chongmelaxme, B., Sruamsiri, R., Dilokthornsakul, P., Dhippayom, T., Kongkaew, C., Saokaew, S., Chuthaputti, A. and Chaiyakunapruk, N. 2017. Clinical effects of *Zingiber cassumunar* (Plai): A systematic review. *Complementary Therapies in Medicine*. 35: 70–77.

Dai, J.R., Cardellina, J.H., Mahon, J.B.M. and Boyd, M.R. 1997. Zerumbone, an human HIV-inhibitory and cytotoxic sesquiterpene of *Zingiber aromaticum. Evidence-based Complementary and Alternative Medicine*. 10(2): 115–118.

da Silva, T.M., Pinheiro, C.D., Orlandi, P.P., Pinheiro, C.C. and Pontes, G.S. 2018. Zerumbone from *Zingiber zerumbet* (L.) smith: A potential prophylactic and therapeutic agent against the cariogenic bacterium Streptococcus mutans. *BMC Complementary and Alternative Medicine*. 18(1): 1–9.

Hasimun, P., Sulaeman, A., Mulyani, Y., Islami, W.N. and Lubis, F.A.T. 2019. Antihyperlipidemic activity and HMG CoA reductase inhibition of ethanolic extract of Zingiber cassumunar roxb in fructose-induced hyperlipidemic wistar rats. *Journal of Pharmaceutical Sciences and Research*. 11(5): 1897–1901.

Huang, G.C., Chien, T.Y., Chen, L.G. and Wang, C.C. 2005. Antitumor effects of zerumbone from *Zingiber zerumbet* in P-388D1 cells *in vitro* and *in vivo*. *Planta Medica*. 71(3): 219–224.

Keong, Y.S., Alitheen, N.B., Mustafa, S., Aziz, S.A., Rahman, M.A. and Ali, A.M. 2010. Immunomodulatory effects of zerumbone isolated from roots of *Zingiber zerumbet. Pakistan Journal of Pharmaceutical Sciences*. 23(1): 75–82.

Kishore, N. and Dwivedi, R.S. 1992. Zerumbone: A potential fungitoxic agent isolated from *Zingiber cassumunar* Roxb. *Mycopathologia*. 120: 155–159.

Kubo, M., Gima, M., Baba, K., Nakai, M., Harada, K., Suenaga, M., Matsunaga, Y., Kato, E., Hosoda, S. and Fukuyama, Y. 2015. Novel neurotrophic phenylbutenoids from Indonesian ginger Bangle, *Zingiber purpureum*. *Bioorganic & Medicinal Chemistry Letters*. 25(7): 1586–1591.

Kumar, S.S., Srinivas, P., Negi, P.S. and Bettadaiah, B.K. 2013. Antibacterial and antimutagenic activities of novel zerumbone analogues. *Food Chemistry*. 141(2): 1097–1103.

Matsui, N., Kido, Y., Okada, H., Kubo, M., Nakai, M., Fukuishi, N., Fukuyama, Y. and Akagi, M. 2012. Phenylbutenoid dimers isolated from *Zingiber purpureum* exert neurotrophic effects on cultured neurons and enhance hippocampal neurogenesis in olfactory bulbectomized mice. *Neuroscience Letters*. 513(1): 72–77.

Ozaki, Y., Kawahara, N. and Harada, M. 1991. Anti-inflammatory effect of Zingiber cassumunar Roxb. and its active principles. *Chemical and Pharmaceutical Bulletin*. 39(9): 2353–2356.

Shamoto, T., Matsuo, Y., Shibata, T., Tsuboi, K., Nagasaki, T., Takahashi, H., Funahashi, H., Okada, Y. and Takeyama, H. 2014. Zerumbone inhibits angiogenesis by blocking NF-κB activity in pancreatic cancer. *Pancreas*. 43(3): 396–404.

Sharifah Sakinah, S.A., Tri Handayani, S. and Azimahtol Hawariah, L.P. 2007. Zerumbone induced apoptosis in liver cancer cells via modulation of Bax/Bcl-2 ratio. *Cancer Cell International*. 7(1): 1–11.

Shin, D.S. and Eom, Y.B. 2019. Zerumbone inhibits *Candida albicans* biofilm formation and hyphal growth. *Canadian Journal of Microbiology*. 65(10): 713–721.

Somchit, M.N., Mak, J.H., Bustamam, A.A., Zuraini, A., Arifah, A.K., Adam, Y. and Zakaria, Z.A. 2012. Zerumbone isolated from *Zingiber zerumbet* inhibits inflammation and pain in rats. *Journal of Medicinal Plants Research*. 6(2): 177–180.

Tandirogang, N., Anitasari, S., Arung, E.T., Paramita, S. and Shen, Y.K. 2022. Evaluations of antibacterial properties *of Zingiber purpureum* essential oil against 13 different gram-positive and gram-negative bacteria. *The Indonesian Biomedical Journal*. 14(3): 303–308.

Wang, Y., You, C.X., Yang, K., Wu, Y., Chen, R., Zhang, W.J., Liu, Z.L., Du, S.S., Deng, Z.W., Geng, Z.F. and Han, J. 2015. Bioactivity of essential oil of *Zingiber purpureum* rhizomes and its main compounds against two stored product insects. *Journal of Economic Entomology*. 108(3): 925–932.

Zingiber zerumbet **(L.) Roscoe ex Sm.**

[From the Greek *ziggiberis* = ancient spice known to Arabs and the ancient name of shampoo ginger = *zurunbad*]

Published in *Exotic Botany* 2: 105. 1806

Synonyms: *Amomum sylvestre* (L.) Lam.; *Amomum zerumbet* L.; *Zingiber spurium* J. Koenig
Local names: *limpuyang* (Tindung); *lempoyang* (Bisaya); *benggalai* (Murut)
Common names: bitter ginger, shampoo ginger
Habitat: cultivated
Geographical distribution: from India to the Philippines
Botanical description: this ginger reaches a height of about 1 m. Rhizomes: yellow inside, aromatic. Leaves: simple, alternate. Petioles: tiny. Ligules: entire, 1.5–3.5 cm long, papery. Blades: emerge from culms, lanceolate to oblong, tapering at base, acuminate at apex, 3–8 cm × 15–40 cm. Spikes: somewhat dark red, 10–15 cm long, fleshy, glossy, ovate, on about 10–30 cm long peduncles. Bracts: about 4 cm long. Calyx: white, membranous, tubular, about 1.2–2 cm long, open at one side, and trilobed. Corolla: tubular, 5 cm long, light yellow, membranous. Labellum: suborbicular, about 2.5 cm across, undulate, emarginate at apex. Staminodes: present. Stamens: 1, about 8 mm long, curved like some kind of beak. Capsules: small and globose.
Traditional therapeutic indications: wounds (Murut); fatigue (Tindung)
Pharmacology and phytochemistry: the plant produces the sesquiterpene zerumbone (see *Zingiber purpureum* Roscoe).
Toxicity, side effects, and drug interaction: not known

Comment: The Dusun and Kadazan employ plants of the genus *Zingiber* Martinov (1820) as medicines: "*tolidus*" for toothaches and "*dangalai taragang*" for diarrhea and vomiting.

4.1.5 Superorder Ranunculanae Takht. ex Reveal (1992), the Eudicots

4.1.5.1 Order Ranunculales Juss. ex Bercht. & J. Presl (1820)

4.1.5.1.1 Family Menispermaceae Juss. (1789), the Moonseed Family

The family Menispermaceae consists of about 70 genera and 500 species of climbing plants, often growing from massive tubers. The cross sections of stems show characteristic patterns of bicycle-wheel-like medullary rays. Wood: often bright yellow, bitter. Leaves: simple, alternate, exstipulate. Petioles: often straight, thin, often swollen at both ends. Blades: often cordate and peltate, with secondary nerves emerging from the base. Flowers: tiny. Sepals: 2–10. Petals: 1–6 or none. Stamens: 1–40, short. Carpels: often 1–30, free, sessile, containing each a single ovule. Fruits: drupes. Seeds: horseshoe-shaped or somewhat like bending slugs, verrucose.

***Coscinium fenestratum* (Gaertn.) Colebr.**

[From Greek *koskinion* = little sieve and Latin *fenestratum* = with little windows]

Published in *Transactions of the Linnean Society of London* 13: 65. 1822

Synonyms: *Coccinium usitatum* Pierre; *Coscinium miosepalum* Diels; *Coscinium peltatum* Merr.; *Coscinium usitatum* Pierre; *Coscinium wallichianum* Miers; *Coscinium wightianum* Miers ex Diels; *Pereiria medica* Lindl.

Local name: *babas lingungan* (Murut)

Common names: false calumba, knotted plant, tree turmeric

Habitat: forests

Geographical distribution: from India to Borneo

Botanical description: this woody climbing plant grows up to a length of about 15 m. Bark: ash colored, rough. Wood: bright yellow, bitter, porous.. Stems: terete with bicycle-like medullary rays in transversal section. Leaves: simple, alternate, peltate, exstipulate. Petioles: about 4–7 cm long, thin, straight, swollen at base and apex. Blades: dark green, smooth, glossy above, glaucous beneath, broadly cordate to broadly lanceolate, with up to about 5–7 pairs of secondary nerves emerging from the base, coriaceous, about 5–10 cm × 10–20 cm, cordate or rounded at base, acuminate at apex. Fascicles: cauliflorous, hairy, umbelloid, up to about 5 cm long. Flowers: tiny. Sepals: 3. Petals: 3. Stamens: 6. Drupes: up to about 3 cm across, hairy, globose, light brown, like Annonaceae ripe carpels.

Traditional therapeutic indications: jaundice (Murut)

Pharmacology and phytochemistry: human head and neck cancer cells (HN31) (Potikanond et al. 2015), antiplasmodial (Le Tran et al. 2003), antibacterial (Chomnawang et al. 2009), antidiabetic (Yibchok-anun et al. 2009).

Antibacterial and antiplasmodial protoberberine alkaloids (berberine, jatrorrhirizine) (Pinho et al. 1992; Singburaudom et al. 2015). Berberin is toxic to liver cancer cells (HepG2) (Hwang et al. 2006) and antiviral (Warowicka et al. 2020).

Toxicity, side effects, and drug interaction: aqueous extract of stems administered orally at the dose of 2.5 g/kg/day to rats for 90 days was found to cause no signs of toxicity (Wongcome et al. 2007).

REFERENCES

Chomnawang, M.T., Trinapakul, C. and Gritsanapan, W. 2009. *In vitro* antigonococcal activity of *Coscinium fenestratum* stem extract. *Journal of Ethnopharmacology*. 122(3): 445–449.

Hwang, J.M., Kuo, H.C., Tseng, T.H., Liu, J.Y. and Chu, C.Y. 2006. Berberine induces apoptosis through a mitochondria/caspases pathway in human hepatoma cells. *Archives of Toxicology*. 80: 62–73.

Le Tran, Q., Tezuka, Y., Ueda, J.Y., Nguyen, N.T., Maruyama, Y., Begum, K., Kim, H.S., Wataya, Y., Tran, Q.K. and Kadota, S. 2003. *In vitro* antiplasmodial activity of antimalarial medicinal plants used in Vietnamese traditional medicine. *Journal of Ethnopharmacology*. 86(2–3): 249–252.

Pinho, P.M., Pinto, M.M., Kijjoa, A., Pharadai, K., Díaz, J.G. and Herz, W. 1992. Protoberberine alkaloids from *Coscinium fenestratum*. *Phytochemistry*. 31(4): 1403–1407.

Potikanond, S., Chiranthanut, N., Khonsung, P. and Teekachunhatean, S. 2015. Cytotoxic effect of *Coscinium fenestratum* on human head and neck cancer cell line (HN31). *Evidence-Based Complementary and Alternative Medicine*. 2015: 1–8.

Singburaudom, N. 2015. The alkaloid berberine isolated from *Coscinium fenestratum* is an inhibitor of phytopathogenic fungi. *Journal of Biopesticides*. 8(1): 28.

Warowicka, A., Nawrot, R. and Goździcka-Józefiak, A. 2020. Antiviral activity of berberine. *Archives of Virology*. 165: 1935–1945.

Wongcome, T., Panthong, A., Jesadanont, S., Kanjanapothi, D., Taesotikul, T. and Lertorasertsuke, N. 2007. Hypotensive effect and toxicology of the extract from *Coscinium fenestratum* (Gaertn.) Colebr. *Journal of Ethnopharmacology*. 111(3): 468–475.

Yibchok-anun, S., Jittaprasatsin, W., Somtir, D., Bunlunara, W. and Adisakwattana, S. 2009. Insulin secreting and α-glucosidase inhibitory activity of *Coscinium fenestratum* and postorandial hyperglycemia in normal and diabetic rats. *Journal of Medicinal Plants Research*. 3: 646–651.

Fibraurea tinctoria **Lour.**

[From Latin *fibra* = fiber, *aureus* = golden, and *tinctor* = a dyer]

Published in *Flora Cochinchinensis, denuo in Germania edita* 679. 1793

Synonyms: *Fibraurea laxa* Miers; *Fibraurea trotteri* Watt ex Diels
Local names: *takop, tapa buawang, tapa bohuang, tapa bawang, tonsisilou* (Dusun, Kadazan); *tolungon, war birar* (Murut); *babas* (Lundayeh)
Common name: yellow roots
Habitat: forests
Geographical distribution: India, Bangladesh, Southeast Asia
Botanical description: this woody climbing plant grows up to a length of about 40 m. Bark: dark gray. Stems: ligneous, yielding a white latex once incised. Leaves: simple, spiral, exstipulate. Petioles: about 4–10 cm long, thin, swollen to curved at base, somewhat straight. Blades: almost peltate, elliptic, to ovate, to somewhat tongue-shaped, rounded at base, acuminate at apex, dark green, smooth glossy above, somewhat fleshy, 10–20 cm × 5–15 cm, with about 6–8 pairs of the secondary nerves emerging from the base. Panicles: up to about 30 cm long, axillary or cauliflorous, lax, thin, with numerous fragrant flowers. Sepals: 6, light yellowish, about 4 mm long, oblong. Petals: none. Stamens: 6, with thick filaments, vaguely phallus-shaped. Staminodes: 6. Carpels: 3. Drupes: light yellowish to somewhat orangish, about 2 cm long, ellipsoid, eaten by monkeys, including *Presbitys femoralis* (Figure 4.15).

FIGURE 4.15 *Fibraurea tinctoria* Lour.

Traditional therapeutic indications: fever, headaches (Dusun); hypertension, wounds (Murut); chest pain, jaundice, fatigue, stomach aches (Dusun, Kadazan); malaria (Dusun, Kadazan, Lundayeh, Murut); eczema (Lundayeh)

Pharmacology and phytochemistry: antiplasmodial (Nguyen-Pouplin et al. 2007; Fikriah & Sawitri 2020). The plant yields anti-inflammatory furanoditerpenes (Su et al. 2008) as well as protoberberine alkaloids (berberine, palmatine, and jatrorrhizine) (Su et al. 2007; Purwaningsih et al. 2023) with broad-spectrum antibacterial and antifungal effects (Deng et al. 2012). Berberine decreases blood pressure (Suadoni & Atherton 2021), toxic to human hepatoma (HepG2) (Hwang et al. 2006), and antiviral (Warowicka et al. 2020).

Toxicity, side effects, and drug interaction: not known

REFERENCES

Deng, Y., Zhang, M. and Luo, H. 2012. Identification and antimicrobial activity of two alkaloids from traditional Chinese medicinal plant *Tinospora capillipes*. *Industrial Crops and Products*. 37(1): 298–302.

Fikriah, I. and Sawitri, E. 2020. *In vivo* antimalarial effect of yellow root stem (*Fibraurea tinctoria* Lour) on Plasmodium berghei. *Systematic Reviews in Pharmacy*. 11(6): 380–383.

Hwang, J.M., Kuo, H.C., Tseng, T.H., Liu, J.Y. and Chu, C.Y. 2006. Berberine induces apoptosis through a mitochondria/caspases pathway in human hepatoma cells. *Archives of Toxicology*. 80: 62–73.

Nguyen-Pouplin, J., Tran, H., Tran, H., Phan, T.A., Dolecek, C., Farrar, J., Tran, T.H., Caron, P., Bodo, B. and Grellier, P. 2007. Antimalarial and cytotoxic activities of ethnopharmacologically selected medicinal plants from South Vietnam. *Journal of Ethnopharmacology*. 109(3): 417–427.

Purwaningsih, I., Maksum, I.P., Sumiarsa, D. and Sriwidodo, S. 2023. A review of Fibraurea tinctoria and its component, berberine, as an antidiabetic and antioxidant. *Molecules*. 28(3): 1294.

Su, C.R., Chen, Y.F., Liou, M.J., Tsai, H.Y., Chang, W.S. and Wu, T.S. 2008. Anti-inflammatory activities of furanoditerpenoids and other constituents from *Fibraurea tinctoria*. *Bioorganic & Medicinal Chemistry*. 16(21): 9603–9609.

Su, C.R., Ueng, Y.F., Dung, N.X., Vijaya Bhaskar Reddy, M. and Wu, T.S. 2007. Cytochrome P3A4 inhibitors and other constituents of Fibraurea tinctoria. *Journal of Natural Products*. 70(12): 1930–1933.

Suadoni, M.T. and Atherton, I. 2021. Berberine for the treatment of hypertension: A systematic review. *Complementary Therapies in Clinical Practice*. 42: 101287.

Warowicka, A., Nawrot, R. and Goździcka-Józefiak, A. 2020. Antiviral activity of berberine. *Archives of Virology*. 165: 1935–1945.

***Pycnarrhena tumefacta* Miers**

[Greek *pyknos* = close-packed and *arrhen* = male and Latin *tumefactus* = tumescent]

Published in *Annals and Magazine of Natural History, ser. 3, 20: 12*. 1867

Synonyms: *Cocculus celebicus* Boerl.; *Pycnarrhena balabacensis* Yamamoto; *Pycnarrhena borneensis* Diels; *Pycnarrhena celebica* (Boerl.) Diels; *Pycnarrhena castanopsidifolia* Yamam; *Pycnarrhena merrillii* Diels; *Pycnarrhena membranifolia* Merr.;
Local names: *apa* (Murut); *fatagah* (Lundayeh)
Habitat: forests around Mount Kinabalu
Geographical distribution: Borneo, Indonesia, the Philippines, Papua New Guinea, the Solomon Islands
Botanical description: this shrub reaches a height of about 3 m. Leaves: simple, spiral, exstipulate. Petioles: thin, straight, 1–6 cm long, curved and swollen at apex. Blades: elliptic to oblong, dark green and glossy above, somewhat asymmetrical, rounded at base, acuminate at apex, 3.5–15 cm ×10–35 cm, with 4–9 pairs of secondary nerves. Flower fascicles: cauliflorous, 1–3 cm long, with numerous fragrant and tiny flowers. Sepals: outer 2–4, inner 4–6, hooded, yellowish-green. Petals: 3–6, obovate. Stamens: 6–18 in clusters. Carpels: 3–4. Drupes: yellow to red, globose to obovate, 1.5–3 cm long. Seeds: 1, moon-shaped, verrucose.
Traditional therapeutic indication: pimples (Lundayeh)
Pharmacology and phytochemistry: α-glucosidase inhibition (Ramadhan et al. 2020). Antiplasmodial (Fernandez 2010).
Toxicity, side effects, and drug interaction: not known

Comments:

(i) The Murut use the leaves to flavor dishes, as well as the natives of Sarawak, who call the plant "*tubu*". In East Kalimantan, it is used for diabetes and is locally called "*bekai*".
(ii) The bright yellow wood may have prompted this plant's use against jaundice. Menispermaceae produce alkaloids, often intense yellow in color and with antibacterial activity. This explains the use of these plants for the treatment of wounds, pimples and others bacterial infections.

REFERENCES

Fernandez, L.S. 2010. *Identification of Novel Natural Product Antimalarial Compounds* (PhD thesis, Griffith University).
Ramadhan, R., Phuwapraisirisan, P., Kusuma, I.W. and Amirta, R. 2020. Ethnopharmacological evaluation of selected east Kalimantan flora for diabetes therapy: The isolation of lupane triterpenoids as α-glucosidase inhibitors from ceriops tagal (perr) cb robb. *Rasayan Journal of Chemistry*. 13(3): 1727–1734.

Tinospora crispa **(L.) Hook.f. & Thoms.**

[From Greek *teino* = to stretch, *spora* = seed, and Latin *crispus* = curly]

Published in *Flora Indica: being a systematic account of the plants*. 1: 183. 1855

Synonyms: *Cocculus crispus* DC; *Menispermum crispum* L.; *Tinospora gibbericaulis* Hand.-Mazz.; *Tinospora mastersii* Diels; *Tinospora rumphii* Boerl.; *Tinospora thorelii* Gagnep.

Local names: *patawali* (Bajau); *ubat it mato* (Lundayeh); *sapai* (Dusun, Kadazan)

Common name: Chinese tinospora

Habitat: forests

Geographical distribution: Southeast Asia, South China

Botanical description: this woody climbing plant grows up to a length of about 15 m. Stems: tuberculate, somewhat sappy and glossy, with a membranous bark. Leaves: simple, spiral, exstipulate. Petioles: 5–20 cm long, thin. Blades: broadly ovate, 6–13 cm × 6–15 cm, base cordate, apex acuminate, with 2–3 pairs of secondary nerves emerging from the base. Flower fascicules: thin, cauliflorous, 2–10 cm long. Sepals: 6. Petals: 3–6, yellow, tiny. Stamens: 6. Carpels: 3. Drupes: orange, subglobose, to 2 cm. Seeds: 1, horseshoe-shaped.

Traditional therapeutic indications: sore eyes (Dusun, Kadazan, Lundayeh); hypertension (Bajau, Dusun, Kadazan); malaria (Murut); heart diseases, scabies, wounds, diabetes (Bajau); malaria (Dusun, Murut, Kadazan); insect bites, intestinal worms (Dusun, Kadazan)

Pharmacology and phytochemistry: antiplasmodial (Nguyen-Pouplin et al. 2007; Ramadani et al. 2018), antibacterial, antifungal (Md et al. 2011), antidiabetic (Noor & Ashcroft 1989), toxic to head and neck squamous cell carcinoma (Phienwej et al. 2015).

The plant produces antimicrobial aporphine alkaloids (magnoflorine, *N*-formylannonaine, *N*-formylnornuciferine), as well as the protoberberine alkaloids (dihydrodiscretamine and columbamine) (Choudhary et al. 2010; Yusoff et al. 2014).

Toxicity, side effects, and drug interaction: in volunteers, intake of 250 mg of powder twice a day for 2 months was safe (Sriyapai et al. 2009).

Comment: A plant of the genus *Tinospora* Miers (1851) is used by the Dusun and Kadazan for stomach aches and eye infections (local name: *wakau*).

REFERENCES

Choudhary, M.I., Ismail, M., Ali, Z., Shaari, K., Lajis, N.H. and Rahman, A.U. 2010. Alkaloidal constituents of *Tinospora crispa. Natural Product Communications*. 5: 1747–1750.

Md, H.A., Sm, I.A. and Mohammad, S. 2011. Antimicrobial, cytotoxicity and antioxidant activity of *Tinospora crispa. Journal of Pharmaceutical and Biomedical Sciences (JPBMS)*. 12(12): 1–4.

Nguyen-Pouplin, J., Tran, H., Tran, H., Phan, T.A., Dolecek, C., Farrar, J., Tran, T.H., Caron, P., Bodo, B. and Grellier, P. 2007. Antimalarial and cytotoxic activities of ethnopharmacologically selected medicinal plants from South Vietnam. *Journal of Ethnopharmacology*. 109(3): 417–427.

Noor, H. and Ashcroft, S.J. 1989. Antidiabetic effects of *Tinospora crispa* in rats. *Journal of Ethnopharmacology*. 27(1–2): 149–161.

Phienwej, H., Swasdichira, I.S., Amnuoypol, S., Pavasant, P. and Sumrejkanchanakij, P. 2015. *Tinospora crispa* extract inhibits MMP-13 and migration of head and neck squamous cell carcinoma cell lines. *Asian Pacific Journal of Tropical Biomedicine*. 5(9): 738–743.

Ramadani, A.P., Paloque, L., Belda, H., Tamhid, H.A., Augereau, J.M., Valentin, A., Wijayanti, M.A. and Benoit-Vical, F. 2018. Antiprotozoal properties of Indonesian medicinal plant extracts. *Journal of Herbal Medicine*. 11: 46–52.

Sriyapai, C., Dhumma-upakorn, R., Sangwatanaroj, S., Kongkathip, N. and Krittiyanunt, S. 2009. Hypoglycemic effect of *Tinospora crispa* dry powder in outoatients with metabolic syndrome at king chulalongkorn memorial hospital. *Journal of Health Research*. 23(3): 125–133.

Yusoff, M., Hamid, H. and Houghton, P. 2014. Anticholinesterase inhibitory activity of quaternary alkaloids from *Tinospora crispa. Molecules*. 19(1): 1201–1211.

***Stephania corymbosa* Walp.**

[From Greek *stephanos* = crown and Latin *corymbosa* = with corymb]

Published in *Repert. Bot. Syst.* 1: 96. 1842

Synonyms: *Clypea corymbosa* Bl.; *Stephania borneensis* Yamam.; *Stephania calosepala* Diels; *Stephania cauliflora* Becc.; *Stephania ramosii* Merr.; *Stephania ramuliflora* Miers
Local names: *penaki*, *purut* (Murut)
Habitat: forests, thickets
Distribution: Malaysia, Indonesia, the Philippines
Botanical description: this woody climbing plant grows up to a length of about 10 m. Tubers: massive. Bark: light brown, rough. Leaves: simple, spiral, exstipulate. Petioles: 3–19 cm long, somewhat swollen at base. Blades: broadly lanceolate to triangular, membranous, peltate, 3.5–13 cm × 3–11 cm, somewhat glaucous beneath, dark green and glossy above, truncate at base, acute at apex, with about 3 pairs of secondary nerves emerging from the base. Cymes: axillary or cauliforous, umbelliform, on thin and up to about 10 cm long peduncles. Sepals: 6. Petals: 1–3, papillose. Stamens: fascicled. Drupes: olive-shaped, glossy, turning orange, about 1 cm long.
Traditional therapeutic indications: poison antidote (Murut)
Pharmacology and phytochemistry: alkaloids (protostephanine) (Barbosa-Filho et al. 2000).
Toxicity, side effects, and drug interaction: not known

Comment: Compared with other Asian areas, the medicinal plants of Sabah are often used as poison antidotes and food The Murut, who might be the oldest ethnic group of Sabah, have deep knowledge about poisonous plants and their antidotes.

REFERENCE

Barbosa-Filho, J., da-Cunha, E.V.L. and Gray, A.I. 2000. Alkaloids of the menispermaceae. *The Alkaloids: Chemistry and Biology*. 54: 1–190.

4.1.6 Superorder Proteanae Takht. (1967), the Eudicots

4.1.6.1 Order Proteales Juss. ex Bercht. & J. Presl (1820)

4.1.6.1.1 Family Proteaceae Juss. (1789), the Protea Family

The family Proteaceae consists of about 80 genera and 1,700 species of shrubs or trees. Leaves: simple or pinnate, alternate, exstipulate. Blades: coriaceous. Inflorescences: spikes. Sepals: 4. Stamens: 4, free. Carpels: 1, forming a unilocular ovary containing 3–many ovules. Styles: 1. Stigma: clavate. Fruits: follicles, achenes, or drupes. Seeds: 1 or 2, winged or not.

Helicia serrata **Bl.**

[From Greek *helikoiedes* = helix and Latin *serrata* = serrate]

Published in *Annales des Sciences Naturelles (Paris)* 2(1): 215. 1834

Synonym: *Roupala serrata* R.Br.
Local names: *andaun motuka* (Murut), *lepu punchu* (Dusun)
Habitat: forests
Geographical distribution: from the Andaman Islands to Indonesia
Botanical description: this tree reaches a height of about 30 m. Stems: hairy, stout. Leaves: simple, spiral, exstipulate. Petioles: 1.5–2 cm long, stout. Blades: elliptic, lanceolate to obovate, base attenuate, apex acuminate, coriaceous, serrate, 5–8.5 cm × 12–20 cm, with 8–10 pairs of secondary nerves. Spikes: axillary or cauliflorous, 9–15 cm long. Perianth: about 8 mm long, tubular, thin, whitish, 4-lobed, the lobes revolute. Disc: present, in the form of glands. Stamens: 4. Ovary: ovoid, hairy. Styles: 1, thin. Stigma: clavate. Drupes: pod-like, greenish, 3–4 cm long (Figure 4.16).

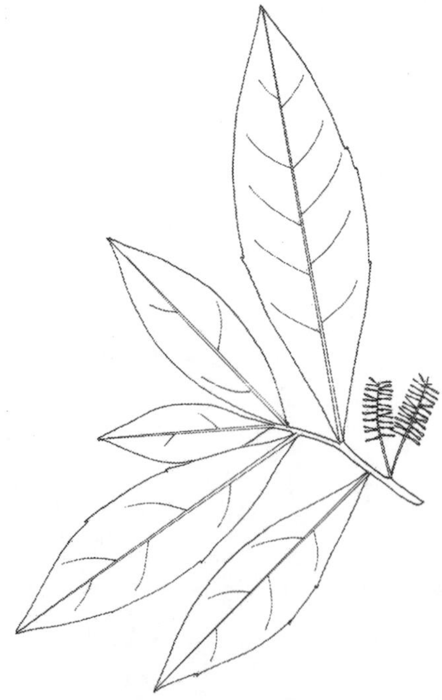

FIGURE 4.16 *Helicia serrata* Bl.

Traditional therapeutic indications: medicinal (Murut)
Pharmacology and phytochemistry: not known
Toxicity, side effects, and drug interaction: not known

Comments:

(i) The young leaves are used as food in Java.
(ii) A plant of the genus *Heliciopsis* Sleumer (1955) is used by the Dusun and Kadazan for coughs, fatigue, and skin diseases.

4.1.7 Superorder Dillenianae Takht. ex Doweld (2001), the Core Eudicots

4.1.7.1 Order Dilleniales DC. ex Bercht. & J. Presl (1820)

4.1.7.1.1 Family Dilleniaceae Salisb. (1807), the Elephant Apple Family

The family Dilleniaceae consists of about 12 genera and 400 species of trees, shrubs, and woody climbing plants. Leaves: simple, spiral, exstipulate. Blades: coriaceous, serrate, marked by clearly visible straight secondary nerves. Inflorescences: cymes, racemes, or solitary. Sepals: 4–5, fleshy, often hooded, imbricate. Petals: 3–5, membranous, caducous. Stamens: numerous. Carpels: 2–7, each with 1–100 ovules. Fruits: follicles, berries, capsules enclosed in accrescent fleshy sepals. Seeds: 1 to many, often arillate

***Dillenia excelsa* (Jack) Martelli**

[After the 17th century German botanist Johann Jakob Dillen and Latin *excelsa* = lofty]

Published in *Malesia Raccolta*. . . 3(3): 164. 1887

Synonym: *Wormia excelsa* Jack
Local names: *doingins* (Dusun, Kadazan); *pampan kazu* (Rungus)
Habitat: forests near streams
Geographical distribution: Thailand, Malaysia, Indonesia, the Philippines
Botanical description: this timber tree reaches a height of about 40 m. Trunk: straight. Bark: grayish red, scaly. Leaves: simple, spiral, exstipulate. Petioles: stout, 2–5 cm long. Blades: oblong to elliptic, serrate, glossy, coriaceous, 7–10 cm × 15–30 cm, with 10–13 pairs of clearly visible secondary nerves, acute and somewhat asymmetrical at base, acuminate to rounded at apex. Racemes: few-flowered, showy, 12–20 cm across, terminal. Sepals: 5, rounded, imbricate, about 2 cm long, fleshy, dull green, coriaceous. Petals: 5, yellow, obovate to oblong, somewhat spoon-shaped, membranous at the margin, 4–5 cm long. Stamens: numerous, about 2 mm long, linear. Styles: 8–12, pinkish, ensiform, recurved. Berries: globose, heavy, about 20 cm across, with persistent sepals, somewhat glossy to dull green.
Traditional therapeutic indications: bloody stools, chest pain (Dusun, Kadazan); foul body odor (Rungus)
Pharmacology and phytochemistry: antibacterial (Abdulah et al. 2017), antidiabetic (Matusin et al. 2021).
Toxicity, side effects, and drug interaction: not known

REFERENCES

Abdulah, R., Milanda, T., Sugijanto, M., Barliana, M.I., Diantini, A., Supratman, U. and Subarnas, A. 2017. Antibacterial properties of selected plants consumed by primates against *Escherichia coli* and *Bacillus subtilis*. *The Southeast Asian Journal of Tropical Medicine and Public Health*. 48: 109–116.

Matusin, A.H.A., Abd Ghani, N.I. and Ahmad, N. 2021. Pancreatic islet regenerative capability of *Dillenia excelsa* in alloxan-induced diabetic rats. *Journal of Applied Pharmaceutical Science*. 11(3): 121–129.

Dillenia grandifolia Wall. ex Hook.f. & Thomson

[After the 17th century German botanist Johann Jakob Dillen and Latin *grandifolia* = large leaf]

Published in *Flora Indica; or descriptions of Indian Plants* 1: 71. 1855

Synonyms: *Dillenia crassisepala* Martelli; *Dillenia eximia* Miq.; *Dillenia rhizophora* Boerl. & Koord.-Schum.; *Dillenia scortechinii* (King) Ridl.; *Wormia scortechinii* King
Local names: *dudungin, tembakau* (Murut)
Habitat: forests
Geographical distribution: from Thailand to Borneo
Botanical description: this timber tree reaches a height of about 40 m. Trunk: straight. Bark: grayish red, scaly. Leaves: simple, spiral, exstipulate. Petioles: stout, 2.5–5 cm long, hairy. Blades: elliptic, coriaceous, 15–30 cm × 18–60 cm, with 11–37 pairs of clearly visible secondary nerves, acute and somewhat asymmetrical at base, somewhat wavy and dentate to entire, hairy beneath, acute to rounded at apex. Racemes: terminal, pubescent, few-flowered, the flowers about 15 cm across. Sepals: 5, elliptic, about 3 cm long, thick. Petals: 5, bright yellow, obovate to oblong, 5–6 cm long, somewhat membranous at margin. Stamens: numerous, up to 2 cm long. Styles: 8–12, yellowish, linear. Berries: globose with persistent fleshy sepals, about 6 cm across, yellowish. Seeds: black.
Traditional therapeutic indications: stomach aches (Murut)
Pharmacology and phytochemistry: not known
Toxicity, side effects, and drug interaction: not known

Dillenia indica L.

[After the 17th century German botanist Johann Jakob Dillen and Latin *indica* = from India]

Published in *Species Plantarum* 1: 535. 1753

Synonym: *Dillenia speciosa* Thunb.
Local names: *morotud, pampan* (Rungus)
Common name: elephant apple
Habitat: forests near streams and rivers, cultivated
Geographical distribution: from India to the Philippines
Botanical description: this magnificent tree reaches a height of about 40 m. Trunk: straight. Bark: reddish-brown, flaky. Leaves: simple, spiral, exstipulate. Petioles: narrowly winged, stout, 2–6 cm long. Blades: obovate-oblong, coriaceous, thin, 15–40 cm × 7–20 cm, with 30–40 pairs of clearly visible secondary nerves, tapering to rounded at base, serrate, glossy, acuminate to rounded at apex. Flowers: showy, 12–20 cm across, terminal, solitary. Sepals: 5, rounded, imbricate, 4–6 cm across, fleshy, dull green, coriaceous. Petals: 5, pure white, obovate-oblong, membranous, 7–9 cm long. Stamens: numerous. Styles: up to 20, linear, lanceolate, pure white. Berries: globose, heavy, massive, 10–25 cm across, with persistent sepals, somewhat glossy to dull green.
Traditional therapeutic indications: medicinal (Murut); mouthwash, toothaches (Rungus)
Pharmacology and phytochemistry: antibacterial and antifungal (Alam et al. 2011), anti-inflammatory (Yeshwante et al. 2009), anti-leukemic (Kumar et al. 2010), antidiabetic and hypolipidemic (Kumar et al. 2011), antiviral (Goswami et al. 2016). The plant produces proanthocyanidins, condensed tannins (Fu et al. 2015), as well as triterpenes (betulinic acid) (Ghosh et al. 2014).
Toxicity, side effects, and drug interaction: not known

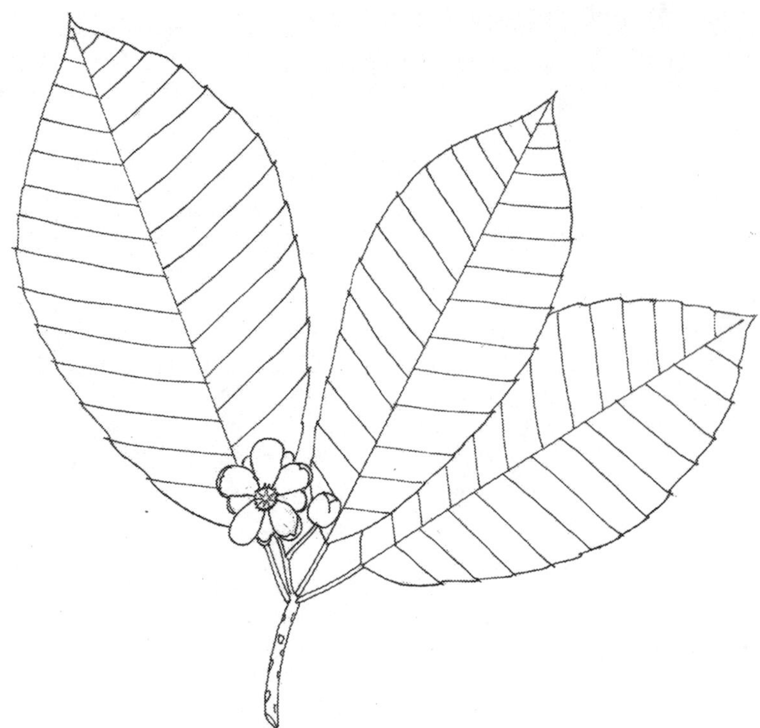

FIGURE 4.17 *Dillenia sumatrana* Miq

Comments:

(i) A plant of the genus *Dillenia* L. (1753) is used by the Rungus as styptic (local name: *rungin*).
(ii) *Dillenia suffruticosa* (Griff. ex Hook. f. & Thomson) Martelli is used by the Dusun as styptic (local name: *simpur*).
(iii) *Dillenia sumatrana* Miq. (Figure 4.17) is used by the Dusun for rheumatisms.

REFERENCES

Alam, M.B., Chowdhury, N.S., Mazumder, M.E.H. and Haque, M.E. 2011. Antimicrobial and toxicity study of different fractions of *Dillenia indica* Linn. bark extract. *International Journal of Pharmaceutical Sciences and Research*. 2(4): 860.

Fu, C., Yang, D., Peh, W.Y.E., Lai, S., Feng, X. and Yang, H. 2015. Structure and antioxidant activities of proanthocyanidins from elephant apple (*Dillenia indica* Linn.). *Journal of Food Science*. 80(10): 2191–2199.

Ghosh, P.S., Sarma, I.S., Sato, N., Harigaya, Y. and Dinda, B. 2014. Lupane-triterpenoids from stem bark of *Dillenia indica*. *Indian Journal of Chemistry*. 53B: 1284–1287.

Goswami, D., Mukherjee, P.K., Kar, A., Ojha, D., Roy, S. and Chattopadhyay, D. 2016. Screening of ethnomedicinal plants of diverse culture for antiviral potentials. *Indian Journal of Traditional Knowledge*. 15(3): 474–481.

Kumar, D., Mallick, S., Vedasiromoni, J.R. and Pal, B.C. 2010. Anti-leukemic activity of *Dillenia indica* L. fruit extract and quantification of betulinic acid by HPLC. *Phytomedicine*. 17(6): 431–435.

Kumar, S., Kumar, V. and Prakash, O. 2011. Antidiabetic, hypolipidemic and histopathological analysis of *Dillenia indica* (L.) leaves extract on alloxan induced diabetic rats. *Asian Pacific Journal of Tropical Medicine*. 4(5): 347–352.

Yeshwante, S.B., Juvekar, A.R., Nagmoti, D.M., Wankhede, S.S., Shah, A.S., Pimprikar, R.B. and Saindane, D.S. 2009. Anti-inflammatory activity of methanolic extracts of *Dillenia indica* L. leaves. *Journal of Young Pharmacists*. 1(1): 63.

Tetracera akara **Merr.**

[From Latin *tetracera* = 4 horns and from the Malay *akar* = root]

Published in *Philippine Journal of Science* 19: 366. 1921

Synonyms: *Calophyllum akara* Burm.f.; *Tetracera axillaris* Martelli

Tetracera rheedii DC.; *Tetracera sericea* Bl.; *Tetracera sylvestris* Ridl.
Local names: *daun ampelas, tampan kuning* (Bajau, Dusun, Kadazan)
Habitat: forests
Geographical distribution: from India to the Philippines
Botanical description: this climbing plant grows up to a length of about 25 m. Stems: bright brown, hairy at apex. Leaves: simple, spiral, exstipulate. Petioles: 5–7 mm long. Blades: oblong to lanceolate, 1.5–10.5 cm × 5–22 cm, acute at base, acuminate at apex, dull green above, rough or not, entire or serrate, with 6–8 pairs of clearly visible secondary nerves. Panicles: axillary or terminal, up to about 8 cm long. Sepals: 4, hairy, about 8 mm long. Petals: 4, pure white, spathulate, about 1.2 cm long. Stamens: numerous, about 8 mm long, yellowish. Carpels: 3, free. Follicles: somewhat globose, about 1 cm across, horned at apex, on persistent calyx. Seeds: ovoid, black, glossy, about 4 mm long, arillate.
Traditional therapeutic indications: medicinal.
Pharmacology and phytochemistry: hepatoprotective (Nair et al. 2019).
Toxicity, side effects, and drug interaction: the median lethal dose (LD_{50}) of ethanol extract administered to rats was above 5 g/kg (Nair et al. 2017).

REFERENCES

Nair, R.R., Suja, S.R., Aneeshkumar, A.L., Vilash, V., Kumar, B.B., Rajasekharan, S. and Latha, P.G. 2017. Evaluation of acute and sub-acute oral toxicity of ethanolic root extract of Tetracera akara (Burm.f.) Merr., an ethnomedicinal plant used by the Kani tribe of Kerala. *Journal of Traditional and Folk Practices*. 5(2).

Nair, R.R., Suja, S.R., Vilash, V., Aneeshkumar, A.L., Rajasekharan, S. and Kumar, B.B. 2019. Protective effect of ethanol extract of roots of tetracera akara (Burm.f.) Merr., on carbon tetrachloride-induced hepatic injury in wistar rats. *Indian Journal of Pharmaceutical Sciences*. 81(5).

***Tetracera indica* (Christm. & Panz.) Merr.**

[From Latin *tetracera* = 4 horns and *indica* = from India]

Published in *An Interpretation of Rumphius's Herbarium Amboinense* 367. 1917

Synonyms: *Assa indica* Christm. & Panz; *Tetracera assa* DC.; *Tetracera laevis* Vahl
Local names: *pampan* (Dusun); *mempelas* (Bajau)
Habitat: at the edges of forests, roadsides
Geographical distribution: India, Bangladesh, Myanmar, Cambodia, Laos, Vietnam, Thailand, Malaysia, Indonesia, Borneo
Botanical description: this climbing plant grows up to a length of about 2 m. Stems: purplish and smooth when young. Leaves: simple, spiral, somewhat in fascicles, exstipulate. Petioles: 6 mm–1 cm long, channeled. Blades: ovate to oblong, 3–5 cm × 6–10 cm, acute at base, obtuse to acute at apex, rough or not, entire or serrate, dark green, sometimes glossy above, with about 6–9 pairs of secondary nerves. Panicles: terminal, up to 8 cm long. Sepals: 4, about 1 cm long, glabrous, hooded, glossy. Petals: 4, reddish to pure white, spathulate to wedge-shaped, bilobed, about 1–1.5 cm long. Stamens: numerous, light red. Carpels: 3–4, free. Follicles: 3, somewhat globose, glossy, on a persistent calyx, about 1 cm across, burgundy, dehiscent, horned at apex. Seeds: black, glossy, ovoid, 4 mm long, in a bright red aril.
Traditional therapeutic indications: refresher, anti-inflammatory, to increase appetite, headaches (Dusun); coughs (Dusun, Kadazan)
Pharmacology and phytochemistry: antidiabetic (Qamar et al. 2012).
Sulfated flavones (Alhassan et al. 2019). Flavones (quercetin, norwogonin, techtochrysin, wogonin) (Harrison et al. 1994; Hasan et al. 2017). Wogonin is anti-inflammatory (Chi et al. 2003; Lee & Park 2015), vasodilatory (Qu et al. 2015), and protects mice against coughs induced by *Mycoplasma pneumoniae* (Liang et al. 2022).
Toxicity, side effects, and drug interaction: the median lethal dose (LD_{50}) of aqueous extract of leaves administered to rats was above 5 g/kg (Ahmed et al. 2012).

Comment: The word "*mempelas*" means sandpaper because the leaves are used for this purpose by natives.

REFERENCES

Ahmed, Q.U., Dogarai, B.B., Amiroudine, M.Z.A.M., Taher, M., Latip, J., Umar, A. and Muhammad, B.Y. 2012. Antidiabetic activity of the leaves of Tetracera indica Merr. (Dilleniaceae) *in vivo* and *in vitro*. *Journal of Medicinal Plants Research*. 6(49): 5912–5922.

Alhassan, A.M., Ahmed, Q.U., Latip, J. and Shah, S.A.A. 2019. A new sulphated flavone and other phytoconstituents from the leaves of Tetracera indica Merr. and their alpha-glucosidase inhibitory activity. *Natural Product Research*. 33(1): 1–8.

Chi, Y.S., Lim, H., Park, H. and Kim, H.P. 2003. Effects of wogonin, a plant flavone from Scutellaria radix, on skin inflammation: *In vivo* regulation of inflammation-associated gene expression. *Biochemical Pharmacology*. 66(7): 1271–1278.

Harrison, L.J., Sia, G.L. and Sim, K.Y. 1994. 5, 7-Dihydroxy-8-methoxyflavone from Tetracera indica. *Planta Medica*. 60(5): 493–494.

Hasan, M.M., Ahmed, Q.U., Soad, S.Z.M., Latip, J., Taher, M., Syafiq, T.M.F., Sarian, M.N., Alhassan, A.M. and Zakaria, Z.A. 2017. Flavonoids from Tetracera indica Merr. induce adipogenesis and exert glucose uptake activities in 3T3-L1 adipocyte cells. *BMC Complementary and Alternative Medicine*. 17: 1–14.

Lee, J.Y. and Park, W. 2015. Anti-inflammatory effect of wogonin on RAW 264.7 mouse macrophages induced with polyinosinic-polycytidylic acid. *Molecules*. 20(4): 6888–6900.

Liang, M., Meng, Y., Wang, X., Wang, L., Tang, G. and Wang, W. 2022. The effectiveness of wogonin on treating cough mice with mycoplasma pneumoniae infection. *Frontiers in Molecular Biosciences*. 9: 803842.

Qamar, U.A., Bashar, B.D., Mohamed, Z.A.M.A., Muhammad, T., Jalifah, L., Abdulrashid, U. and Bala, Y.M. 2012. Antidiabetic activity of the leaves of Tetracera indica Merr. (Dilleniaceae) *in vivo* and *in vitro*. *Journal of Medicinal Plants Research*. 6(49): 5912–5922.

Qu, J.T., Zhang, D.X., Liu, F., Mao, H.P., Ma, Y.K., Yang, Y., Li, C.X., Qiu, L.Z., Geng, X., Zhang, J.M. and Gao, X.M. 2015. Vasodilatory effect of wogonin on the rat aorta and its mechanism study. *Biological and Pharmaceutical Bulletin*. 38(12): 1873–1878.

Tetracera scandens **(L.) Merr.**

[From Latin *tetracera* = 4 horns and *scandens* = climbing]

Published in *An Interpretation of Rumphius's Herbarium Amboinense* 365. 1917

Synonyms: *Delima scandens* (L.) Burkill; *Tetracera loureiroi* (Finet & Gagnep.); Pierre ex Craib *Tragia scandens* L.

Local names: *agupit, kerub kerub, pan pan* (Bajau); *riyop* (Lundayeh); *akar ampalas* (Bajau); *papan* (Sungai); *tambar* (Dusun)

Habitat: roadsides, at the edges of forests

Geographical distribution: from India to the Philippines

Botanical description: this climbing plant grows up to a length of about 30 m. Stems: smooth and purplish at apex. Leaves: simple, spiral, somewhat in fascicles, exstipulate. Petioles: 6 mm–1.2 cm long, channeled. Blades: oblong to obovate, 3–7 cm × 6–15 cm, acute to tapering at base, serrate, sometimes dark glossy above, with about 15–18 pairs of showy secondary nerves sunken above. Panicles: terminal, many-flowered, up to about 20 cm long. Sepals: 4, about 3 mm long, ovate. Petals: 2–3, pure white, caducous, spathulate, about 4 mm long. Stamens: numerous, white. Carpels: 2. Follicles: 1, ovoid to pear-shaped, burgundy, somewhat hairy, about 1 cm long, dehiscent, horned at apex, on persistent calyx. Seeds: 1–2, arillate, about 4 mm long.

Traditional therapeutic indication: coughs (Dusun)

Pharmacology and phytochemistry: hypoglycemic (Umar et al. 2010), antiviral (HIV) (Kwon et al. 2012), active against *Babesia gibsoni* (Matsuura et al. 2004).

Triterpenes (Nguyen & Nguyen 2013). Hypoglycemic prenylated isoflavones (3′,5′-diprenylgenistein) (Chung-Hee et al. 2008; Lee et al. 2009).

Toxicity, side effects, and drug interaction: not known

Comments:

(i) The young leaves are used as food by the Lundayeh.
(ii) *Tetracera macrophylla* Hook.f. & Thomson is used by the Dusun (local: *pampan indu*) for diarrhea.
(iii) *Tetracera fagifolia* Bl. Thomson is used by the Dusun for diarrhea and sore eyes (local names: *pampan*, *pampan mianai*).

REFERENCES

Chung-Hee, K.I.M., Myung-Sun, L.E.E., Manh, H.D., Mee-ree, K.I.M., Bo-Yeon, K.I.M. and Seog, A.J. 2008. 3',5'-Diprenylgenistein isolated from *Tetracera scandens* enhances muscle glucose uptake via AMP-activated Kinase. 추계총회 및 학술대회: 209–209.

Kwon, H.S., Park, J., Kim, J.H. and You, J.C. 2012. Identification of anti-HIV and anti-reverse transcriptase activity from *Tetracera scandens*. *BMB Reports*. 45(3): 165–170.

Lee, M.S., Kim, C.H., Hoang, D.M., Kim, B.Y., Sohn, C.B., Kim, M.R. and Ahn, J.S. 2009. Genistein-derivatives from *Tetracera scandens* stimulate glucose-uptake in L6 myotubes. *Biological and Pharmaceutical Bulletin*. 32(3): 504–508.

Matsuura, H., Yamasaki, M., Yamato, O., Maede, Y., Katakura, K., Suzuki, M. and Yoshihara, T. 2004. Effects of central Kalimantan plant extracts on intraerythrocytic *Babesia gibsoni* in culture. *Journal of Veterinary Medical Science*. 66(7): 871–874.

Nguyen, M.T.T. and Nguyen, N.T. 2013. A new lupane triterpene from *Tetracera scandens* L., xanthine oxidase inhibitor. *Natural Product Research*. 27(1): 61–67.

Umar, A., Ahmed, Q.U., Muhammad, B.Y., Dogarai, B.B.S. and Soad, S.Z.B.M. 2010. Anti-hyperglycemic activity of the leaves of *Tetracera scandens* Linn. Merr. (Dilleniaceae) in alloxan-induced diabetic rats. *Journal of Ethnopharmacology*. 131(1): 140–145.

4.1.8 Superorder Myrothamnanae Takht. (1997), the Core Eudicot

4.1.8.1 Order Saxifragales Bercht. & J. Presl (1820)

4.1.8.1.1 Family Crassulaceae J.St.-Hil. (1805), the Stonecrop Family

The family Crassulaceae consists of about 40 genera and 1,500 species of herbaceous plants or shrubs, which are often fleshy and growing among stones in dry and sunny places. Leaves: simple or pinnate, alternate, opposite, or verticillate, exstipulate. Inflorescences: terminal or axillary cymes, spikes, or solitary. Sepals: 4–6, free or connate. Corolla: 4–6 petals or tubular. Stamens: 3–30. Carpels: 4–6, free, each with 5–50 ovules. Fruits: follicles.

***Kalanchoe pinnata* (Lam.) Pers.**

[The word *kalanchoe* might come from an ancient Chinese name for a plant of the genus *Kalanchoe* Adans and Latin *pinnata* = pinnate]

Published in *Synopsis Plantarum* 1: 446. 1805

Synonyms: *Bryophyllum calycinum* Salisb.; *Bryophyllum germinans* Blanco; *Bryophyllum pinnatum* (Lam.) Kurz; *Calanchoe pinnata* Pers.; *Cotyledon calycina* Roth; *Cotyledon calyculata* Solander; *Cotyledon pinnata* Lam.; *Cotyledon rhizophilla* Roxb.; *Crassula pinnata* L. f.; *Crassuvia floripenula* Comm.; *Kalanchoe calycina* Salisb.; *Sedum madagascariense* Clus.; *Verea pinnata* (Lam.) Spreng.

Local names: *dingan dingan*, *kapal* (Dusun); *sedingin* (Bajau), *setawar* (Bajau, Kedayan); *tanom tombiog* (Murut)

Common name: air plant

Habitat: dry and stony soils, cultivated

Geographical distribution: tropical

Botanical description: this herbaceous plant reaches a height of about 1.5 m and has an aura of dryness. Stems: terete, fleshy, purplish. Leaves: imparipinnate, decussate, exstipulate. Blades: 10–30 cm long. Petiolules: 2–4 cm long. Folioles: 3–5, oblong to elliptic, 5–20 cm × 2.5–12 cm, crenate, base rounded, fleshy, dull green, obtuse at apex. Panicles: axillary or terminal, 10–40 cm long, many-flowered. Flowers: pendulous and showy. Calyx: tubular, cylindrical, purplish or light green, 2–4 cm long, 4-lobed, the lobes triangular. Corolla: of a heavenly light purple, lantern-shaped, up to 5 cm long, 4-lobed, the lobes ovate-lanceolate. Stamens: 8. Nectar scales: present. Ovary: ovoid, tiny. Styles: thin. Follicles: 4, about 1.5 cm long, included in remnant calyx and corolla. Seeds: numerous, tiny.

Traditional therapeutic indications: headaches (Dusun, Kadazan), wounds, cuts (Murut); "hot body" (Kedayan); stomach aches, fever (Bajau, Kedayan)

Pharmacology and phytochemistry: this plant has been the subject of numerous pharmacological studies (Subrata et al. 2011), which have highlighted, among other things, antilithiatic Sohgaura et al. 2018), gastroprotective (de Araújo et al. 2018), hepatoprotective (Yadav & Dixit 2003), anticonvulsant (Mora-Perez & Hernández-Medel 2016), hypotensive (Bopda et al. 2014), antibacterial (Romulo et al. 2018), analgesic, anti-inflammatory (Matthew et al. 2013), and antipyretic activity (Olajide et al. 1998).

Antibacterial and antifungal flavone glycosides (kaempferol 3,7-dirhamnoside, kaempferol 3-*O*-α-L-(4-acetyl) rhamnopyranoside-7- *O*-α -L-rhamnopyranoside, afzelin, α-rhamnoisorobin) (Tatsimo et al. 2012). Leishmanicidal flavone glycoside (quercitrin) (Muzitano et al. 2006). Antiviral lignans and phenol glycosides (coronavirus, herpes simplex virus, vaccinia virus) (Schwarz et al. 2014; Cryer et al. 2017). Bufadienolides (bryophillin C) (Supratman et al. 2000).

Toxicity, side effects, and drug interaction: the median lethal dose (LD_{50}) of ethanol extract of leaves administered to rats was above 2 g/kg (Saravanan et al. 2021).

REFERENCES

Bopda, O.S.M., Longo, F., Bella, T.N., Edzah, P.M.O., Taïwe, G.S., Bilanda, D.C., Tom, E.N.L., Kamtchouing, P. and Dimo, T. 2014. Hypotensive activities of the aqueous extract of *Kalanchoe pinnata* (Crassulaceae) in high salt-loaded rats. *Journal of Ethnopharmacology*. 153(2): 400–407.

Cryer, M., Lane, K., Greer, M., Cates, R., Burt, S., Andrus, M., Zou, J., Rogers, P., Hansen, M.D., Burgado, J. and Satheshkumar, P.S. 2017. Isolation and identification of compounds from *Kalanchoe pinnata* having human alphaherpesvirus and vaccinia virus antiviral activity. *Pharmaceutical Biology*. 55(1): 1586–1591.

de Araújo, E.R.D., Guerra, G.C.B., de Souza Araújo, D.F., de Araújo, A.A., Fernandes, J.M., de Araújo Júnior, R.F., da Silva, V.C., de Carvalho, T.G., de Santis Ferreira, L. Zucolotto, S.M. 2018. Gastroprotective and antioxidant activity of *Kalanchoe brasiliensis* and *Kalanchoe pinnata* leaf juices against indomethacin and ethanol-induced gastric lesions in rats. *International Journal of Molecular Sciences*. 19: 1265.

Matthew, S., Jain, A.K., James, M., Matthew, C. and Bhowmik, D. 2013. Analgesic and anti-inflammatory activity of *Kalanchoe pinnata* (lam.) pers. *Journal of Medicinal Plants*. 1(2): 24–28.

Mora-Perez, A. and Hernández-Medel, M.D.R. 2016. Anticonvulsant activity of methanolic extract from *Kalanchoe pinnata* (Lam.) stems and roots in mice: A comparison to diazepam. *Neurología (English Edition)*. 31(3): 161–168.

Muzitano, M.F., Cruz, E.A., de Almeida, A.P., Da Silva, S.A., Kaiser, C.R., Guette, C., Rossi-Bergmann, B. and Costa, S.S. 2006. Quercitrin: An leishmanicidalflavonoid glycoside from *Kalanchoe pinnata*. *Planta Medica*. 72(1): 81–83.

Olajide, O.A., Awe, S.O. and Makinde, J.M. 1998. Analgesic, anti-inflammatory and antipyretic effects of *Bryophyllum pinnatum*. *Fitoterapia*. 69(3): 249–252.

Romulo, A., Zuhud, E.A., Rondevaldova, J. and Kokoska, L. 2018. Screening of *in vitro* antimicrobial activity of plants used in traditional Indonesian medicine. *Pharmaceutical Biology*. 56(1): 287–293.

Saravanan, V., Murugan, S.S., Navaneetha Krishnan, K.R., Mohana, N., Sakthive, K. and Sathya, T.N. 2021. Toxicological assessment of ethanolic leaves extract of *Kalanchoe pinnata* in Rats. *Indian Journal of Forensic Medicine & Toxicology*. 15(2): 1148.

Schwarz, S., Sauter, D., Wang, K., Zhang, R., Sun, B., Karioti, A., Bilia, A.R., Efferth, T. and Schwarz, W. 2014. Kaempferol derivatives as antiviral drugs against the 3a channel protein of coronavirus. *Planta Medica*. 80(02/03): 177–182.

Sohgaura, A.K., Bigoniya, P. and Shrivastava, B., 2018. In vitro antilithiatic potential of Kalanchoe pinnata, Emblica officinalis, Bambusa nutans, and Cynodon dactylon. *Journal of Pharmacy and Bioallied Sciences*. 10(2): 83–89.

Subrata, K.B., Anusua, C., Joysree, D., Riaz, U. and Md, S.R. 2011. Literature review on pharmacological potentials of *Kalanchoe pinnata* (Crassulaceae). *African Journal of Pharmacy and Pharmacology*. 5(10): 1258–1262.

Supratman, U., Fujita, T., Akiyama, K. and Hayashi, H. 2000. New insecticidal bufadienolide, bryophyllin C, from *Kalanchoe pinnata*. *Bioscience, Biotechnology, and Biochemistry*. 64(6): 1310–1312.

Tatsimo, S.J.N., de Dieu Tamokou, J., Havyarimana, L., Csupor, D., Forgo, P., Hohmann, J., Kuiate, J.R. and Tane, P. 2012. Antimicrobial and antioxidant activity of kaempferol rhamnoside derivatives from *Bryophyllum pinnatum*. *BMC Research Notes*. 5(1): 158.

Yadav, N.P. and Dixit, V.K. 2003. Hepatoprotective activity of leaves of *Kalanchoe pinnata* Pers. *Journal of Ethnopharmacology*. 86(2–3): 197–202.

4.1.9 Superorder Rosanae Takht. (1967), the Rosids

4.1.9.1 Order Vitales Juss. ex Bercht. & J. Presl (1820)

4.1.9.1.1 Family Vitaceae Juss. (1789), the Grape Family

The family Vitaceae consists of about 15 genera and 800 species of woody climbing plants. Leaves: simple or compound, spiral, sometimes stipulate. Inflorescences: panicles, cymes, or spikes, which are often leaf-opposite, pseudoterminal, or axillary. Flowers: tiny. Calyx: 4- or 5-lobed. Petals: 4–5. Disc: present. Stamens: 4–5. Carpels: 2–8 forming a 2–8 locular ovary, each locule with 1 ovule. Fruits: drupes or berries. Seeds: up to 4.

Leea indica **(Burm.f.) Merr.**

[After the 18th century nurseryman James Lee and Latin *indica* = from India]

Published in *Philippine Journal of Science* 14(2): 245. 1919

Synonyms: *Aquilicia sambucina* L.; *Leea sambucina* Willd. *Staphylea indica* Burm.f.;
Common name: bandicoot berry
Habitat: forests
Geographical distribution: from India to Pacific Islands
Botanical description: this shrub reaches a height of about 5 m. Stems: terete, longitudinally and deeply striated. Leaves: pinnate, alternate, stipulate. Stipules: broadly obovate, 2.5–4.5 cm long, rounded at apex. Petioles: 13–23 cm long. Petiolules 0.2–5 cm long. Folioles: elliptic to lanceolate, 5–8 cm × 6–32 cm, rounded at base, glossy, crenate to serrate, apex acuminate, with 6–11 pairs of clearly visible secondary nerves. Cymes: opposite to leaves. Calyx: tubular, somewhat 5-lobed. Petals: 5, somewhat fleshy, convolute, elliptic, about 2 mm long, white. Stamens: 5, about 2 mm long, within a tube. Ovary: globose. Styles: tiny, white. Stigma: expanded. Berries: somewhat globose to oblate, about 1 cm across, dull black. Seeds: 4–6.
Traditional therapeutic indications: sprains (Dusun, Kadazan, Murut); fatigue, headaches (Dusun, Kadazan)
Pharmacology and phytochemistry: toxic to prostate cancer cells (Ghagane et al. 2017) and Ehrlich ascites carcinoma cells (Raihan et al. 2012), hypoglycemic, hypolipidemic (Dalu et al. 2014), anti-inflammatory (Sakib et al. 2021), analgesic (Emran et al. 2012), antifungal (Ramesh et al. 2015).
Toxicity, side effects, and drug interaction: not known

REFERENCES

Dalu, D., Duggirala, S. and Akarapu, S. 2014. Anti-hyperglycemic and hypolipidemic activity of *Leea indica*. *International Journal of Bioassays*. 3(7): 3155–3159.

Emran, T.B., Rahman, M.A., Hosen, S.Z., Rahman, M.M., Islam, A.M.T., Chowdhury, M.A.U. and Uddin, M.E. 2012. Analgesic activity of *Leea indica* (Burm.f.) Merr. *Phytopharmacology*. 3(1): 150–157.

Ghagane, S.C., Puranik, S.I., Kumbar, V.M., Nerli, R.B., Jalalpure, S.S., Hiremath, M.B., Neelagund, S. and Aladakatti, R. 2017. *In vitro* antioxidant and anticancer activity of *Leea indica* leaf extracts on human prostate cancer cell lines. *Integrative Medicine Research*. 6(1): 79–87.

Raihan, M.O., Tareq, S.M., Brishti, A., Alam, M.K., Haque, A. and Ali, M.S. 2012. Evaluation of antitumor activity of *Leea indica* (Burm.f.) Merr. extract against Ehrlich ascites carcinoma (EAC) bearing mice. *American Journal of Biomedical Sciences*. 4(2): 143–152.

Ramesh, D., Ramesh, D.Y., Kekuda, T.R., Onkarappa, R., Vinayaka, K.S. and Raghavendra, H.L. 2015. Antifungal and radical scavenging activity of leaf and bark of *Leea indica* (Burm.f.) Merr. *Journal of Chemical and Pharmaceutical Research*. 7(1): 105–110.

Sakib, S.A., Tareq, A.M., Islam, A., Rakib, A., Islam, M.N., Uddin, M.A., Rahman, M.M., Seidel, V. and Emran, T.B. 2021. Anti-inflammatory, thrombolytic and hair-growth promoting activity of the n-hexane fraction of the methanol extract of *Leea indica* leaves. *Plants*. 10(6): 1081.

***Tetrastigma leucostaphylum* (Dennst.) Alston**

[From Latin *tetrastigma* = 4 stigmas, the Greek *leuko* = white, and *staphyle* = grape]

Published in *Taxon* 26: 539. 1977

Synonyms: *Cissus lanceolaria* Rooxb.; *Cissus leucostaphyla* Dennst.; *Tetrastigma lanceolarium* Planch. *Tetrastigma muricatum* Gamble; *Vitis lanceolaria* Wall.; *Vitis muricata* Wall. ex Wight & Arn.

Common name: Indian chestnut vine

Habitat: forests

Geographical distribution: from India to Papua New Guinea

Botanical description: this climbing plant grows up to a length of about 10 m. Stems: flattish, tuberculate, with tendrils. Tendrils: simple. Leaves: pedate, alternate, stipulate. Petioles: 3–9 cm long. Petiolules: 0.5–3 cm long. Folioles: lanceolate, coriaceous, sometimes glossy, serrate, 4–6 cm × 13–15 cm, acute at base, acuminate at apex, with 8–10 pairs of secondary nerves. Cymes: axillary on 4 cm long peduncles. Flowers: tiny. Calyx: 4-lobed. Petals: 4, whitish green. Disc: present. Stamens: 4, whitish. Ovary: cone-shaped. Stigma: 4-lobed. Berries: globose, about 1 cm across, somewhat glossy, orange. Seeds: 2–4, obovoid.

Traditional therapeutic indications: fatigue, fever, jaundice (Dusun, Kadazan); chest pain (Murut)

Pharmacology and phytochemistry: toxic for brine shrimps, anti-inflammatory, thrombolytic, anthelmintic (Rudra et al. 2020), anxiolytic, toxic for malignant glioblastoma cells (Rudra et al. 2023).

Toxicity, side effects, and drug interaction: not known

REFERENCES

Rudra, S., Sawon, S.U., Emon, N.U., Alam, S., Tareq, S.M., Islam, M.N., Uddin, M.R., Md Sazid, A., Hasbe, A.N., Shakil, M. and Sakib, S.A. 2020a. Biological investigations of the methanol extract of *Tetrastigma leucostaphylum* (Dennst.) Alston ex Mabb. (Vitaceae): *In vivo* and *in vitro* approach.

Rudra, S., Faruque, M.O., Tahamina, A., Emon, N.U., Al Haidar, I.K. and Uddin, S.B. 2023. Neuropharmacological and antiproliferative activity of Tetrastigma leucostaphyllum (Dennst.) Alston: Evidence from in-vivo, in-vitro and in-silico approaches. *Saudi Pharmaceutical Journal*. 31(6): 929–941.

Rudra, S., Tahamina, A., Emon, N.U., Adnan, M., Shakil, M., Chowdhury, M.H.U., Barlow, J.W., Alwahibi, M.S., Soliman Elshikh, M., Faruque, M.O. and Uddin, S.B. 2020. Evaluation of various solvent extracts of Tetrastigma leucostaphylum (Dennst.) Alston leaves, a Bangladeshi traditional medicine used for the treatment of Diarrhea. *Molecules*. 25(21): 4994.

Tetrastigma diepenhorstii (Miq.) Latiff

[From Latin *tetrastigma* = 4 stigmas and after the 19th century Dutch botanist H. Diepenhorst]

Published in *Folia Malaysiana* 2: 185. 2001

Synonyms: *Cissus diepenhorstii* Miq.; *Vitis diepenhorstii* Miq.
Local names: *daramatin* (Murut); *pupus, pumpaos* (Dusun)
Habitat: forests
Geographical distribution: Indonesia, Borneo, the Philippines
Botanical description: this climbing plant grows up to a length of about 20 m. Stems: pustulate, terete, striated, with tendrils. Tendrils: furcate, 2.8–26 cm long. Leaves: trifoliate, alternate, stipulate. Stipules: triangular, about 5 mm long. Petioles: 14–28 cm long. Petiolules: 0.7–6.5 cm long. Folioles: coriaceous, glossy or not, brownish beneath, sometimes loosely serrate, 3–10 cm × 7–23 cm, lanceolate to elliptic, acute at base, acuminate at apex, with 6–11 pairs of clearly visible secondary nerves. Cymes: 4.5 cm across, on 0.5–2 cm long peduncles. Flowers: tiny. Calyx: vaguely 4-lobed. Petals: 4, whitish green, obovate. Disc: present. Stamens: 4. Ovary: cone-shaped. Stigma: 4-lobed. Drupes: oblong, reddish, about 8 mm long, sometimes smooth. Seeds: oblong, about 8 mm long, curved.
Traditional therapeutic indication: sprains (Murut)
Pharmacology and phytochemistry: not known
Toxicity, side effects, and drug interaction: not known

Vitis trifolia L.

[From Latin *vitis* = vine and *trifolia* = trifoliate]

Published in *Species Plantarum* 1: 203. 1753

Synonyms: *Causonis trifolia* (L.) Mabb. & J. Wen; *Cayratia trifolia* (L.) Domin; *Cissus carnosa* Lam.; *Cissus trifolia* (L.) K.Schum.; *Columella trifolia* (L.) Merr.
Local name: *susumoloi* (Murut)
Common names: bush grape, fox grape, three-leaved cayratia
Habitat: at the edges of forests
Distribution: from India to the Pacific Islands
Botanical description: this climbing plant grows up to a length of about 20 m. Stems: striated, sometimes purplish when young, with tendrils. Tendrils: 4–5, furcate, curly, with adhesive discs at apex. Leaves: trifoliate, alternate, stipulate. Stipules: triangular, about 5 mm long. Petioles: about 1–6 cm long, sometimes purplish. Petiolules: 0.3–1.5 cm long, sometimes purplish. Folioles: of a peculiar kind of green, sometimes loosely serrate, 2.5–9 cm × 3–9 cm, elliptic, attenuate at base, acute to rounded at apex, with 5–6 pairs of clearly visible secondary nerves, the midrib sometimes purplish. Cymes: showy, on about 10 cm long peduncles. Flowers: tiny. Calyx: cup-shaped, sinuate. Petals: 4, green, about 2 mm long, cuculate at apex. Disc: lobed. Stamens: 4, white. Styles: 1 saltshaker-shaped, of a magnificent bright red color. Ovary: 4-lobed. Stigma: oblate. Berries: oblate, dull gray to blackish, glossy, about 1–2 cm across. Seeds: 2–4, triangular, about 6 mm long.
Traditional therapeutic indications: medicinal (Murut)
Pharmacology and phytochemistry: this plant has been the subject of numerous pharmacological studies (Kumar et al. 2011) which have highlighted, among other things, antibacterial (Rahman & Khatun 2010), analgesic (Ahmed et al. 2007), and antiulcer properties (Gupta et al. 2012). Cytotoxic triterpenes (epifriedelanol) (Kundu et al. 2000). Antibacterial triterpenes (friedelane-3-β-ol), amide alkaloids (trans-*N*-caffeoyltyramine) (Patnaik et al. 2008).
Toxicity, side effects, and drug interaction: not known

FIGURE 4.18 *Ampelocissus ochracea* (Teijsm. & Binn.) Merr.

Comments:

(i) Other plants of the genus *Tetrastigma* (Miq.) Planch. (1887) used as medicines in Sabah are "*torumum dakon*" for acne, dandruff, and toothaches by the Dusun whilst "*lipoi*" is used by the Dusun and Kadazan for pancreatitis (owed to alcoholism, in general).
(ii) *Ampelocissus ochracea* (Teijsm. & Binn.) Merr. (Figure 4.18) is used by the Dusun and Kadazan (local name: *tabai*) for beriberi.
(iii) *Ampelocissus imperialis* (Miq.) Planch is used by the Dusun for bloody vomiting (local name: *kamburat*).
(iv) *Ampelocissus polita* (Miq.) Pelser is used by the Dusun as flavoring agent (local name: *mban ambuk*).

REFERENCES

Ahmed, F., Rahman, K.M., Alam, S.M.M. and Masud, M.M. 2007. Antinociceptive activity of *Vitis trifolia*. *Dhaka University Journal of Pharmaceutical Sciences*. 6(2): 129–130.

Gupta, J., Kumar, D. and Gupta, A. 2012. Evaluation of gastric anti-ulcer activity of methanolic extract of *Cayratia trifolia* in experimental animals. *Asian Pacific Journal of Tropical Disease*. 2(2): 99–102.

Kumar, D., Kumar, S., Gupta, J., Arya, R. and Gupta, A. 2011. A review on chemical and biological properties of *Cayratia trifolia* Linn. (Vitaceae). *Pharmacognosy Reviews*. 5(10): 184.

Kundu, J.K., Rouf, A.S.S., Hossain, M.N., Hasan, C.M. and Rashid, M.A. 2000. Antitumor activity of epifriedelanol from *Vitis trifolia*. *Fitoterapia*. 71(5): 577–579.

Patnaik, T., Dey, R.K. and Gouda, P. 2008. Antimicrobial activity of friedelan-3 [beta]-ol and trans-N-caffeoyltyramine isolated from the root of *Vitis trifolia*. *Asian Journal of Chemistry*. 20(1): 417.

Rahman, M. and Khatun, A. 2010. Antioxidant, antimicrobial and cytotoxic activities of *Vitis trifolia* linn. *Journal of Dhaka International University*. 1(1): 181–183.

4.1.10 Superorder Rosanae Takht. (1967), the Fabids

4.1.10.1 Order Cucurbitales Juss. ex Bercht. & J. Presl (1820)

4.1.10.1.1 Family Anisophylleaceae Ridl. (1922), the Anisophyllea Family

The family Anisophylleaceae consists of about 4 genera and 36 species of trees and shrubs. Leaves: simple, alternate, exstipulate. Inflorescences: axillary racemes, spikes, or panicles. Sepals: 4. Petals: 4. Stamens: 4-8. Carpels: 4 forming 4 locular ovary, each locule with 1 ovule. Fruits: drupes or samaras.

***Anisophyllea disticha* Baill.**

[from Greek *anisos* = unequal, *phullon* = leaf and Latin *disctichus* = made of 2 lines]

Published in *Adansonia* 11: 311. 1875

Local names: *lapad tulang* (Lundayeh); *sapad* (Dusun)
Common name: leechwood
Habitat: forests
Geographical distribution: Malaysia, Indonesia, Borneo
Botanical description: this treelet reaches a height of about 7 m. Stems: hairy, fissured, brown, and zigzag-shaped. Leaves: simple, sessile, alternate, exstipulate, look from a distance like the many alternate and compactly arranged folioles of a pinnate leaf. Blades: strongly asymmetrical, glossy, of a peculiar kind of grayish green, with a visible sunken midrib, about 6 mm × 2 mm. Flowers: tiny, axillary, solitary. Sepals: 4, triangular. Petals: 4, deeply divided. Stamens: 4. Styles: 4. Drupes: shaped like some kind of tiny cacao capsules, about 1 cm × 2 cm, glossy, light green at first then turning red, similar to that of the fruits of *Ruscus aculeatus* L.
Traditional therapeutic indications: joints pain (Lundayeh); fatigue (Dusun)
Pharmacology and phytochemistry: antidiabetic (Almurdani et al. 2020), antimicrobial (Bari et al. 2021), leishmanicidal (Roshan-Jahn et al. 2018). Ellagitannins (Lowry 1968).
Toxicity, side effects, and drug interaction: not known

REFERENCES

Almurdani, M., Fikriyah, M., Zamri, A., Nugroho, T.T., Jasril, J., Eryanti, Y. and Teruna, H.Y. 2020, October. Antidiabetic activity of some extracts from *Anisophyllea disticha* leaves. *Journal of Physics: Conference Series*. 1655(1): 012032.

Bari, N.A.A., Ferdosh, S. and Sarker, M.Z.I. 2021. Antimicrobial and antioxidant activities of a malaysian medicinal plant *Anisophyllea disticha* (Jack) baill. and quantification of its phenolic constituents. *Bangladesh Journal of Botany*. 50(3): 515–521.

Lowry, J.B. 1968. The Geographical distribution and potential taxonomic value of alkylated ellagic acids. *Phytochemistry*. 7(10): 1803–1813.

Roshan-Jahn, M.S., Getha, K., Mohd-Ilham, A., Norhayati, I., Muhammad-Haffiz, J. and Amyra, A.S. 2018. *In vitro* anti-leismanial activity of Malaysian medicinal and forest plant species. *Journal of Tropical Forest Science*. 30(2): 234–241.

4.1.10.1.2 Family Cucurbitaceae Juss. (1789), the Gourd Family

The family Cucurbitaceae consists of about 140 genera and 700 species of climbing plants. Stems: often hairy, fleshy, with tendrils. Leaves: simple, alternate, exstipulate. Blades: often palmately lobed, hairy. Flowers: axillary, often showy. Calyx: tubular, 5-lobed. Corolla: often funnel-shaped, membranous, 5-lobed. Stamens: 3–5. Carpels: often 3, merged into a trilocular ovary, each locule with 3–many ovules. Styles: often 1–3. Stigma: enlarged or bifid. Fruits: berries, capsules. Seeds: numerous, flattened.

***Benincasa hispida* (Thunb.) Cogn.**

[After the 16th century Italian botanist Giuseppe Benincasa and Latin *hispida* = hispid]

Published in *Monographiae Phanerogamarum* 3: 513. 1881

Synonyms: *Benincasa cerifera* Savi; *Cucurbita hispida* Thunb.
Local names: *buah kundru* (Bajau); *kundur* (Murut)
Common names: ash gourd, wax gourd
Habitat: cultivated
Geographical distribution: tropical
Botanical description: this herbaceous plant grows up to a length of about 8 m. Stems: angular, hairy. Leaves: simple, alternate, exstipulate. Petioles: stout, hairy, 5–20 cm long. Blades: reniform-orbicular, 15–30 cm across, cordate at base, acute at apex, 5–7-lobed, the lobes broadly triangular or ovate. Inflorescences: axillary, solitary on 5–15 cm long, hairy peduncles. Calyx: tubular, hairy, 5-lobed, the lobes about 1 cm long. Corolla: meadow buttercup yellow, membranous, about 6 cm long. Stamens: 3, the anthers about 5 mm long. Staminodes: present. Ovary: 2–4 cm long, hairy. Styles: 1, 2 mm long. Berries: massive, covered with some kind of sticky dust, of a strange light, almost glaucous, green, somewhat oblong, up to 25 cm × 60 cm. Seeds: numerous in a white pulp.
Traditional therapeutic indications: jaundiced babies (Bajau)
Pharmacology and phytochemistry: this plant has been the subject of numerous pharmacological studies (Al-Snafi 2013) which have highlighted, among other things, hepatoprotective activity (Das & Roy 2012).
Toxicity, side effects, and drug interaction: the median lethal dose (LD_{50}) of ethanol extract of fruit pulp administered orally to rats was above 2 g/kg (Shakya et al. 2020).

REFERENCES

Al-Snafi, A.E. 2013. The Pharmacological importance of *Benincasa hispida*. A review. *International Journal of Pharma Sciences and Research*. 4(12): 165–170.

Das, S.K. and Roy, C. 2012. The protective role of *Benincasa hispida* on diclofenac sodium induced hepatotoxicity in albino rat model. *International Journal of Pharmaceutical Research and Development*. 3(11): 171–179.

Shakya, A., Chaudhary, S.K., Bhat, H.R. and Ghosh, S.K. 2020. Acute and sub-chronic toxicity studies of *Benincasa hispida* (Thunb.) cogniaux fruit extract in rodents. *Regulatory Toxicology and Pharmacology*. 118: 104785.

***Luffa cylindrica* M. Roem.**

[From Arabic *lufah* = *Luffa acutangula* Roxb. and from Latin *cylindrica* = cylindrical]

Published in *Familiarum Naturalium Regni Vegetabilis Synopses Monographicae* 2: 63. 1846

Synonyms: *Cucumis fricatorius* Sessé & Moc.; *Luffa acutangula* var. *subangulata* (Miq.) Cogn.; *Luffa aegyptiaca* Mill.; *Luffa subangulata* Miq.; *Melothria touchanensis* H. Lév.; *Momordica cylindrica* L.; *Momordica luffa* L.

Local names: *kosula* (Dusun), *petola* (Bajau); *pucula* (Lundayeh)

Common name: sponge gourd

Habitat: cultivated

Geographical distribution: tropical

Botanical description: this climbing plant grows up to a length of about 10 m. Stems: angular, hairy, with tendrils. Tendrils: 2–4-fid. Leaves: simple, alternate, exstipulate. Petioles: about 10 cm long. Blades: papery, 5–7–lobed, cordate at base, acute at apex, up to about 20 cm across. Inflorescences of male flowers: axillary racemes, 15–20 flowered, up to about 15 cm long. Inflorescences of female flowers: solitary, on 2 cm–10 cm long peduncles. Calyx: hairy, about 2 cm long, tubular, 5-lobed. Corolla: funnel-shaped, 5–9 cm across, membranous, yellow. Stamens: 5, about 1 cm long. Ovary: cylindrical. Berries: smooth, cylindrical, somewhat sausage-shaped, up to about 45 cm long. Seeds: numerous, smooth, black.

Traditional therapeutic indication: hemorrhoids (Lundayeh)

Pharmacology and phytochemistry: anticancer (Abdel-Salam et al. 2019), analgesic (Sultana et al. 2014), anti-inflammatory, hypoglycemic (Al-Snafi 2019), hepatoprotective (Balakrishnan & Huria 2011), antibacterial, antifungal (Devi et al. 2009), atherosclerosis (Raut et al. 2021), and hypoglycemic (Akther et al. 2014).

Toxicity, side effects, and drug interaction: aqueous and alcoholic extracts of fruits were poisonous to mice at an oral dose above 2 g/kg (Al-Snafi 2019).

Comments:

(i) In the Philippines, Cebuanos use the plant to induce sleep and to treat hypertension (local name: *sikwa*).
(ii) Dried fruits afford a net of dense fibers used to make sponges.

REFERENCES

Abdel-Salam, I.M., Awadein, N.E.S. and Ashour, M. 2019. Cytotoxicity of *Luffa cylindrica* (L.) M. Roem. extract against circulating cancer stem cells in hepatocellular carcinoma. *Journal of Ethnopharmacology*. 229: 89–96.

Akther, F., Rahman, A., Proma, J.J., Kabir, Z., Paul, P.K. and Rahmatullah, M. 2014. Methanolic extract of *Luffa cylindrica* fruits show anti hyperglycemic potential in Swiss albino mice. *ANAS*. 8: 62–65.

Al-Snafi, A.E. 2019. Constituents and pharmacology of *Luffa cylindrica*-A review. *IOSR Journal of Pharmacy*. 9(9): 68–79.

Balakrishnan, N. and Huria, T. 2011. Protective effect of *Luffa cylindrica* L. fruit in Paracetamol induced hepatotoxicity in rats. *International Journal of Pharmaceutical & Biological Archive*. 2(6): 1761–1764.

Devi, G.S., Muthu, A.K., Kumar, D.S. and Rekha, S. 2009. Studies on the antibacterial and antifungal activities of the ethanolic extracts of *Luffa cylindrica* (Linn) fruit. *International Journal of Drug Development and Research*. 1(1): 105–109.

Raut, P., Dhawale, S., Kulkarni, D., Pekamwar, S., Shelke, S., Panzade, P. and Paliwal, A. 2021. Pharmacodynamic findings for the usefulness of *Luffa cylindrica* (L.) leaves in atherosclerosis therapy with supporting antioxidant potential. *Future Journal of Pharmaceutical Sciences*. 7(1): 1–8.

Sultana, J., Sarker, M.M.H., Rahman, A., Monalisa, M.N., Marzan, M., Faisal, M. and Rahmatullah, M. 2014. Analgesic activity evaluation of methanol extract of Luffa cylindrica fruits. *Advances in Natural and Applied Sciences*. 8(9): 13–17.

***Momordica charantia* L.**

[From Latin *mordere* = bite, Greek *kallos* = beautiful, and *anthos* = flower]

Published in *Species Plantarum* 2: 1009. 1753

Synonyms: *Cucumis argyi* H. Lév.; *Momordica indica* L.; *Momordica sinensis* Spreng.

Local names: *poporia* (Dusun); *peria* (Bajau, Dusun, Murut); *feria* (Lundayeh); *poporia* (Murut); *peria katak* (Bajau)

Common name: bitter gourd

Habitat: cultivated

Geographical distribution: tropical

Botanical description: this climbing plant reaches a length of about 4 m. Stems: hairy, thin, with tendrils. Tendrils: opposite the leaves, 1-6 cm long. Leaves: simple, alternate, exstipulate. Petioles: 1–2 cm long, hairy. Blades: thin, mottled with very small blackish blotches beneath, hairy, about 3.5 cm across, deeply 5–7-lobed, loosely toothed. Flowers: axillary, solitary, yellow, on 4–10 cm long peduncles. Calyx: tubular, 5-lobed, hairy, 0.8–1 cm long, the lobes acute elliptic. Corolla: yellowish, 5-lobed, the lobes 1.6–3 cm long. Stamens: 3. Ovary: fusiform, muricate. Stigma: trifid. Berries: fusiform, muricate, bright orange when ripe, 5–15 cm long. Seeds: 0.8–1.3 cm long, compressed, embedded into a crimson aril.

Traditional therapeutic indications: hemorrhoids (Lundayeh); smallpox (Murut); hypertension (Bajau, Dusun, Kadazan); yellow fever (Bajau, Rungus); high cholesterol, coughs, diabetes, postpartum (Bajau)

Pharmacology and phytochemistry: this plant has been the subject of numerous pharmacological studies which have highlighted, among other things, hypoglycemic (Ahmed et al. 2001; Welihinda et al. 1982), hypocholesterolemic (Kinoshita & Ogata. 2018), wound healing (Teoh et al. 2009), anticancer (Güneş et al. 2019), and antibacterial activity (Costa et al. 2010). Hypoglycemic saponins (Oishi et al. 2007).

Toxicity, side effects, and drug interaction: eating too much bitter gourd causes alarming weakness and can be life-threatening due to low blood sugar levels. Fatal comatose hypoglycemia has been caused in children by the consumption of tea made from the leaves, while the consumption of products based on this plant has caused potentially fatal poisoning in adults (Demmers et al 2022). The use of any herbal remedy or medicinal plant extract in toddlers and children should be strictly prohibited. Medical and pharmacy students no longer receive serious lessons on medicinal plants. How many people will have to be poisoned and die before medical and pharmacy schools integrate medicinal and poisonous plant into their curricula again?

Comment: In the Philippines, the plant affords a remedy for diabetes (local name: *parya*).

REFERENCES

Ahmed, I., Lakhani, M.S., Gillett, M., John, A. and Raza, H. 2001. Hypotriglyceridemic and hypocholesterolemic effects of anti-diabetic *Momordica charantia* (karela) fruit extract in streptozotocin-induced diabetic rats. *Diabetes Research and Clinical Practice*. 51(3): 155–161.

Costa, J.G.M., Nascimento, E.M., Campos, A.R. and Rodrigues, F.F. 2010. Antibacterial activity of *Momordica charantia* (Curcubitaceae) extracts and fractions. *Journal of Basic and Clinical Pharmacy*. 2(1): 45.

Güneş, H., Alper, M. and Çelikoğlu, N. 2019. Anticancer effect of the fruit and seed extracts of *Momordica charantia* L. (Cucurbitaceae) on human cancer cell lines. *Tropical Journal of Pharmaceutical Research*. 18(10): 2057–2065.

Kinoshita, H. and Ogata, Y. 2018. Effect of bitter melon extracts on lipid levels in Japanese subjects: A randomized controlled study. *Evidence-Based Complementary and Alternative Medicine*. 2018: 1–6.

Kola, V., Mondal, P., Thimmaraju, M.K., Mondal, S. and Rao, N.V. 2018. Antiarthritic potential of aqueous and ethanolic fruit extracts of "*Momordica charantia*" using different screening models. *Pharmacognosy Research*. 10(3): 258–264.

Oishi, Y., Sakamoto, T., Udagawa, H., Taniguchi, H., Kobayashi-Hattori, K., Ozawa, Y. and Takita, T. 2007. Inhibition of increases in blood glucose and serum neutral fat by *Momordica charantia* saponin fraction. *Bioscience, Biotechnology, and Biochemistry*. 71(3): 735–740.

Teoh, S.L., Latiff, A.A. and Das, S. 2009. The effect of topical extract of *Momordica charantia* (bitter gourd) on wound healing in nondiabetic rats and in rats with diabetes induced by streptozotocin. *Clinical and Experimental Dermatology: Experimental Dermatology*. 35(7): 815–822.

Welihinda, J., Arvidson, G., Gylfe, E., Hellman, B. and Karlsson, E. 1982. The insulin-releasing activity of the tropical plant *Momordica charantia*. *Acta Biologica et Medica Germanica*. 41(12): 1229–1240.

***Trichosanthes cucumerina* L.**

[From Greek *trichos* = hair, *anthos* = flower]

Published in *Species Plantarum* 2: 1008. 1753

Synonyms: *Trichosanthes anguina* L.; *Trichosanthes brevibracteata* Kundu; *Trichosanthes pachyrrhachis* Kundu
Local names: *molisun*, *mamula* (Murut)
Common name: snake gourd
Habitat: cultivated
Geographical distribution: from Pakistan to Australia
Botanical description: this climbing plant grows up to a length of about 5 m. Stems: angular, thin, hairy. Leaves: simple, alternate, exstipulate. Petioles: 3–7 cm long. Blades: reniform or broadly ovate, 7–16 cm × 8–18 cm, membranous, denticulate, deeply 5–7-lobed, the lobes triangular or rhombic. Inflorescences of male flowers: axillary racemes, up to about 20 cm long. Inflorescences of female flowers: solitary, axillary. Calyx: tubular, about 1.5 cm long, 5-lobed. Corolla: funnel-shaped, pure white, 5-lobed, about 3 cm across, finely cut like lace. Berries: cylindrical, up to about 1.5 m long, smooth, having the resemblance of hideous giant dangling leeches, edible. Seeds: numerous, oblong, up to 1.7 cm long, flat.
Traditional therapeutic indications: body swellings (Murut)
Pharmacology and phytochemistry: this plant has been the subject of numerous pharmacological studies (Bobade et al. 2022) which have highlighted, among other things, antibacterial, antifungal (Ali et al. 2011), antidiabetic (Arawwawala et al. 2009; Kirana & Srinivasan 2008), anti-inflammatory, antiulcers (Arawwawala et al. 2010, 2010a), and hepatoprotective activity (Kumar et al. 2009).
Anti-inflammatory steroids (cucurbitacin B, isocucurbitacin B) (Suebsakwong et al. 2020).
Toxicity, side effects, and drug interaction: the median lethal dose (LD_{50}) of aqueous extract administered to rats was 1.1 g/kg (Kirana & Srinivasan 2008).

Comments:

(i) Cucumber (*Cucumis sativus* L.) is used by the Bajau for gout (local name: *timun*).
(ii) Pumpkin (*Cucurbita pepo* L.) is called "*tawadak*" by the Dusun and "*labu*" by the Bajau and Murut. It is used for flatulence by the Dusun and scalds by the Dusun and Kadazan.
(iii) Chayote (*Sechium edule* (Jacq.) Sw.) is used for asthma by the Bugis (local name: *lawo*). In the Philippines, the Cebuanos use it for hypertension (local name: *sayote*).
(v) In the family Begoniaceae C. Agardh (1824), plants in the genus *Begonia* L. (1753) are used as medicines such as "*tonsom onsom*" for fungal infection by Dusun and Kadazan and "*sompotungu*" for vomiting by the Dusun. The Sabah rainforest was once home to a stunning array of endemic begonia species.

REFERENCES

Ali, M.A., Sayeed, M.A., Islam, M.S., Yeasmin, M.S., Khan, G.R.M.A.M. and Muhamad, I.I. 2011. Physicochemical and antimicrobial properties of *Trichosanthes anguina* and *Swietenia mahagoni* seeds. *Bulletin of the Chemical Society of Ethiopia*. 25(3): 227–436.

Arawwawala, M., Thabrew, I. and Arambewela, L. 2009. Antidiabetic activity of *Trichosanthes cucumerina* in normal and streptozotocin—induced diabetic rats. *International Journal of Biological and Chemical Sciences*. 3(2): 287–296.

Arawwawala, M., Thabrew, I. and Arambewela, L. 2010a. Gastroprotective activity of Trichosanthes cucumerina in rats. *Journal of Ethnopharmacology*. 127(3): 750–754.

Arawwawala, M., Thabrew, I., Arambewela, L. and Handunnetti, S. 2010b. Anti-inflammatory activity of *Trichosanthes cucumerina* Linn. in rats. *Journal of Ethnopharmacology*. 131(3): 538–543.

Bobade, A.A., Thatte, C.V. and Tijare, R.B. 2022. *Trichosanthes cucumerina*: A perspective on various medicinal uses or activities. *GSC Biological and Pharmaceutical Sciences*. 20(3): 141–147.

Kirana, H. and Srinivasan, B.P. 2008. *Trichosanthes cucumerina* Linn. improves glucose tolerance and tissue glycogen in non insulin dependent diabetes mellitus induced rats. *Indian Journal of Pharmacology*. 40(3): 103.

Kumar, S.S., Kumar, B.R. and Mohan, G.K. 2009. Hepatoprotective effect of *Trichosanthes cucumerina* var *cucumerina* L. on carbon tetrachloride induced liver damage in rats. *Journal of Ethnopharmacology*. 123(2): 347–350.

Suebsakwong, P., Chulrik, W., Chunglok, W., Li, J.X., Yao, Z.J. and Suksamrarn, A. 2020. New triterpenoid saponin glycosides from the fruit fibers of *Trichosanthes cucumerina* L. *RSC Advances*. 10(18): 10461–10470.

4.1.10.1.3 Family Tetramelaceae Airy Shaw (1965), the Tetramela Family

The family Tetramelaceae consists of about 2 genera and 2 species of trees. Buttresses: sometimes present. Leaves: simple, alternate, or spiral, exstipulate. Petioles: often elongated. Blades: palmately 3–5-veined. Inflorescences: spikes. Calyx: tubular, 4–8-lobed, or made of 4–8 sepals. Petals: none or 6–8. Stamens: 4–8. Carpels: 4–8, forming a unilocular ovary, with 2–many ovules. Styles: 4–8, short, stout. Stigma: sometimes capitate. Fruits: dehiscent capsules. Seeds: numerous, ovoid, or fusiform, tiny.

***Octomeles sumatrana* Miq.**

[From Greek *octo* = eight, *melos* = part, and Latin *Sumatrana* = from Sumatra]

From *Flora van Nederlandsch Indië, Eerste Bijvoegsel* (Supplement) (2): 336. 1860

Synonym: *Octomeles moluccana* Warb.
Local names: *binuang* (Bajau); *tiniwaung* (Dusun, Kadazan)
Common names: benuang, ilimo tree
Habitat: forests, cultivated
Geographical distribution: from Indonesia to the Solomon Islands
Botanical description: this magnificent tree reaches a height of about 60 m. Bark: fissured, grayish. Buttresses: up to 6 m tall. Stems: stout, rough, with leaf scars. Leaves: simple, spiral, exstipulate. Petioles: stout, smooth, swollen at base, 8–30 cm long, sometimes purplish. Blades: broadly cordate, somewhat glossy, with 5–9 pairs of secondary nerves, sometimes purplish, membranous, the first pairs emerging from the base, acuminate at apex, marked beneath with tiny discoid domatia. Spikes: 15–60 cm long, pendulous, axillary. Calyx: tubular, tiny, 8-lobed. Petals: none. Stamens: 8. Styles: 5–8 inserted on the throat of an urn-shaped calyx tube. Stigma: capitate. Capsules: about 8 mm long, splitting from the apex, with persistent crown of styles, the whole structure weird. Seeds: numerous, ovoid, or fusiform, tiny (Figure 4.19).
Traditional therapeutic indications: bone aches, cuts, chest pain, fatigue, wounds (Dusun, Kadazan)
Pharmacology and phytochemistry: hypoglycemic (Azahar et al. 2012). Flavonols (kaempferol), flavonol glycosides (quercetin 3-*O*-glucoside) (Bohm 1988).
Toxicity, side effects, and drug interaction: not known

REFERENCES

Azahar, M.A., Al-Naqeb, G., Hasan, M. and Adam, A. 2012. Hypoglycemic effect of *Octomeles sumatrana* aqueous extract in streptozotocin—induced diabetic rats and its molecular mechanisms. *Asian Pacific Journal of Tropical Medicine*. 5(11): 875–881.

Bohm, B.A. 1988. Flavonoid systematics of the datiscaceae. *Biochemical Systematics and Ecology*. 16(2): 151–155.

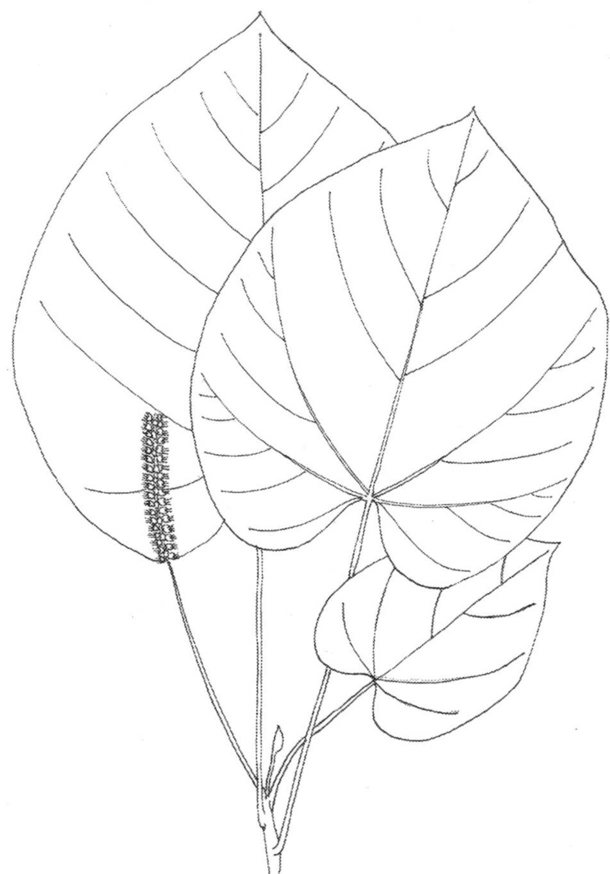

FIGURE 4.19 *Octomeles sumatrana* Miq.

4.1.10.2 Order Fabales Bromhead (1838)

4.1.10.2.1 Family Fabaceae Lindley (1836), the Pea Family

The family Fabaceae is a vast taxon consisting of about 650 genera and 12,000 species of trees, shrubs, climbing plants, or herbaceous plants. Leaves: simple or compound, alternate or spiral, stipulate. Inflorescences: racemes, corymbs, spikes, heads, or panicles. Calyx: 5-lobed or 5 sepals. Petals: 5, either similar or organized into a standard, a pair of wings, and a keel. Stamens: often 10, free or partially merged, diadelphous. Carpels: often 1, forming a unilocular ovary with 2–100 ovules. Fruits: pods.

***Airyantha borneensis* (Oliv.) Brummitt**

[After the 20th century English botanist Herbert Kenneth Airy Shaw and Latin *borneensis* = from Borneo]

Published in *Kew Bulletin* 22(3): 378. 1968

Synonyms: *Baphia borneensis* Oliv.; *Baphiastrum borneense* (Oliv.) Yakovlev
Local names: *barayung* (Dusun); *molisun, matamis* (Murut)
Geographical distribution: Borneo, the Philippines

Botanical description: this climbing plant grows up to a length of about 8 m. Stems: terete, hairy at apex. Leaves: simple, alternate, stipulate. Stipules: 2 mm long, lanceolate. Petioles: 1–3.5 cm long, somewhat straight, thin, hairy. Blades: coriaceous, 8.5–22 cm × 3.5–10.5 cm, oblong-elliptic to obovate, with about 7–9 pairs of straight secondary nerves, acute to rounded, or truncate at base, acuminate at apex. Racemes: axillary, about 5 cm long. Calyx: tubular, about 2 mm long, 2–3 dentate. Corolla: cream-colored and tinted in some places light pink. Standard: orbiculate, about 1 cm long, with a yellow blotch. Pods: hooked, hairy, acuminate. Seeds: red (Figure 4.20).

Traditional therapeutic indications: fever, hypertension, toothaches (Murut); female fatigue (Dusun)

Pharmacology and phytochemistry: not known

Toxicity, side effects, and drug interaction: not known

Cassia alata L.

[From the Hebrew word *qase' ah* = some kind of cinnamon bark and Latin *alata* = winged]

Published in *Species Plantarum* 1: 378. 1753

Synonyms: *Cassia bracteata* L. f.; *Cassia herpetica* Jacq.; *Herpetica alata* (L.) Raf.; *Senna alata* (L.) Roxb.

Local names: *emon* (Murut); *kurubau* (Rungus); *daung galingkan* (Bugis); *hihinggos* (Kadazan); *gombirang, gombiring, raun suluk* (Dusun); *kayabau, manggarut* (Dusun, Kadazan)

Common name: candle bush

Habitat: roadsides, around villages, wastelands, public damps

Geographical distribution: tropical

Botanical description: this weird shrub reaches a height of about 4 m. Leaves: spiral, paripinnate, stipulate. Stipules: about 1–2 cm long, triangular, persistent. Petioles: angular, somewhat yellowish to light pinkish brown. Folioles: sessile, dull green, coriaceous, 6–12

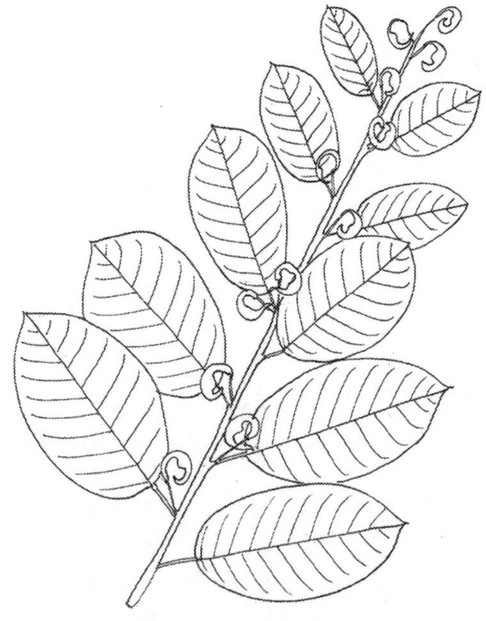

FIGURE 4.20 *Airyantha borneensis* (Oliv.) Brummitt.

pairs, parallel to almost perfection, oblong or obovate, 6–15 cm × 3.5–7.5 cm, obliquely truncate to almost cordate at base, cuspidate at apex, drying black in herbarium samples. Racemes: axillary, erect, about 60 cm tall. Sepals: 5. Petals: 5, bright yellow, ovate-orbicular, 1.5–2.5 cm long, forming a somewhat globose corolla. Stamens: 10. Ovary: hairy. Pods: woody, straight, winged, tetragonal, drying black, dehiscent, 10–20 cm long. Seeds: about 50, deltoid to somewhat toothlike, woody.

Traditional therapeutic indications: stomach aches (Dusun, Rungus); beriberi, fatigue, asthma, jaundice, ringworms, scabies (Dusun, Kadazan); skin diseases (Bugis, Dusun, Kadazan)

Pharmacology and phytochemistry: this plant has been the subject of numerous pharmacological studies (Suresh Babu et al. 2012) which have highlighted, among other things, antibacterial, antifungal, cytotoxic (Thaina et al. 2008; Fatmawati et al. 2020; Bharathi et al. 2022), antiulcer (Bharathi et al. 2022), anti-inflammatory (Babatunde et al. 2019), and bronchorelaxant activity (Ouédraogo et al. 2013). Toxic to breast cancer cells (MCF-7) (Chahardehi et al. 2021). Anthraquinones (aloe-emodin, emodin, physcion) (Promgool et al. 2014).

Toxicity, side effects, and drug interaction: the median lethal dose (LD_{50}) of hydroalcoholic extract of leaves administered to mice was 18.5 g/kg (Pieme et al. 2006).

Comment: A plant of the genus *Cassia* L. (1753) is used by the Dusun and Kadazan (local name: *wollu*) for asthma.

REFERENCES

Babatunde, J.O. and Kayode, O.K. 2019. Inhibitory action of dried leaf of Cassia alata (Linn.) Roxb against lipoxygenase activity and nitric oxide generation. *Scientia Agricola*. 10: 185–190.

Bharathi, D.R., Kumar Mani, R., Ajay, B.V., Pooja, R.C., Kumar, K. and Mahesh, C. 2022. *Cassia alata*: Phytopharmacological, traditional, and medicinal considerations. *World Journal of Current Medical and Pharmaceutical Research*: 147–150.

Chahardehi, A.M., Arsad, H., Ismail, N.Z. and Lim, V. 2021. Low cytotoxicity, and antiproliferative activity on cancer cells, of the plant *Senna alata* (Fabaceae). *Revista de Biología Tropical*. 69(1): 317–330.

Fatmawati, S., Purnomo, A.S. and Bakar, M.F.A. 2020. Chemical constituents, usage and pharmacological activity of Cassia alata. *Heliyon*. 6(7): 1–11.

Ouédraogo, M., Da, F.L., Fabré, A., Konaté, K., Dibala, C.I., Carreyre, H., Thibaudeau, S., Coustard, J.M., Vandebrouck, C., Bescond, J. and Belemtougri, R.G. 2013. Evaluation of the bronchorelaxant, genotoxic, and antigenotoxic effects of *Cassia alata* L. *Evidence-Based Complementary and Alternative Medicine*. 2013: 1–11.

Pieme, C.A., Penlap, V.N., Nkegoum, B., Taziebou, P.C.L., Tekwu, E.M., Etoa, F.X. and Ngongang, J. 2006. Evaluation of acute and subacute toxicities of aqueous ethanolic extract of leaves of *Senna alata* (L.) Roxb (Ceasalpiniaceae). *African Journal of Biotechnology*. 5(3): 283–289.

Promgool, T., Pancharoen, O. and Deachathai, S. 2014. Antibacterial and antioxidative compounds from *Cassia alata* Linn. *Songklanakarin Journal of Science and Technology*. 36(4): 459–463.

Suresh Babu, V.V., Narayana, S.V., Naik, N., Geethanjali, B., Yamini, K., Sultana, N. and Malothu, R. 2012. Evaluation of antiulcer activity of cassia alata linn leaves. *International Journal of Pharmacy & Therapeutics*. 3(2): 1–4.

Thaina, P., Panichayupakaranant, P., Wongnawa, M. and Bumrungwong, N. 2008. Comparative study on the laxative effects of anthraquinone glycosides and aglycones extracted from Senna alata (Cassia alata) leaves and possible involvement of nitric oxide in the laxative action. In *The Second Joint PSU—UNS International Conference on BioScience: Food, Agriculture and the Environment June 22–24*, Novi Sad, Serbia (p. 195).

***Millettia nieuwenhuis* J.J. Smith**

[After 19th century plant collector Charles Millet and after the 19th century Dutch explorer Anton Willem Nieuwenhuis]

Published in *Species Plantarum* 1: 378. 1753

Synomyms: *Adinobotrys myrianthus* Dunn; *Adinobotrys nieuwenhuisii* (J. J. Smith) Dunn; *Callerya nieuwenhuisii* (J. J. Smith) Schot; *Millettia cuspidata* Ridl.; *Whitfordiodendron nieuwenhuisii* Dunn
Local name: *ramus* (Murut)
Habitat: forests
Geographical distribution: Borneo
Botanical description: this enormous climbing plant grows up to a length of about 10 m. Leaves: alternate, imparipinnate, stipulate. Stipules: triangular, up to about 4 mm long. Rachis: 15–25 cm long. Petiolules: 5–9 mm long. Folioles: 2–6 pairs, ovate to elliptic, 4–12 cm × 7.5–25 cm, rounded to cuneate at base, cuspidate at apex, marked with 4–6 pairs of secondary nerves. Racemes: cauliflorous, up to about 1 m long. Calyx: tubular, about 4 mm long, 5-toothed. Standard: elliptic, about 1 cm long, dull red, with a yellowish-green middle line. Wings and keel: dull red, about 1 cm long. Disc: present. Stamens: 10, about 1 cm long. Ovary: linear, about 7 mm long. Styles: 1, ciliate. Pods: ovate to obovate, 2–4.5 cm long, pendulous, brownish-pink somewhat resembling peanuts. Seeds: 1–3, white, in a white pulp.
Traditional therapeutic indications: thrush (Murut)
Pharmacology and phytochemistry: not known
Side effects and toxicity: not known

Comment: One could search for antifungal principles in this plant before it becomes extinct.

***Mimosa pudica* L.**

[From Latin *mimus* = mime and *pudica* = modest]

Published in *Species Plantarum* 1: 518. 1753

Synonyms: *Mimosa hispidula* Kunth; *Mimosa tetrandra* Humb. & Bonpl. ex Willd.; *Mimosa unijuga* Duchass. & Walp.
Local names: *sikot mou, tenom molu* (Murut); *rumput semalu* (Kedayan; Dusun); *togop togop* (Dusun); *puteri malu* (Bajau, Kedayan)
Common names: sensitive mimosa, sensitive plant
Habitat: roadsides, lawns, flower pots in gardens, wastelands
Botanical description: this creeping herbaceous plant grows up to a length of about 1 m. Stems: purplish, ligneous, armed with sharp and curved thorns with a length of 3 mm. Leaves: bipinnate, spiral, stipulate. Stipules: lanceolate, up to about 1 cm long. Blades: 2–8 cm long, with the incredible ability to close immediately after being touched. Folioles: 5–26 pairs, narrowly elliptic, 0.6–1.6 cm × 1.5–3 mm, purple-margined. Flowers: very small, arranged in delicate and beautiful pinkish heads. Calyx: tiny. Corolla: about 2 mm long, tubular. Stamens: 4, about 5 mm long, pink. Styles: 1, thin. Pods: 1–2 cm × 4 mm, linear, slightly constricted between the seeds.
Traditional therapeutic indications: abscesses, insect bites, skin diseases, excessive menses, malaria, swellings (Kedayan); stomach aches (Kedayan, Murut); hypertension, cuts, diarrhea, anxiety, spitting blood (Dusun); asthma, wounds (Kedayan, Dusun). Other uses include remedies for dizziness and diarrhea.

Pharmacology and phytochemistry: this plant has been the subject of numerous pharmacological studies (Adurosakin et al. 2023) which have highlighted, among other things, analgesic (Vikram et al. 2012), antibacterial (Le Thoa et al. 2015), antiplasmodial (Aarthi & Murugan. 2011), anxiolytic (Shaikh et al. 2016), antiasthma (Yang et al. 2011), diuretic (Kalabharathi et al. 2015), hypotensive (Hanumanthappa et al. 2022), and wound healing activity (Kokane et al. 2019). Tannins (Kokane et al. 2019).

Side effects and toxicity: the plant produces a poisonous nonprotein amino acid: mimosine (Crounse et al. 1962).

REFERENCES

Aarthi, N. and Murugan, K. 2011. Antimalarial activity and phytochemical screening of ethanolic leaf extract of *Phyllanthus niruri* and *Mimosa pudica. International Journal of Pharmaceutical Research and Development.* 3(3): 198–205.

Adurosakin, O.E., Iweala, E.J., Otike, J.O., Dike, E.D., Uche, M.E., Owanta, J.I., Ugbogu, O.C., Chinedu, S.N. and Ugbogu, E.A. 2023. Ethnotraditional therapeutic indications, phytochemistry, pharmacological activities and toxicological effects of *Mimosa pudica*-A review. *Pharmacological Research.* 7: 100241.

Crounse, R.G., Maxwell, J.D. and Blank, H. 1962. Inhibition of growth of hair by mimosine. *Nature.* 194(4829): 694–695.

Hanumanthappa, S.K., Harris, J.H., Sudhakar, N.P., Ashoka, P. and Hanumanthappa, M. 2022. Hypotensive and cardioprotective property of *Mimosa pudica* using zebra fish. *Pharmacognosy Research.* 14(3): 333–337.

Kalabharathi, H.L., Shruthi, S.L., Vaibhavi, P.S., Pushpa, V.H., Satish, A.M. and Sibgatullah, M. 2015. Diuretic activity of ethanolic root extract of *Mimosa pudica* in albino rats. *Journal of Clinical and Diagnostic Research: JCDR.* 9(12): FF05.

Kokane, D.D., More, R.Y., Kale, M.B., Nehete, M.N., Mehendale, P.C. and Gadgoli, C.H. 2009. Evaluation of wound healing activity of root of *Mimosa pudica. Journal of Ethnopharmacology.* 124(2): 311–315.

Le Thoa, N.T., Nam, P.C. and Nhat, D.M. 2015. Antibacterial activities of the extracts of *Mimosa pudica* L. an *in-vitro* study. *International Journal on Advanced Science, Engineering and Information Technology.* 5(5): 358–361.

Shaikh, Z., Roy, S.P., Patel, P. and Gohil, K. 2016. Medicinal value of Mimosa pudica as an anxiolytic and antidepressant: A comprehensive review. *World Journal of Pharmacy and Pharmaceutical Sciences.* 5(3): 420–432.

Vikram, P.K., Malvi, R. and Jain, D.K. 2012. Evaluation of analgesic and anti-inflammatory potential of *Mimosa pudica* Linn. *International Journal of Current Pharmaceutical Research.* 4(4): 49–52.

Yang, E.J., Lee, J.S., Yun, C.Y., Ryang, Y.S., Kim, J.B. and Kim, I.S. 2011. Suppression of ovalbumin-induced airway inflammatory responses in a mouse model of asthma by *Mimosa pudica* extract. *Phytotherapy Research.* 25(1): 59–66.

***Vigna unguiculata* (L.) Walp.**

[After the 17th century Italian botanist Dominico Vigna and Latin *unguiculata* = with a small claw]

Published in *Repertorium Botanices Systematicae.* 1(5): 779. 1843

Synonyms: *Dolichos biflorus* L.; *Dolichos catjang* L.; *Dolichos monachalis* Brot.; *Dolichos sesquipedalis* L.; *Dolichos sinensis* L.; *Dolichos sphaerospermus* (L.) DC.; *Dolichos unguiculatus* L.; *Vigna catjang* (L.) Walp.; *Vigna cylindrica* (L.) Skeels; *Vigna sesquipedalis* (L.) Fruwirth; *Vigna sinensis* (L.) Endl. ex Hassk.
Local names: *balatong* (Dusun); *kacang panjang* (Bajau)
Common names: black-eyed pea, cowpea, catjang
Habitat: cultivated
Geographical distribution: subtropical, tropical
Botanical description: this climbing plant grows up to a length of about 3 m. Leaves: pinnate, spiral, stipulate. Stipules: lanceolate, about 1 cm long. Petioles: 5–25 cm long. Folioles: 3, ovate to rhomboid, dark dull green, 5–15 cm × 4–6 cm, lateral ones asymmetrical, acute to rounded at base, acute at apex. Racemes: axillary, about 20 cm long. Calyx: tubular, bilobed, about 1 cm long. Corolla: light purplish. Standard: somewhat broadly bilobed, up to about 3 cm long. Wings: about 2 cm long. Keel: boat-shaped, subequal to wings. Stamens: 10. Styles: 1, filiform. Pods: terete, linear, up to 90 cm long. Seeds: numerous, kidney-shaped, up to 1 cm long, whitish, marked with a black spot around hilum.
Traditional therapeutic indications: feverish children, headaches (Dusun); stomach aches (Bajau)
Pharmacology and phytochemistry: hypocholesterolemic (Alah et al. 2017), anti-inflammatory (Ojwang et al. 2015), antibacterial (Sandeep 2014), vasoprotector (Haryati et al. 2021). Flavonols (Aziagba et al. 2017).
Side effects and toxicity: not known

Comments:

(i) Plants of the genus *Bauhinia* L. (1753) are used for boils by the Dusun and Kadazan: "*kukuak*" and "*tuturukon*".
(ii) A plant of the genus *Dalbergia* L.f. (1781) is used by the Dusun and Kadazan for joint pains and asthma but has the side effect of causing dizziness.
(iii) A plant of the genus *Albizzia* Durazz. (1772) is used by the Dusun and Kadazan for fatigue (local name: *sapang*).
(iv) *Bauhinia kockiana* Korth. is used by the Murut for magic rituals (local: *kulih akah*).
(v) A plant of the genus *Sindora* Miq. (1860) is medicinal for the Murut (local name: *talikakasam*).
(vi) *Archidendron clypearia* (Jack) I.C. Nielsen is used by the Dusun and Kadazan for toothaches, dandruff, and thrush. The Dusun call this plant "*sogo*" and use it for itchiness.
(vii) *Archidendron ellipticum* (Blume) I.C. Nielsen is used by the Dusun and Kadazan for dandruff, as antidote for poisons whilst it is used by the Dusun for pancreas pain (local name: *sabano*).
(viii) *Caesalpinia bonduc* (L.) Roxb. is used by the Bajau for chicken pox and malaria (local name: *mentayang*).
(ix) *Caesalpinia sappan* L. (1753) is used by the Bajau (local name: *sapang*) for anemia, asthma, body pains, chest pain, colds, wounds, and postpartum.
(x) *Crotalaria pallida* Aiton is used by the Dusun for flu (local names: *kiri kiri*, *ngrik ngrik*).
(xi) *Dalbergia parviflora* Roxb. is used by the Dusun and Kadazan (local name: *tampan*

kalabau) for fatigue, jaundice, and postpartum.

(xii) *Desmodium heterocarpum* DC. is used by the Dusun and Kadazan (local name: *mampan sokot*) at postpartum. They also use "*rupot rupot*" or "*rupet rupet*".

(xiii) *Intsia palembanica* Miq. is used by the Dusun and Kadazan as styptic (local name: *tupin*).

(xiv) *Koompassia malaccensis* Maing. is used by the Bajau (local name: *raja kayu*) for stomach aches, dysentery, bloating, allergy, toothaches, swollen gums, asthma, convulsions, gastritis, and body pains.

(xv) *Parkia singularis* Miq. is used by the Murut (local name: *kundai*) for kidney detoxification.

(xvi) *Parkia speciosa* Hassk. (1842) is used by the Bajau (local name: *petai*) for flu, hypertension, kidney diseases, and indigestion.

(xvii) *Sesbania grandiflora* L. is used by the Kedayan (local name: *pokok geti*) for canker sores, coughs, diarrhea, phlegm, and stomach aches. The Bajau call this plant "*daun turi*".

(xviii) *Spatholobus gyrocarpus* Benth, is used by the Murut (local name: *ramus*) for thrush. *Spatholobus ferrugineus* (Zoll. & Moritzi) Benth. is used by the Dusun as an aphrodisiac. The Murut and Dusun use plants of the genus *Spatholobus* Hassk. (1852) as medicines: "*belohu*" for cough and diarrhea whilst "*lipoi*" is given for fever and cold. Another medicinal plant in this genus is "*gingor*" sold by root gatherers in the wet markets of Sabah.

(xix) *Entada rheedei* Spreng. is used by the Dusun for itchiness.

(xx) A plant of the genus *Erythrina* L. (1753) is used by the Dusun for fever (local name: *radap*).

(xxi) *Neptunia oleracea* Lour. is used as vegetable by the locals (local name: *semalu air*) (Dr. Pauline Yap, personal communication).

REFERENCES

Alah, N.S.K., Eltayeb, I.M. and Hamad, A.E.H. 2017. Phytochemical screening and hypolipidemic activity of extracts from seeds and leaves of *Vigna unguiculata* growing in Sudan. *Journal of Pharmacognosy and Phytochemistry*. 6(3): 488–491.

Aziagba, B.O., Okeke, C.U., Ezeabara, A.C., Ilodibia, C.V., Ufele, A.N. and Egboka, T.P. 2017. Determination of the flavonoid composition of seven varieties of *Vigna unguiculata* (L.) walp as food and therapeutic values. *Universal Journal of Applied Science*. 5(1): 1–4.

Haryati, N.P.S., Kurniawati, E.D., Lestary, T.T., Norahmawati, E., Wiyasa, I.W.A., Hidayati, D.Y.N. and Nurseta, T. 2021. Cowpea (*Vigna unguiculata*) extract reduce malondialdehyde levels and prevent aortic endothelial cell decline in ovariectomized rats. *Medical Laboratory Technology Journal*. 7(2): 132–143.

Ojwang, L.O., Banerjee, N., Noratto, G.D., Angel-Morales, G., Hachibamba, T., Awika, J.M. and Mertens-Talcott, S.U. 2015. Polyphenolic extracts from cowpea (*Vigna unguiculata*) protect colonic myofibroblasts (CCD18Co cells) from lipopolysaccharide (LPS)-induced inflammation—modulation of microRNA 126. *Food & Function*. 6(1): 145–153.

Sandeep, D. 2014. Evaluation of antibacterial activity of seed extracts of *Vigna unguiculata*. *International Journal of Pharmacy and Pharmaceutical Sciences*. 6(1): 75–77.

4.1.10.2.2 Family Polygalaceae Hoffmanns. & Link (1809), the Milkwort Family

The family Polygalaceae consists of about 45 genera and 1,000 species of herbaceous plants, shrubs, or trees. Leaves: simple, alternate, opposite, or whorled, exstipulate. Sepals: 3–5, free or connate, sometimes all alike. Petals: 3–5, forming a somewhat fabaceous corolla. Stamens: 6–8, free or merged at base. Carpels: 2, forming a bilocular ovary, each locule with 1 ovule. Styles: 3. Fruits: capsules, samaras, or drupes.

***Polygala paniculata* L.**

[From Greek *polus* = many, *gala* = milk, and Latin *paniculata* = paniculate]

Published in *Systema Naturae, Editio Decima* 2: 1154. 1759

Synonym: *Polygala tenella* Willd.
Local names: *gosok* (Dusun); *bunga tali tali* (Dusun, Bajau); *mentimagas* (Dusun, Kadazan)
Habitat: lawns, roadsides, open lands
Geographical distribution: tropical
Botanical identification: this delicate and diffuse herbaceous plant reaches a height of about 50 cm. Stems: terete, thin. Leaves: simple, sessile, alternate, exstipulate. Blades: linear, 1–4 mm × 5–20 mm, attenuate at base, dull green, acute at apex. Racemes: terminal or opposite to leaves, up to 15 cm long. Sepals: 3, the inner 2 purple, petaloid, elliptic-oblong, and 2 mm long. Petals: 3, white to purplish, forming an elongated corolla. Stamens: 8, merged at base into a sheath. Stigma: cup-shaped. Capsules: oblong, 2 mm long, dehiscent. Seeds: 2, oblong, blackish
Traditional therapeutic indications: flatulence (Dusun); coughs, fever, gastritis, hypertension, toothaches (Dusun, Kadazan)
Pharmacology and phytochemistry: antifungal (Hamburger et al. 1985), hypotensive (da Rocha-Lapa et al. 2011), antidepressant (Bettio et al. 2011), analgesic (Lapa et al. 2009), neuroprotective (Farina et al. 2005).
Prenylated coumarins (aurapten, phebalosin, murrangatin and 7-methoxy-8-(1,4-dihydroxy-3-methyl-2-butenyl) coumarin) (Hamburger et al. 1985). Other constituents are xanthones (Cristiano et al. 2003) and flavone glycosides (rutin) (Lapa et al. 2009).
Toxicity, side effects, and drug interaction: not known

Comment: A plant of the genus *Polygala* Hoffmanns. & Link (1809) is used for heart diseases by the Dusun and Kadazan.

REFERENCES

Bettio, L.E., Machado, D.G., Cunha, M.P., Capra, J.C., Missau, F.C., Santos, A.R., Pizzolatti, M.G. and Rodrigues, A.L.S. 2011. Antidepressant-like effect of extract from Polygala paniculata: Involvement of the monoaminergic systems. *Pharmaceutical Biology*. 49(12): 1277–1285.

Cristiano, R., Pizzolatti, M.G., Monache, F.D., Rezende, C.M. and Branco, A. 2003. Two xanthones from *Polygala paniculata* and confirmation of the 1-hydroxy-2, 3, 5-trimethoxy-xanthone at trace level by HRGC-MS. *Zeitschrift für Naturforschung C*. 58(7–8): 490–494.

da Rocha Lapa, F., Soares, K.C., Rattmann, Y.D., Crestani, S., Missau, F.C., Pizzolatti, M.G., Marques, M.C.A., Rieck, L. and Santos, A.R.S. 2011. Vasorelaxant and hypotensive effects of the extract and the isolated flavonoid rutin obtained from *Polygala paniculata* L. *Journal of Pharmacy and Pharmacology*. 63(6): 875–881.

Farina, M., Franco, J.L., Ribas, C.M., Meotti, F.C., Dafré, A.L., Santos, A.R., Missau, F. and Pizzolatti, M.G. 2005. Protective effects of *Polygala paniculata* extract against methylmercury-induced neurotoxicity in mice. *Journal of Pharmacy and Pharmacology*. 57(11): 1503–1508.

Hamburger, M., Gupta, M. and Hostettmann, K. 1985. Coumarins from *Polygala paniculata*. *Planta Medica*. 51(3): 215–217.

Lapa, F.D.R., Gadotti, V.M., Missau, F.C., Pizzolatti, M.G., Marques, M.C.A., Dafré, A.L., Farina, M., Rodrigues, A.L.S. and Santos, A.R. 2009. Antinociceptive properties of the hydroalcoholic extract and the flavonoid rutin obtained from *Polygala paniculata* L. in mice. *Basic & Clinical Pharmacology & Toxicology*. 104: 306–315.

***Xanthophyllum excelsum* (Blume) Miq.**

[From the Greek *xantho* = yellow, *phullon* = leaf, and Latin *excelsum* = lofty]

Published in *Flora van Nederlandsch Indië* 1(2): 9. 1858

Synonyms: *Banisterodes affine* (Korth. ex Miq.) Kuntze; *Banisterodes excelsum* (Blume) Kuntze; *Jackia excelsa* Bl.; *Kaulfussia geminiflora* Dennst.; *Monnina excelsa* (Blume) Spreng.; *Xanthophyllum adenopodum* Miq.; *Xanthophyllum affine* Korth. ex Miq.; *Xanthophyllum angustifolium* Wight; *Xanthophyllum arnottianum* Wight; *Xanthophyllum floriferum* Elmer; *Xanthophyllum geminiflorum* (Dennst.) Alston; *Xanthophyllum glandulosum* Merr.; *Xanthophyllum loheri* Merr.; *Xanthophyllum multiramosum* Elmer; *Xanthophyllum obliquum* Craib; *Xanthophyllum pallidum* Ridl.; *Xanthophyllum roxburghianum* Wight; *Xanthophyllum sarawakensis* Chodat; *Xanthophyllum siamense* Craib; *Xanthophyllum undulatum* Wight; *Xanthophyllum virescens* D.Dietr.

Local name: *lapad atag* (Lundayeh)

Habitat: forests

Geographical distribution: from India to the Philippines

Botanical description: this tree reaches a height of about 30 m. Buttresses: present. Bark: grayish to greenish brown, somewhat smooth. Sapwood: dark straw colored. Stems: terete, sometimes hairy at apex. Leaves: simple, aternate, exstipulate. Petioles: about 1–1.5 cm long. Blades: beautiful, dark green and glossy above, light ochre in herbarium samples, of a peculiar kind of light green beneath, elliptic to oblong-elliptic, 2.8–10 cm × 6–27 cm, with 4–12 pairs of secondary nerves sunken above and somewhat looping at the margin, tertiary veins scalariform, apex acuminate, base attenuate. Panicles: terminal, hairy, up to about 20 cm long. Sepals: 5, hairy. Petals: 5, white to yellow, the upper petal bilobed and yellow at base. Stamens: 8. Drupes: about 2 cm across, dirty white to green, somewhat globose, apiculate. The flowers and fruits can make one think of a plant in the Rutaceae (Figure 4.21).

FIGURE 4.21 *Xanthophyllum excelsum* (Blume) Miq.

Traditional therapeutic indications: gastritis (Lundayeh)
Pharmacology and phytochemistry: antiplasmodial (Horgen et al. 2001).
Toxicity, side effects, and drug interaction: not known

Comment: *Xanthophyllum reticulatum* Chodat is endemic and used for magic rituals by the Dusun (local name: *ngkruab*).

REFERENCE

Horgen, F.D., Edrada, R.A., De Los Reyes, G., Agcaoili, F., Madulid, D.A., Wongpanich, V., Angerhofer, C.K., Pezzuto, J.M., Soejarto, D.D. and Farnsworth, N.R. 2001. Biological screening of rain forest plot trees from Palawan Island (Philippines). *Phytomedicine*. 8(1): 71–81.

4.1.10.3 Order Fagales Engl. (1892)

4.1.10.3.1 Family Casuarinaceae R.Br. (1814), the Casuarina Family

The family Casuarinaceae consists of about 4 genera and 65 species of trees or shrubs, some of them looking like pine trees. Leaves: simple, linear, whorled, exstipulate. Inflorescences: spikes or heads. Perianth: reduced to 1–2 scales or none. Stamens: 1. Carpels: 2 forming a unilocular or bilocular ovary, each locule with 2 ovules. Styles: 1. Stigma: bifid. Fruits: tiny nuts, samaras. Woody syncarps.

Casuarina sumatrana **Jungh. ex de Vriese**

[Probably from the Moluccan *kasuwari* = cassowary bird and Latin *sumatrana* = from Sumatra]

Published in *Tijdschrift voor Natuurlijke Geschiedenis en Physiologie* 11: 115. 1844

Synonym: *Gymnostoma sumatranum* (Jungh. ex de Vriese) L.A.S. Johnson
Local name: *aru* (Bajau)
Common name: Sumatran ru
Habitat: seashores, cultivated
Geographical distribution: from Malaysia to Papua New Guinea
Botanical description: this tree reaches a height of about 10 m. Trunk: not straight. Bark: deeply fissured, light brown, very dry. Stems: thin, terete. Leaves: filiform, articulate, each article about 1 mm × 5 mm, pointed at both ends and grooved. Inflorescences: heads of female flowers and spikes of male flowers. Syncarps: somewhat elliptic, woody, coriaceous, about 6 cm × 4 cm, with some sorts of beaks opening to release little winged seeds.
Traditional therapeutic indications: hair loss (Bajau)
Pharmacology and phytochemistry: not known
Toxicity, side effects, and drug interaction: not known

Comment: A legend says that sitting under this tree during the full moon is a means to know the future.

4.1.10.4 Order Malpighiales Juss, ex Bercht. & J. Presl (1820)

4.1.10.4.1 Family Clusiaceae Lindley (1836), the Garcinia Family

The family Clusiaceae consists of about 50 genera and 1,200 species of trees, climbing plants, or herbaceous plants, exuding a sticky yellow gum resin once incised. Leaves: simple, opposite exstipulate. Petioles: often clasping the stem at base. Flowers: often showy in terminal cymes or racemes or, less often, solitary. Sepals: 2–5. Petals: 3–6. Stamens: often numerous, in bundles. Carpels: 1–13, forming a 1–13 locular ovary, each locule with 2-many. Styles: 1–5. Stigma: lobed or peltate. Fruits: berries, drupes, capsules.

Garcinia mangostana **L.**

[After the 18th century French botanist Laurent Garcin and from the latinized Malay *manggis* = mangosteen]

Published in *Species Plantarum* 1: 443–444. 1753

Synonym: *Mangostana garcinia* Gaertn.
Local name: *timpurog* (Murut)
Common name: mangosteen
Habitat: cultivated
Geographical distribution: tropical Asia

Botanical description: this tree reaches a height of about 10 m. Bark: yellowish brown, yielding an opaque yellow latex once incised. Leaves: simple, decussate, exstipulate. Petioles: 1–2 cm long, stout, cracked, clasping the stems. Blades: ovate to ovate-oblong, coriaceous, 3.5–5.5 cm × 6.5–9.5 cm, attenuate at base, acute at apex, with numerous secondary nerves, which are thin, parallel, 2 mm apart, and slightly joining into an intramarginal nerve. Inflorescences: racemes. Sepals: 4, fleshy, about 1.7 cm long. Petals: 4, fleshy, pinkish, about 2.5 cm long. Stamens: 6–20. Stigma: stout, 6-10-lobed. Berries: 5 cm across, ripening dark purple, with a thick and somewhat sappy pericarp yielding yellow latex, marked by flat stigma divided in wedge-shaped lobes. Seeds: few, embedded into a thick, whitish, and delicious pulp.

Traditional therapeutic indications: blood vomiting (Murut)

Pharmacology and phytochemistry: this plant has been the subject of numerous pharmacological studies (Rizaldy et al. 2021). In regards to pharmacological activities that can be related to medicinal uses, an extract of fruits administered to rats orally at the dose of 1 g/kg/day for 15 days has demonstrated some levels of protection against indomethacin-induced gastric ulcers (Mahendran et al. 2002).

Anti-inflammatory and anti-ulcerogenic prenylated xanthones (α-mangostin and γ-mangostin) (Chen et al. 2008; Sidahmed et al. 2013).

Toxicity, side effects, and drug interaction: the median lethal dose (LD_{50}) of ethylacetate extract of pericaps administered orally to rats was above 15.4 g/kg (Rahmayanti et al. 2016).

Comments:

(i) The Dusun and Kadazan use a plant of the genus *Calophyllum* L. (1753) for back and waist pains, pimples, and for kidney diseases (local name: *lawong*).

(ii) *Garcinia parvifolia* (Miq.) Miq. (Figure 4.22) is used by the Dusun for postpartum (local name: *kandis*).

(iii) *Hypericum japonicum* Thunb. ex Murr. is used by the Dusun for ringworms and shingles (local name: *tungkedem*).

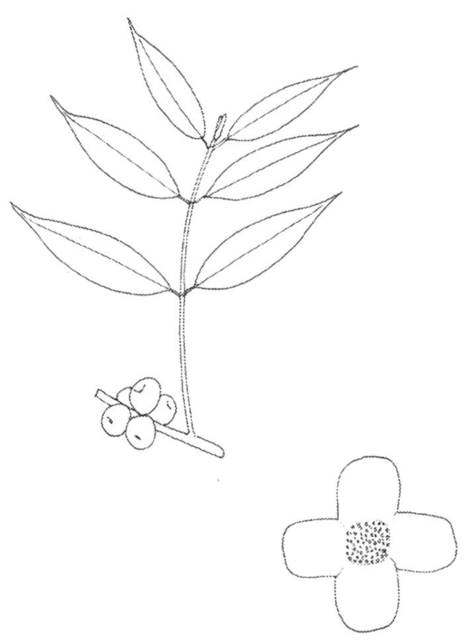

FIGURE 4.22 *Garcinia parvifolia* (Miq.) Miq.

REFERENCES

Chen, L.G., Yang, L.L. and Wang, C.C. 2008. Anti-inflammatory activity of mangostins from *Garcinia mangostana*. *Food and Chemical Toxicology*. 46(2): 688–693.

Mahendran, P., Vanisree, A.J. and Shyamala Devi, C.S. 2002. The antiulcer activity of *Garcinia cambogia* extract against indomethacin-induced gastric ulcer in rats. *Phytotherapy Research*. 16(1): 80–83.

Rahmayanti, F., Sastradipura, D.F.S., Mas'Ud, Z.A., Bachtiar, B.M., Wimardhani, Y.S. and Permana, G. 2016. Acute oral toxicity testing of ethyl acetate fraction from *Garcinia mangostana* Linn extract in sprague-dawley rats. *Research Journal of Medicinal Plant*. 10(3): 261–264.

Rizaldy, D., Hartati, R., Nadhifa, T. and Fidrianny, I. 2021. Chemical compounds and pharmacological activities of mangosteen (*Garcinia mangostana* L.)—Updated review. *Biointerface Research in Applied Chemistry*. 12: 2503–2516.

Sidahmed, H., Abdelwahab, S.I., Mohan, S., Abdulla, M.A., Mohamed Elhassan Taha, M., Hashim, N.M., Hadi, A.H.A., Vadivelu, J., Loke Fai, M., Rahmani, M. and Yahayu, M. 2013. α-Mangostin from *Cratoxylum arborescens* (Vahl) Blume demonstrates anti-ulcerogenic property: A mechanistic study. *Evidence-Based Complementary and Alternative Medicine*. 2013: 1–10.

4.1.10.4.2 Family Dichapetalaceae Baill. (1886), the Dichapetalum Family

The family Dichapetalaceae consists of about 4 genera and 200 species of treelets or shrubs. Leaves: simple, alternate, stipulate. Inflorescences: axillary cymes or capitula. Sepals: 5, free, or merged at base. Corolla: 5 petals or tubular. Stamens: 5, free, or merged. Disc: lobed. Carpels: 2–3 forming a bilocular or trilocular ovary each locule with 2 ovules per locule. Styles: 1–2. Fruits: drupes.

***Dichapetalum gelonioides* (Roxb.) Engl.**

[From Greek *dicha* = two, *petalon* = petal, and Greek *gelonioides* = resembling *Gelonium*]

Published in *Die Natürlichen Pflanzenfamilien* 3(4): 348. 1896

Synonyms: *Chailletia gelonioides* (Roxb.) Hook.f.; *Moacurra gelonioides* Roxb.
Common name: gelonium poison-leaf
Local name: *kokos* (Bajau)
Habitat: forests
Geographical distribution: from India to the Philippines
Botanical description: this shrub reaches a height of about 5 m. Stems: terete, zigzag-shaped at apex. Leaves: simple, alternate, stipulate. Stipules: needlelike, about 3 mm long, hairy. Petioles: ligneous, 3–5 mm long, channeled. Blades: elliptic or oblong-elliptic, 2–6 cm × 6–16 cm, dark green and glossy above, with 5 or 6 pairs of secondary nerves looping (like smiles?), somewhat wavy, attenuate at base, acuminate to almost caudate at apex (with resemblance to the leaves of *Gelonium multiflorum* (A. Juss) Baill). Cymes: axillary, up to about 1 cm long. Sepals: 5, oblong, about 2 mm long, yellowish. Petals: 5, spathulate, somewhat lobed, pure white, about 4 mm long. Stamens: 5. Disc glands: 5, rounded. Ovary: yellow-brown, hairy. Syles: 1. Stigma: trifid. Drupes: obcordate to elliptic, or bilobed, somewhat compressed, about 1.7 cm across, dull green to glaucous, somewhat velvety.
Traditional therapeutic indications: dysmenorrhea, postpartum, gastritis, colds, deafness, hypertension, diabetes (Bajau); aphrodisiac (Dusun)
Pharmacology and phytochemistry: antiplasmodial (Zamil et al. 2022), antifungal, nematicidal, anti-inflammatory triterpenes (dichapetalins) (Jing et al. 2014; Zhang et al. 2021). Dichapetalin A, I, and J are toxic to human ovarian cancer cells (SW626) (Jing et al. 2014). Diterpenes (Fang et al. 2006).
Toxicity, side effects, and drug interaction: plants of the genus *Dichapetalum* Thouars (1806) produce poisonous monofluoroacetate derivatives (Vickery & Vickery 1975).

Comments:

(i) Parts of the plant are sold in the wet markets of Sabah as medicine.
(ii) A plant of the genus *Dichapetalum* Thouars (1806) is used by the Murut for fatigue (local name: *akar urat*).

REFERENCES

Fang, L., Ito, A., Chai, H.B., Mi, Q., Jones, W.P., Madulid, D.R., Oliveros, M.B., Gao, Q., Orjala, J., Farnsworth, N.R. and Soejarto, D.D. 2006. Cytotoxic constituents from the stem bark of *Dichapetalum gelonioides* collected in the Philippines. *Journal of Natural Products*. 69(3): 332–337.

Jing, S.X., Luo, S.H., Li, C.H., Hua, J., Wang, Y.L., Niu, X.M., Li, X.N., Liu, Y., Huang, C.S., Wang, Y. and Li, S.H. 2014. Biologically active dichapetalins from *Dichapetalum gelonioides*. *Journal of Natural Products*. 77(4): 882–893.

Vickery, B. and Vickery, M.L. 1975. The synthesis and defluorination of monofluoroacetate in some *Dichapetalum* species. *Phytochemistry*. 14(2): 423–427.

Zamil, M.F., Sazed, S.A., Hossainey, M.R.H., Biswas, A., Alam, M.S., Khanum, H. and Barua, P. 2022. Antimalarial investigation of *Acorus calamus, Dichapetalum gelonioides,* and *Leucas aspera* on *Plasmodium falciparum* strains. *The Journal of Infection in Developing Countries*. 16(11): 1768–1772.

Zhang, D.L., Li, M., Xu, W.F., Yu, H., Jin, P.F., Li, S.Y. and Tang, S.A. 2021. Nine new dichapetalin-type triterpenoids from the twigs of *Dichapetalum gelonioides* (Roxb.) Engl. *Fitoterapia*. 151: 104868.

4.1.10.4.3 Family Euphorbiaceae Juss. (1789), the Spurge Family

The family Euphorbiaceae consists of about 300 genera and about 8,000 species of trees, shrubs, herbaceous plants, climbing plants, or even cactuses, often exuding a milky poisonous latex once incised. Leaves: simple or compound, alternate, sometimes stipulate. Numerous sorts of inflorescences occur in this family. Flowers: small, unisexual. Sepals: none or, 2–6, or tubular. Petals: none or 3–5. Stamens: 1-many. Disc: present. Carpels: 3, forming a 3-locular ovary, each locule with 1–2 ovules. Styles: 3. Fruits: trilocular capsules, berries, or drupes.

***Antidesma montanum* Bl.**

[From Greek *anti* = against, the Greek *desma* = chain, and Latin *montanum* = mountainous]

Published in *Bijdragen tot de flora van Nederlandsch Indië* (17): 1124. 1826

Local name: *damat mandalom* (Murut)
Common name: mountain currant tree
Habitat: forests, on the banks of rivers
Geographical distribution: from India to Australia
Botanical description: this tree reaches a height of about 20 m. Stems: hairy at apex. Leaves: simple, alternate, stipulate. Stipules: linear, about 1 cm long. Petioles: 2 mm–1 cm long, hairy. Blades: coriaceous, oblong, elliptic, or lanceolate, 1.5–10 cm × 3–25 cm, with 6–12 pairs of secondary nerves, acute at base, acuminate at apex. Spikes: about 10 cm long. Perianth: cupular and 3- to 4-lobed. Disc: present. Stamens: 3–5. Ovary: glabrous. Stigma: 2–6-lobed. Drupes: somewhat globose, about 5 mm across, and with a magnificent range of yellow to red to dark blue (the whole structure can be reminiscent of a spike of black pepper fruits).
Traditional therapeutic indications: chest pain (Murut)
Pharmacology and phytochemistry: diuretic triterpene (friedelin), anti-inflammatory alkane (n-titriacontane), triterpenes (antidesmol A, canophyllal, cacnophyllol) (Rizvi et al. 1980).
Toxicity, side effects, and drug interaction: not known

REFERENCE

Rizvi, S.H., Shoeb, A., Kapil, R.S. and Popli, S.P. 1980. Antidesmanol-a new pentacyclic triterpenoid from *Antidesma menasu* Miq. ex. Tul. *Experientia*. 36: 146–147.

Baccaurea lanceolata **(Miq.) Müll.Arg.**

[From Latin *bacca* = berry, *aurea* = golden, and *lancea* = lance]

Published in *Prodromus Systematis Naturalis Regni Vegetabilis* 15(2): 457. 1866

Synonyms: *Baccaurea glabriflora* Pax & K.Hoffm.; *Baccaurea pyrrhodasya* (Miq.) Müll. Arg.; *Pierardia pyrrhodasya* Miq.
Local names: *limposu* (Murut); *lepasu, lipasu, liposu, nipassu* (Dusun)
Habitat: forests
Geographical distribution: Thailand, Malaysia, Indonesia, the Philippines
Botanical description: this tree reaches a height of about 30 m. Bark: grayish, more or less scaly. Stems: hairy. Leaves: simple, alternate, stipulate. Stipules: acuminate, about 1 cm long and caducous. Petioles: 2–10 cm long, straight. Blades: broadly elliptic to lanceolate, membranous, somewhat asymmetrical, wavy, acute at base, acute to acuminate at apex, 3–30 cm × 7–40 cm, with 6–13 pairs of secondary nerves. Spikes: cauliflorous, up to 30 cm long, made of innumerable tiny flowers, pendulous. Sepals: 4–5, spathulate. Petals: 3–5. Stamens: 4–5. Ovary: oblong. Stigma: bifid. Berries: about 5 cm across, dull light green. Seeds: embedded in an edible, white, and translucent aril (Figure 4.23).
Traditional therapeutic indications: abdominal pain, stomach aches (Murut); cuts, bone aches (Dusun, Kadazan)
Pharmacology and phytochemistry: antibacterial (Fitriansyah et al. 2018).
Toxicity, side effects, and drug interaction: not known

REFERENCE

Fitriansyah, S.N., Putri, Y.D., Haris, M. and Ferdiansyah, R. 2018. Aktivitas antibakteri ekstrak etanol buah, daun, dan kulit batang limpasu (*Baccaurea lanceolata* (Miq.) Müll.Arg.) dari Kalimantan Selatan. *PHARMACY: Jurnal Farmasi Indonesia (Pharmaceutical Journal of Indonesia)*. 15(2): 111–119.

FIGURE 4.23 *Baccaurea lanceolata* (Miq.) Müll.Arg.

***Bischofia javanica* Bl.**

[After the 19th century German botanist Gottleib Wilhelm Bischoff and Latin *javanica* = from Java]

Published in *Bijdragen tot de flora van Nederlandsch Indië* (17): 1168. 1826

Synonyms: *Andrachne trifoliata* Roxb.; *Bischofia cumingiana* Decne.; *Bischofia leptopoda* Müll.Arg.; *Bischofia oblongifolia* Decne.; *Bischofia roeperiana* Decne. ex Jacquem.; *Bischofia toui* Decne.; *Bischofia trifoliata* (Roxb.) Hook.; *Microelus roeperianus* (Decne. ex Jacquem.) Wight & Arn.; *Stylodiscus trifoliatus* (Roxb.) Benn
Local names: *kapas kapas, tungo* (Dusun); *tongon* (Dusun, Kadazan)
Common name: bishopwood tree
Habitat: forests, around villages
Geographical distribution: from India to Pacific Islands
Botanical description: this tree reaches a height of about 40 m and has some kind of sinister aura. Trunk: not straight, buttressed. Bark: brownish, rough, cracked. Inner bark: reddish. Leaves: trifoliate, shaped like rubber tree leaves, spiral, stipulate. Stipules: about 5 mm long. Petioles: thin, straight, about 5.5–17.5 cm long. Folioles: elliptic to oblanceolate, 2.5–10 cm × 5–18 cm, coriaceous, dull green to glossy, serrate, attenuate at base, acuminate at apex, with 7–8 pairs of secondary nerves. Panicles: lax, axillary, up to about 30 cm long. Sepals: 5, about 4 mm long, rounded, hooded (male flowers), or lanceolate (female flowers). Stamens: 5, merged at base. Ovary: tiny, ovoid. Stigma: trifid, linear, about 5 mm long. Drupes: numerous, claimed to be edible (?), somewhat globose, light dull brown, about 1 cm across. Seeds: curved, about 4 mm long (Figure 4.24).

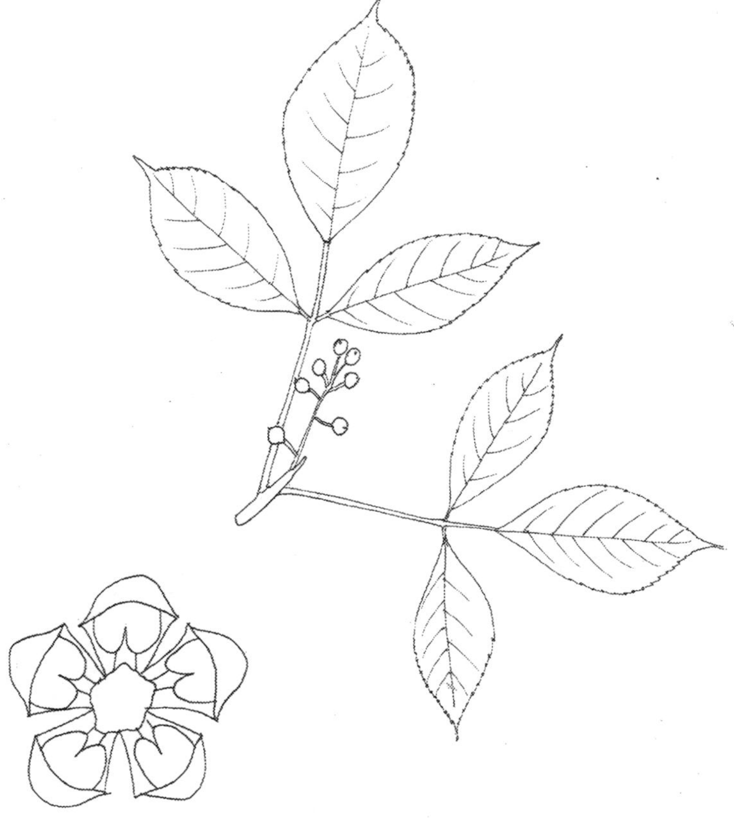

FIGURE 4.24 *Bischofia javanica* Bl.

Traditional therapeutic indications: blood circulation, diarrhea (Dusun); stomach aches (Dusun, Kadazan)

Pharmacology and phytochemistry: antiviral (poliovirus) (Lipipun et al. 2003), toxic to leukemia cells (U937, K562, HL60) (Lingadurai et al. 2011), anti-inflammatory (Pangondian et al. 2020), antibacterial (Khan et al. 2001).

Tannins (geraniin, corilagin, punicalagin, procyanidin B-1) (Tanaka et al. 1995). Triterpernes inhibitors of topoisomerase II (betulinic acid, 3 β-*O*-(*Z*)-coumaroylbetulinic acid, and 3 β-*O*-(*E*)-coumaroylbetulinic) (Wada et al. 2005).

Toxicity, side effects, and drug interaction: not known

Comment: Compared with other medicinal practices in Southeast Asia, the Dusunic traditional system of medicine in Sabah seems to employ more baths and inhalations made of plant decoctions.

REFERENCES

Khan, M.R., Kihara, M. and Omoloso, A.D. 2001. Anti-microbial activity of *Bidens pilosa, Bischofia javanica, Elmerillia papuana* and *Sigesbekia orientalis*. *Fitoterapia*. 72(6): 662–665.

Lingadurai, S., Roy, S., Joseph, R.V. and Nath, L.K. 2011. Antileukemic activity of the leaf extract of *Bischofia javanica* blume on human leukemic cell lines. *Indian Journal of Pharmacology*. 43(2): 143.

Lipipun, V., Kurokawa, M., Suttisri, R., Taweechotipatr, P., Pramyothin, P., Hattori, M. and Shiraki, K. 2003. Efficacy of Thai medicinal plant extracts against herpes simplex virus type 1 infection *in vitro* and *in vivo*. *Antiviral Research*. 60(3): 175–180.

Pangondian, A., Nainggolan, M. and Dalimunthe, A. 2020. Characterization and anti-inflammatory activity of ethanol extract of sikkam (*Bischofia javanica* Blume) stem bark. *Asian Journal of Pharmaceutical Research and Development*. 8(4): 16–20.

Tanaka, T., Nonaka, G.I., Nishioka, I., Kouno, I. and Ho, F.C. 1995. Bischofianin, a dimeric dehydroellagitannin from *Bischofia javanica*. *Phytochemistry*. 38(2): 509–513.

Wada, S.I. and Tanaka, R. 2005. Betulinic acid and its derivatives, potent DNA topoisomerase II inhibitors, from bark of *Bischofia javanica*. *Chemistry & Biodiversity*. 2(5): 689–694.

***Bridelia stipularis* (L.) Bl.**

[After the 19th century Swiss bryologist Samuel Elisée Bridel-Brideri and Latin *stipularis* = with stipules]

Published in *Bijdragen tot de flora van Nederlandsch Indië* (12): 597. 1825

Synonyms: *Bridelia retusa* (L.) A. Juss.; *Bridelia scandens* (Roxb.) Willd.; *Clutia scandens* Roxb.; *Clutia stipularis* L.

Local names: *belingkut, bingkarut* (Dusun); *kenidai babi, kenidai sanak, cenerai gajah, kernam* (Bajau); *bolingkut* (Murut); *kutu, kandrikebo* (Lundayeh)

Habitat: forests, roadsides

Geographical distribution: from India to the Philippines

Botanical description: this shrub reaches a height of about 10 m tall. Stems: zigzag-shaped, hairy at apex. Leaves: simple, alternate, stipulate. Stipules: triangular, up to about 1 cm long. Petioles: up to 1.3 cm long, hairy. Blades: broadly elliptic, 4–17 cm × 2–11 cm, obtuse to rounded at base, hairy beneath, dull green, acute to obtuse at apex, somewhat wavy, with 10–14 pairs of secondary nerves. Spikes: axillary. Sepals: 5, triangular, about 5 mm long. Petals: 5, spathulate, dentate at apex, about 2 mm long. Disc: present. Stamens: 5, merged at base. Ovary: ovoid. Styles: 2. Drupes: globose, 7 mm–1.3 cm long, green, glossy, bilocular. Seeds: about 8 mm long, yellowish brown, eaten by birds, notably the magnificent pink-necked green pigeon.

Traditional therapeutic indications: diabetes (Dusun, Kadazan, Murut); fever, postpartum (Dusun, Kadazan); thrush (Dusun, Kadazan, Murut)

Pharmacology and phytochemistry: antibacterial (Anjum et al. 2011), antidiabetic (Khan et al. 2018).

Toxicity, side effects, and drug interaction: not known

REFERENCES

Anjum, A., Haque, M.R., Rahman, M.S., Hasan, C.M., Haque, M.E. and Rashid, M.A. 2011. *In vitro* antibacterial, antifungal and cytotoxic activity of three Bangladeshi Bridelia species. *International Research of Pharmacy and Pharmacology*. 1(7): 149–154.

Khan, S.A., Islam, R., Alam, J.M. and Rahman, M.M. 2018. Investigation of anti-diabetic properties of ethanol leaf extract of *Bridelia stipularis* L. on alloxan induced type-2 diabetic rats. *Journal of Advances in Medical and Pharmaceutical Sciences*. 18(4): 1–9.

Euphorbia hirta L.

[After the 1st-century BC Greek physician Euphorbus and Latin *hirta* = hairy]

Published in *Species Plantarum* 1: 454. 1753

Synonyms: *Anisophyllum piluliferum* (L.) Haw.; *Chamaesyce gemella* (Lag.) Small; *Chamaesyce hirta* (L.) Millsp.; *Chamaesyce karwinskyi* (Boiss.) Millsp.; *Chamaesyce microcephala* (Boiss.) Croizat; *Chamaesyce pilulifera* (L.) Small; *Chamaesyce rosei* Millsp.; *Euphorbia bancana* Miq.; *Euphorbia capitata* Lam.; *Euphorbia chrysochaeta* W. Fitzg.; *Euphorbia gemella* Lag.; *Euphorbia globulifera* Kunth; *Euphorbia karwinskyi* Boiss.; *Euphorbia microcephala* Boiss.; *Euphorbia nodiflora* Steud.; *Euphorbia obliterata* Jacq.; *Euphorbia pilulifera* L.; *Euphorbia verticillata* Vell.; *Tithymalus pilulifer* (L.) Moench; *Tithymalus piluliferus* (L.) Moench

Local names: *rumput susu* (Dusun); *gelang susu*, *ara tanah* (Bajau); *rumput susu kambing* (Kedayan); *kohonsizud* (Dusun, Kadazan)

Common names: hairy spurge, asthma-plant

Habitat: abandoned lands, parking lots, roadsides, villages

Geographical distribution: subtropical, tropical

Botanical description: this strange herbaceous plant reaches a height of about 25 cm. It produces an irritating milky latex. Stems: hairy, reddish, subglabrous, scorpioid. Leaves: simple, opposite, stipulate. Stipules: triangular, tiny, caducous. Petioles: up to about 3 mm long. Blades: asymmetrical, serrate, lanceolate to oblong, hairy, obtuse at apex, green with purplish margins, 2–4 cm × 0.8–1.5 cm. Cymes: somewhat capitate, axillary. Cyathia: tiny. Stamens: 1. Ovary: trilobed, hairy. Stigma: bifid. Capsules: tiny, trilobed, smooth.

Traditional therapeutic indications: asthma, headaches (Kedayan); sore eyes, broken bones (Lundayeh); itchiness, swellings, boils (Dusun, Kadazan)

Pharmacology and phytochemistry: this plant has been the subject of numerous pharmacological studies which have highlighted, among other things, nematicidal (Ndjonka et al. 2011), antiplasmodial (Attah et al. 2013; Tona et al. 1999), antibacterial (Sudhakar et al. 2006; Vijaya et al. 1995), antifungal (Sudhakar et al. 2006), antiviral (HIV) (Hamidi et al. 1996), cytotoxic (Duez et al. 1991), analgesic, antipyretic, and anti-inflammatory activity (Lanhers et al. 1991).

Antiviral flavan (epicatechin 3-*O*-gallate) (Perumal et al. 2015). Antiviral (hepatitis B virus) and antibacterial hydroxycinnamic acid derivatives (caffeic acid, chlorogenic acid) (Perumal et al. 2015; Wang et al. 2009). Antiviral (influenza virus) tannin (1,3,4,6-tetra-*O*-galloyl-β-D-glucopyranoside) (Chang et al. 2016). Phenolic compounds (ellagic acid, gallic acid) (Chang et al. 2016; Weng et al. 2019).

Toxicity, side effects, and drug interaction: the median lethal dose (LD_{50}) of methanol extract administered orally to rats was above 6 g/kg (Yuet Ping et al. 2013).

REFERENCES

Attah, S.K., Ayeh-Kumi, P.F., Sittie, A.A., Oppong, I.V. and Nyarko, A.K. 2013. Extracts of *Euphorbia hirta* Linn. (Euphorbiaceae) and *Rauvolfia vomitoria* Afzel (Apocynaceae) demonstrate activities against *Onchocerca volvulus* microfilariae *in vitro*. *BMC Complementary and Alternative Medicine*. 13(1): 66.

Chang, S.Y., Park, J.H., Kim, Y.H., Kang, J.S. and Min, J.Y. 2016. A natural component from *Euphorbia humifusa* Willd displays novel, broad-spectrum anti-influenza activity by blocking nuclear export of viral ribonucleoprotein. *Biochemical and Biophysical Research Communications*. 471(2): 282–289.

Duez, P., Livaditis, A., Guissou, P.I., Sawadogo, M. and Hanocq, M. 1991. Use of an *Amoeba proteus* model for *in vitro* cytotoxicity testing in phytochemical research. Application to *Euphorbia hirta* extracts. *Journal of Ethnopharmacology*. 35(2–3): 235–246.

Hamidi, J.A., Ismaili, N.H., Ahmadi, F.B. and Lajisi, N.H. 1996. Antiviral and cytotoxic activities of some plants used in Malaysian indigenous medicine. *Pertanika Journal of Tropical Agricultural Science*. 19(2/3): 129–136.

Lanhers, M.C., Fleurentin, J., Dorfman, P., Mortier, F. and Pelt, J.M. 1991. Analgesic, antipyretic and anti-inflammatory properties of *Euphorbia hirta*. *Planta Medica*. 57(3): 225–231.

Ndjonka, D., Agyare, C., Lüersen, K., Djafsia, B., Achukwi, D., Nukenine, E.N., Hensel, A. and Liebau, E. 2011. *In vitro* activity of Cameroonian and Ghanaian medicinal plants on parasitic (*Onchocerca ochengi*) and free-living (*Caenorhabditis elegans*) nematodes. *Journal of Helminthology*. 85(3): 304–312.

Perumal, S., Mahmud, R. and Ramanathan, S. 2015. Anti-infective potential of caffeic acid and epicatechin 3-gallate isolated from methanol extract of *Euphorbia hirta* (L.) against *Pseudomonas aeruginosa*. *Natural Product Research*. 29(18): 1766–1769.

Sudhakar, M., Rao, C.V., Rao, P.M., Raju, D.B. and Venkateswarlu, Y. 2006. Antimicrobial activity of *Caesalpinia pulcherrima, Euphorbia hirta* and *Asystasia gangeticum*. *Fitoterapia*. 77(5): 378–380.

Tona, L., Ngimbi, N.P., Tsakala, M., Mesia, K., Cimanga, K., Apers, S., De Bruyne, T., Pieters, L., Totte, J. and Vlietinck, A.J. 1999. Antimalarial activity of 20 crude extracts from nine African medicinal plants used in Kinshasa, Congo. *Journal of Ethnopharmacology*. 68(1–3): 193–203.

Vijaya, K., Ananthan, S. and Nalini, R. 1995. Antibacterial effect of theaflavin, polyphenon 60 (*Camellia sinensis*) and *Euphorbia hirta* on *Shigella* spp.—a cell culture study. *Journal of Ethnopharmacology*. 49(2): 115–118.

Wang, G.F., Shi, L.P., Ren, Y.D., Liu, Q.F., Liu, H.F., Zhang, R.J., Li, Z., Zhu, F.H., He, P.L., Tang, W. and Tao, P.Z. 2009. Anti-hepatitis B virus activity of chlorogenic acid, quinic acid and caffeic acid *in vivo* and *in vitro*. *Antiviral Research*. 83(2): 186–190.

Weng, J.R., Lin, C.S., Lai, H.C., Lin, Y.P., Wang, C.Y., Tsai, Y.C., Wu, K.C., Huang, S.H. and Lin, C.W. 2019. Antiviral activity of *Sambucus f ormosana* Nakai ethanol extract and related phenolic acid constituents against human coronavirus NL63. *Virus Research*, 273: 197767.

Yuet Ping, K., Darah, I., Chen, Y., Sreeramanan, S. and Sasidharan, S. 2013. Acute and subchronic toxicity study of *Euphorbia hirta* L. methanol extract in rats. *BioMed Research International*. 2013: 1–14.

***Euphorbia prostrata* Aiton**

[After the 1st century BC Greek physician Euphorbus and Latin *prostratus* = prostrate]

Published in *Hortus Kewensis; or, a catalogue*. . . 2: 139. 1789

Synonyms: *Chamaesyce prostrata* (Aiton) Small; *Euphorbia callitrichoides* Kunth
Local names: *nipon nipon betina, nipon nipon merah, galung galung, sinting anak, rolapan* (Murut); *dukun anak betina* (Bajau); *meniran, gendong anak* (Lundayeh)
Common names: ground spurge, prostrate sandmat
Habitat: roadsides, wastelands, villages
Geographical distribution: tropical
Botanical description: this prostrate herbaceous plant grows up to a length of about 20 cm. Roots: fibrous. Stems: thin, smooth, reddish. Leaves: simple, sometimes sessile, opposite, stipulate. Stipules: triangular, tiny, caducous. Petioles: tiny or absent. Blades: elliptic to obovate, membranous, 3–7 mm × 2–5 mm, somewhat asymmetrical at base, rounded at apex. Cyathia: tiny, axillary. Tepals: 5, petal-like with triangular appendages. Stamens: 1. Stigma: bifid. Capsules: trilobed, green, smooth.
Traditional therapeutic indications: yellow fever (Murut); fever, hypertension, malaria (Dusun)
Pharmacology and phytochemistry: antiviral (HIV) (Hussein et al. 1999), antibacterial (Voukeng et al. 2017). Ellagitannin oligomers (prostratin A, B, and C) (Yoshida et al. 1990).
Toxicity, side effects, and drug interaction: methanol extract administered orally to rats at the dose of 400 mg/kg induced atrophy of reproductive organs (Bataineh and Mohammad 2012).

REFERENCES

Bataineh, H.N. and Mohammad, M.A. 2012. Effects of oral administration of *Euphorbia prostrata* extract on the reproductive system of male albino rats: A histometric and biochemical study. *Comparative Clinical Pathology*. 21: 433–439.

Hussein, G., Miyashiro, H., Nakamura, N., Hattori, M., Kawahata, T., Otake, T., Kakiuchi, N. and Shimotohno, K. 1999. Inhibitory effects of Sudanese plant extracts on HIV-1 replication and HIV-1 protease. *Phytotherapy Research: An International Journal Devoted to Pharmacological and Toxicological Evaluation of Natural Product Derivatives*. 13(1): 31–36.

Voukeng, I.K., Beng, V.P. and Kuete, V. 2017. Multidrug resistant bacteria are sensitive to *Euphorbia prostrata* and six others Cameroonian medicinal plants extracts. *BMC Research Notes*. 10(1): 1–8.

Yoshida, T., Namba, O., Chen, L., Liu, Y. and Okuda, T. 1990. Ellagitannin monomers and oligomers from *Euphorbia prostrata* AIT. and oligomers from *Loropetalum chinense* OLIV. *Chemical and Pharmaceutical Bulletin*. 38(12): 3296–3302.

Glochidion macrostigma Hook.f

[From Greek *glochis* = barb of an arrow, *makros* = large, and *stigma* = a mark]

Published in *The Flora of British India* 5: 313. 1887

Synonyms: *Diasperus macrostigma* (Hook.f.) Kuntze; *Glochidion capitatum* J.J.Sm.
Local name: *sondot laling* (Murut)
Habitat: forests
Geographical distribution: Malaysia, Indonesia, Borneo
Botanical description: this tree reaches a height of about 5 m. Stems: terete, hairy at apex. Leaves: simple, closely alternate, stipulate. Stipules: triangular, tiny, caducous. Petioles: fissured, up to 5 mm long, glabrous. Blades: elliptic to oblong, 7.5–10 cm long, asymmetrical, somewhat cordate at base, acute at apex, glaucous beneath, with 5–10 pairs of secondary nerves. Flower fascicles: axillary, about 1 cm across. Sepals: 6, merged at base, triangular to ovate, yellow. Stamens: 5. Ovary: ribbed. Stigma: 5-6-lobed.
Traditional therapeutic indications: feverish colds (Murut)
Pharmacology and phytochemistry: not known
Toxicity, side effects, and drug interaction: not known

Comment: *Glochidion rubrum* Bl. is used by the Dusun for bloody stools (local name: *dampul*).

Homalanthus populneus (Geiseler) Pax & Prantl

[From Greek *homalos* = even, *anthos* = flower, and Latin *populneus* = poplar-shaped]

Published in *Die Natürlichen Pflanzenfamilien* 3(5): 96. 1892

Synonyms: *Carumbium populneum* (Geiseler) Müll.Arg.; *Stillingia populnea* Geiseler
Local names: *dayang mato, mato* (Dusun); *boto boto* (Dusun, Kadazan), *dolimato* (Dusun, Kadazan); *sipapaloi* (Murut)
Habitat: roadsides, forests
Geographical distribution: Thailand, Malaysia, Indonesia, the Philippines
Botanical description: this handsome tree reaches a height of about 10 m. Bark: grayish, yields a white latex once incised. Leaves: simple, spiral, stipulate. Stipules: about 1–2 cm long. Petioles: thin, reddish, up to 15 cm long. Blades: rounded to acute at base, acuminate at apex, lanceolate to diamond-shaped to somewhat poplar-shaped, 1.5–20 cm × 1–5 cm, with 9–15 pairs of secondary nerves. Cymes: somewhat conical, up to 30 cm long, with numerous tiny flowers. Sepals: 2. Stamens: 8–10 in male flowers. Stigma: bifid. Capsules: dull green to almost somewhat glaucous to dark purple, smooth, globose, about 1 cm long. Seeds: 2.
Traditional therapeutic indications: muscle cramps, swollen feet (Dusun); rheumatisms (Kadazan); fatigue, fever, headaches, bone aches, rheumatisms, sprains (Dusun, Kadazan)
Pharmacology and phytochemistry: antibacterial (Ismayati et al. 2021), antiviral (influenza virus, HIV) (Dewi 2014; Kartika 2015).
Toxicity, side effects, and drug interaction: not known

Comment: Plants of the genus *Homalanthus* Wittst. (1852) are used by the Dusun and Kadazan as medicines: "*moropingan*" for toothaches and "*tombubuto*" for pancreatitis.

REFERENCES

Dewi, N.W.E.S. 2014. *Potensi Ekstrak Etanolik Kareumbi (Homalanthus Populneus (Giesel.) Pax.) Dalam Penghambatan Ekspresi Reseptor Sel T: Studi Penghambatan Infeksi HIV* (Doctoral dissertation, Universitas Gadjah Mada).

Ismayati, M., Zulfiana, D., Himmi, S.K., Tarmadi, D., Meisyara, D., Zulfitri, A. and Kartika, T. 2021. Antimicrobial activity of ten extractives from toba, north sumatra and Mt. Merapi National Park Regions, Indonesia. *Jurnal Sylva Lestari*. 9(1): 76–85.

Kartika, N. 2015. *Aktivitas Antivirus Ekstrak Etanolik Batang Tutup Abang (Homalanthus Populneus (Giesel.) Pax) Terhadap Virus Avian Influenza H5N1* (Doctoral dissertation, Universitas Gadjah Mada).

Jatropha curcas **L.**

[From Greek *iatros* = physician, *trophe* = food, and *kuraki* = raven]

Published in *Species Plantarum* 2: 1006. 1753

Castiglionia lobata Ruiz & Pav.; *Curcas adansonii* Endl.; *Curcas curcas* (L.) Britton & Millsp.; *Curcas drastica* Mart.; *Curcas indica* A. Rich.; *Curcas purgans* Medic.; *Curcas purgans* Medik.; *Jatropha acerifolia* Salisb.; *Jatropha afrocurcas* Pax; *Jatropha condor* Wall.; *Jatropha edulis* Cerv.; *Jatropha moluccana* Wall.; *Jatropha tuberosa* Elliot; *Jatropha yucatanensis* Briq.; *Manihot curcas* (L.) Crantz; *Ricinus americanus* Mill.; *Ricinus jarak* Thunb.

Local names: *jarak* (Dusun, Murut); *jarak betina* (Kedayan); *pai pai* (Bugis)

Common names: physic nut, Barbados nut

Habitat: cultivated, villages

Geographical distribution: tropical Asia

Botanical description: this shrub reaches a height of about 3 m. Stems: smooth, covered with very small whitish lenticels, produce an abundant and poisonous milky latex once incised. Leaves: simple, spiral, exstipulate. Petioles: 9–12 cm long, smooth, somewhat straight. Blades: thin, 5-lobed, cordate, wavy, of a peculiar kind of dull green, with 5–10 pairs of secondary nerves, the first 2 emerging from the base. Cymes: terminal, about 8 cm long, with numerous green, tiny flowers. Sepals: 5, lanceolate, about 3 mm long. Petals: 5, free, yellowish-green, about 5 mm long, hairy. Stamens: 10. Ovary: ovoid. Stigma: bifid. Capsules: somewhat globose, fleshy, about 2.5 cm long, dull green turning blackish, dehiscent. Seeds: 6, brownish black, somewhat glossy, oleiferous, about 1.8 cm long.

Traditional therapeutic indications: arthritis, rheumatisms, toothaches (Kedayan); cuts (Dusun), gastritis (Murut); abscesses, convulsions, piles (Bugis); medicinal (Bajau)

Pharmacology and phytochemistry: wound healing (Esimone et al. 2008), amoebicidal (Tona et al. 1998; Tona et al. 2000), nematicidal (Monteiro et al. 2011), antibacterial (Rachana et al. 2012), antiviral (HIV) (Matsuse et al. 1998).

Lignan (isoamericanol A) toxic to breast cancer cells (MCF-7) (Katagi et al. 2016). Antibacterial phorbol esters (Devappa et al. 2012).

Toxicity, side effects, and drug interaction: promotes tumor growth (Horiuchi et al. 1987).

Comment: The presence of tumor promoters and anticancer agents in different parts of this plant exemplifies the fact that a plant often produces both a poison and its antidote as believed by Asian medicinal traditions. There is always a duality or polarity as explained thousands of years ago by the sages Ancient Egypt.

REFERENCES

Devappa, R.K., Rajesh, S.K., Kumar, V., Makkar, H.P. and Becker, K. 2012. Activities of *Jatropha curcas* phorbol esters in various bioassays. *Ecotoxicology and Environmental Safety,* 78: 57–62.

Esimone, C.O., Nworu, C.S. and Jackson, C.L. 2008. Cutaneous wound healing activity of a herbal ointment containing the leaf extract of *Jatropha curcas* L. (Euphorbiaceae). *International Journal of Applied Research in Natural Products.* 1(4): 1–4.

Horiuchi, T., Fujiki, H., Hirota, M., Suttajit, M., Suganuma, M., Yoshioka, A., Wongchai, V., Hecker, E. and Sugimura, T. 1987. Presence of tumor promoters in the seed oil of Jatropha curcas L. from Thailand. *Japanese Journal of Cancer Research GANN.* 78(3): 223–226.

Katagi, A., Sui, L., Kamitori, K., Suzuki, T., Katayama, T., Hossain, A., Noguchi, C., Dong, Y., Yamaguchi, F. and Tokuda, M. 2016. Inhibitory effect of isoamericanol A from *Jatropha curcas* seeds on the growth of MCF-7 human breast cancer cell line by G2/M cell cycle arrest. *Heliyon.* 2(1): e00055.

Matsuse, I.T., Lim, Y.A., Hattori, M., Correa, M. and Gupta, M.P. 1998. A search for anti-viral properties in Panamanian medicinal plants.: The effects on HIV and its essential enzymes. *Journal of Ethnopharmacology*. 64(1): 15–22.

Monteiro, M.V.B., Bevilaqua, C.M., Morais, S.M., Machado, L.K.A., Camurça-Vasconcelos, A.L.F., Campello, C.C., Ribeiro, W.L. and Mesquita, M.D.A. 2011. Anthelmintic activity of *Jatropha curcas* L. seeds on *Haemonchus contortus*. *Veterinary Parasitology*. 182(2–4): 259–263.

Rachana, S., Tarun, A., Rinki, R., Neha, A. and Meghna, R. 2012. Comparative analysis of antibacterial activity of *Jatropha curcas* fruit parts. *Journal of Pharmaceutical and Biomedical Sciences (Jpbms)*. 15(15): 1–4.

Tona, L., Kambu, K., Ngimbi, N., Cimanga, K. and Vlietinck, A.J. 1998. Antiamoebic and phytochemical screening of some Congolese medicinal plants. *Journal of Ethnopharmacology*. 61(1): 57–65.

Tona, L., Kambu, K., Ngimbi, N., Mesia, K., Penge, O., Lusakibanza, M., Cimanga, K., De Bruyne, T., Apers, S., Totte, J. and Pieters, L. 2000. Antiamoebic and spasmolytic activities of extracts from some antidiarrhoeal traditional preparations used in Kinshasa, Congo. *Phytomedicine*. 7(1): 31–38.

Jatropha podagrica **Hook.**

[From Greek *iatros* = physician, *trophe* = food, and Latin *podagra* = gout]

Published in *Botanical Magazine* 74: pl. 4376. 1848

Local names: *jarak, segima* (Dusun); *jarak hutan* (Bajau)
Common names: Buddha belly plant, bottle euphorbia, gout plant
Habitat: cultivated, indoor plant
Geographical distribution: tropical
Botanical description: this shrub reaches a height of about 2 m. The main stem is somewhat bottle-shaped and yields a poisonous latex once incised. Leaves: simple, spiral, stipulate. Stipules: needlelike. Petioles: straight, 8–10 cm long. Blades: peltate, lobed, 8–18 cm × 7–16 cm, base and apex obtuse, wavy, with 3–4 pairs of secondary nerves. Cymes: bright red, fleshy, up to about 25 cm long, green, straight, on smooth peduncles. Sepals: 5, tiny, triangular. Petals: 5, oblong, about 6 mm long. Stamens: 6–8, about 8 mm long, merged. Ovary: elliptic. Stigma: capitate. Capsules: dull green, trilobed, dehiscent, 1.5 cm across. Seeds: ellipsoid, up to 6 mm long.
Traditional therapeutic indications: jaundice (Dusun)
Pharmacology and phytochemistry: antibacterial diterpenes (Aiyelaagbe et al. 2007). Antiviral diterpenes (hepatitis C virus) (Falodun et al. 2014).
Toxicity, side effects, and drug interaction: the plant produces poisonous amide alkaloids including tetramethylpyrazine (Ojewole & Odebiyi 1980).

REFERENCES

Aiyelaagbe, O.O., Adesogan, K., Ekundayo, O. and Gloer, J.B. 2007. Antibacterial diterpenoids from *Jatropha podagrica* Hook. *Phytochemistry*. 68(19): 2420–2425.

Falodun, A., Imieje, V., Erharuyi, O., Ahomafora, J.J., Akunyuli, C., Udu-Cosi, A.A., Theophilus, O., Ali, I., Albadry, M., Fasinu, P. and Hamann, M.T. 2014. Isolation of Diterpenoids from *Jatropha podagrica* against Hepatitis C virus. *Journal of African Association of Physiological Sciences*. 2(1): 21–25.

Ojewole, J.A.O. and Odebiyi, O.O. 1980. Neuromuscular and cardiovascular actions of tetramethylpyrazine from the stem of *Jatropha podagrica*. *Planta Medica*. 38(4): 332–338.

***Macaranga gigantea* (Zoll.) Müll.Arg.**

[After the Malagasy name of a plant of the genus *Macaranga* Thouars and Latin *gigantea* = gigantic]

Published in *Prodromus Systematis Naturalis Regni Vegetabilis* 15(2): 995. 1866

Synonyms: *Macaranga incisa* Gage; *Mappa macrophylla* Kurz ex Teijsm. & Binnend.; *Mappa gigantea* Rchb.f. & Zoll.; *Mappa megalophylla* Müll.Arg.; *Mappa rugosa* Müll.Arg.
Common names: elephant's ear, giant mahang
Habitat: forests
Geographical distribution: Thailand, Malaysia, Indonesia, Borneo
Botanical description: this tree reaches a height of about 30 m. Stems: rough, stout, marked with clearly visible leaf scars, exude a clear latex once incised. Leaves: simple, spiral, stipulate. Stipules: ovate, hairy, up to 6 cm long. Petioles: stout, lanceolate, up to 50 cm long. Blades: somewhat trilobed, peltate, enormous (up to 1 m across), somewhat cordate to rounded at base, dentate, acute at apex, irregular, with 5–8 pairs of secondary nerves, with scalariform tertiary nerves. Panicles: up to about 30 cm long, axillary, many-flowered. Male flowers: tiny. Sepals: 3. Stamens: 2–3. Female flowers: tiny, hairy, urceolate. Styles: 2. Capsules: about 1 cm long, 4-lobed, eaten by birds (gray-breasted spiderhunter). Seeds: about 5 mm across, embedded in a purplish aril (Figure 4.25)

FIGURE 4.25 *Macaranga gigantea* (Zoll.) Müll.Arg.

Traditional therapeutic indications: thrush (Murut, Dusun); diarrhea (Kadazan)

Pharmacology and phytochemistry: antifungal (Kusuma et al. 2016), antibacterial (Lim et al. 2009). Prenylated flavonols (macagigantin, glyasperin A) toxic to murine leukemia cells (P-388) (Tanjung et al. 2009).

Toxicity, side effects, and drug interaction: not known

Comments:

(i) *"Merakubong"* means something like flying lemurs because the leaves look like flying lemurs, which are in parenthese on the verge of extinction.

(ii) In Sarawak, the Iban use the plant for canker sores and apply the latex from fresh young stems to wounds. In Sabah, there are still a small population of Iban in the South East.

(iii) The Kedayan carry amulets made with the latex of this plant which they wear to protect themselves from evil spirits and curses.

REFERENCES

Kusuma, I.W., Sari, N.M., Murdiyanto and Kuspradini, H. 2016, July. Anticandidal activity of numerous plants used by Bentian tribe in East Kalimantan, Indonesia. In *AIP Conference Proceedings* (1755, No. 1: 040002). AIP Publishing LLC.

Lim, T.Y., Lim, Y.Y. and Yule, C.M. 2009. Evaluation of antioxidant, antibacterial and anti-tyrosinase activities of four Macaranga species. *Food Chemistry*. 114(2): 594–599.

Tanjung, M., Hakim, E.H., Mujahidin, D., Hanafi, M. and Syah, Y.M. 2009. Macagigantin, a farnesylated flavonol from *Macaranga gigantea*. *Journal of Asian Natural Products Research*. 11(11): 929–932.

***Macaranga gigantifolia* Merr.**

[After the Malagasy name of a plant of the genus *Macaranga* Thouars and Latin *gigantifolia* = with giant leaves]

Published in *Philippine Journal of Science* 7: 391. 1912

Local name: *binawong* (Murut)
Habitat: forests, roadsides, villages
Geographical distribution: Borneo, the Philippines
Botanical description: this tree reaches a height of about 30 m. Stems: rough, stout, with clearly visible leaf scars, exude a clear latex once incised. Leaves: simple, spiral, stipulate. Stipules: ovate, hairy, up to 6 cm long. Petioles: stout, up to 40 cm long. Blades: broadly ovate to trilobed, rounded at base, up to about 40 cm × 60 cm, loosely dentate, acute at apex, with 4–7 pairs of secondary nerves, with scalariform tertiary nerves. Panicles: up to about 25 cm long, axillary, hairy, many-flowered. Flowers: tiny. Sepals: 3. Stamens: 2–4. Capsules: somewhat broadly obovate. Seeds: a few, embedded in a pale purple aril.
Traditional therapeutic indications: diarrhea (Dusun, Kadazan); thrush (Murut)
Pharmacology and phytochemistry: flavonol glycosides (Primahana & Darmawan 2017), prenylated flavonols (Darmawan et al. 2015).
Toxicity, side effects, and drug interaction: not known

REFERENCES

Darmawan, A., Suwarso, W.P., Kosela, S., Kardono, L.B. and Fajriah, S. 2015. Macarangin, a geranylated flavonoid and anticancer active compound isolated from ethyl acetate fraction of *Macaranga gigantifolia* leaves. *Indonesian Journal of Pharmacy*. 26(1): 52.

Primahana, G. and Darmawan, A. 2017. A Flavonoid glycoside compound isolated from *Macaranga gigantifolia* Merr leaves. *The Journal of Pure and Applied Chemistry Research*. 6(1): 22.

***Macaranga tanarius* (L.) Müll.Arg.**

[After the Malagasy name of a plant of the genus *Macaranga* Thouars = *macaranga*]

Published in *Prodromus Systematis Naturalis Regni Vegetabilis* 15(2): 997. 1866

Synonyms: *Macaranga glabra* (Juss.) Pax & Hoffm; *Macaranga vulcanica* Elmer ex Merr.; *Mappa glabra* Juss.; *Mappa tanarius* (L.) Bl.; *Ricinus mappa* Roxb.; *Ricinus tanarius* L.
Local names: *daun bayangan* (Murut); *muyung, sedaman buta buta* (Kedayan)
Common names: macaranga, parasol tree
Habitat: at the edges of forests, roadsides, wastelands, construction sites
Geographical distribution: from India to Pacific Islands
Botanical description: this tree reaches a height of about 10 m. Stems: glaucous, somewhat articulate, with clearly visible leaf scars, strongly constricted at nodes, often inhabited with red ants. Leaves: simple, spiral, stipulate. Stipules: membranous, hairy, oblong, 1–3 cm long, dentate, caducous. Petioles: straight, up to 30 cm long. Blades: somewhat triangular, peltate, membranous, truncate to almost cordate at base, acute at apex, irregular, with 7–11 pairs of secondary nerves, with scalariform tertiary nerves, finely dentate. Panicles: up to about 15 cm long. Male flowers: tiny. Stamens: 3–10. Female flowers: tiny. Perianth: 2- to trilobed. Ovary: trilobed. Capsules: 1–1.5 cm long, glaucous green, with some kind of sinister aura, covered with fleshy spines.
Traditional therapeutic indications: fatigue, for blind animals (Kedayan)
Pharmacology and phytochemistry: antiplasmodial prenylated flavanones (nymphaeol C, solophenol D, nymphaeol A, and nymphaeol B) (Marliana et al. 2018). Cytotoxic prenylated stilbenes (Péresse et al. 2017) and prenylated flavanones (Kawakami et al. 2008). Antibacterial and antifungal prenylated flavanone (propolin D) (Lee et al. 2019).
Toxicity, side effects, and drug interaction: not known

Comments:

(i) The Murut use the leaves to wrap rice.
(ii) The Dusun and Kadazan use the sap of an unidentified plant of the genus *Macaranga* Thouars (1806) for trush (local name: *limbukon*).

REFERENCES

Kawakami, S., Harinantenaina, L., Matsunami, K., Otsuka, H., Shinzato, T. and Takeda, Y. 2008. Macaflavanones A– G, prenylated flavanones from the leaves of *Macaranga tanarius*. *Journal of Natural Products*. 71(11): 1872–1876.

Lee, J.H., Kim, Y.G., Khadke, S.K., Yamano, A., Woo, J.T. and Lee, J. 2019. Antimicrobial and antibiofilm activities of prenylated flavanones from *Macaranga tanarius*. *Phytomedicine*. 63: 153033.

Marliana, E., Hairani, R., Tjahjandarie, T.S. and Tanjung, M. 2018, April. Antiplasmodial activity of flavonoids from *Macaranga tanarius* leaves. In *IOP Conference Series: Earth and Environmental Science* (144, No. 1: 012011). IOP Publishing.

Péresse, T., Jézéquel, G., Allard, P.M., Pham, V.C., Huong, D.T., Blanchard, F., Bignon, J., Lévaique, H., Wolfender, J.L., Litaudon, M. and Roussi, F. 2017. Cytotoxic prenylated stilbenes isolated from *Macaranga tanarius*. *Journal of Natural Products*. 80(10): 2684–2691.

***Mallotus paniculatus* (Lam.) Müll.Arg.**

[From Greek *mallos* = wool and Latin *paniculatus* = paniculate]

Published in *Linnaea* 34: 189. 1865

Synonyms: *Croton paniculatus* Lam.; *Echinus trisulcus* Lour.; *Mallotus chinensis* Müll.Arg.; *Mallotus cochinchinensis* Lour.; *Mallotus formosanus* Hayata; *Rottlera paniculata* (Lam.) A. Juss.

Local name: *dauk* (Dusun)

Common name: turn-in-the-wind

Habitat: roadsides, open forests, wastelands

Geographical distribution: from Bangladesh to Australia

Botanical description: this tree reaches a height of about 10 m. Stems: hairy. Leaves: simple, spiral, stipulate. Stipules: tiny. Petioles: 2–15 cm long, hairy, somewhat straight, thin. Blades: ovate, lanceolate to rhombic to trilobed, 3–12 cm × 5–15 cm, somewhat almost peltate, glaucous beneath, acute at base, with a pair of glands at base, acuminate at apex, with about 3–6 pairs of secondary nerves, with scalariform tertiary nerves. Racemes: axillary, up to 25 cm long, hairy. Calyx: 3–5-lobed, the lobes ovate, hairy, about 2 cm long. Stamens: up to about 60. Stigma: trilobed, linear. Capsules: trilobed, covered with soft spines, green. Seeds: 3, 4 mm long, blackish.

Traditional therapeutic indications: cuts, scabies (Dusun)

Pharmacology and phytochemistry: leishmanicidal (Monzote et al. 2014). Steroids (Wang et al. 2013).

Toxicity, side effects, and drug interaction: not known

REFERENCES

Monzote, L., Piñón, A. and Setzer, W.N. 2014. Leishmanicidal potential of tropical rainforest plant extracts. *Medicines*. 1(1): 32–55.

Wang, W.J., Jiang, J.H. and Chen, Y.G. 2013. Steroids from *Mallotus paniculatus*. *Chemistry of Natural Compounds*. 49: 577–578.

***Manihot esculenta* Crantz**

[The word manihot is a latinized version of the South American name of cassava and Latin *esculenta* = edible]

Published in *Institutiones Rei Herbariae* 1: 167. 1766

Synonyms: *Janipha manihot* (L.) Kunth.; *Jatropha manihot* L.; *Jatropha stipulata* Vell.; *Manihot edulis* A. Rich.; *Manihot utilissima* Pohl

Local names: *daung lamey* (Bugis); *dikayu* (Lundayeh); *kasila tatanum, mundok, bayag kayu* (Dusun); *lui* (Murut); *ubi kayu* (Bajau, Dusun, Kedayan, Murut)

Common names: cassava, tapioca

Habitat: cultivated

Geographical distribution: tropical

Botanical description: this shrub reaches a height of about 5 m. Tubers: elongated, pure white inside, rough and brownish outside, which can make one think of bunches of monstrous carrots, heavy, up to 60 cm long. Leaves: palmate, spiral, stipulate. Petioles: 6–35 cm long, thin, reddish, straight. Stipules: lanceolate, about 6 mm long. Blades: 3–9-lobed, 5–20 cm across, the lobes oblanceolate, 8–18 cm × 1.5–4 cm, finely marked with 8–9 pairs of secondary nerves, acuminate at apex, of a sinister kind of dull green above, membranous. Racemes: terminal or axillary, 5–8 cm long. Calyx: 5-lobed, light greenish-yellow to white with purple lines, the lobes tongue-shaped and about 4 mm–1 cm long. Disc: 10-lobed. Stamens: 5, about 7 mm long, the filaments thin. Ovary: somewhat ovoid, somewhat hexagonal. Stigma: trifid, fleshy, pure white. Capsules: somewhat globose, dehiscent when dry, glaucous, fleshy at first, up to about 1.5 cm long, 6-winged. Seeds: 3.

Traditional therapeutic indications: measles, child rashes (Lundayeh); flatulence, headaches, festering wounds, fever, diarrhea (Dusun); anemia in pregnant women (Bugis); fever, headaches (Kedayan)

Pharmacology and phytochemistry: antibacterial (Mustarichie et al. 2020), cytotoxic (Diana et al. 2018), increase of red blood cell count (Suzanne et al. 2020).

Triterpenes (maesculentins A and B) toxic to gastric cancer cells (HGC-27) (Pan et al. 2015).

Toxicity, side effects, and drug interaction: tubers and leaves contain hydrocyanic acid and are therefore deadly poisonous (Gondwe 1974).

REFERENCES

Diana, W., Heri, M.A., Herlina, E., Warnasih, S. and Yudhie, S. 2018. Cytotoxic effects of cassava (Manihot esculenta Crantz), Adira-2, karikil and sao pedro petro varieties against P-388 murine leukemia cells. *Research Journal of Chemistry and Environment*. 22: 206–208.

Gondwe, A.T.D. 1974. Studies on the hydrocyanic acid contents of some local varieties of cassava (*Manihot esculenta* Crantz) and some traditional cassava food products. *East African Agricultural and Forestry Journal*. 40(2): 161–167.

Mustarichie, R., Sulistyaningsih, S. and Runadi, D. 2020. Antibacterial activity test of extracts and fractions of cassava leaves (*Manihot esculenta* Crantz) against clinical isolates of *Staphylococcus epidermidis* and *Propionibacterium acnes* causing acne. *International Journal of Microbiology*. 2020: 1–9.

Pan, Y.M., Zou, T., Chen, Y.J., Chen, J.Y., Ding, X., Zhang, Y., Hao, X.J. and He, H.P. 2015. Two new pentacyclic triterpenoids from Stems of Manihot esculenta. *Phytochemistry Letters*. 12: 273–276.

Suzanne, B.B., Adeline, F.Y., Theodora, K.K., Dairou, H., Pradel, K.L., Aristide, K.M., Mathieu, N., Gabriel, A.A., Anne, N.N. and Clerge, T. 2020. Hemopoietic effects of some herbal extracts used in treatment of infantile anemia in Cameroon. *World Journal of Pharmaceutical Research*. 6(1): 147–155.

***Phyllanthus niruri* L.**

[From Greek *phullon* = leaf and *anthos* = flower]

Published in *Species Plantarum* 2: 981–982. 1753

Synonyms: *Diasperus niruri* (L.) Kuntze; *Phyllanthus amarus* Schum. & Thonn; *Phyllanthus asperulatus* Hutch.; *Phyllanthus filiformis* Pavon ex Baillon; *Phyllanthus fraternus* G.L. Webster; *Phyllanthus lathyroides* Kunth; *Phyllanthus microphyllus* Mart.; *Phyllanthus urinaria* L.

Local names: *galung galung, koronipon, piasau piasau, sinting anak* (Dusun); *amin buah, keman jolok* (Bajau); *kararu* (Lundayeh); *nipon nipon* (Dusun, Kadazan); *pilujala, rolapan* (Murut); *koronippon* (Tobilung); *dukung anak* (Bajau, Kedayan)

Common names: niruri, stonebreaker

Habitat: parking lots, roadsides, gardens, wastelands

Geographical distribution: tropical Asia, Pacific Islands

Botanical description: this delicate herbaceous plant reaches a height of about 30 cm. Stems: glabrous, ligneous, flexuous, reddish. Leaves: simple, closely alternate, subsessile, stipulate. Stipules: tiny, acute. Blades: 5–9 mm × 3–4 mm, oblong-elliptic, somewhat asymmetrical, rounded at base and apex, membranous, dull green. Flowers: tiny, axillary, solitary. Sepals: 5, ovate. Stamens: 3, merged into a column. 'Disc: 5-lobed, flat. Ovary: lobed. Stigma: trifid. Capsules: tiny, depressed, globose, trilobed, on remnant calyx, smooth. Seeds: longitudinally ribbed.

Traditional therapeutic indications: diabetes, bloody stools, malaria (Dusun, Kadazan); fever (Dusun, Kadazan, Lundayeh, Tobilung); coughs (Murut); cuts (Kedayan, Dusun); diarrhea (Kedayan, Dusun, Kadazan, Lundayeh); swellings (Kedayan, Dusun); jaundice (Kedayan, Dusun, Murut); hypertension (Kedayan, Dusun, Kadazan); ulcers, wounds, teething, fatigue, body heat, phlegm (Dusun).

Pharmacology and phytochemistry: this plant has been the subject of numerous pharmacological studies which have highlighted, among other things, antibacterial (Valle et al. 2015), hypotensive (Bello et al. 2020), hepatoprotective (Syamasundar et al. 1985), diuretic, antihyperglycemic (Srividya & Periwal 1995), hypolipidemic (Khanna et al. 2002), antiplasmodial (Kabiru et al. 2013), parasiticidal (de Oliveira et al. 2017), and antiviral activity (dengue) (Sood et al. 2015).

Lignans (niranthin) (Chowdhury et al. 2012). Antiviral (hepatitis B virus) lignans (nirtetralin B and niranthin) (Liu et al. 2014). Antiviral phenolic glycoside (niruruside) (Qian-Cutrone et al. 1996). Prenylated flavanones (Shakil et al. 2008). Antiviral (coxsackievirus) ellagitannin (corilagin) (Yeo et al. 2015). Antiviral (hepatitis C virus) apocarotenoid lactone (loliolide) (Chung et al. 2016).

Toxicity, side effects, and drug interaction: the median lethal dose (LD_{50}) of aqueous extract administered to mice was 2.5 g/kg (Singh et al. 2016).

Comments:

(i) In Sarawak, the Iban call this plant "*rumpit blis*" and use it for cuts. Mixed with *Melastoma malabathricum* L., it is used for infant jaundice. In Peninsular Malaysia, the Chinese people use this herb for liver detoxification.

(ii) A plant of the genus *Phyllanthus* L. (1753) is used by the Dusun and Kadazan for foul body odor (local name: *mongkolongkoi*).

REFERENCES

Bello, I., Usman, N.S., Dewa, A., Abubakar, K., Aminu, N., Asmawi, M.Z. and Mahmud, R. 2020. Blood pressure lowering effect and vascular activity of *Phyllanthus niruri* extract: The role of NO/cGMP signaling pathway and β-adrenoceptor mediated relaxation of isolated aortic rings. *Journal of Ethnopharmacology*. 250: 112461.

Chowdhury, S., Mukherjee, T., Mukhopadhyay, R., Mukherjee, B., Sengupta, S., Chattopadhyay, S., Jaisankar, P., Roy, S. and Majumder, H.K. 2012. The lignan niranthin poisons *Leishmania donovani* topoisomerase IB and favours a Th1 immune response in mice. *EMBO Molecular Medicine*. 4(10): 1126–1143.

Chung, C.Y., Liu, C.H., Burnouf, T., Wang, G.H., Chang, S.P., Jassey, A., Tai, C.J., Tai, C.J., Huang, C.J., Richardson, C.D. and Yen, M.H. 2016. Activity-based and fraction-guided analysis of *Phyllanthus urinaria* identifies loliolide as a potent inhibitor of hepatitis C virus entry. *Antiviral Research*. 130: 58–68.

de Oliveira, C.N.F., Frezza, T.F., Garcia, V.L., Figueira, G.M., Mendes, T.M.F. and Allegretti, S.M. 2017. *Schistosoma mansoni*: *In vivo* evaluation of *Phyllanthus amarus* hexanic and ethanolic extracts. *Experimental Parasitology*. 183: 56–63.

Kabiru, A.Y., Abdulkadir, A., Timothy, A. and Gbodi, U.M. 2013. Evaluation of haematological changes in *Plasmodium-berghei*-infected mice administered with aqueous extract of *Phyllantus amarus*. *Pakistan Journal of Biological Sciences*. 16(11): 510.

Khanna, A.K., Rizvi, F. and Chander, R. 2002. Lipid lowering activity of *Phyllanthus niruri* in hyperlipemic rats. *Journal of Ethnopharmacology*. 82(1): 19–22.

Liu, S., Wei, W., Shi, K., Cao, X., Zhou, M. and Liu, Z. 2014. *In vitro* and *in vivo* anti-hepatitis B virus activities of the lignan niranthin isolated from *Phyllanthus niruri* L. *Journal of Ethnopharmacology*. 155(2): 1061–1067.

Qian-Cutrone, J., Huang, S., Trimble, J., Li, H., Lin, P.F., Alam, M., Klohr, S.E. and Kadow, K.F. 1996. Niruriside, a new HIV REV/RRE binding inhibitor from *Phyllanthus niruri*. *Journal of Natural Products*. 59(2): 196–199.

Shakil, N.A., Kumar, J., Pandey, R.K. and Saxena, D.B. 2008. Nematicidal prenylated flavanones from *Phyllanthus niruri*. *Phytochemistry*. 69(3): 759–764.

Singh, T., Kumar, R. and Singh, J.K. 2016. Acute toxicity study of *Phyllanthus niruri* and its effect on the cyto-architectural structure of nephrocytes in Swiss albino mice *Mus-musculus*. *Pharmacognosy Journal*. 8(1): 77–80.

Sood, R., Raut, R., Tyagi, P., Pareek, P.K., Barman, T.K., Singhal, S., Shirumalla, R.K., Kanoje, V., Subbarayan, R., Rajerethinam, R. and Sharma, N., 2015. Cissampelos pareira Linn: natural source of potent antiviral activity against all four dengue virus serotypes. *PLoS Neglected Tropical Diseases*. 9(12): e0004255.

Srividya, N.A. and Periwal, S. 1995. Diuretic, hypotensive and hypoglycaemic effect of *Phyllanthus amarus*. *Indian Journal of Experimental Biology*. 33(11): 861–864.

Syamasundar, K.V., Singh, B., Thakur, R.S., Husain, A., Yoshinobu, K. and Hiroshi, H. 1985. Antihepatotoxic principles of Phyllanthus niruri herbs. *Journal of Ethnopharmacology*. 14(1): 41–44.

Valle Jr, D.L., Andrade, J.I., Puzon, J.J.M., Cabrera, E.C. and Rivera, W.L. 2015. Antibacterial activities of ethanol extracts of Philippine medicinal plants against multidrug-resistant bacteria. *Asian Pacific Journal of Tropical Biomedicine*. 5(7): 532–540.

Yeo, S.G., Song, J.H., Hong, E.H., Lee, B.R., Kwon, Y.S., Chang, S.Y., Kim, S.H., Lee, S.W., Park, J.H. and Ko, H.J. 2015. Antiviral effects of *Phyllanthus urinaria* containing corilagin against human enterovirus 71 and Coxsackievirus A16 in vitro. *Archives of Pharmacal Research*. 38: 193–202.

***Sauropus androgynus* (L.) Merr.**

[From Latin *saurus* = reptile, *pus* = foot, and *androgynus* = hermaphrodyte]

Published in *Philipp. Bur. For. Bull.* 1: 30. 1903

Synonyms: *Clutia androgyna* L.; *Phyllanthus strictus* Roxb.; *Sauropus indicus* Wight; *Sauropus sumatranus* Miq.

Local names: *cangkuk manis* (Kedayan); *sayur manis* (Bajau; Murut); *totopus teropuk* (Dusun)

Common name: Sabah vegetable

Habitat: cultivated, villages

Geographical distribution: from India to Papua New Guinea

Botanical description: this shrub reaches a height of about 3 m. Stems: thorny. Leaves: compound, spiral, stipulate. Stipules: triangular, about 2 mm long. Petioles: 2–3 mm long. Folioles: somewhat lanceolate, 3–10 cm × 1.5–3.5 cm, membranous, fleshy, base and apex acute, with 6–10 pairs of secondary nerves, dull green, soft. Flowers: axillary, solitary. Sepals: 6, obovate, up to 5 mm long, with a beautiful burgundy color. Disc: 6-lobed. Stamens: 3. Ovary: globose. Styles: 3, forked. Berries: of a peculiar kind of ivory white to somewhat pinkish, smooth, vaguely trilobed, about 1 cm across, on persistent calyx. Seeds: numerous, tiny, black.

Traditional therapeutic indications: cuts (Dusun); fever, sore eyes (Kedayan); medicinal (Murut)

Pharmacology and phytochemistry: antibacterial, cytotoxic, wound healing, hypocholesterolamic, antidiabetic, anti-inflammatory (Fikri & Purnama 2020).

Toxicity, side effects, and drug interaction: consumption of this plant leads to serious lung damage (Lin et al. 1996).

Comments:

(i) *Breynia coronata* Hook.f. is a remedy for bloody stools and postpartum for the Dusun and Kadazan.

(ii) *Codiaeum variegatum* (L.) Blume is used for gastric ulcers by the Dusun.

(iii) *Mallolus macrostachyus* (Miq.) Müll.Arg. is used to kill germs and for stomach aches by the Dusun and Kadazan whilst it is used for cuts by the Rungus. The plant is called "*dahu*" by the Dusun, Kadazan, and Rungus. The Dusun use this plant for healing and call it "*jabai*".

(iv) *Mallotus miquelianus* (Scheff.) Boerl. is used for cirrhosis by the Lundayeh.

(v) *Pedilanthus tithymaloides* (L.) Poit. is medicinal for the Murut (local name: *tatapis tindukon*).

(vi) *Ricinus communis* L. is a remedy for excessive menstruation and fever for the Kedayan (local name: *jarak jantan*).

(vii) *Sapium discolor* (Champ. ex Benth.) Müll.Arg. is used for itchiness by the Lundayeh (local name: *simbobolou*).

(viii) *Trigonopleura malayana* Hook.f. leaves are occasionally chewed with betel nut by the Dusun (local name: *saripa*).

(ix) *Trigonostemon polyanthus* Merr. is used by the Dusun as an antidote for poison (local name: *ambuk sagubang kayu*).

(x) Plants of the genus *Croton* L. (1753) are medicinal in Sabah. The Dusun and Kadazan use "*tolotok*" and "*kalayon*" to give strength to sickly children whilst "*rolok taragan*" is a remedy for fatigue for the Rungus.

REFERENCES

Fikri, F. and Purnama, M.T.E. 2020. Pharmacology & phytochemistry overview on *Sauropus androgynus*. *Systematic Reviews in Pharmacy*. 11(6): 124–128.

Lin, T.J., Lu, C.C., Chen, K.W. and Deng, J.F. 1996. Outbreak of obstructive ventilatory impairment associated with consumption of *Sauropus androgynus* vegetable. *Journal of Toxicology: Clinical Toxicology*. 35(1): 1–8.

4.1.10.4.4 Family Flacourtiaceae Rich ex DC. (1824), the Flacourtia Family

The family Flacourticeae consists of about 90 genera and about 1,000 species of trees and shrubs. Stems: often armed with thorns. Leaves: alternate, simple, sometimes stipulate. Blades: often serrate. Inflorescences: spikes, cymes, panicles, fascicles, racemes. Flowers: tiny. Sepals: 3–6, free or connate. Petals: 3–8 or none. Disc: present. Stamens: up to 100. Carpels: 2–10, merged in a unilocular or 2–10 locular ovary with 15–100 ovules. Styles: up to 10. Fruits: capsules, berries, drupes. Seeds: often embedded in an aril, or hairy, or winged.

***Casearia grewiifolia* Vent.**

[After the 17th century Dutch clergyman Johanes Caesarius and Latin *grewiifolia* = with *Grewia*-like leaves]

Published in *Choix de Plantes, Dont la Plupart Sont Cultivées dans le Jardin de Cels*. 48. 1808

Synonym: *Samyda grewiifolia* (Vent.) Poir.
Local name: *salokdan* (Murut)
Habitat: forests
Geographical distribution: from Vietnam to the Solomon Islands
Botanical description: this tree reaches a height of about 20 m. Trunk: buttressed. Bark: light gray, rough. Stems: angular, hairy at apex. Leaves: simple, alternate, stipulate. Stipules: tiny, triangular. Petioles: 5 mm–1 cm long. Blades: 3.5–6 cm × 8–10 cm, oblong, coriacous, crenate, rounded and somewhat asymmetrical at base, acuminate at apex, hairy beneath, with 10–14 pairs of secondary nerves. Flowers: about 7 mm across, arranged in cauliflorous fascicles. Calyx: 5-lobed, the lobes about 3 mm long, hairy. Stamens: 8. Ovary: tiny. Stigma: capitate. Capsules: about 2.5 cm long, oblong, glossy, yellow to orange, 3-valved. Seeds: numerous, tiny, in a red aril (the whole fruit has a kind of myristicaceous appearance).
Traditional therapeutic indication: pancreatitis (Murut)
Pharmacology and phytochemistry: diterpenes (caseagrewifolin B) toxic to human carcinoma cells (KB) and liver cancer cells (HepG2) (Kanokmedhakul et al. 2005, 2007; Nguyen et al. 2015).
Phenolics (Rayanil et al. 2012), phenolic glycosides (Nhoek et al. 2021).

Comments:

(i) *Casearia rugulosa* Bl. (Figure 4.26) is endemic to Sabah and used by the Dusun for skin infection (local name: *keh lupor*). In Sarawak, this plant is called *"ayam hantu sebayan"* by the Iban.
(ii) *Flacourtia rukam* Zoll. & Moritzi is used by the Dusun and Kadazan for colic and headaches (local name: *peripot*).
(iii) *Homalium foetidum* (Roxb.) Benth. is medicinal for the Murut (local name: *lulumada*).
(iv) In the family Achariaceae Harms (1897), *Trichadenia philippinensis* Merr. (Figure 4.27) is used by the Murut for magic (local name: *ulok ulok*) whilst the seeds of *Pangium edule* Reinw, although toxic, are used by the Dusun (local name: *pangi*) as preservation ingredients for certain traditional fermented meat recipes.
(v) In the family Passifloraceae Juss. ex Roussel (1806), *Adenia macrophylla* (Blume) Koord. is used by the Lundayeh for epilepsy (local name: *war ruai*) whilst *Passiflora foetida* L. is taken for hypertension by the Dusun (local name: *lapak lapak*) and for fever by the Lundayeh (local name: *timun belanda*).

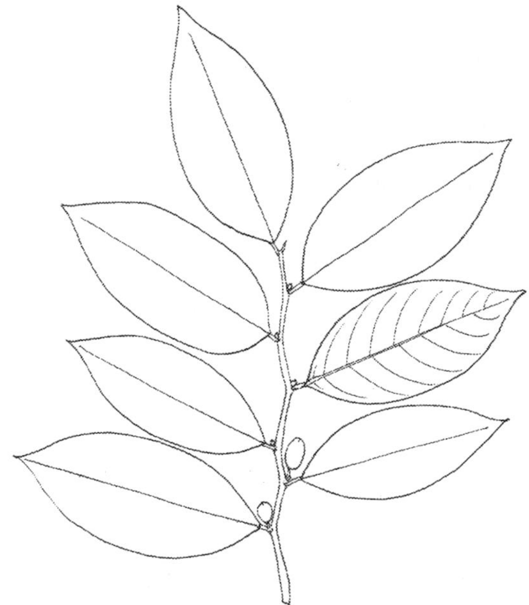

FIGURE 4.26 *Casearia rugulosa* Bl.

FIGURE 4.27 *Trichadenia philippinensis* Merr.

REFERENCES

Kanokmedhakul, S., Kanokmedhakul, K. and Buayairaksa, M. 2007. Cytotoxic clerodane diterpenoids from fruits of Casearia grewiifolia. *Journal of Natural Products*. 70(7): 1122–1126.

Kanokmedhakul, S., Kanokmedhakul, K., Kanarsa, T. and Buayairaksa, M. 2005. New bioactive clerodane diterpenoids from bark off *Casearia grewiifolia. Journal of Natural Products*. 68(2): 183–188.

Nguyen, H.T., Truong, N.B., Doan, H.T., Litaudon, M., Retailleau, P., Do, T.T., Nguyen, H.V., Chau, M.V. and Pham, C.V. 2015. Cytotoxic clerodane diterpenoids from the leaves of *Casearia grewiifolia. Journal of Natural Products*. 78(11): 2726–2730.

Nhoek, P., Ahn, S., Park, I.G., Pel, P., Huh, J., Kim, H.W., Ahn, J., Khiev, P., Choi, Y.H., Lee, K. and Noh, M. 2021. Salicinoyl Quinic acids and their prostaglandin E2 production inhibitory activities from the fruits of *Casearia grewiifolia. Journal of Natural Products*. 84(9): 2437–2446.

Rayanil, K.O., Nimnoun, C. and Tuntiwachwuttikul, P. 2012. New phenolics from the wood of *Casearia grewiifolia. Phytochemistry Letters*. 5(1): 59–62.

4.1.10.4.5 Family Rhizophoraceae Pers. (1806), the Rhizophora Family

The family Rhizophoraceae consists of about 16 genera and 120 species of shrubs and trees. Stilt-roots: often present. Leaves: simple, entire, opposite, stipulate. Stipules: interpetiolar. Inflorescences: solitary or cymes. Hypanthium: present. Sepals: 3–16. Petals: 3–16. Stamens: 8–40. Disc: present. Carpels: often 2–5 forming a 1–6 locular, each locule with 2- 25 ovules. Styles: 1. Stigma: capitate. Fruits: drupes, berries. Seeds: often viviparous.

Bruguiera parviflora **Wight**

[After the 18th century French physician J. G. Bruguieres and Latin *parviflora* = with small flowers]

Published in *Trans. Med. Soc. Calcutta* 8: 10. 1836

Synonym: *Bruguiera parviflora* (Roxb.) Wight & Arn. ex Griff.
Local names: *bakau, langadoi* (Rungus)
Common name: thin-fruit orange mangrove
Habitat: mangroves
Geographical distribution: from India to Pacific Islands
Botanical description: this massive tree reaches a height of about 30 m. Stilt roots: present. Trunk: not straight. Bark: ash colored, smooth. Leaves: simple, spiral, stipulate. Stipules: up to about 4.5 cm long. Petioles: thin, up to about 2 cm long. Blades: elliptic to spathulate, coriaceous, 7.5–13 cm × 2–4.5 cm. Inflorescences: terminal cymes of yellowish-green flowers. Hypantium: about 9 mm long, yellowish. Calyx: tubular, 8-lobed, the lobes about 2 mm long, slightly curved at apex. Petals: 8, hairy, bilobed. Hypocotyls: cylindrical, up to about 25 cm long, smooth, pendulous.
Traditional therapeutic indication: fatigue (Rungus)
Pharmacology and phytochemistry: anti-inflammatory (Bui et al. 2022). Antiplasmodial triterpene (3-(Z)-caffeoyllupeol) (Chumkaew et al. 2005).
Toxicity, side effects, and drug interaction: not known

Comment: The fruits are used as food by the Bajau.

REFERENCES

Bui, T.T., Nguyen, K.P.T., Nguyen, P.P.K., Le, D.T. and Nguyen, T.L.T. 2022. Anti-inflammatory and α-glucosidase inhibitory activities of chemical constituents from *Bruguiera parviflora* leaves. *Journal of Chemistry*. 2022: 1–9.

Chumkaew, P., Kato, S. and Chantrapromma, K. 2005. A new triterpenoid ester from the fruits of *Bruguiera parviflora*. *Chemical and Pharmaceutical Bulletin*. 53(1): 95–96.

***Rhizophora apiculata* Bl.**

[From Greek *rhiza* = root, *phoros* = bearing, and Latin *apiculata* = ending abruptly in a short point]

Published in *Enumeratio Plantarum Javae* 1: 91. 1827

Synonyms: *Rhizophora candelaria* DC.; *Rhizophora conjugata* L.
Local names: *bakau, bangkita* (Bajau); *tongog* (Rungus)
Common names: mangroves, tall-stilted mangrove
Habitat: mangroves
Geographical distribution: From India to Pacific Islands
Botanical description: this tree reaches a height of about 15 m. Aerial roots: coming out of the mud vertically (pneumatophores). Bark: grayish brown, cracked. Stems: terete, stout, rough. Leaves: simple, opposite, stipulate, at apex of stems. Stipules: interpetiolar, reddish, lanceolate, about 4–8 cm long. Petioles: stout, yellowish-green to reddish, about 1–2.5 cm long. Blades: coriaceous, elliptic, smooth, glossy, tapering at base, apiculate at apex, without apparent secondary nerves, 6–14 cm × 3–6 cm. Flowers: cauliflorous, solitary, or in pairs. Calyx lobes: 4, woody, stout, about 1.5 cm × 9 mm, light greenish-yellow arranged into a cross. Petals: 4, narrowly lanceolate, caducous, membranous, white, about 1 cm long. Stamens: 10–12, sessile, about 8 mm long. Disc: present. Ovary: conical, yellowish. Styles: 1, tiny. Drupes: somewhat pear-shaped, 2.5 cm long, brownish. Hypocotyl: linear, curved at base about 33 cm long (Figure 4.28).

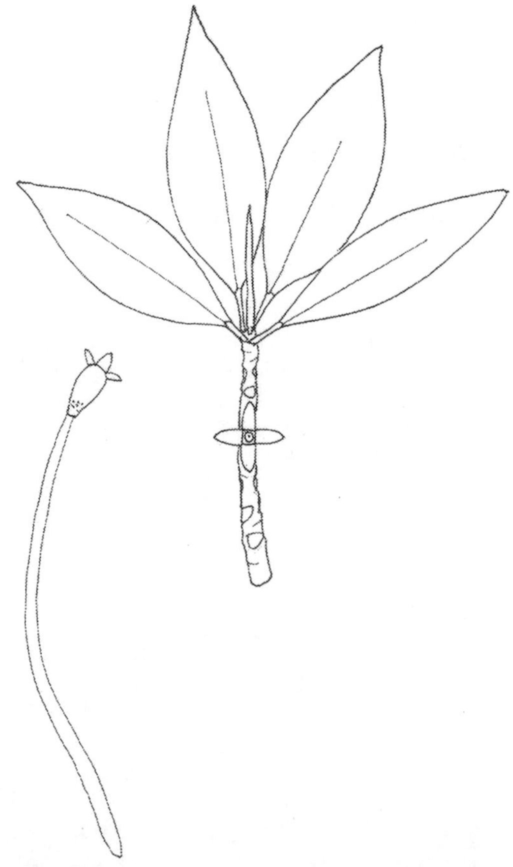

FIGURE 4.28 *Rhizophora apiculata* Bl.

Traditional therapeutic indications: laxative, bad breath (Bajau)

Pharmacology and phytochemistry: anti-inflammatory, cytotoxic (Prabhu & Guruvayoorappan 2012).

Antibacterial tannins (Lim et al. 2006). Triterpenes (taraxeryl cis-*p*-hydroxycinnamate) (Kokpol et al. 1990). Arylnaphthalene lignans glycosides (Gao & Xiao 2012).

Toxicity, side effects, and drug interaction: not known

Comments:

(i) The plant is used in Sabah to make a red dye.

(ii) *Ceriops tagal* (Perr.) C. B. Rob. is used by the Bajau (local name: *tengar*) for sore throats and to make a drink.

(iii) In the family Violaceae Batsch (1802), a plant of the genus *Rinorea* Aubl. (1775) is used by the Dusun and Kadazan for coughs (local name: *posiou*).

(iv) In the family Pandaceae Engl. & Gilg (1913), *Galearia fulva* Zoll. & Moritzi is used by the Dusun (local name: *sanggara*) as medicine and food.

REFERENCES

Gao, M. and Xiao, H. 2012. Activity-guided isolation of antioxidant compounds from Rhizophora apiculata. *Molecules*. 17(9): 10675–10682.

Kokpol, U., Chavasiri, W., Chittawong, V. and Miles, D.H. 1990. Taraxeryl cis-p-hydroxycinnamate, a novel taraxeryl from *Rhizophora apiculata*. *Journal of Natural Products*. 53(4): 953–955.

Lim, S.H., Darah, I. and Jain, K. 2006. Antimicrobial activities of tannins extracted from *Rhizophora apiculata* barks. *Journal of Tropical Forest Science*.: 59–65.

Prabhu, V.V. and Guruvayoorappan, C. 2012. Anti-inflammatory and anti-tumor activity of the marine mangrove *Rhizophora apiculata*. *Journal of Immunotoxicology*. 9(4): 341–352.

4.1.10.5 Order Oxalidales Bercht. & J. Presl. (1820)

4.1.10.5.1 Family Connaraceae R. Br. (1818), the Kana Family

The family Connaraceae consists of about 18 genera and 350 species of treelets, shrubs, or climbing plants. Leaves: compound, alternate, exstipulate. Inflorescences: terminal or axillary panicles or racemes. Sepals: 5. Petals: 5. Stamens: 5–10. Disc: sometimes present. Carpels: 1–8, forming a 1–8 locular ovary, each locule with 2 ovules. Fruits: follicles. Seeds: arillate.

***Cnestis platantha* Griff.**

[From Greek *knestis* = a rasp, Latin *plat* = flat, and *anthos* = flower]

Published in *Notulae ad Plantas Asiaticas* 4: 434. 1854

Synonyms: *Rourea dasyphylla* Miq.; *Santalodes dasyphyllum* O. Ktze
Local name: *lingem* (Dusun)
Habitat: forests
Geographical distribution: Malaysia, Indonesia
Botanical description: this climbing plant grows up to a length of about 20 m. Stems: terete, stout, hairy at apex. Leaves: imparipinnate, spiral, at apex of stems, exstipulate. Blades: up to about 50 cm long. Rachis: hairy. Folioles: of up to about 24 pairs, hairy beneath, rounded to somewhat cordate at base, oblong to lanceolate, rounded to acute at apex, about 1.5–2.8 cm × 4.5–10.2 cm, sessile. Racemes: axillary, hairy, the flowers with a pungent odor. Sepals: 5, lanceolate, about 5 mm long, hairy. Petals: 5, lanceolate, whitish to pink, about 3 mm long. Stamens: 10. Carpels: 5. Follicles: obovoid, hairy, dehiscent, about 5 cm × 8 cm. Seeds: back, glossy, partially embedded in an aril.
Traditional therapeutic indications: coughs, stomach aches (Dusun); fever, measles, sprains, fatigue, flu (Dusun, Kadazan)
Pharmacology and phytochemistry: not known
Toxicity, side effects, and drug interaction: poisonous (Garon et al. 2007).

Comments:

(i) *Agelaea macrophylla* (Zoll.) Leenh. is used by the Murut for fatigue (local name: *kalam malam*).
(ii) *Ellipanthus tomentosus* Kurz (Figure 4.29) is used by the Dusun and Kadazan for coughs (local name: *bidon*).
(iii) In the family Family Oxalidaceae R.Br. (1818), *Averrhoa bilimbi* L. is used by the Dusun and Kadazan for fatigue (local name: *tulod ulod*). *Averrhoa carambola* L. is used for viral fever by the Murut (local name: *belimbing*) whilst the Chinese people of Sabah employ it as a home remedy to reduce plasma triglycerides.
(iv) *Rourea mimosoides* Planch. is used by the Dusun for bloody stools (local name: *udang udang*).

REFERENCE

Garon, D., Chosson, E., Rioult, J.P., de Pecoulas, P.E., Brasseur, P. and Vérité, P. 2007. Poisoning by *Cnestis ferruginea* in Casamance (Senegal): An etiological approach. *Toxicon*. 50(2): 189–195.

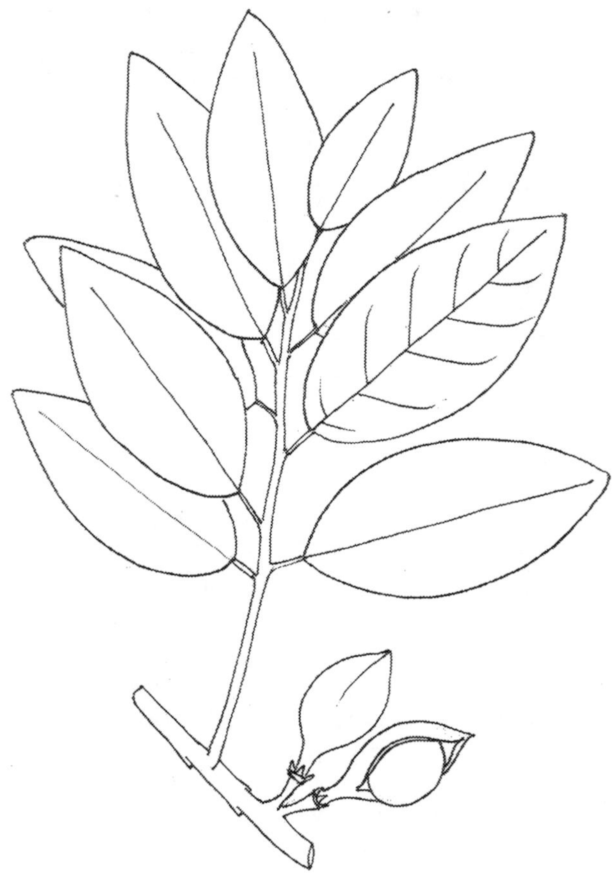

FIGURE 4.29 *Ellipanthus tomentosus* Kurz.

4.1.10.5.2 Family Elaeocarpaceae Juss. ex DC. (1816), the Elaeocarpus Family

The family Elaeocarpaceae consists of about 9 genera and 40 species of trees or shrubs. Leaves: simple, alternate, sometimes stipulate. Petioles: often straight, thin, somewhat geniculate and swollen at apex, Inflorescences: axillary or terminal, racemes, panicles, corymbs, the flowers often dangling and graceful. Sepals: 4–5, free or merged at base. Petals: 4–5, often finely fringed. Stamens: 8-many. Disc: present. Carpels: 2–10, forming a 1–many locular ovary, each locule with 2–50 ovules. Fruits: drupes often resembling small olives.

Elaeocarpus clementis **Merr.**

[From the Greek *elaio* = olive and *karpos* = fruit]

Published in *Journal of the Straits Branch of the Royal Asiatic Society* 77: 195. 1917

Local name: *timbarazung* (Rungus)
Habitat: forests on the banks of rivers
Geographical distribution: Borneo
Botanical description: this tree reaches a height of about 25 m. Bark: green. Sapwood: white. Leaves: simple, spiral, stipulate. Stipules: linear, up to about 4 mm long. Petioles: straight, 4–8 cm long, thin, somewhat geniculate and swollen at apex. Blades: elliptic to oblong, 2–10 cm × 5–20 cm, with about 8–11 pairs of secondary nerves, acute to somewhat attenuate at base, crenate, long acuminate at apex. Racemes: 10–20 cm long, many-flowered. Sepals: 5, ovate to elliptic, up to about 5 mm long, somewhat hairy. Petals: 5, fleshy, ovate to elliptic, about 5 mm long, fringed at apex. Stamens: up to about 50. Disc: present. Ovary: tiny, hairy. Styles: 1. Drupes: olive-shaped, green-purple, up to about 2.5 cm long (Figure 4.30).
Traditional therapeutic indications: medicinal (Rungus)
Pharmacology and phytochemistry: not known
Toxicity, side effects, and drug interaction: not known

Comment: The roots are said to taste like licorice.

FIGURE 4.30 *Elaeocarpus clementis* Merr.

4.1.10.6 Order Rosales Bercht. & J. Presl. (1820)

4.1.10.6.1 Family Moraceae Link (1831), the Mulberry Family

The family Moraceae consists of about 40 genera and 1,400 species of trees, shrubs, climbing plants, or herbaceous plants. Stems: often somewhat articulate and secreting latex when incised. Leaves: simple, spiral, stipulate. Stipules: often form a cap over the leaf buds. Blades: often coriaceous, glossy, serrate. Inflorescences: spikes, heads, or fleshy and more or less hollowed recepacles (hypanthodium). Flowers: tiny. Sepals: 3–5, merged at base. Petals: none. Stamens: 1–4. Carpels: 2, forming a unilocular or bilocular ovary, each locule with 1 ovule. Styles: 2. Fruits: drupes, nuts, achenes.

***Artocarpus elasticus* Reinw ex Bl.**

[From Greek *artos* = bread, *karpos* = fruit, and *elastikos* = springy]

Published in *Bijdragen tot de flora van Nederlandsch Indië* (9): 481. 1825

Synonyms: *Artocarpus blumei* Trécul; *Artocarpus kunstleri* King; *Artocarpus pubescens* Bl.; *Saccus blumei* (Trécul) Kuntze; *Saccus elasticus* (Reinw. ex Blume) Kuntze; *Saccus kunstleri* (King) Kuntze

Local names: *kikian, puputul* (Murut); *togop* (Dusun)

Common name: wild breadfruit

Habitat: forests, roadsides, wastelands, around villages

Geographical distribution: Thailand, Malaysia, Indonesia, the Philippines

Botanical description: this tree reaches a height of about 40 m. Trunk: straight. Buttresses: about 3 m tall. Bark: dark gray, secretes a white and sticky latex (used as bird trap) once incised. Stems: stout, rugose, somewhat articulate, hairy at apex. Leaves: simple, spiral, stipulate. Stipules: lanceolate, hairy, 6–15 cm long. Petioles: 4–10 cm long, stout. Blades: coriaceous, dark green and glossy above, wavy, heavy, elliptic, 10–35 cm × 12–60 cm, rounded to attenuate at base, crenate to deeply incised, acuminate at apex, hairy beneath with 12–14 pairs of clearly visible secondary nerves. Flower heads: oblong, fleshy, 6–15 cm long. Perianth: tubular, bilobed. Stamens: 1. Syncarps: spiny, about 15 cm long, on approximately 10 cm long stout peduncles. Seeds: numerous, about 1 cm long.

Traditional therapeutic indications: medicinal (Murut)

Pharmacology and phytochemistry: antibacterial and cytotoxic prenylated dihydrochalcone (elastichalcone C) (Ramli et al. 2013; Daus et al. 2017). Prenylated flavonol (artonol A) toxic for non-small cell lung cancer cells (A549) (Ko et al. 2005; Cidade et al. 2001).

Toxicity, side effects, and drug interaction: not known

Comment: The fruits are used as food by the Dusun.

REFERENCES

Cidade, H.M., Nacimento, M.S.J., Pinto, M.M., Kijjoa, A., Silva, A.M. and Herz, W. 2001. Artelastocarpin and carpelastofuran, two new flavones, and cytotoxicities of prenyl flavonoids from *Artocarpus elasticus* against three cancer cell lines. *Planta Medica*. 67(9): 867–870.

Daus, M., Chaithada, P., Phongpaichit, S., Watanapokasin, R., Carroll, A.R. and Mahabusarakam, W. 2017. New prenylated dihydrochalcones from the leaves of *Artocarpus elasticus*. *Phytochemistry Letters*. 19: 226–230.

Ko, H.H., Lu, Y.H., Yang, S.Z., Won, S.J. and Lin, C.N. 2005. Cytotoxic prenylflavonoids from *Artocarpus elasticus*. *Journal of Natural Products*. 68(11): 1692–1695.

Ramli, F., Rahmani, M., Kassim, N.K., Hashim, N.M., Sukari, M.A., Akim, A.M. and Go, R. 2013. New diprenylated dihyrochalcones from leaves of *Artocarpus elasticus*. *Phytochemistry Letters*. 6(4): 582–585.

Artocarpus tamaran Becc.

[From Greek *artos* = bread, *karpos* = fruit, and from the local name of the plant = *tamaran*]

Published in *Nelle Foreste di Borneo* 626. 1902

Local names: *tarap tempunan* (Bajau); *timbagan* (Dusun)
Common name: elephant Jack
Habitat: forests
Geographical distribution: Borneo
Botanical description: this tree reaches a height of about 40 m. Buttresses: present. Bark: gray, brown, lenticelled. Stems: stout, rugose, somewhat articulate, hairy at apex, yield a white latex once incised. Leaves: simple, spiral, stipulate. Stipules: lanceolate, hairy, 3–9 cm long. Petioles: 3.5–4 cm long, stout. Blades: coriaceous, dark green, glossy, wavy, heavy, elliptic, 11–17 cm × 20–35 cm, rounded at base, crenate to deeply incised, acuminate at apex, hairy beneath, with 17–23 pairs of clearly visible secondary nerves (with a vague resemblance to leaves of *Dillenia*). Heads: oblong, fleshy, about 7 cm long. Perianth: tubular, bilobed. Stamens: 1. Syncarps: spiny, which can make one think of a furry durian fruit, oblong, about 10 cm long seeds on an approximately 10 cm long stout peduncle. Seeds: somewhat globose and 2 cm across.
Traditional therapeutic indications: medicinal
Pharmacology and phytochemistry: not known
Toxicity, side effects, and drug interaction: not known

Comment: The fruits are used as food in Sabah and Sarawak.

Ficus deltoidea Jack

[From Latin *ficus* = fig and *deltoidea* = deltoid]

Published in *Malayan Miscellanies* 2(7): 71. 1822

Synonyms: *Ficus diversifolia* Bl.; *Ficus diversifolia* var. *deltoidea* (Jack) Ridl.; *Ficus ovoidea* Jack
Local names: *agolauran* (Murut); *daun sampit sampit*, *mas cotek* (Bajau, Bugis)
Common names: mistletoe fig, mistletoe rubber-plant
Habitat: forests, cultivated
Geographical distribution: Thailand, Malaysia, Indonesia, the Philippines
Botanical description: this shrub reaches a height of about 2 m. Bark: whitish gray. Stems: articulate. Leaves: simple, spiral, stipulate. Stipules: linear-lanceolate, 8–12 mm long, convolute. Petioles: stout, about 8 mm long. Blades: coriaceous, thick, spathulate, elliptic, or deltoid, smooth, 1.5–5 cm × 2.5–7.5 cm, truncate, rounded, obtuse, or notched at apex. Hypanthodium: solitary or paired, axillary, ovoid, globose or pyriform, 5–10 mm across, on 5–20 mm long peduncles. Sepals: 3–4. Stamens: 1. Styles: 1, thin. Figs: 1 cm across, smooth, somewhat globose, yellow to orange-red.
Traditional therapeutic indications: medicinal (Murut); itchiness (Bugis); sore eyes, colds (Bajau); to make women barren (Dusun, Kadazan)
Pharmacology and phytochemistry: wound healing (Abdulla et al. 2010), anti-inflammatory (Abdullah et al. 2009), antidiabetic (Ilyanie et al. 2011; Haslan et al. 2021), hypotensive (Azis et al. 2019), antibacterial (Othman et al. 2012).
Cytotoxic flavonol *C*-glycosides (vitexin, isovitexin) (Choo et al. 2012; He et al. 2016).
Toxicity, side effects, and drug interaction: the median lethal dose (LD_{50}) of ethanol extract administered to rats was 2 g/kg (Nugroho et al. 2020).

REFERENCES

Abdulla, M.A., Ahmed, K.A.A., Abu-Luhoom, F.M. and Muhanid, M. 2010. Role of *Ficus deltoidea* extract in the enhancement of wound healing in experimental rats. *Biomedical Research.* 21(3): 241–245.

Abdullah, Z., Hussain, K., Ismail, Z. and Ali, R.M. 2009. Anti-inflammatory activity of standardised extracts of leaves of three varieties of *Ficus deltoidea. International Journal of Pharmaceutical and Clinical Research.* 1(3): 100–105.

Azis, N.A., Agarwal, R., Ismail, N.M., Ismail, N.H., Kamal, M.S.A., Radjeni, Z. and Singh, H.J. 2019. Blood pressure lowering effect of *Ficus deltoidea* var *kunstleri* in spontaneously hypertensive rats: Possible involvement of renin—angiotensin—aldosterone system, endothelial function and anti-oxidant system. *Molecular Biology Reports.* 46: 2841–2849.

Choo, C.Y., Sulong, N.Y., Man, F. and Wong, T.W. 2012. Vitexin and isovitexin from the leaves of *Ficus deltoidea* with *in-vivo* α-glucosidase inhibition. *Journal of Ethnopharmacology.* 142(3): 776–781.

Haslan, M.A., Samsulrizal, N., Hashim, N., Zin, N.S.N.M., Shirazi, F.H. and Goh, Y.M. 2021. *Ficus deltoidea* ameliorates biochemical, hormonal, and histomorphometric changes in letrozole-induced polycystic ovarian syndrome rats. *BMC Complementary Medicine and Therapies.* 21: 1–13.

He, M., Min, J.W., Kong, W.L., He, X.H., Li, J.X. and Peng, B.W. 2016. A review on thepharmacological effects of vitexin and isovitexin. *Fitoterapia.* 115: 74–85.

Ilyanie, Y., Wong, T.W. and Choo, C.Y. 2011. Evaluation of hypoglycemic activity and toxicity profiles of the leaves of *Ficus deltoidea* in rodents. *Journal of Complementary and Integrative Medicine.* 8(1): 1–16.

Nugroho, R.A., Aryani, R., Manurung, H., Rudianto, R., Prahastika, W., Juwita, A., Alfarisi, A.K., Pusparini, N.A.O. and Lalong, A. 2020. Acute and subchronic toxicity study of the ethanol extracts from *Ficus deltoidea* leaves in male mice. *Open Access Macedonian Journal of Medical Sciences.* 8(A): 76–83.

Othman Abd, S., Nur Tarwiyah Ahmad, Z. and Abu Bakar, S. 2012. Antimicrobial activity of *Ficus deltoidea* Jack [Mas Cotek]. *Pakistan Journal of Pharmaceutical Sciences.* 25(3): 675–678.

Ficus elliptica Hook ex Miq.

[From Latin *ficus* = fig and *elliptica* = elliptic]

Published in *F.W.H von Humboldt, A.J.A Bonpland, & C.S. Kunth, Nov. Gen.* Sp. 2: 46. 1817

Synonyms: *Ficus retusa* L.; *Perula retusa* (L.) Raf.; *Urostigma retusum* (L.) Gasp.; *Ficus truncata* (Miq.) Miq.; *Urostigma truncatum* Miq.

Local names: *hintotobu* (Dusun, Kadazan); *silabon rondoh* (Murut)

Habitat: cultivated, forests

Geographical distribution: from India to the Philippines

Botanical identification: this strangler fig tree reaches a height of about 35 m. Stems: hairy, articulate. Leaves: simple, spiral, stipulate. Stipules: linear-lanceolate, 0.2–1.5 cm long, hairy. Petioles: stout, 0.5–1.5 cm long. Blades: coriaceous, oblong to ovate, smooth, 0.5–5 cm × 1.9–5.5 cm, attenuate at base, rounded at apex, with 3–6 pairs of secondary nerves. Hypanthodium: solitary or paired, axillary, ovoid, globose or pyriform, 3–5 cm across, on about 2 cm long peduncles. Sepals: red. Figs: globose, 1 cm across, smooth, yellow to orange-red.

Traditional therapeutic indications: shivers (Dusun, Kadazan); medicinal (Murut)

Pharmacology and phytochemistry: not known

Toxicity, side effects, and drug interaction: not known

Ficus lepicarpa Bl.

[From Greek *lepis* = scaly and *karpos* = fruit]

Published in *Bijdragen tot de flora van Nederlandsch Indië* (9): 459. 1825

Synonyms: *Covellia didynama* Miq.; *Covellia lepicarpa* (Blume) Miq.; *Covellia volkameriifolia* Miq.; *Ficus cuneifolia* Hook. ex Miq.; *Ficus volkameriifolia* (Miq.) Wall. ex Miq.

Local names: *lintotobow* (Murut); *kelupang gajah* (Bajau); *ombuwasak* (Rungus); *tombuwasak* (Dusun)

Common name: saraca fig

Habitat: forests

Geographical distribution: from Myanmar to the Philippines

Botanical description: this tree reaches a height of about 10 m. Buttresses: present. Bark: light gray. Leaves: spiral, simple, stipulate. Stems: articulated, angular. Stipules: lanceolate, 1–2 cm long. Petioles: stout, 0.5–5 cm long. Blades: membranous, oblanceolate to obovate, dark green, glossy, smooth, 3–11.5 cm × 9–25 cm, with about 7–10 pairs of secondary nerves, acuminate at apex. Hypanthodium: solitary, axillary, sessile, ovoid, about 1 cm long. Perianth: tubular. Figs: somewhat globose, 1 cm across, smooth, glossy, green turning somewhat dark purple, covered with tiny whitish bodies.

Traditional therapeutic indications: fever, fatigue, ringworms (Dusun)

Pharmacology and phytochemistry: hepatoprotective (Vun-Sang et al. 2022).

Toxicity, side effects, and drug interaction: not known

Comments:

(i) The leaves are used as food by the natives.

(ii) The Murut use the latex of *Antiaris toxicaria* Lesch (local name: *paliu*) to poison blowgun darts.

REFERENCE

Vun-Sang, S., Rodrigues, K.F., Dsouza, U.J. and Iqbal, M. 2022. Suppression of oxidative stress and proinflammatory cytokines is a potential therapeutic action of *Ficus lepicarpa* B. (Moraceae) against carbon tetrachloride (CCl4)-induced hepatotoxicity in rats. *Molecules*. 27(8): 2593.

Ficus septica **Burm.f.**

[From Latin *ficus* = fig and *septicus* = pertaining to putrefaction]

Published in *Flora Indica. . . nec non Prodromus Florae Capensis* 226. 1768

Synonyms: *Ficus haulii* Blanco; *Ficus kaukauensis* Hayata; *Ficus leucantatoma* Poir.; *Ficus oldhamii* Hance

Local names: *lintatobu*, *lintotobu taragang*, *hintotobu* (Dusun); *hintotobow*, *lintotobu* (Dusun, Kadazan)

Common name: hauli tree

Habitat: seashores, wastelands, roadsides, at the edges of forests

Geographical distribution: from Malaysia to Pacific Islands

Botanical identification: this tree reaches a height of about 20 m. Trunk: not straight. Bark: light gray. Stems: stout, terete, with an acrid sap, somewhat articulate. Leaves: simple, spiral, stipulate. Stipules: ovate to lanceolate, up to about 3 cm long. Petioles: stout, 2–8 cm long. Blades: lanceolate, 10–14 cm × 15–26 cm, fleshy, glossy, attenuate at base, acuminate at apex, with 6–12 pairs of secondary nerves. Hypothandium: axillary, solitary, or paired, with 8–12 longitudinal ridges as well as whitish lenticels, oblate, about 1.2–2.5 cm across. Flowers: tiny. Calyx: 2-3-lobed. Stamens: 1. Stigma: clavate. Achenes: ovoid, tiny, embedded by in a repulsive fig.

Traditional therapeutic indications: flatulence, gastritis, fatigue, stomach aches (Dusun); fever, headaches, postpartum (Dusun, Kadazan)

Pharmacology and phytochemistry: analgesic, anti-inflammatory (Muaña et al. 2013), antibacterial, antifungal, antiprotozoal, cytotoxic (Vital et al. 2010; Sudirga et al. 2014), neuroactive (Chung et al. 2005), toxic to non-small cell lung cancer cells (A549) and colon carcinoma cells (HCT-116) (Canoy et al. 2010).

Antibacterial and antifungal phenanthroindolizidine alkaloid (antofine) (Baumgartner et al. 1989; Baumgartner et al. 1990; Damu et al. 2005; Damu et al. 2009). Antiplasmodial phenanthroindolizidine alkaloids (dehydroantofine) (Kubo et al. 2016).

Toxicity, side effects, and drug interaction: not known

Comments:

(i) This plant is sold in the wet markets of Sabah as medicine.

(ii) The Dusun and Kadazan use plants of the genus *Ficus* L. (1753) as medicines: "*tambunan*" for swollen knees, "*togung*" for stomach aches, "*sintotobou kusai*" for postpartum, "*tongkungkop*" for coughs and cuts, and "*lintotobutopurak*" for headaches and for postpartum.

(iii) The Murut use a plant of the genus *Ficus* L. (1753) as medicine (local name: *mamponoh*).

(iv) *Artocarpus heterophyllus* Lam. is used by the Dusun for skin infections and sore throats (local name: *nanko*). In places, villagers protect the fruits by covering them with rags, a sight which can be surprising.

(v) *Artocarpus odoratissimus* Blanco (Figure 4.31) is used for rough skin by the Murut (local name: *tarap*).

(vi) *Ficus racemosa* var *elongata* (King) M.F. Barrett medicinal for the Murut (local name: *tandilan*).

(vii) *Ficus fulva* Reinw. is used for cuts, stomach aches and wounds by the Dusun and Kadazan.

(viii) Dusun believe that spirits dwell in "*banyan*" (*Ficus benghalensis* L.).

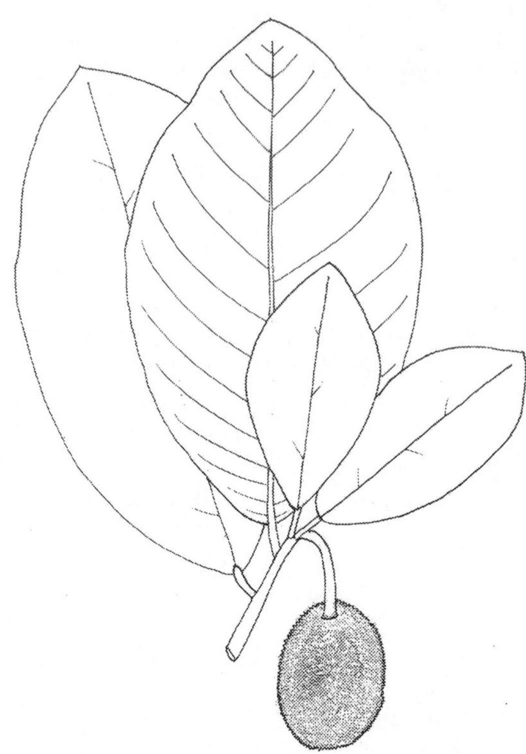

FIGURE 4.31 *Artocarpus odoratissimus* Blanco.

REFERENCES

Baumgartner, B., Erdelmeier, C.A., Wright, A.D., Rali, T. and Sticher, O. 1989. Antofine, a strong antifungal alkaloid from *Ficus septica* leaves. *Planta Medica*. 55(7): 652–653.

Baumgartner, B., Erdelmeier, C.A., Wright, A.D., Rali, T. and Sticher, O. 1990. An antimicrobial alkaloid from *Ficus septica*. *Phytochemistry*. 29(10): 3327–3330.

Canoy, R.J.C., Lomanta, J.M.J.C., Ballesteros, P.M., Chun, E.A.C., Dator, R.P. and Jacinto, S.D. 2010. Cancer chemotherapeutic potential of endemic and indigenous plants of Kanawan, Morong, Bataan Province, Philippines. *Asia Life Sciences-The Asian International Journal of Life Sciences*. 20(2): 331–339.

Chung, L.Y., Yap, K.F., Mustafa, M.R., Goh, S.H. and Imiyabir, Z. 2005. Muscarinic receptor activity of some Malaysian plant species. *Pharmaceutical Biology*. 43(8): 672–682.

Damu, A.G., Kuo, P.C., Shi, L.S., Li, C.Y., Kuoh, C.S., Wu, P.L. and Wu, T.S. 2005. Phenanthroindolizidine alkaloids from Stems: Of *Ficus septica*. *Journal of Natural Products*. 68(7): 1071–1075.

Damu, A.G., Kuo, P.C., Shi, L.S., Li, C.Y., Su, C.R. and Wu, T.S. 2009. Cytotoxic phenanthroindolizidine alkaloids from the roots of *Ficus septica*. *Planta Medica*. 75(10): 1152–1156.

Kubo, M., Yatsuzuka, W., Matsushima, S., Harada, K., Inoue, Y., Miyamoto, H., Matsumoto, M. and Fukuyama, Y. 2016. Antimalarial phenanthroindolizine alkaloids from *Ficus septica*. *Chemical and Pharmaceutical Bulletin*. 64(7): 957–960.

Muaña, C.G., Belga, M.L.L., Castillo, R.J.C., Dimaano, K.I.A., Fuentes, R.L.M. and Odasco, D.D.A. 2013. Potential analgesic and anti-inflammatory properties of the dried leaf extract of lagnub (*Ficus septica*). *Root Gatherers*. 5(1): 1–1.

Sudirga, S.K., Suprapta, D.N., Sudana, I.M. and Wirya, I.G.N.A.S. 2014. Antifungal activity of leaf extract of *Ficus septica* against *Colletotrichum acutatum* the cause of anthracnose disease on chili pepper. *Journal of Biology Agriculture and Healthcare*. 4(28): 47–52.

Vital, P.G., Velasco, R.N., Demigillo, J.M. and Rivera, W.L. 2010. Antimicrobial activity, cytotoxicity and phytochemical screening of *Ficus septica* Burm and *Sterculia foetida* L. leaf extracts. *Journal of Medicinal Plants Research*. 4(1): 58–63.

4.1.10.6.2 Family Rhamnaceae Juss. (1789), the Buckthorn Family

The family Rhamnaceae consists of about 50 genera and 900 species of trees, shrubs, or climbing plants. Stems: often thorny. Leaves: simple, alternate, stipulate. Blades: often coriaceous. Inflorescences: axillary cymes, corymbs, racemes, spikes, panicles or fascicles. Flowers: tiny. Calyx: 5 sepals or tubular. Petals: 5. Stamens: 5. Disc: present. Carpels: often 5, forming a 1–5 locular ovary, each locule with 1 ovule. Styles: 1. Stigma: often 5-lobed. Fruits: drupes, schizocarps.

Alphitonia incana **(Roxb.) Teijsm. & Binn. ex Kurz**

[From Greek *alphiton* = baked barley meal and Latin *incana* = grayish]

Published in *Journal of Botany, British and Foreign* 11: 208. 1873

Synonyms: *Alphitonia excelsa* (Fenzl) Reissek ex Endl.; *Alphitonia philippinensis* Braid; *Ceanothus excelsus* (Fenzl) Steud.; *Colubrina excelsa* Fenzl; *Rhamnus incana* Roxb.

Local names: *balịk anjin* (Bajau); *bolotion* (Murut)*; patiyata, pakodita, pokudata* (Dusun); *pukudita* (Kadazan)

Common name: cooper's wood

Geographical distribution: Indonesia, Malaysia, the Philippines, Papua New Guinea, Australia

Botanical description: this tree reaches a height of about 30 m. Bark and wood: fragrant. Bark: gray. Stems: terete, hairy and somewhat zigzag-shaped at apex. Leaves: simple, alternate, stipulate. Stipules: tiny, triangular, caducous. Petioles: about 1 cm long, hairy. Blades: 3–20 cm × 2–6 cm, rounded to somewhat truncate at base, with 7–14 pairs of secondary nerves, acuminate at apex, somewhat membranous, with a unique kind of golden color beneath, dark brown above in herbarium samples. Cymes: cauliflorous, terminal, or axillary, with tiny flowers that are light green to yellowish. Sepals: 5, hairy, tiny. Petals: 5, acuminate, about 3 mm long. Stamens: 5. Disc: present. Drupes: dark blue (which can make one think of blueberries), globose, about 1 cm across, on a cupular accrescent calyx. Seeds: in a red and glossy aril (Figure 4.32).

Traditional therapeutic indications: skin diseases, itchiness, pancreatitis, jaundice, sprains, headaches (Dusun, Kadazan)

Pharmacology and phytochemistry: not known

Toxicity, side effects, and drug interaction: not known

Comment: A plant of the genus *Alphitonia* Reissek ex Endl. (1840) is used by the Dusun for itchiness (local name: *pakodita*).

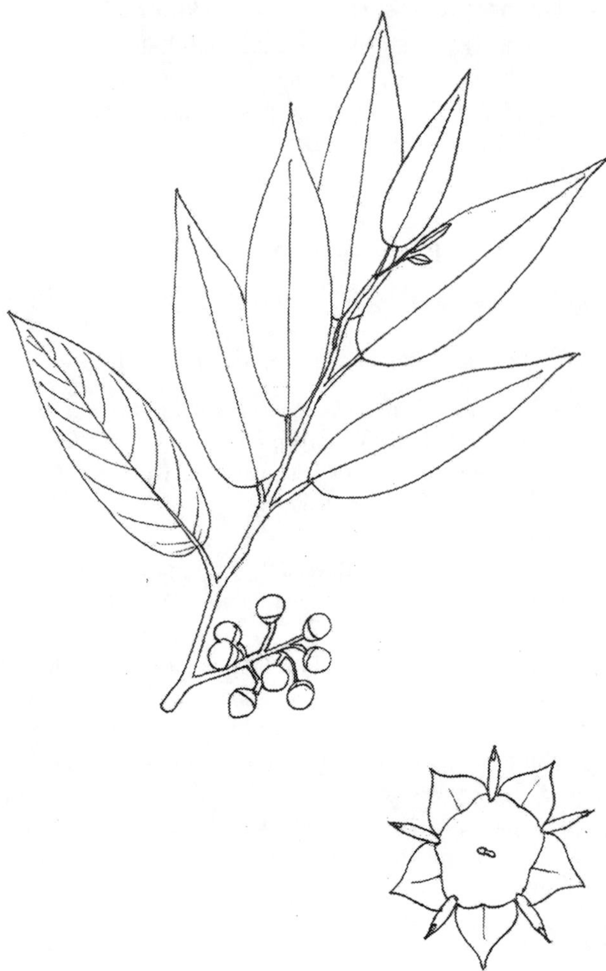

FIGURE 4.32 *Alphitonia incana* (Roxb.) Teijsm. & Binn. ex Kurz.

***Ziziphus horsfieldii* Miq.**

[From Persian *zaizafun* = jujube and after the 19th century American naturalist Thomas Horsfield]

Published in *Flora van Nederlandsch Indië* 1: 643. 1855

Synonym: *Ziziphus palawanensis* Elmer
Local names: *longkowit* (Rungus); *jingjing kulit* (locals of Javanese descent)
Habitat: forests
Geographical distribution: Malaysia, Indonesia, Borneo
Botanical description: this woody climbing plant grows up to a length of about 30 m. Stems: spiny, zigzag-shaped at apex. Leaves: simple, alternate, stipulate. Stipules: tiny, spiny, hooked, about 3 mm long. Petioles: about 5 mm long. Blades: elliptic, somewhat asymmetrical, finely serrate, about 1.2–4.5 cm × 3.5–10 cm, acute base, with a pair of longitudinal secondary nerves, acuminate to caudate at apex. Cymes: axillary, with tiny light green to light yellow flowers. Sepals: 5. Petals: 5, triangular, about 3 mm long. Stamens: 5. Disc: yellowish, pentagonal. Styles: 1. Stigma: bifid. Drupes: red, globose, 1–2 cm across.
Traditional therapeutic indication: medicinal (locals of Javanese descent)
Pharmacology and phytochemistry: not known
Toxicity, side effects, and drug interaction: not known

Comment: The Rungus use this plant for magic rituals.

4.1.10.6.3 Family Rosaceae Juss. (1789), the Rose Family

The family Rosaceae consists of about 100 genera and 3,000 species of trees, shrubs, herbaceous plants, and climbing plants. Leaves: simple or compound, alternate or spiral, often serrate, stipulate. Inflorescences: solitary or racemes. Hypanthium: present. Sepals: 5. Petals: 5. Stamens: up to 100. Carpels: 1–50, forming a 1–5 locular ovary each locule with 1–10 ovules per locule. Styles: 2–5. Fruits: drupes, berries.

***Prunus arborea* (Bl.) Kalkman**

[From Latin *prunus* = plum tree and *arbor* = tree]

Published in *Blumea* 13: 90. 1965

Synonyms: *Polydontia arborea* Bl.; *Pygeum arboretum* (Bl.) Bl., *Pygeum parviflorum* Teijsm. & Binnend., *Pygeum stipulaceum* King.
Local name: *kalanos* (Rungus)
Common name: current laurel
Habitat: forests
Geographical distribution: From India to Papua New Guinea
Botanical description: this magnificent tree reaches a height of about 40 m. Buttresses: present. Bark: smooth, odoriferous, brownish. Stems: terete, sometimes hairy. Leaves: simple, alternate, stipulate. Stipules: about 5 mm long. Petioles: 5 mm–1.5 cm long, sometimes hairy. Blades: lanceolate, attenuate to rounded at base, acuminate to somewhat caudate at apex, with about 9–13 pairs of secondary nerves, 3–6 cm × 6–15 cm. Racemes: axillary, about 2.5–6 cm long, sometimes hairy, with numerous tiny flowers. Sepals: 5. Petals: 5, whitish to pink. Stamens: about 50, 8 mm long. Ovary: hairy. Styles: 5 mm long. Drupes: about 1 cm across, compressed, which can make one think of a pair of buttocks, glossy, dark purple when ripe (Figure 4.33).

FIGURE 4.33 *Prunus arborea* (Blume) Kalkman.

Traditional therapeutic indications: veterinary medicine (Rungus)
Pharmacology and phytochemistry: not known
Toxicity, side effects, and drug interaction: not known

Comments:

(i) *Prunus arborea* (Blume) Kalkman and many other trees with medicinal value are disappearing. Until the 1980s, about 80% of Sabah's surface was covered in virgin forest.
(ii) *Rubus moluccanus* L. is used by the Dusun for sore eyes.

4.1.10.6.4 Family Urticaceae Juss. (1789), the Nettle Family

The family Urticaceae consists of about 50 genera and 550 species of herbaceous plants, climbing plants, shrubs, or trees. Leaves: simple, alternate, or opposite, stipulate, sometimes hairy. Inflorescences: axillary cymes or fascicles of tiny flowers. Calyx: tubular, 4–5-lobed. Corolla: none. Stamens: 4–5. Carpels: 1, with 1 ovule. Styles: 1. Fruits: achenes.

***Dendrocnide elliptica* (Merr.) Chew**

[From Greek *dendron* = tree, *knide* = stinging nettle, and from Latin *elliptica* = elliptical]

Published in *Gardens' Bulletin, Singapore* 21(2): 203. 1965

Synonym: *Laportea elliptica* Merr.
Local name: *ohopoi* (Rungus)
Habitat: forests near rivers
Geographical distribution: Borneo, the Philippines
Botanical description: this tree reaches a height of about 10 m. Stems: terete, stout, with extremely vesicant hairs. Leaves: simple, spiral, stipulate. Petioles: stout, to about 2 cm long. Blades: broadly oblanceolate to elliptic, somewhat wavy, tapering at base, acuminate at apex, 11–15 cm × 15–30 cm, with about 15 pairs of secondary nerves. Flower fascicles: thin, pendulous, terminal. Flowers: tiny. Calyx: tubular, thin, 4-lobed. Stamens: 4. Ovary: ovoid. Stigma: filiform. Achenes: somewhat asymmetrical.
Traditional therapeutic indications: medicinal (Rungus)
Pharmacology and phytochemistry: cytotoxic (Singchai & Jantawe 2020), antibacterial (Mariani et al. 2014).
Toxicity, side effects, and drug interaction: not known

Comments:

(i) In West Java, *Dendrocnide stimulans* (L.fil.) Chew is used for numbness (local name: *pulus*).
(ii) In Papua New Guinea, the Maenge use a plant of the genus *Laportea* Gaudich. (1826) for wounds, sore throats, sores, and magic rituals.
(iii) In the Moluccas, *Laportea decumana* (Roxb.) Wedd. (local name: *sinan*) is used for stomach aches.
(iv) *Leucosyke capitellata* Wedd. is used by the Dusun for body pains and postpartum (local name: *mandahasi*). They also employ *Pipturus arborescens* (Link) C.B. Rob. for flu and fever (local name: *bayug*).

REFERENCES

Mariani, R., Sukandar, E.Y. and Suganda, A.G. 2014. Antimicrobial activity from Indonesian urticaceae. *International Journal of Pharmacy and Pharmaceutical Sciences*. 6(4): 191–193.
Singchai, B. and Jantawe, R. 2020. Isolated compound and its biological activities from dendrocnide stimulans. *Burapha Science Journal (วารสาร วิทยาศาสตร์ บูรพา)*. 25(1): 1–12.

***Poikilospermum cordifolium* (Barg.-Petr.) Merr.**

[From *Greek poikilos* = spotted and Latin *cordifolium* = heart-shaped leaves]

Published in *Contributions from the Arnold Arboretum of Harvard University* 8: 49. 1934

Synonym: *Conocephalus cordifolius* Barg.-Petr.
Local name: *sodingkalan* (Dusun, Kadazan, Rungus)
Habitat: forests near streams and waterfalls, on trees or limestone rocks
Geographical distribution: Thailand, Malaysia, Borneo, the Philippines
Botanical description: this stout climbing plant grows up to a length of about 10 m. Bark: scaly, brown, rugose, fissured. Stems: terete, stout. Leaves: simple, spiral, stipulate. Stipules: 4–6 cm long, purplish. Petioles: about 10–40 cm long, stout, smooth. Blades: enormous (30–50 cm × 20–35 cm), broadly ovate, oblong, or elliptic, coriaceous, rounded to cordate at base, obtuse at apex, wavy, fleshy, with 10–12 pairs of secondary nerves, the first pairs emerging from the base. Cymes: cauliflorous, stout, about 4–10 cm long, purplish, with fluffy and purplish heads, which are about 2 cm across. Flowers: tiny. Sepals: 4, about 4 mm long. Ovary and style white. Achenes: 5 mm long (Figure 4.34).
Traditional therapeutic indication: fainting (Rungus)
Pharmacology and phytochemistry: not known
Toxicity, side effects, and drug interaction: not known

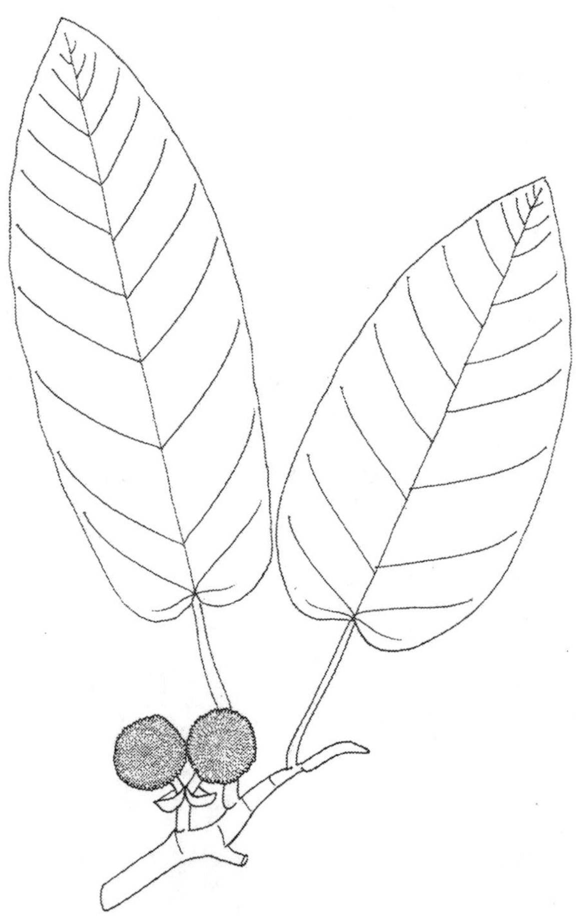

FIGURE 4.34 *Poikilospermum cordifolium* (Barg.-Petr.) Merr.

Comment: A plant of the genus *Poikilospermum* Zipp. ex (1864) is used by the Dusun for postpartum, flatulence, fatigue, and wounds (local name: *saringkalang*).

***Poikilospermum suaveolens* (Bl.) Merr.**

[From Greek *poikilos* = spotted and Latin *suaveolens* = fragrant]

Published in *Contributions from the Arnold Arboretum of Harvard University* 8: 49. 1934

Synonyms: *Conocephalus sinensis* C.H. Wright; *Conocephalus suaveolens* Bl.; *Poikilospermum sinense* (C.H. Wright) Merr.
Local names: *bunatol* (Murut); *gunaton* (Dusun)
Habitat: forests
Geographical distribution: Thailand, Malaysia, Borneo, the Philippines
Botanical description: this stout climbing plant grows up to a length of about 10 m. Bark: scaly, fissured. Leaves: simple, spiral, moraceous, stipulate. Stipules: 2–4 cm long. Petioles: about 4–14 cm long, smooth, somewhat yellowish brown. Blades: broadly lanceolate, elliptic to ovate, 6–10 cm × 25–40 cm, coriaceous, rounded, attenuate, or cordate at base, obtuse at apex, with 12–14 pairs of secondary nerves. Cymes: cauliflorous, with purplish heads, about 3–7 cm across. Male flowers: 4 tepals, 4 stamens. Female flowers: perianth tubular, thin, 4-lobed, about 2.5 mm long. Achenes: tiny.
Traditional therapeutic indications: postpartum (Murut); sore eyes (Dusun)
Pharmacology and phytochemistry: antiprotozoal (Matsuura et al. 2004), antibacterial (Hapid et al. 2021), antifungal (Kusuma et al. 2016), wound healing effects (Ahmad et al. 2022).
Toxicity, side effects, and drug interaction: not known

Comments:

(i) The young leaves are used as food by the Murut.
(ii) In Sulawesi, the Kaili use the plant (local name: *bajakah*) for cancer and fatigue.
(iii) A plant of the genus *Urtica* L. (1753) is used by the Dusun (local name: *mandahasi*) for coughs.
(iv) A plant of the genus *Pouzolzia* Gaudich. (1826) is used by the Dusun and Kadazan to expedite childbirth (local name: *komburiong*).

REFERENCES

Ahmad, H., Ahmad, M.H., Ali, A., Pagarra, H., Salempa, P., Salleh, L.M. and Passitta, M. 2022. Evaluation of antioxidant, antimicrobial and wound healing activity of *Poikilospermum suaveolens*. *Jurnal Teknologi*. 84(1): 41–48.

Hapid, A., Napitupulu, M. and Zubair, M.S. 2021. Phytochemical screening, GC-MS analysis, toxicity and antimicrobial properties of extracts outer shell *Poikilospermum suaveolens* (Blume) Merr. *International Journal of Research and Innovation in Applied Science*. 6(9): 111–117.

Kusuma, I.W., Sari, N.M., Murdiyanto, M. and Kuspradini, H. 2016, July. Anticandidal activity of numerous plants used by Bentian tribe in East Kalimantan, Indonesia. In *AIP Conference Proceedings* (1755, No. 1). AIP Publishing.

Matsuura, H., Yamasaki, M., Yamato, O., Maede, Y., Katakura, K., Suzuki, M. and Yoshihara, T. 2004. Effects of central Kalimantan plant extracts on intraerythrocytic *Babesia gibsoni* in culture. *Journal of Veterinary Medical Science*. 66(7): 871–874.

4.1.11 Superorder Rosanae Takht. (1967), the Malvids

4.1.11.1 Order Brassicales Bromhead (1838)

4.1.11.1.1 Family Caricaceae Dumort. (1829), the Papaya Family

The family Caricaceae comprises about 5 genera and 55 species of treelets. Stems: secrete a latex once incised. Leaves: simple, spiral, exstipulate. Corolla: tubular, 5-lobed. Blades: palmate or palmatifid. Inflorescences: axillary cymes. Sepals: 5, free, or merged at base. Corolla: tubular, 5-lobed. Stamens: 5–10. Carpels: 5, forming a 1–5 locular ovary, each locule with 30–100 ovules. Styles: 15. Stigma: 5-lobed. Fruits: berries.

***Carica papaya* L.**

[From Latin *carica* = a fig and Central American local name of *Carica papaya* L. = *papaya*]

Published in *Species Plantarum* 2: 1036. 1753

Local names: *betik* (Bajau, Kedayan, Murut); *kepayas, tapayas* (Bajau); *tepayas* (Dusun); *pellek keniki* (Bugis)

Common names: papaya, melon tree

Habitat: roadsides, around villages

Geographical distribution: tropical

Botanical description: this treelet reaches a height of about 5 m. Trunk: soft wooded, regularly marked with clearly visible leaf scars, secretes a latex once incised. Leaves: simple, spiral, terminal, exstipulate. Petioles: about 30 cm long, fleshy, somewhat straight. Blades: palmatifid, incised, up to about 30 cm across, dull green. Male flowers: long, pendulous, and lax panicles. Female flowers: short cymes. Calyx: 5-lobed, the lobes about 1 cm long. Corolla: tubular, 5-lobed, creamy yellow, the lobes about 5.5 cm long. Stamens: 10. Ovary: ovoid. Styles: 1. Stigma 5-lobed. Berries: fleshy, ovoid to oblong, greenish-yellow, heavy, smooth, up to about 15 cm × 30 cm. Seeds: numerous, black, glossy, in a reddish, sappy, and delicious flesh.

Traditional therapeutic indications: flatulence, postpartum, increases breast milk, weak uterus, constipation (Dusun); sinusitis (Bajau); malaria (Bajau, Sungai); hypertension (Bajau, Dusun, locals of Javanese descent, Murut, Rugus); abscesses, swellings, stomach aches warts (Kedayan); gout (Bugis); hemorrhoids, gonorrhea, to promote lactation, to prevent menstruation (Dusun, Kadazan); diabetes (Bajau, Bugis); fever (Kedayan, Irannun); painful joins (Iranum); headaches, kidney diseases, hypertension (Bajau); dengue (Suluk)

Pharmacology and phytochemistry: this plant has been the subject of numerous pharmacological studies which have highlighted, among other things, antiviral (Dengue) (Hettige 2008; Ahmad et al. 2011; Subenthiran et al. 2013), antibacterial (Rakholiya et al. 2014), antifungal (Chávez-Quintal et al. 2011), hypotensive (Eno et al. 2000), hypoglycemic (Juárez-Rojop et al. 2012), and anthelmintic activity (Satrija et al. 1995).

Antiamoebal, antiplasmodial, antithrombocytopenic piperidine alkaloids (carpaine) (Zunjar et al. 2016; Cimanga et al. 2018; Teng et al. 2019). Flavonols (kaempferol), flavonol glycosides (quercetin 3-rutinoside, myricetin 3-rhamnoside) (Nugroho et al. 2017). Antiviral (influenza) flavonols (quercetin, isoquercetin) (Kim et al. 2010).

Toxicity, side effects, and drug interaction: the latex and seeds are dangerous in pregnancy (Gopalakrishnan & Rajasekharasetty 1978). The latex dissolves proteins and causes severe esophageal injuries when used to treat food impaction (Weiner et al. 1978).

REFERENCES

Ahmad, N., Fazal, H., Ayaz, M., Abbasi, B.H., Mohammad, I. and Fazal, L. 2011. Dengue fever treatment with *Carica papaya* leaves extracts. *Asian Pacific Journal of Tropical Biomedicine*. 1(4): 330.

Chávez-Quintal, P., González-Flores, T., Rodríguez-Buenfil, I. and Gallegos-Tintoré, S. 2011. Antifungal activity in ethanolic extracts of *Carica papaya* L. cv. Maradol leaves and seeds. *Indian Journal of Microbiology*. 51(1): 54–60.

Cimanga, K.R., Makila, B.M.F., Kambu, K.O., Tona, L.G., Vlietinck, A.J. and Pieters, L. 2018. *In vitro* amoebicidal activity of aqueous extracts and their fractions from some medicinal plants used in traditional medicine as antidiarrheal agents in Kinshasa-Democratic Republic of Congo against entamoeba histolytica. *European Journal of Biomedical and Pharmaceutical Sciences*. 5(7): 103–114.

Eno, A.E., Owo, O.I., Itam, E.H. and Konya, R.S. 2000. Blood pressure depression by the fruit juice of *Carica papaya* (L.) in renal and DOCA-induced hypertension in the rat. *Phytotherapy Research: An International Journal Devoted to Pharmacological and Toxicological Evaluation of Natural Product Derivatives*. 14(4): 235–239.

Gopalakrishnan, M. and Rajasekharasetty, M.R. 1978. Effect of papaya (*Carica papaya* linn) on pregnancy and estrous cycle in albino rats of Wistar strain. *Indian Journal of Physiology and Pharmacology*. 22(1): 66–70.

Hettige, S. 2008. Salutary effects of *Carica papaya* leaf extract in dengue fever patients—a pilot study. *Sri Lankan Family Physician*. 29(1): 17–19.

Juárez-Rojop, I.E., Díaz-Zagoya, J.C., Ble-Castillo, J.L., Miranda-Osorio, P.H., Castell-Rodríguez, A.E., Tovilla-Zárate, C.A., Rodríguez-Hernández, A., Aguilar-Mariscal, H., Ramón-Frías, T. and Bermúdez-Ocaña, D.Y. 2012. Hypoglycemic effect of *Carica papaya* leaves in streptozotocin-induced diabetic rats. *BMC Complementary and Alternative Medicine*. 12: 1–11.

Kim, Y., Narayanan, S. and Chang, K.O. 2010. Inhibition of influenza virus replication by plant-derived isoquercetin. *Antiviral Research*. 88(2): 227–235.

Nugroho, A., Heryani, H., Choi, J.S. and Park, H.J. 2017. Identification and quantification of flavonoids in *Carica papaya* leaf and peroxynitrite-scavenging activity. *Asian Pacific Journal of Tropical Biomedicine*. 7(3): 208–213.

Rakholiya, K., Kaneria, M. and Chanda, S. 2014. Inhibition of microbial pathogens using fruit and vegetable peel extracts. *International Journal of Food Sciences and Nutrition*. 65(6): 733–739.

Satrija, F., Nansen, P., Murtini, S. and He, S. 1995. Anthelmintic activity of papaya latex against patent *Heligmosomoides polygyrus* infections in mice. *Journal of Ethnopharmacology*. 48(3): 161–164.

Subenthiran, S., Choon, T.C., Cheong, K.C., Thayan, R., Teck, M.B., Muniandy, P.K., Afzan, A., Abdullah, N.R. and Ismail, Z. 2013. *Carica papaya* leaves juice significantly accelerates the rate of increase in platelet count among patients with dengue fever and dengue haemorrhagic fever. *Evidence-Based Complementary and Alternative Medicine*. 2013: 1–7.

Teng, W.C., Chan, W., Suwanarusk, R., Ong, A., Ho, H.K., Russell, B., Rénia, L. and Koh, H.L. 2019. *In vitro* antimalarial evaluations and cytotoxicity investigations of *Carica papaya* leaves and carpaine. *Natural Product Communications*. 14(1): 33–36.

Weiner, B., Curtis, R., Lovejoy, D. and Rheinlander, H.F. 1978. Three case reports. *Drug Intelligence & Clinical Pharmacy*. 12(8): 458–460.

Zunjar, V., Dash, R.P., Jivrajani, M., Trivedi, B. and Nivsarkar, M. 2016. Antithrombocytopenic activity of carpaine and alkaloidal extract of *Carica papaya* Linn. leaves in busulfan induced thrombocytopenic Wistar rats. *Journal of Ethnopharmacology*. 181: 20–25.

4.1.11.1.2 Family Cleomaceae Bercht. & J. Presl (1820), the Spider Flower Family

The family Cleomaceae consists of about 8 genera and 275 species of herbaceous plants. Leaves: alternate, opposite, simple or trifoliate, or often palmately compound. Stipules: absent or tiny. Inflorescences: racemes, or solitary. Gynophore or androgynophore: present. Sepals: 4. Petals: 4. Stamens: up to 100, showy. Carpels: 2, forming a unilocular or bilocular ovary, each locule with 10–100 ovules. Styles: 1. Fruits: capsules. Seeds: reniform or angular.

***Cleome chelidonii* L.f.**

[Probably from Greek *kleos* = glory and Latin *chelidonii* = chelidonium-like]

Published in *Supplementum Plantarum* 300. 1781

Synonyms: *Aubion chelidonii* (L. f.) Raf.; *Corynandra chelidonii* (L. f.) Cochrane & Iltis ex Spreng.; *Polanisia chelidonii* (L. f.) DC.
Local name: *janggut kucing* (Murut)
Common name: celandine spider flower
Habitat: wastelands, roadsides, villages
Geographical distribution: from India to Indonesia
Botanical description: this magnificent herbaceous plant reaches a height of about 30 cm. Stems: thin, terete. Leaves: palmate, spiral, exstipulate. Petioles: thin, about 3–7 cm long. Blades: includes 3–7 folioles. Folioles: membranous, elliptic to spathulate, 1–1.5 cm × 1.5–4 cm, acute at apex, tapering at base, somewhat asymmetrical and wavy. Racemes: axillary, with beautiful flowers. Sepals: 4, 4 mm long. Petals: 4, pink, about 1.5 cm long, broadly lanceolate, acuminate at apex, cross-shaped, somewhat recurved. Stamens: numerous, filamentous, about 1 cm long. Ovary: oblong. Stigma: capitate. Capsules: dehiscent, 4–7 cm long, narrowly oblong to cylindrical. Seeds: numerous, tiny, somewhat nautilus-shaped.
Traditional therapeutic indications: stomach aches (Murut)
Pharmacology and phytochemistry: antimicrobial (Sridhar et al. 2014), anti-inflammatory, analgesic, antipyretic (Parimalakrishnan et al. 2007).
Flavonol glycosides toxic to human hepatoma cells (Nguyen et al. 2017).
Toxicity, side effects, and drug interaction: not known

Comments:

(i) *Gynandropsis gynandra* (L.) Briq is used by the Bajau (local name: *maman hantu*) for hemorrhoids and rheumatisms.
(ii) In the family Moringaceae Martinov (1820), *Moringa oleifera* Lam. is used for skin cleansing by the Bugis (local name: *daung kelor*), whilst it is a medicine for asthma, coughs, diarrhea, and fever for the Bajau (local name: *marrungai*).

REFERENCES

Nguyen, T.P., Mai, D.T., Do, T.H.T. and Phan, N.M. 2017. Flavonoids with hepatoprotective activity from the leaves of Cleome chelidonii. *Natural Product Communications*. 12(7): 2587–2592.

Parimalakrishnan, S., Dey, A., Smith, A.A. and Manavalan, R. 2007. Evaluation of anti-inflammatory, antinociceptive and antipyretic effects of methanol extract of *Cleome chelidonii*. *International Journal of Biological and Chemical Sciences*. 1(3): 223–228.

Sridhar, N., Sasidhar, D.T. and Kanthal, L.K. 2014. *In vitro* antimicrobial screening of methanolic extracts of *Cleome chelidonii* and *Cleome gynandra*. *Bangladesh Journal of Pharmacology*. 9(2): 161–166.

4.1.11.2 Order Malvales Juss. ex Bercht. & J. Presl (1820)

4.1.11.2.1 Family Bixaceae Kunth (1822), the Achiote Family

The family Bixaceae consists of the single genus *Bixa* L. (1753)

***Bixa orellana* L.**

[From the ancient Central American name of achiote = *biche* and after the 16th century Spanish explorer Francisco de Orellana]

Published in *Species Plantarum* 1: 512. 1753

Synonyms: *Bixa acuminata* Bojer; *Bixa americana* Poir.; *Bixa odorata* Ruiz & Pav. ex G.Don; *Bixa platycarpa* Ruiz & Pav. ex G.Don; *Bixa tinctoria* Salisb.; *Bixa upatensis* Ram. Goyena; *Bixa urucurana* Willd.; *Orellana americana* Kuntze; *Orellana orellana* (L.) Kuntze

Local names: *kesumba, puloh, suliabai* (Murut)

Common names: achiote, annatto, lipstick tree

Habitat: cultivated

Geographical distribution: tropical

Botanical description: this shrub reaches a height of about 5 m. Leaves: simple, spiral, stipulate. Petioles: 5–7.5 cm long. Blades: 6–12.5 cm × 10–20 cm, dark green, ovate, acute, or acuminate at apex, truncate or subcordate at base. Flowers: 5 cm across, showy, arranged in terminal panicles. Sepals: 5, of irregular length, up to 1 cm long. Petals: 5, pinkish, red, or white, obovate, membranous, up to 3 cm long. Stamens: numerous. Capsules: dehiscent, bright red, ovoid, or subglobose, spiny, become hideous when they ripen. Seeds: 15–20, trigonous, in a bright red pulp.

Traditional therapeutic indications: gastritis, pain, white hair (Murut)

Pharmacology and phytochemistry: this plant has been the subject of numerous pharmacological studies (Vilar et al. 2014) which have highlighted, among other things, gastroprotective activity (Huamán et al. 2009).

Antinociceptive and anti-inflammatory carotenoids (bixin) (Pacheco et al. 2019).

Toxicity, side effects, and drug interaction: 750 mg of leaf powder given daily for 6 months to volunteers did not cause serious adverse effects (Stohs 2014).

REFERENCES

Huamán, O., Sandoval, M., Arnao, I. and Béjar, E. 2009, June. Efecto antiulceroso del extracto hidroalcohólico liofilizado de hojas de *Bixa orellana* (achiote), en ratas. In *Anales de la Facultad de Medicina* (70, No. 2, pp. 97–102). UNMSM. Facultad de Medicina.

Pacheco, S.D.G., Gasparin, A.T., Jesus, C.H.A., Sotomaior, B.B., Ventura, A.C.S.S.B., Redivo, D.D.B., de Almeida Cabrini, D., Dias, J.D.F.G., Miguel, M.D., Miguel, O.G. and da Cunha, J.M. 2019. Antinociceptive and anti-inflammatory effects of bixin, a carotenoid extracted from the seeds of *Bixa orellana*. *Planta Medica*. 85(16): 1216–1224.

Stohs, S.J. 2014. Safety and efficacy of *Bixa orellana* (achiote, annatto) leaf extracts. *Phytotherapy Research*. 28(7): 956–960.

Vilar, D.D.A., Vilar, M.S.D.A., Raffin, F.N., Oliveira, M.R.D., Franco, C.F.D.O., de Athayde-Filho, P.F., Diniz, M.D.F.F.M. and Barbosa-Filho, J.M. 2014. Traditional uses, chemical constituents, and biological activities of Bixa orellana L.: A review. *The Scientific World Journal*. 2014: 1–11.

4.1.11.2.2 Family Bombacaceae Kunth (1822), the Bombax Family

The family Bombacaceae consists of about 30 genera and 180 species of trees. Trunk: sometimes somewhat bottle-shaped to enormously swollen, sometimes with broad conical woody thorns. Buttresses: sometimes present. Leaves: simple or palmate, alternate, stipulate. Stipules: tiny. Petioles: often swollen at base, straight. Flowers: often showy, axillary, solitary. Epicalyx: sometimes present. Calyx: 5 sepals or tubular. Corolla: 5 petals or tubular. Stamens: up to 100, showy, merged at base. Carpels: 2–5, forming a 2–5 locular ovary each locule with 2–6 ovules. Fruits: capsules. Seeds: numerous.

***Bombax ceiba* L.**

[From Greek *bombyx* = silk and the ancient local Central American name of a plant in the Bombaceae family = *ceiba*]

Published in *Species Plantarum* 511. 1753

Synonyms: *Bombax heptaphyllum* L.; *Bombax malabaricum* DC.; *Gossampinus malabarica* (DC.) Merr.; *Salmalia malabarica* (DC.) Schott & Endl.

Local name: *kapok* (Bajau, Dusun, Murut)

Common name: red silk cotton tree

Habitat: roadsides, around villages

Geographical distribution: tropical Asia

Botanical description: this massive tree reaches a height of about 25 m. Trunk: with woody thorns. Butresses: present. Leaves: spiral, palmate, stipulate. Stipules: tiny. Petioles: thin, up to 20 cm long. Folioles: 5–7, oblong, 3.5–5.5 cm × 10–15 cm, with 15–17 pairs of secondary nerves, soft, acuminate at apex. Flowers: solitary, showy, terminal. Calyx: cup-shaped, fleshy, green, glossy, 2–3 cm long, somewhat 5-lobed, the lobes triangular. Petals: 5, oblong, 8–10 cm × 3–4 cm, of a peculiar kind of dull red. Stamens: showy, numerous. Capsules: oblong, 4.5–5 cm × 10–15 cm, open to release a bulky cottonous mass and numerous seeds.

Traditional therapeutic indications: fatigue (Dusun, Kadazan); blood vomiting (Murut)

Pharmacology and phytochemistry: this plant has been the subject of numerous pharmacological studies (Chaudhary & Khadabadi 2012; Rajput 2022) which have highlighted, among other things, antibacterial (Digge et al. 2015; Masood-ur-Rehman et al. 2017), antiviral (herpes simplex virus) (Rajbhandari 2001), antidiabetic (Guang-Kai et al. 2017), hypotensive (Saleem et al. 2003), and diuretic activity (Jalalpure & Gadge 2011).

Triterpenes (lupeol and lupeol acetate), flavonols, and flavonol glycosides (Saleem et al. 2003). Xanthones (mangiferin) with antiviral (HIV) activity (Faizi & Ali 1999; Shahat et al. 2003; Wang et al. 2011). Antiviral (hepatitis B virus) lignans (bombasinol A, 5,6-dihydroxymatairesinol, (+) pinoresinol, maitaresinol) (Wang et al. 2013). Antiviral (respiratory syncytial virus) hydroxycinnamic acid derivatives (Zhang et al. 2015).

Toxicity, side effects, and drug interaction: the median lethal dose (LD_{50}) of ethanol extract of fruits administered to rats was 2 g/kg (Jalalpure & Gadge 2011).

REFERENCES

Chaudhary, P.H. and Khadabadi, S.S. 2012. *Bombax ceiba* Linn.: Pharmacognosy, ethnobotany and phytopharmacology. *Pharmacognosy Communications*. 2(3): 2–9.

Digge, V.G., Kuthar Sonali, S., Hogade Maheshwar, G., Poul, B.N. and Jadge Dhanraj, R. 2015. Screening of antibacterial activity of aqueous bark extract of *Bombax ceiba* against some gram positive and gram negative bacteria. *Screening*. 5: 8.

Faizi, S. and Ali, M. 1999. Shamimin: A new flavonol C-glycoside from leaves of *Bombax ceiba*. *Planta Medica*. 65(4): 383–385.

Guang-Kai, X.U., Xiao-Ying, Q.I.N., Guo-Kai, W.A.N.G., Guo-Yong, X.I.E., Xu-Sen, L.I., Chen-Yu, S.U.N., Bao-Lin, L.I.U. and Min-Jian, Q.I.N. 2017. Antihyperglycemic, antihyperlipidemic and antioxidant effects of standard ethanol extract of *Bombax ceiba* leaves in high-fat-diet- and streptozotocin-induced Type 2 diabetic rats. *Chinese Journal of Natural Medicines*. 15(3): 168–177.

Jalalpure, S.S. and Gadge, N.B. 2011. Diuretic effects of young fruit extracts of *Bombax ceiba* L. in rats. *Indian Journal of Pharmaceutical Sciences*. 73(3): 306.

Masood-ur-Rehman, N.A. and Mustafa, R. 2017. Antibacterial and antioxidant potential of stem bark extract of *Bombax ceiba* collected locally from south Punjab Area of Pakistan. *African Journal of Traditional, Complementary, and Alternative Medicines*. 14(2): 9.

Rajbhandari, M., Wegner, U., Jülich, M., Schoepke, T. and Mentel, R. 2001. Screening of Nepalese medicinal plants for antiviral activity. *Journal of Ethnopharmacology*. 74(3: 251–255.

Rajput, R.T. 2022. Ethnomedicine and pharmacology of semal (*Bombax ceiba* L.)-A Indian medicinal plant: A review. *Agricultural Reviews*. 43(2): 145–153.

Saleem, R., Ahmad, S.I., Ahmed, M., Faizi, Z., Zikr-ur-Rehman, S., Ali, M. and Faizi, S. 2003. Hypotensive activity and toxicology of constituents from *Bombax ceiba* stem bark. *Biological and Pharmaceutical Bulletin*. 26(1): 41–46.

Shahat, A.A., Hassan, R.A., Nazif, N.M., Van Miert, S., Pieters, L., Hammuda, F.M. and Vlietinck, A.J. 2003. Isolation of mangiferin from *Bombax malabaricum* and structure revision of shamimin. *Planta Medica*. 69(11): 1068–1070.

Wang, G.K., Lin, B.B., Rao, R., Zhu, K., Qin, X.Y., Xie, G.Y. and Qin, M.J. 2013. A new lignan with anti-HBV activity from the roots of *Bombax ceiba*. *Natural Product Research*. 27(15): 1348–1352.

Wang, R.R., Gao, Y.D., Ma, C.H., Zhang, X.J., Huang, C.G., Huang, J.F. and Zheng, Y.T. 2011. Mangiferin, an anti-HIV-1 agent targeting protease and effective against resistant strains. *Molecules*. 16(5): 4264–4277.

Zhang, Y.B., Wu, P., Zhang, X.L., Xia, C., Li, G.Q., Ye, W.C., Wang, G.C. and Li, Y.L. 2015. Phenolic compounds from the flowers of *Bombax malabaricum* and their antioxidant and antiviral activities. *Molecules*. 20(11): 19947–19957.

***Ceiba pentandra* (L.) Gaertn.**

[From the ancient local Central American name of a plant in the Bombaceae family = *ceiba*, the Greek *pente* = five, and *andros* = male]

Published in *De Fructibus et Seminibus Plantarum*. . . . 2(2): 244, pl. 133, f. 1. 1791

Synonyms: *Bombax pentandrum* L.; *Ceiba casearia* Medik.; *Eriodendron anfractuosum* DC.; *Eriodendron caribaeum* G.Don; *Eriodendron occidentale* (Spreng.) G.Don; *Gossampinus rumphii* Schott; *Xylon pentandrum* (L.) Kuntze

Local names: *kapok kapok* (Murut); *kapek, kekawu* (Lundayeh); *kekabu* (Bajau)

Common name: kapok tree

Habitat: roadsides, villages

Geographical distribution: tropical Asia

Botanical description: this massive tree reaches a height of about 40 m. Buttresses: present. Bark: smooth, greenish, with woody conical thorns. Wood: whitish, soft. Stems: terete, smooth. Leaves: palmate, spiral, stipulate. Petioles: 7–14 cm long. Blades: 5–9 folioles. Folioles: sessile, 1.5–6.5 cm × 5–20 cm, narrowly elliptic to oblanceolate, somewhat glossy, wedge-shaped at base, acuminate at apex, with about 10 pairs of secondary nerves. Flowers: cauliflorous, solitary, or in fascicles. Calyx: tubular, 5-lobed, about 1.5 cm, the lobes broadly lanceolate Petals: 5, pink, somewhat fleshy, hooded, obovate, about 3 cm long, of a peculiar kind of yellowish color. Stamens: 5, filaments pinkish, slightly curved, protuding. Ovary: glabrous. Styles: 1. about 3 cm long, white, often twisted. Stigma: clavate. Capsules: green, fleshy, fusiform, 8–14 cm long. Seeds: numerous, up to 8 mm long in a cottony mass.

Traditional therapeutic indications: feverish children (Dusun)

Pharmacology and phytochemistry: antiviral (Dewi et al. 2019), antipyretic (Saptarini & Deswati 2015), antimicrobial (Singh et al. 2010). Anticancer (Kumar et al. 2016), anti-inflammatory, analgesic (Anosike et al. 2016), antidiabetic (Odoh et al. 2016), and anti-Alzheimer flavonolignans (Abouelela et al. 2020). Antibacterial and antifungal sesquiterpene lactones (Rao et al. 1993).

Toxicity, side effects, and drug interaction: not known

Comment: The plant needs to be studied for its possible analgesic effects, and flavonolignans could undergo further studies for the treatment of Alzheimer's disease.

REFERENCES

Abouelela, M.E., Orabi, M.A., Abdelhamid, R.A., Abdelkader, M.S., Darwish, F.M., Hotsumi, M. and Konno, H. 2020. Anti-Alzheimer's flavanolignans from *Ceiba pentandra* aerial parts. *Fitoterapia*. 143: 104541.

Anosike, C.A., Okagu, I.U., Amaechi, K.C. and Nweke, V.C. 2016. *In vivo* anti-inflammatory and analgesic potentials of methanol extract of *Ceiba pentandra* stem back. *American Journal of Research Communication*. 4(9): 116–129.

Dewi, B.E., Angelina, M., Ardiantara, S., Prakoso, A.R., Desti, H. and Sudiro, T.M. 2019, December. Antiviral activity of *Ceiba pentandra* and Eugenia uniflora leaf extracts to dengue virus serotype-2 in Huh 7it-1 cell line. In *AIP Conference Proceedings* (2193, No. 1: 030003). AIP Publishing LLC.

Kumar, R., Kumar, N., Ramalingayya, G.V., Setty, M.M. and Pai, K.S.R. 2016. Evaluation of *Ceiba pentandra* (L.) Gaertner bark extracts for *in vitro* cytotoxicity on cancer cells and *in vivo* antitumor activity in solid and liquid tumor models. *Cytotechnology*. 68(5): 1909–1923.

Odoh, U.E., Onugha, V.O. and Chukwube, V.O. 2016. Evaluation of antidiabetic effect and hematotological profile of methanol extract of *Ceiba pentandra* G (Malvaceae) stem bark on alloxan-induced diabetic rats. *African Journal of Pharmacy and Pharmacology*. 10(28): 584–590.

Rao, K.V., Sreeramulu, K., Gunasekar, D. and Ramesh, D. 1993. Two new sesquiterpene lactones from *Ceiba pentandra*. *Journal of Natural Products*. 56(12): 2041–2045.

Saptarini, N.M. and Deswati, D.A. 2015. The Antipyretic activity of leaves extract of *Ceiba pentandra* better than gossypium arboreum. *Journal of Applied Pharmaceutical Science*. 5(7): 118–121.

Singh, M., Khatoon, S., Singh, S., Kumar, V., Rawat, A.K.S. and Mehrotra, S. 2010. Antimicrobial screening of ethnobotanically important stem bark of medicinal plants. *Pharmacognosy Research*. 2(4): 254.

Durio zibethinus **Rumph. ex Murray**

[From the Malay *duri* = thorn and the Latin *zibethinus* = in reference to the civet cat *Viverra zibetha* L.]

Published in *Systema Vegetabilium. Editio decima tertia* 581. 1774

Local names: *durian* (Kedayan, Murut); *ratu* (Dusun); *lampun* (Murut); *hampun* (Dusun, Murut)
Common name: durian tree
Habitat: cultivated
Geographical distribution: Thailand, Malaysia, Indonesia, the Philippines
Botanical description: this massive tree reaches a height of about 40 m. Buttresses: present. Bark: grayish. Wood: yellowish orange. Leaves: simple, spiral, exstipulate. Petioles: 1.5–2 cm long, straight, channeled. Blades: coriaceous, rounded at base, somewhat coriaceous, acuminate at apex, covered beneath with glossy golden stellate scales, elliptic-oblong, 10–15 cm × 3.5–4 cm, with 12–16 pairs of secondary nerves. Flower fascicles: cauliflorous, with caducous flowers. Calyx: tubular, deeply 5-lobed, about 2 cm long. Petals: 5, pure white, spathulate, about 2.5–5 cm long. Stamens: whitish, numerous, about as long as the petals, in some kinds of bundles, with curly anthers. Ovary: 5-ribbed. Styles: 1. somewhat as long as the petals, white. Stigma: oblate. Capsules: up to 25 cm. 20 cm, heavy, greenish-yellow, thorny, dehiscent. Seeds: 3–4 per valve, smooth, angular, about 5 cm long, embedded in a creamy horribly stinking, yellowish, but edible pulp.
Traditional therapeutic indications: stomach aches, canker sores (Dusun)
Pharmacology and phytochemistry: anti-inflammatory triterpenes (Feng et al. 2018; Charoenphun & Klangbud 2022), antimicrobial (Chigurupati et al. 2017), anticancer (Saminathan & Doraiswamy 2020).
Alkyl sulfurs (1-propanethiol, ethyl-2-methylbutanoate, and 1, 1 diethoxane) (Baldry et al. 1972; Moser et al. 1980; Weenen et al. 1996).
Toxicity, side effects, and drug interaction: durian fruit abound with saturated fatty acids, such as palmitic acid (Moser et al. 1980; Swangpol et al. 2020) (found in palm oil), which favors the development of hypertension and cardiovascular diseases (Grimsgaard et al. 1999).

Comment: The fruits of *Durio graveolens* Becc. are used as food by the Murut (local name: *ruyan*) and the fruits of *Durio grandiflorus* (Mast.) Kosterm. & Soegeng are used as food by the Dusun (local name: *durian mantui*).

REFERENCES

Baldry, J., Dougan, J. and Howard, G.E. 1972. Volatile flavoring constituents of durian. *Phytochemistry*. 11: 2081–2084.
Charoenphun, N. and Klangbud, W.K. 2022. Antioxidant and anti-inflammatory activities of durian (*Durio zibethinus* Murr.) pulp, seed and peel flour. *PeerJ*. 10: p.e12933.
Chigurupati, S., Mohammad, J.I., Vijayabalan, S., Vaipuri, N.D., Selvarajan, K.K. and Nemala, A.R. 2017. Quantitative estimation and antimicrobial potential of ethanol extract of *Durio zibethinus* Murr. Leaves. *Asian Journal of Pharmaceutical and Clinical Research*. 10(9): 1–4.
Feng, J., Yi, X., Huang, W., Wang, Y. and He, X. 2018. Novel triterpenoids and glycosides from durian exert pronounced anti-inflammatory activities. *Food Chemistry*. 241: 215–221.
Grimsgaard, S., Bønaa, K.H., Jacobsen, B.K. and Bjerve, K.S. 1999. Plasma saturated and linoleic fatty acids are independently associated with blood pressure. *Hypertension*. 35(3): 478–483.
Moser, R., Düvel, D. and Greve, R. 1980. Volatile constituents and fatty acid composition of lipids in *Durio zibethinus*. *Phytochemistry*. 19(1): 79–81.
Saminathan, V. and Doraiswamy, R. 2020. Phytochemical analysis, antioxidant and anticancer activities of durian (*Durio zibethinus* Murr.) fruit extract. *Journal of Research in Pharmacy*. 24: 882–892.
Swangpol, S., Charoenkiatkul, S., Sridonpai, P., Thiyajai, P., Kulpradit, K. and Judprasong, K. 2020. Nutritional composition of indigenous durian varieties. *Malaysian Journal of Nutrition*. 26(1): 93.
Weenen, H., Koolhaas, W.E. and Apriyantono, A. 1996. Sulfur-containing volatiles of durian fruits (*Durio zibethinus* Murr.). *Journal of Agricultural and Food Chemistry*. 44(10): 3291–3293.

4.1.11.2.3 Family Dipterocarpaceae Blume (1825), the Dipterocarp Family

The family Dipterocarpaceae consists of about 20 genera and 580 species of magnificent giant timber trees. Trunk: straight. Buttresses: present. Bark: often secretes an aromatic resin once incised. Leaves: simple, alternate, stipulate. Blades: often coriaceous. Inflorescences: racemes or panicles. Calyx: 5 sepals or tubular. Petals: 5, contorted. Stamens: numerous. Carpels: 2–many forming a 2–3 locular ovary, each locule with 2–4 ovules. Styles: 1–3. Fruits: nuts, winged nor not.

***Shorea macroptera* Dyer**

[After 18th century British East India Company official John Shore, from the Greek *makros* = large, and *pteron* = wing]

Published in *The Flora of British India* 1: 308. 1874

Synonyms: *Rubroshorea macroptera* (Dyer) P.S. Ashton & J. Heck.; *Shorea auriculata* Scort. ex Foxw.; *Shorea baillonii* Heim; *Shorea sandakanensis* Sym.
Local name: *omnompik* (Murut)
Habitat: forests
Geographical distribution: Thailand, Malaysia, Indonesia, Borneo
Botanical description: this magnificent timber tree reaches a height of about 50 m. Bark: fissured, secretes a resin once incised. Buttresses: up to 2.5 m tall. Wood: very hard, dense, heavy. Leaves: simple, alternate, stipulate. Stipules: small, lanceolate. Petioles: about 1 cm long. Blades: 2.5–6 cm × 10–18 cm, narrowly oblong, glossy above, base obtuse, apex acuminate, with about 10–15 pairs of secondary nerves. Panicles: terminal. Sepals: 5. Corolla: 5-lobed, the lobes contorted, light creamy yellow to pinkish, linear. Stamens: numerous, tiny. Ovary: ovoid, pubescent. Nuts: about 2 cm long, with a set of 5 wings ranging in length from about 1–15 cm long, the longest of these are somewhat pink at first, then become woody.
Traditional therapeutic indication: antidote for food poisoning (Murut)
Pharmacology and phytochemistry: antibacterial and cytotoxic oligostilbene (davidiol A) (Nur et al. 2012).
Toxicity, side effects, and drug interaction: not known

REFERENCE

Nur, A., Norizan, A., Mashita, A., Norrizah, J.S. and Syed, A. 2012. Antioxidant, antimicrobial and cytotoxic activities of resveratrol oligomers of *Shorea macroptera* dyer. *Australian Journal of Basic and Applied Sciences*. 6(8): 431–436.

Shorea parvistipulata F. Heim

[After 18th century British East India Company official John Shore, from Latin *parvistipulata* = with small stipules]

Published in *Bull. Mens. Soc. Linn. Paris* 2: 974. 1981

Synonyms: *Rubroshorea parvistipulata* (F.Heim) P.S. Ashton & J. Heck.; *Shorea cristata* Brandis; *Shorea nebulosa* Meijer
Local names: *seraya lupa* (Bajau); *roloi* (Murut)
Habitat: forests
Geographical distribution: Borneo
Botanical description: this magnificent timber tree reaches a height of about 70 m. Bark: scaly, secretes a resin once incised. Wood: very hard, dense, heavy. Leaves: simple, alternate, stipulate. Stipules: about 1.5 cm long. Petioles: hairy, about 1.5 cm long. Blades: 3–9 cm × 6–20 cm, oblong, glossy above, hairy beneath, cordate, attenuate, or obtuse at base, acuminate at apex. with about 13–21 pairs of secondary nerves and scalariform tertiary nerves. Panicles: terminal, hairy, up to about 15 cm long. Sepals: 5. Corolla: tubular, 5-lobed, light creamy yellow to pinkish, the lobes linear. Stamens: 15, tiny. Ovary: ovoid, pubescent. Nuts: about 2.5 cm long, ovoid, with a set of 5 wings ranging in length from about 1–20 cm long, the longest wings woody and spathulate (Figure 4.35).
Traditional therapeutic indications: fatigue (Murut)
Pharmacology and phytochemistry: not known
Toxicity, side effects, and drug interaction: not known

Comment: *Parashorea malaanonan* Merr. is medicinal for the Murut (local name: *melapi*).

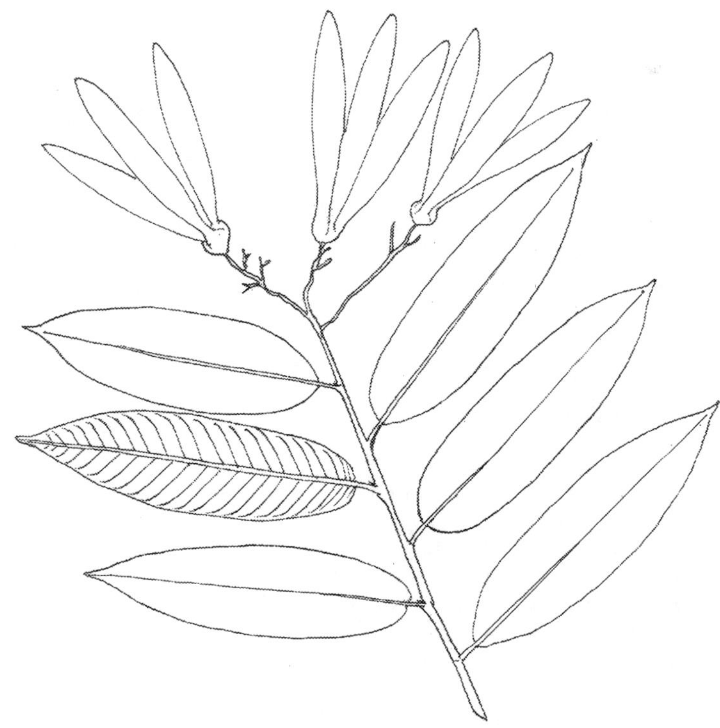

FIGURE 4.35 *Shorea parvistipulata* F.Heim

4.1.11.2.4 Family Malvaceae Juss. (1789), the Mallow Family

The family Malvaceae consists of about 100 genera and 1,000 species of herbaceous plants, shrubs, or trees. Stems: terete, often hairy, fibrous. Leaves: simple, alternate or spiral, stipulate. Petioles: often thin and long. Inflorescences: solitary, cymes, racemes, or panicles. Flowers: often showy. Epicalyx: often present. Calyx: 5 sepals or tubular. Petals: 5. Stamens: 5–100, partially forming a column. Carpels: often 5, merged in a 5-locular ovary, each locule with 1–50 ovules. Styles: numerous. Fruits: loculicidal capsules or schizocarps. Seeds: numerous, reniform.

***Abelmoschus esculentus* L.**

[from Arabic *ḥabb al-misk* = seed of musk and Latin *esculenta* = edible]

Published in *Methodus Plantas Horti Botanici et Agri Marburgensis, a staminum situ describendi* 2: 617. 1794

Synonym: *Hibiscus esculentus* L.
Local names: *kacang bendi* (Bajau); *bendi* (Murut)
Common name: lady's fingers
Habitat: cultivated
Geographical distribution: tropical
Botanical description: this erect herbaceous plant reaches a height of about 2 m. Stems: hairy, fibrous. Leaves: simple, spiral, stipulate. Stipules: linear, about 1 cm long. Petioles: 7–15 cm long, straight, hairy. Blades: 10–30 cm across, palmately 3–7-lobed, the lobes somewhat oblong to obovate and dentate. Flowers: solitary, axillary, magnificent, on 1–2 cm long peduncles. Epicalyx: 7-10-lobed, the lobes linear and about 2 cm long. Calyx: tubular, 2–3 cm long, hairy. Petals: 5, immaculate yellow, dark purple at base, obovate, about 5 cm long. Stamens: numerous, merged into a column, about 2 cm long. Stigma: 5-lobed. Capsules: pentagonal, fibrous, up to 20 cm long, beaked. Seeds: numerous, globose, about 5 mm across.
Traditional therapeutic indications: fever, flu (Bajau)
Pharmacology and phytochemistry: this plant has been the subject of numerous pharmacological studies (Chowdhury et al. 2019) which have highlighted, among other things, analgesic, anti-inflammatory (Nesa et al. 2014), and anti-ulcerogenic activity (Ortac et al. 2018). Immunostimulating polysaccharides (Wahyuningsih et al. 2018). Flavonol glycosides (Liao et al. 2012)
Toxicity, side effects, and drug interaction: ethanol extract of aerial parts administered orally to rats at the dose of 5 g/kg was found to cause no signs of toxicity (Ortac et al. 2018).

REFERENCES

Chowdhury, N.S., Jamaly, S., Farjana, F., Begum, N. and Zenat, E.A. 2019. A review on ethnomedicinal, pharmacological, phytochemical and pharmaceutical profile of lady's finger (*Abelmoschus esculentus* L.) plant. *Pharmacology & Pharmacy*. 10(2): 94–108.

Liao, H., Liu, H. and Yuan, K. 2012. A new flavonol glycoside from the Abelmoschus esculentus Linn. *Pharmacognosy Magazine*. 8(29): 12–15.

Nesa, M., Islam, M., Alam, B., Munira, S., Mollika, S., Naher, N. and Islam, M.R. 2014. Analgesic, anti-inflammatory and CNS depressant activities of the methanolic extract of *Abelmoschus esculentus* Linn. seed in mice. *British Journal of Pharmaceutical Research*. 4(7): 849–860.

Ortac, D., Cemek, M., Karaca, T., Büyükokuroğlu, M.E., Özdemir, Z.Ö., Kocaman, A.T. and Göneş, S. 2018. *In vivo* anti-ulcerogenic effect of okra (*Abelmoschus esculentus*) on ethanol-induced acute gastric mucosal lesions. *Pharmaceutical Biology*. 56(1): 165–175.

Wahyuningsih, S.P.A., Pramudya, M., Putri, I.P., Winarni, D., Savira, N.I.I. and Darmanto, W. 2018. Crude polysaccharides from okra pods (*Abelmoschus esculentus*) grown in Indonesia enhance the immune response due to bacterial infection. *Advances in Pharmacological Sciences*. 2018: 1–7.

***Urena lobata* L.**

[From ancient Malabar local name of this plant = *urena* and Latin *lobata* = with lobes]

Published in *Species Plantarum* 2: 692. 1753

Synonyms: *Urena americana* L. f.; *Urena grandiflora* DC.; *Urena reticulata* Cav.; *Urena trilobata* Vell.

Local names: *pulut pulut* (Kedayan); *injilokot* (Murut); *tong gilupang* (Dusun); *pong*, *tongilopan*, *tangirupang* (Dusun, Kadazan)

Common names: aramina, caesar weed, burr mallow

Habitat: grassy places, wastelands

Geographical distribution: tropical

Botanical description: this herbaceous plant reaches a height of about 1 m. Stems: terete, hairy. Leaves: simple, spiral, stipulate. Stipules: filiform, tiny, caducous. Petioles: 1–4 cm long, hairy. Blades: orbicular to ovate, 4–7 cm × 3–6.5 cm, base rounded to cordate, somewhat palmately lobed, hairy. Flowers: solitary, axillary. Calyx: tubular, 5-lobed, hairy. Petals: 5, membranous, about 1.5 cm long, pink, obovate. Stamens: numerous, forming a 1.5 cm long column. Styles: 10. Capsules: 5-lobed, about 1 cm across, spiny, become hideous as they ripen.

Traditional therapeutic indications: boils, fever, stomach aches, flatulence, gastritis, skin diseases, toothaches, coughs (Dusun, Kadazan); constipation, thrush (Murut); jaundice, rashes, swellings (Dusun)

Pharmacology and phytochemistry: antibacterial (Mazumder et al. 2001), antidiabetic (Purnomo et al. 2015), anti-inflammatory (Rajagopal et al. 2018), leishmanicidal (Salamanca et al. 2009), toxic to Ehrlich ascites carcinoma cells (Mathappan et al. 2019).

Lignan glycosides (Luo et al. 2019). Anti-inflammatory megastigmane glycoside (urenalobaside C) (Luo et al. 2019). Antifungal lignan (trachelogenin) (Gao et al. 2015). Antifungal triterpene saponin (clematoside S) (Gao et al. 2015). Megastigmane glycosides (Su et al. 2018).

Toxicity, side effects, and drug interaction: aqueous extract of roots administered orally to rats at the dose of 300 mg/kg for 28 days induced intestinal and hepatic injuries (Mshelia et al. 2013).

Comments:

(i) A plant of the genus *Urena* L. (1753) is used for boils by the Dusun and Kadazan (local name: *tondorupang*).

(ii) A plant of the genus *Gossypium* L. (1753) is used by the Dusun for flatulence (local names: *gapas*).

(iii) *Abutilon indicum* (L.) Sweet is used for sprains by the Dusun and Kadazan (local names: *tindokot, tondorupang*).

(iv) *Hibiscus rosa-sinensis* L. is used for boils (Bajau, Dusun, Kadazan, Murut), coughs, hair loss, dysmenorrhea, phlegm, mumps (Bajau, local name: *bunga raya*), jaundice (Dusun), fever (Kedayan, Murut, Dusun), headaches (Lundayeh), sprains, asthma, wounds, swelling (Dusun, Kadazan, local name: *tongkuango*).

(v) *Hibiscus sabdariffa* L. is used for cancer (Bajau).

(vi) *Sida acuta* Burm.f. is used for stomach aches by the Dusun and Kadazan (local name: *bulitotok*).

(vii) *Sida rhombifolia* L. is used as poison antidote by the Murut (local name: *dalupang*), as laxative, and for body pains by the Bugis (local name: *daung canggadori*), for snake bites by the Lundayeh (local name: *tahong*), for abscesses, swellings, boils, headaches, diarrhea, skin diseases by the Dusun and the Kadazan (local names: *tatak tatak*, *lingkabau*), and fever by Bugis, Kadazan, and Dusun. The Kedayan call the plant *timah timah*.

REFERENCES

Gao, X.L., Liao, Y., Wang, J., Liu, X.Y., Zhong, K., Huang, Y.N., Gao, H., Gao, B. and Xu, Z.J. 2015. Discovery of a potent anti-yeast triterpenoid saponin, clematoside-S from *Urena lobata* L. *International Journal of Molecular Sciences*. 16(3): 4731–4743.

Luo, Y., Su, C., Ding, N., Qi, B., Jia, F., Xu, X., Liu, X., Wang, J., Wang, X., Tu, P. and Shi, S. 2019. Lignan glycosides from *Urena lobata*. *Molecules*. 24(15): 2850.

Mathappan, R., Krishnan Selvarajan, K., Sujeet, S. and Tribedi, S. 2019. Evaluation of antitumor activity of *Urena lobata* against Ehrlich ascites carcinoma treated mice. *Oriental Pharmacy and Experimental Medicine*. 19: 21–26.

Mazumder, U.K., Gupta, M., Manikandan, L. Bhattacharya, S. 2001. Antibacterial activity of *Urena lobata* root. *Fitoterapia*. 72(8): 927–929.

Mshelia, I.Y., Dalori, B.M., Hamman, L.L. and Garba, S.H. 2013. Effect of the aqueous root extract of *Urena lobata* (Linn) on the liver of albino rat. *Research Journal of Applied Sciences, Engineering and Technology*. 5(1): 1–6.

Purnomo, Y., Soeatmadji, D.W., Sumitro, S.B. and Widodo, M.A. 2015. Anti-diabetic potential of *Urena lobata* leaf extract through inhibition of dipeptidyl peptidase IV activity. *Asian Pacific Journal of Tropical Biomedicine*. 5(8): 645–649.

Rajagopal, P.L., Linsha, K.T., Kumar, P.S., Parthasarathy, I.A., Sreejith, K.R. and Aneeshia, S. 2018. Anti-inflammatory activity of the leaves of *Urena lobata* Linn. *World Wide Journal of Multidisciplinary Research and Develop*. 4(11): 59–61.

Salamanca, C., Flores, N., Giménez, A., Ávila, J.A. and Ruiz, G. 2009. Actividad leishmanicida de plantas medicinales de la amazonia Peruana. *Revista Boliviana De Química*. 26(2): 43–48.

Su, C., Qi, B., Wang, J., Ding, N., Wu, Y., Shi, X.P., Zhu, Z.X., Liu, X., Wang, X.H., Zheng, J. and Tu, P.F. 2018. Megastigmane glycosides from *Urena lobata*. *Fitoterapia*. 127: 123–128.

4.1.11.2.5 Family Tiliaceae Juss. (1789), the Jute Family

The family Tiliaceae consists of about 50 genera and 500 species of herbaceous plants, shrubs, or trees. Stems: terete, often hairy. Leaves: simple, alternate or spiral, stipulate. Petioles: often thin. Blades: often serrate, with the first pair of secondary nerves straight and from the base. Inflorescences: cymes, racemes, panicles. Sepals: 5. Petals: 5 or none. Stamens: numerous, filamentous, more or less merged at base. Carpels: 2–100, forming 2–100 celled locular ovaries, each locule with 1–100 ovules. Styles: 1. Fruits: capsules, drupes.

***Microcos antidesmifolia* (King) Burret**

[From Greek *mikros* = small, *kos* = prisoner, and Latin *antidesmifolia* = with *Antidesma*-like leaves]

Published in *Notizbl. Bot. Gart. Berlin-Dahlem* 9: 728. 1926

Synonym: *Grewia antidesmifolia* King
Local name: *kodong* (Rungus)
Habitat: swampy forests
Geographical distribution: Malaysia, Indonesia, Borneo
Botanical description: this tree reaches a height of about 10 m. Stems: terete, somewhat hairy at apex. Leaves: simple, alternate, stipulate. Stipules: about 4 mm long, hairy. Petioles: hairy, about 2 cm long, straight. Blades: oblong, somewhat hairy, 5–11 cm × 7–17 cm, attenuate at base, asymmetrical, acuminate at apex, glossy, the midrib sunken above, with about 4 pairs of secondary nerves and beneath scalariform tertiary nerves. Panicles: terminal axillary. Sepals: 5, oblong, about 5 mm long, hairy. Petals: 5, about 2 mm long, obovate. Stamens: numerous, about 5 mm long, filamentous. Ovary: bottle-shaped. Styles: 1, thin, about 3 mm long. Drupes: somewhat shaped like electric bulbs, about 2 cm long, orange, glossy (Figure 4.36).

FIGURE 4.36 *Microcos antidesmifolia* (King) Burret

Traditional therapeutic indications: magic rituals (Rungus)
Pharmacology and phytochemistry: not known
Toxicity, side effects, and drug interaction: not known

Comments:

(i) *Microcos cinnamomifolia* Burret is used by the Dusun for magic rituals (local name: *ngkodong*).
(ii) In Sumatra, *Grewia acuminata* Juss. is used as an aphrodisiac (local name: *tiga urat*).

4.1.11.2.6 Family Sterculiaceae Vent. (1807), the Cocoa Family

The family Sterculiaceae consists of about 70 genera and 100 species of shrubs or trees. Stems: terete, fibrous. Leaves: simple, alternate, spiral, stipulate. Petioles: often thin and straight. Blades: more or less palmate to deeply incised. Flowers: solitary or arranged in cymes, racemes, or panicles. Sepals: 5, free, or merged at base. Petals: 5 or none. Stamens: 5-many, more or less merged at base. Carpels: 5, forming a 5 locular ovary, each locule with 2–100 ovules. Styles: 1–5. Fruits: capsules, follicles. Seeds: numerous, often reniform.

***Theobroma cacao* L.**

[From Greek *theos* = god, *broma* = food, and from ancient Central American *cacahuati* = cacao]

Published in *Species Plantarum* 2: 782. 1753

Local name: *koko* (Bajau, Dusun, Lundayeh)
Common name: cacao
Habitat: cultivated
Geographical distribution: tropical
Botanical description: this tree reaches a height of about 6 m. Trunk: not straight. Bark: dark gray-brown. Stems: hairy at apex. Leaves: simple, alternate, stipulate. Stipules: linear. Petioles: straight, about 2 cm long, slightly swollen at apex. Blades: obovate-elliptic, 6–10 cm × 15–25 cm, rounded at base, glossy, acuminate at apex, with about 8–9 pairs of secondary nerves, the first pair from the base. Cymes: cauliflorous, with few flowers of an extreme beauty. Calyx: yellowish-green to white, somewhat fleshy, 5-lobed, the lobes triangular and up to 8 mm long. Petals: 5, yellowish-white, about 3 mm long, membranous, spathulate, hooded, then constricted into a sort of spoon. Staminodes: 5, linear, dark purple, about 5 mm long. Stamens: 5. Ovary: obovoid, 5-lobed. Styles: 1, linear. Berries: 10-ribbed, yellowish to purplish, glossy, about 20 cm long. Seeds: numerous, flattish, about 2.5 cm long, in an almost pure white flesh.
Traditional therapeutic indications: skin diseases, itchiness (Lundayeh)
Pharmacology and phytochemistry: this plant has been the subject of numerous pharmacological studies which have highlighted among other things, anti-inflamatory activity (Ono et al. 2003; Oyeleke et al. 2018). The seeds, which are used to make cocoa, contain the xanthine alkaloid theobromine (Matissek 1997).
Toxicity, side effects, and drug interaction: cacao is deadly poisonous to dogs (Bates 2015). Caffeine is teratogenic in rats (Eteng et al. 1997).

Comments:

(i) *Scaphium macropodum* Beumee ex K. Heyne (Figure 4.37) is used as a medicine by the Dusun (local name: *kapayang*).

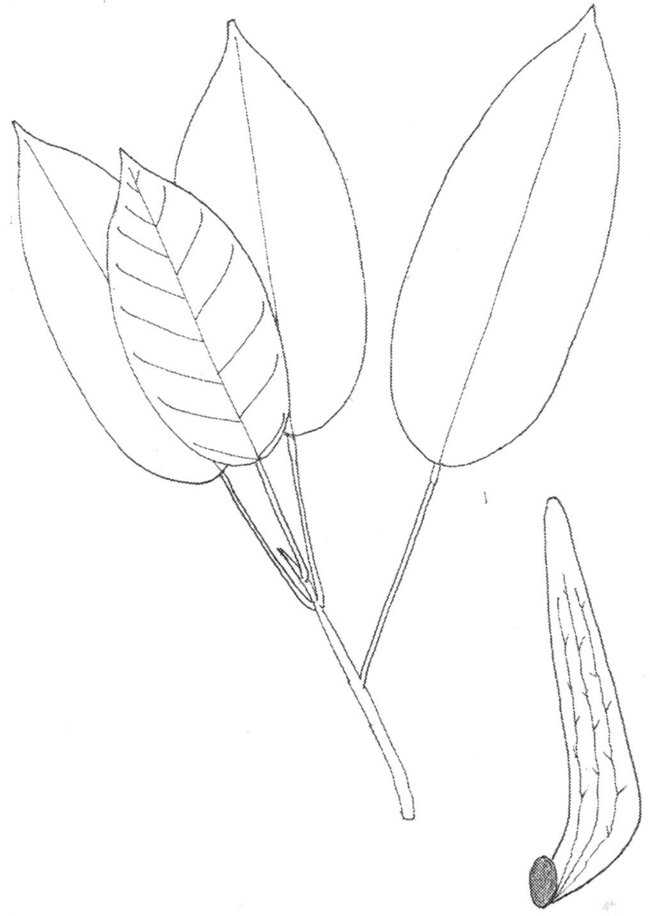

FIGURE 4.37 *Scaphium macropodum* Beumee ex K. Heyne

(ii) *Leptonychia heteroclita* Kurz is used by the Dusun for stomach aches (local name: *tembulang manok*).

REFERENCES

Bates, N. 2015. Chocolate toxicity. *Companion Animal*. 20(10): 579–582.

Cádiz-Gurrea, M.D.L.L., Borrás-Linares, I., Lozano-Sánchez, J., Joven, J., Fernández-Arroyo, S. and Segura-Carretero, A. 2017. Cocoa and grape seed byproducts as a source of antioxidant and anti-inflammatory proanthocyanidins. *International Journal of Molecular Sciences*. 18(2): 376.

Eteng, M.U., Eyong, E.U., Akpanyung, E.O., Agiang, M.A. and Aremu, C.Y. 1997. Recent advances in caffeine and theobromine toxicities: A review. *Plant Foods for Human Nutrition*. 51: 231–243.

Matissek, R. 1997. Evaluation of xanthine derivatives in chocolate—nutritional and chemical aspects. *Zeitschrift für Lebensmitteluntersuchung und-Forschung A*. 205(3): 175–184.

Ono, K., Takahashi, T., Kamei, M., Mato, T., Hashizume, S., Kamiya, S. and Tsutsumi, H. 2003. Effects of an aqueous extract of cocoa on nitric oxide production of macrophages activated by lipopolysaccharide and interferon-γ. *Nutrition*. 19(7–8): 681–685.

Oyeleke, S.A., Ajayi, A.M., Umukoro, S., Aderibigbe, A.O. and Ademowo, O.G. 2018. Anti-inflammatory activity of *Theobroma cacao* L. stem bark ethanol extract and its fractions in experimental models. *Journal of Ethnopharmacology*. 222: 239–248.

4.1.11.2.7 Family Thymeleaceae Juss. (1789), the Eaglewood Family

The family Thymeleaceae consists of about 50 genera and 500 species of herbaceous plants, shrubs, trees. Leaves: simple, opposite, alternate, exstipulate. Inflorescences: racemes, cymes, spikes, umbels, fascicles. Calyx: 4–5 sepals or tubular. Petals: 4–5 or none. Stamens: 2-many. Disc: present. Carpels: 2–5, forming a 2–5 locular ovary, each locule with 1 ovule. Styles: 1, thin. Stigma: capitate to clavate. Fruits: capsules, drupes.

***Wikstroemia androsaemifolia* Hand.-Mazz.**

[After the 19th century Swedish botanist Johan Emanuel Wikstrom and Latin *androsaemifolia* = with leaves like *Androsaemum*]

Published in *Anzeiger der Akademie der Wissenschaften in Wien. Mathematische-naturwissenschaftliche Klasse. Vienna.* 60: 135. 1923

Synonyms: *Daphne lamatsoensis* (Hamaya) Halda; *Daphne pauciflora* Span. ex Walp.; *Daphne viridiflora* Moritzi & Zoll. ex Meisn.; *Eriosolena viridiflora* Zoll. & Moritzi; *Wikstroemia candolleana* Meisn.; *Wikstroemia lamatsoensis* Hamaya; *Wikstroemia junghuhnii* Miq.

Local name: *sinantali* (Dusun)

Common names: male gaharu, red gaharu

Habitat: forests

Geographical distribution: from Cambodia to the Solomon Islands

Botanical description: this shrub reaches a height of about 3 m. Stems: terete, reddish-brown. Leaves: simple, opposite, exstipulate. Petioles: 2–3 mm long. Blades: lanceolate to obovate, thin, acute at base, acuminate to rounded at apex, 1.5–5.5 cm × 4–10 cm, of a strange kind of pale green beneath, finely marked with about 8–11 pairs of secondary nerves. Flower fascicles: terminal, up to 10 cm long. Calyx: tubular, about 1 cm long, yellowish-green, 5-lobed, the lobes oblong and up to about 2 mm long. Disc: present. Stamens: 10. Styles: 1. short. Stigma: globose. Drupes: red, globose, somewhat glossy, about 6 mm across.

Traditional therapeutic indications: headaches (Dusun)

Pharmacology and phytochemistry: antiviral (HIV) diterpenes (Zhang et al. 2021).

Toxicity, side effects, and drug interaction: not known

Comment: A plant of the genus *Wikstroemia* Endl. (1833) is used by the Dusun and Kadazan for body pains (local name: *tindot*).

REFERENCE

Zhang, M., Otsuki, K., Kikuchi, T., Bai, Z.S., Zhou, D., Huang, L., Chen, C.H., Morris-Natschke, S.L., Lee, K.H., Li, N. and Koike, K. 2021. LC-MS identification, isolation, and structural elucidation of anti-HIV tigliane diterpenoids from *Wikstroemia lamatsoensis*. *Journal of Natural Products*. 84(8): 2366–2373.

Wikstroemia ridleyi Gamble

[After the 19th century Swedish botanist Johan Emanuel Wikstrom and after the 20th century British botanist Henry Nicholas Ridley]

Published in *Bulletin of Miscellaneous Information, Royal Gardens, Kew* 4: 200–201. 1912

Synonym: *Daphne ridleyi* (Gamble) Halda
Local names: *tindot* (Dusun); *depu, pelanduk* (Bajau)
Habitat: forests
Geographical distribution: Thailand, Malaysia, Borneo
Botanical description: this shrub reaches a height of about 2 m. Stems: terete. Leaves: simple, opposite, exstipulate. Petioles: 2–3 mm long. Blades: membranous, elliptic, glossy, tapering at base, acute at apex, 2–5 cm × 6–15 cm, finely marked with 7–12 pairs of secondary nerves. Cymes: terminal. Calyx: tubular, 4-lobed, plain yellow, the lobes ovate-oblong, up to about 4 mm long. Stamens: 8. Disc: present. Ovary: ellipsoid. Styles: 1 Stigma: globose. Drupes: red, oblong, about 8 mm long (Figure 4.38).
Traditional therapeutic indications: fatigue after severe sickness (Dusun)
Pharmacology and phytochemistry: not known
Toxicity, side effects, and drug interaction: not known

Comment: *Phaleria papuana* Warb. ex K. Schum. & Lauterb. is a remedy for hypertension, fatigue, gout, asthma, indigestion, allergy, hepatitis, heart diseases, diabetes, kidney problems, and dizziness (local name: *mahkota dewa*).

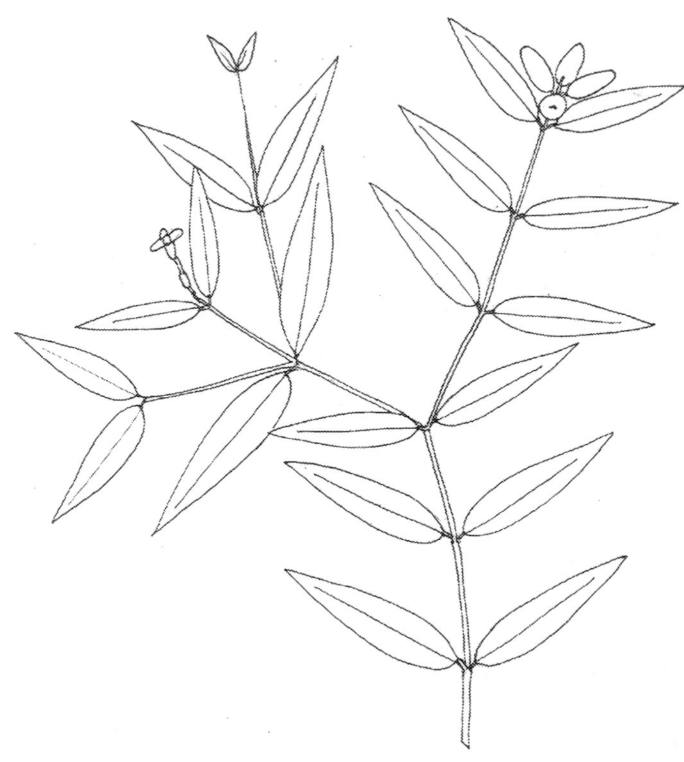

FIGURE 4.38 *Wikstroemia ridleyi* Gamble

4.1.11.3 Order Myrtales Juss. ex Bercht. & J. Presl (1820)

4.1.11.3.1 Family Combretaceae R.Br. (1810), the Indian Almond Family

The family Combretaceae consists of about 20 genera and 600 species of trees, shrubs, or climbing plants. Leaves: opposite, whorled, spiral or alternate, often coriaceous, exstipulate. Inflorescences: spikes, racemes, panicles. Hypanthium: present. Calyx: tubular, 4-5-lobed. Petals: 4–5. Disc: present. Stamens: 4–16. Carpels: 2–5, forming a unilocular ovary with 2–5 ovules. Fruits: capsules, drupes, samaras.

Combretum nigrescens **King**

[The word *combretum* was used by Pliny the Elder to describe a medicinal plant and from Latin *nigrescens* = turning black]

Published in *Journal of the Asiatic Society of Bengal. Part 2. Natural History* 66: 340. 1897

Synonyms: *Combretum adenophorum* Slooten; *Combretum elmeri* Merr.; *Combretum glandulosum* Slooten; *Combretum kunstleri* King; *Combretum scortechinii* King
Local name: *damat dumalarom* (Murut)
Habitat: forests
Geographical distribution: Thailand, Malaysia, Indonesia, Borneo
Botanical description: this climbing plant grows up to a length of about 20 m. Stems: hairy at apex. Leaves: simple, opposite, exstipulate. Petioles: 2–8 mm long, hairy. Blades: membranous, hairy beneath, dark green and glossy above, elliptic to oblong, attenuate to rounded at base, acuminate at apex, with 6–8 pairs of secondary nerves, 3–5 cm × 6–13 cm. Panicles: terminal, about 15 cm long, with numerous tiny white flowers. Calyx: tubular, 4-lobed. Petals: 4. Stamens: 8, 4 mm long. Disc: present. Styles: 1. Drupes: glossy, about 2 cm long.
Traditional therapeutic indications: cuts, internal injuries, wounds (Murut); insect bites, for nails (Dusun, Kadazan)
Pharmacology and phytochemistry: not known
Toxicity, side effects, and drug interaction: not known

Comment: *Lumnitzera littorea* (Jack) Voigt is used by the Bajau as styptic (local name: *santing*).

4.1.11.3.2 Family Lythraceae J. St.-Hil. (1805), the Loosestrife Family

The family Lythraceae consists of about 30 genera and 500 species of herbaceous plants, shrubs, or trees with often some kind of aura of dryness. Leaves: simple, often coriaceous, opposite, exstipulate. Hypanthium: present. Calyx: 4–8 sepals or tubular. Petals: 4–8, often spathulate, membranous. Stamens 8–16. Disc: present. Carpels: 2–4, forming a 1–4 locular ovary, each locule with 5–50 ovules. Fruits: capsules. Seeds: numerous.

Lawsonia inermis **L.**

[After the 17th century Scottish botanist Isaac Lawson and Latin *inermis* = unarmed]

Published in *Species Plantarum* 1: 349. 1753

Synonyms: *Alkanna spinosa* Gaertn.; *Lawsonia alba* Lam.; *Lawsonia speciosa* L.; *Lawsonia spinosa* L.; *Rotantha combretoides* Baker
Local names: *pacar kuku* (Bajau, Dusun); *pacar* (Lundayeh); *inai* (Bajau, Dusun, Murut)

Common name: henna
Habitat: cultivated, roadsides
Geographical distribution: tropical
Botanical description: this delicate shrub reaches a height of about 3 m. Stems: smooth, brownish, red. Leaves: simple, subsessile, opposite, exstipulate. Blades: elliptic, 4–10 mm × 1.5–3 cm, dull green, tapering at base, rounded to acute to somewhat finely apiculate at apex. Panicles: many-flowered, terminal, up to about 20 cm long. Sepals: 4, tiny, ovate. Petals: 4, about 5 mm long, pure white to pink to red. Stamens: 8, about 4 mm long, in 4 pairs. Ovary: obovate, tiny. Capsules: about 7 mm long, dehiscent. Seeds: numerous.
Traditional therapeutic indications: coughs (Dusun); body pains itchiness (Murut)
Pharmacology and phytochemistry: anti-inflammatory, antipyretic, analgesic (Alia et al. 1995), antibacterial, antifungal (Habbal et al. 2005; Rahmoun et al. 2013), antiviral (sindbis virus) (Mouhajir et al. 2001), antibacterial (Jeyaseelan et al. 2012).
Antifungal and cytotoxic naphthoquinone (lawsone) (Rahmoun et al. 2013; Li et al. 2014).
Toxicity, side effects, and drug interaction: not known

Comment: *Sonneratia alba* Sm. is used by the Bajau for diarrhea and fever (local name: *perepat*). They use the fruits and bark of this plant to give taste to their dishes.

REFERENCES

Alia, B.H., Bashir, A.K. and Tanira, M.O.M. 1995. Anti-inflammatory, antipyretic, and analgesic effects of *Lawsonia inermis* L. (henna) in rats. *Pharmacology*. 51(6): 356–363.

Habbal, O.A., Al-Jabri, A.A., El-Hag, A.H., Al-Mahrooqi, Z.H. and Al-Hashmi, N.A. 2005. In-vitro antimicrobial activity of *Lawsonia inermis* Linn (henna). A pilot study on the Omani henna. *Saudi Medical Journal*. 26(1): 69–72.

Jeyaseelan, E.C., Jenothiny, S., Pathmanathan, M.K. and Jeyadevan, J.P. 2012. Antibacterial activity of sequentially extracted organic solvent extracts of fruits, flowers and leaves of Lawsonia inermis L. from Jaffna. *Asian Pacific Journal of Tropical Biomedicine*. 2(10): 798–802.

Li, Q., Gao, W., Cao, J., Bi, X., Chen, G., Zhang, X., Xia, X. and Zhao, Y. 2014. New cytotoxic compounds from flowers of *Lawsonia inermis* L. *Fitoterapia*. 94: 148–154.

Mouhajir, F., Hudson, J.B., Rejdali, M. and Towers, G.H.N. 2001. Multiple antiviral activities of endemic medicinal plants used by Berber peoples of Morocco. *Pharmaceutical Biology*. 39(5): 364–374.

Rahmoun, N., Boucherit-Otmani, Z., Boucherit, K., Benabdallah, M. and Choukchou-Braham, N. 2013. Antifungal activity of the Algerian Lawsonia inermis (henna). *Pharmaceutical Biology*. 51(1): 131–135.

4.1.11.3.3 Family Melastomataceae Juss. (1789), the Melastomes Family

The family Melastomataceae consists of about 200 genera and 4,500 species of herbaceous plants, shrubs, or trees. Leaves: simple, opposite, decussate, exstipulate. Blades: marked by 1–4 pairs of secondary nerves parallel to the midrib and margin as well as scalariform tertiary nerves. Inflorescences: cymes, umbels, corymbs, panicles, spikes. Hypanthium: present. Calyx: tubular, 4-5-lobed. Petals: 4–5. Disc: present. Stamens: 4–10. Carpels: 4–5, forming a 1–5 locular ovary, each locule with 6–50 ovules. Styles: 1. Fruits: capsules, berries.

***Clidemia hirta* (L.) D. Don**

[Possibly after Cleidemus, an author of ancient Greece and Latin *hirtus* = hairy]

Published in *Memoirs of the Wernerian Natural History Society* 4: 309. 1823

Synonyms: *Melastoma elegans* Aubl.; *Melastoma hirtum* L.
Local names: *senduduk paksa* (Bajau); *kahad kahad* (Dusun)
Common names: Koster's curse, soap bush
Habitat: roadsides, at the edges of forests, plantations
Geographical distribution: tropical
Botanical description: this herbaceous plant reaches a height of about 1 m. Stems: terete, somewhat reddish, hairy. Leaves: simple, opposite, exstipulate. Petioles: hairy, up to about 2.5 cm long. Blades: elliptic, to broadly lanceolate, hairy, with 2 pairs of secondary nerves and numerous showy tertiary scalariform nerves (somewhat waffled), up to 15 cm × 8 cm, glossy or not, cordate to rounded at base, acuminate at apex. Racemes: axillary or terminal, hairy, few-flowered. Calyx: campanulate, about 5 mm long, hairy, 5-lobed, the lobes 4 mm long. Petals: 5, oblong, pure white, 8–11 mm long, somewhat recurved. Stamens: 10, pure white, linear, about 5 mm long. Styles: 1, pure white, about 8 mm long. Berries: dark blue, globose, somewhat hairy, glossy, open at apex. Seeds: numerous, about 5 mm across.
Traditional therapeutic indications: heartburn (Bajau); fever, flu (Dusun)
Pharmacology and phytochemistry: hepatoprotective (Amzar & Iqbal 2017), antibacterial (Dianita et al. 2011).
Ellagitannins (El Abdellaoui et al. 2014).
Toxicity, side effects, and drug interaction: tannins in this plant are toxic to goats (Murdiati et al. 1990).

REFERENCES

Amzar, N. and Iqbal, M. 2017. The hepatoprotective effect of *Clidemia hirta* against carbon tetrachloride (CCl 4)– induced oxidative stress and hepatic damage in mice. *Journal of Environmental Pathology, Toxicology and Oncology*. 36(4): 293–307.

Dianita, R., Ramasamy, K. and Ab Rahman, N. 2011. Antibacterial activity of different extracts of Clidemia hirta (L.) D.Don leaves. *Planta Medica*. 77(12): PM11.

El Abdellaoui, S., Destandau, E., Krolikiewicz-Renimel, I., Cancellieri, P., Toribio, A., Jeronimo-Monteiro, V., Landemarre, L., André, P. and Elfakir, C. 2014. Centrifugal partition chromatography for antibacterial bio-guided fractionation of *Clidemia hirta* roots. *Separation and Purification Technology*. 123: 221–228.

Murdiati, T.B., McSweeney, C.S., Campbell, R.S. and Stoltz, D.S. 1990. Prevention of hydrolysable tannin toxicity in goats fed *Clidemia hirta* by calcium hydroxide supplementation. *Journal of Applied Toxicology*. 10(5): 325–331.

Dissochaeta monticola Bl.

[From Greek *dissos* = double, *chaite* = bristle, and Latin *monticola* = inhabiting mountain]

Published in *Flora* 14: 494. 1831

Synonym: *Dissochaeta intermedia* Bl.
Local name: *bina* (Murut)
Habitat: at the edges of forests
Geographical distribution: Thailand, Malaysia, Indonesia, the Philippines
Botanical description: this climbing plant grows up to a length of about 20 m. Stems: terete, hairy. Leaves: simple, opposite, exstipulate. Petioles: hairy, 0.6–2 cm long. Blades: ovate, oblong to lanceolate, 3–4 cm × 6.5–10 cm, rounded at base, with 1–2 pairs of longitudinal nerves, scalariform tertiary nerves, hairy beneath, acuminate at apex. Cymes: terminal, up to 18 cm long, hairy. Calyx: urceolate, rusty, hairy, about 4 mm long, 4-lobed. Petals: 4, ovate to elliptic, light pink to white, 4–6 mm long. Stamens: 4, with light pink anthers. Styles: 1, enlarged at apex, 5–9 mm long. Capsules: urn-shaped, with a conspicuous opening at apex about 5 mm long (Figure 4.39).
Traditional therapeutic indications: poison for blowguns darts (Murut)
Pharmacology and phytochemistry: not known
Toxicity, side effects, and drug interaction: not known

Comment: Poisons for blowgun darts often contain neuroactive natural products of clinical value.

FIGURE 4.39 *Dissochaeta monticola* Bl.

***Melastoma malabathricum* L.**

[From Greek *melanos* = black, *stoma* = mouth, and Latin *malabathricum* = from Malabar]

Published in *Species Plantarum* 1: 390. 1753

Synonyms: *Melastoma affine* D.Don; *Melastoma candidum* D.Don; *Melastoma cavaleriei* H. Lév. & Vaniot; *Melastoma esquirolii* H. Lév.; *Melastoma normale* D.Don; *Melastoma polyanthum* Bl.

Local names: *adok adok duduk gesing, gosing, hosing, kuduk kuduk* (Dusun); *lalarit* (Murut); *kemanden, harendong* (Lundayeh); *senduduk* (Kedayan, Bajau); *gosing gosing* (Dusun, Kadazan)

Common names: Malabar melastome, white melastome

Habitat: roadsides, at the edges of forests, wastelands

Geographical distribution: from India to Pacific Islands

Botanical description: this common shrub reaches a height of about 1.5 m. Stems: terete, rough, hairy, stiff. Leaves: simple, opposite, exstipulate. Petioles: hairy, rough, channeled, 7 mm–1.8 cm long. Blades: broadly lanceolate, hairy, coriaceous, 7–9.5 cm × 4–5.5 cm, with 7 longitudinal nerves, the margin revolute and wavy, the tertiary nerves clearly visible beneath. Cymes: terminal. Calyx: urn-shaped, coriaceous, 5-lobed. Petals: 5–6, obovate, pure white or purplish, about 3 cm long. Disc: present. Stamens: 10, with curved anthers, the whole structure disgusting. Capsules: urn-shaped, hairy, rough, about 1.2 cm long. Seeds: tiny in a black and dismal pulp.

Traditional therapeutic indications: fever, diarrhea (Dusun); cuts, bleeding, stomach aches, measles, wounds (Dusun, Kadazan); blemishes, ulcers (Dusun); postpartum (Dusun, Murut); diarrhea (Bajau, Kedayan, Dusun, Kadazan). Plants with white flowers are used by the Bajau (local name: *senduduk puteh*) for hypertension, diabetes, gout, blood detoxification, and gynecological disorders.

Pharmacology and phytochemistry: this plant has been the subject of numerous pharmacological studies (Joffry et al. 2012) which have highlighted, among other things, antiplasmodial (Subeki 2012), nematicidal (Mackeen et al. 1997; Panda et al. 2017; Suteky 2019), antidiarrheal (Sunilson et al. 2009), wound healing (Anbu et al. 2008), antibacterial (Choudhury et al. 2011), hypoglycemic, and hypolipidemic activity (Balamurugan et al. 2014).

Ellagitannins (Yoshida et al. 1992; Yoshida et al. 2010). Amides alkaloids (auranamide, patriscabratine), triterpenes (α-amyrin), flavonols (quercetin), flavonol glycosides (quercitrin, kaempferol-3- *O* -(2″,6″-di- *O*-*p*-trans -coumaroyl)-β-glucoside) (Sirat et al. 2010).

Toxicity, side effects, and drug interaction: methanol extract of leaves administered orally to rats at the dose of 5 g/kg/day for 14 days was found to cause no signs of toxicity (Kamsani et al. 2019).

Comments:

(i) Plants of the genus *Melastoma* L. (1753) are used by the Dusun and Kadazan as antidotes for poisons, for stomach aches, diarrhea, cuts, and wounds.

(ii) *Sonerila crassiuscula* Stapf. is used by the Lundayeh to repel insects (local name: *bubuk kato*).

(iii) *Pternandra gracilis* (Cogn.) N.P. Nayar (Figure 4.40) is used by the Dusun for food poisoning (local name: *banawar*).

(iv) *Melastoma beccarianum* Cogn. is used by the Dusun to treat blemishes (local name: *duduk abai*).

(v) *Memecylon scolopacinum* Ridl. is used by the Dusun for blood vomiting whilst they employ a plant of the genus *Pternandra* Jack (1822) for diarrhea.

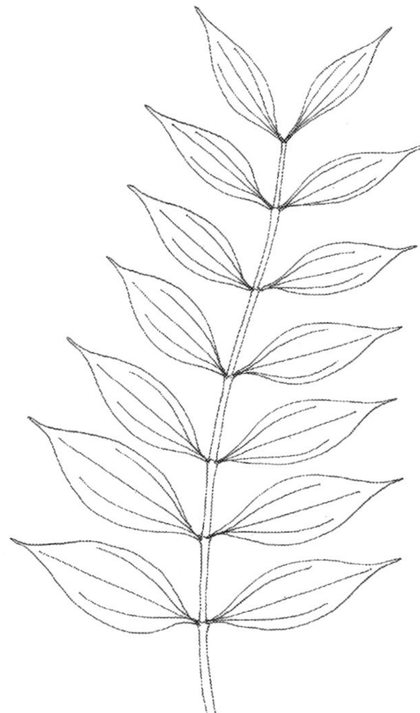

FIGURE 4.40 *Pternandra gracilis* (Cogn.) N.P. Nayar

REFERENCES

Anbu, J., Jisha, P., Varatharajan, R. and Muthappan, M. 2008. Antibacterial and wound healing activities of *Melastoma malabathricum* linn. *African Journal of Infectious Diseases*. 2(2).

Balamurugan, K., Nishanthini, A. and Mohan, V.R. 2014. Antidiabetic and antihyperlipidaemic activity of ethanol extract of *Melastoma malabathricum* Linn. leaf in alloxan induced diabetic rats. *Asian Pacific Journal of Tropical Biomedicine*. 4: S442–S448.

Choudhury, M.D., Nath, D. and Talukdar, A.D. 2011. Antimicrobial activity of *Melastoma malabathricum* L. *Assam University Journal of Science and Technology*. 7(1): 76–78.

Joffry, S.M., Yob, N.J., Rofiee, M.S., Affandi, M.M.R., Suhaili, Z., Othman, F., Akim, A.M., Desa, M.M. and Zakaria, Z.A. 2012. *Melastoma malabathricum* (L.) Smith ethnomedicinal uses, chemical constituents, and pharmacological properties: A review. *Evidence-Based Complementary and Alternative Medicine*. 2012.

Kamsani, N.E., Zakaria, Z.A., Md Nasir, N.L., Mohtarrudin, N. and Mohamad Alitheen, N.B. 2019. Safety assessment of methanol extract of *Melastoma malabathricum* L. leaves following the subacute and subchronic oral consumptions in rats and its cytotoxic effect against the HT29 cancer cell line. *Evidence-Based Complementary and Alternative Medicine*. 2019.

Mackeen, M.M., Ali, A.M., Abdullah, M.A., Nasir, R.M., Mat, N.B., Razak, A.R. and Kawazu, K. 1997. Antinematodal activity of some Malaysian plant extracts against the pine wood nematode, Bursaphelenchus xylophilus. *Pesticide Science*. 51(2): 165–170.

Panda, S.K., Padhi, L., Leyssen, P., Liu, M., Neyts, J. and Luyten, W. 2017. Antimicrobial, anthelmintic, and antiviral activity of plants traditionally used for treating infectious disease in the similipal biosphere reserve, Odisha, India. *Frontiers in Pharmacology*. 8: 658.

Sirat, H.M., Susanti, D., Ahmad, F., Takayama, H. and Kitajima, M. 2010. Amides, triterpene and flavonoids from the leaves of *Melastoma malabathricum* L. *Journal of Natural Medicines*. 64: 492–495.

Subeki, S. 2012. Potency of the Indonesian medicinal plants as antimalarial drugs. *Jurnal Teknologi & Industri Hasil Pertanian*. 13(1): 25–30.

Sunilson, J.A.J., An, K., Kumari, A.V.A.G. and Mohan, S. 2009. Antidiarrhoeal activity of leaves of *Melastoma malabathricum* Linn. *Indian Journal of Pharmaceutical Sciences*. 71(6): 691.

Yoshida, T., Amakura, Y. and Yoshimura, M. 2010. Structural features and biological properties of ellagitannins in some plant families of the order Myrtales. *International Journal of Molecular Sciences*. 11(1): 79–106.

Yoshida, T., Nakata, F., Hosotani, K., Nitta, A. and Okuda, T. 1992. Tannins and related polyphenols of melastomataceous plants. V. Three new complex tannins from *Melastoma malabathricum* L. *Chemical and Pharmaceutical Bulletin*. 40(7): 1727–1732.

4.1.11.3.4 Family Myrtaceae Juss. (1789), the Myrtle Family

The family Myrtaceae consists of about 130 genera and 3,000 species of shrubs and trees. Stems: often somewhat slightly reddish and angular. Leaves: simple, opposite, exstipulate. Blades: often leathery, punctuated with oil glands, the secondary nerves forming or not an intramarginal vein. Inflorescences: cymes. Hypanthium: present. Calyx: 4–5 sepals or tubular, the calyx lobes sometimes merged into a calyptra (operculum). Petals: 4–5. Disc: present. Stamens: 4-many, filamentous, showy. Carpels: 2–5, forming a 2–5 locular ovary, each locule with 2-many ovules. Styles: 1. Fruits: capsules, berries, drupes, marked with disc at apex.

***Psidium guajava* L.**

[From Greek *psidion* = pomegranate and Spanish *guajaba* = guava]

Published in *Species Plantarum* 1: 470. 1753

Synonyms: *Guajava pyrifera* (L.) Kuntze, *Myrtus guajava* (L.) Kuntze, *Psidium guava* Griseb., *Psidium igatemyensis* Barb. Rodr., *Psidium pomiferum* L., *Psidium pumilum* Vahl, *Psidium pyriferum* L.

Local names: *biabas* (Dusun, Kedayan); *kaliabas; pucuk jambu* (Dusun); *liabas, siabas* (Kadazan); *daung jampu* (Bugis); *jambu batu* (Murut); *koliabas* (Dusun, Kadazan); *giabas* (Lundayeh)

Common name: guava tree

Habitat: cultivated

Geographical distribution: tropical

Botanical description: this treelet reaches a height of about 5 m. Trunk: not straight. Bark: smooth. Stems: somewhat angular, smooth, with oil cells, hairy at apex. Internodes: 3–3.5 cm long. Leaves: simple, decussate, exstipulate. Petioles: channeled, hairy, up to about 8 mm long. Blades: elliptic, 6–13 cm × 4–6.5 cm, coriaceous, hairy, with about 16–23 pairs of clearly visible secondary nerves, with oil cells beneath. Flowers: solitary, showy, axillary or terminal. Calyx: up to about 8 mm long, 5-lobed, the lobes triangular, somewhat hairy, up to about 7 mm long. Petals: 5, with oil cells, up to 2 cm long, white. Stamens: numerous, showy. Styles: 1, 1 cm long. Stigma: globose. Berries: pyriform or globose, coriaceous, heavy, greenish, edible, up to about 15 cm across with remnant calyx lobes at apex. Seeds: numerous, in a white or pinkish flesh with a heavenly scent and taste.

Traditional therapeutic indications: diarrhea, dysentery (Bajau; Bugis, Dusun; Kadazan, Lundayeh); stomach aches, indigestion (Bajau, Dusun, Kadazan, Rungus, Sungai, locals of Javanese descent, Kedayan, Iranun); gastroenteritis (Dusun); postpartum, scalds (Bajau); pimples (Kedayan); laxative, fever, chicken pox (Murut)

Pharmacology and phytochemistry: this plant has been the subject of numerous pharmacological studies (Gutiérrez et al. 2008; Ugbogu et al. 2022) which have highlighted, among other things, anti-inflammatory, analgesic, antidiabetic, hypotensive (Olajide et al. 1999; Ojewole 2005), anthelmintic (Tangpu & Yadav 2006), antiplasmodial (Kaushik et al. 2015; Rajendran et al. 2014), antiamoebic (Brandelli et al. 2009), antibacterial (Voravuthikunchai & Kitoipit 2005), antifungal (Morais-Braga et al. 2017), and antiviral (influenza virus) activity (Sriwilaijaroen et al. 2012).

Tannins (guajavin B) (Okuda et al. 1987; Tanaka et al. 1992), antibacterial essential oil (de Souza et al. 2017; Miller et al. 2015).

Toxicity, side effects, and drug interaction: the median lethal dose (LD_{50}) of methanol extract of bark administered orally to rats was above 5 g/kg (Manekeng et al. 2019).

Comments:

(i) In the Philippines, destitute farmers of Mindanao and Cebu use the young leaves to clean teeth and as antiseptic.
(ii) *Rhodomyrtus tomentosa* (Aiton) Hassk. is used for diarrhea by the Dusun (local name: *jilong*).
(iii) *Syzygium aqueum* (Burm.f.) Alston is used by the Bajau for fever (local name: *jambu madu*).
(iv) *Syzygium malaccense* (L.) Merr. & L.M. Perry is used for diarrhea and stomach aches by the Kedayan (local name: *jambu air*).
(v) A plant of the genus *Decaspermum* J.R. Forst. & G. Forst. (1775) is used by the Dusun and Kadazan for the treatment of diarrhea, flatulence, and fever.

REFERENCES

Brandelli, C.L.C., Giordani, R.B., De Carli, G.A. and Tasca, T. 2009. Indigenous traditional medicine: *In vitro* anti-giardial activity of plants used in the treatment of diarrhea. *Parasitology Research*. 104(6): 1345–1349.

de Souza, T.D.S., da Silva Ferreira, M.F., Menini, L., de Lima Souza, J.R.C., Parreira, L.A., Cecon, P.R. and Ferreira, A. 2017. Essential oil of *Psidium guajava*: Influence of genotypes and environment. *Scientia Horticulturae*. 216: 38–44.

Gutiérrez, R.M.P., Mitchell, S. and Solis, R.V. 2008. *Psidium guajava*: A review of its traditional uses, phytochemistry and pharmacology. *Journal of Ethnopharmacology*. 117(1): 1–27.

Kaushik, N.K., Bagavan, A., Rahuman, A.A., Zahir, A.A., Kamaraj, C., Elango, G., Jayaseelan, C., Kirthi, A.V., Santhoshkumar, T., Marimuthu, S. and Rajakumar, G. 2015. Evaluation of antiplasmodial activity of medicinal plants from North Indian Buchpora and South Indian Eastern Ghats. *Malaria Journal*. 14(1): 65.

Manekeng, H.T., Mbaveng, A.T., Ntyam Mendo, S.A., Agokeng, A.J.D. and Kuete, V. 2019. Evaluation of acute and subacute toxicities of *Psidium guajava* methanolic bark extract: A botanical with *in vitro* antiproliferative potential. *Evidence-Based Complementary and Alternative Medicine*. 2019: 1–13.

Miller, A.B., Cates, R.G., Lawrence, M., Soria, J.A.F., Espinoza, L.V., Martinez, J.V. and Arbizú, D.A. 2015. The antibacterial and antifungal activity of essential oils extracted from Guatemalan medicinal plants. *Pharmaceutical Biology*. 53(4): 548–554.

Morais-Braga, M.F., Carneiro, J.N., Machado, A.J., Sales, D.L., dos Santos, A.T., Boligon, A.A., Athayde, M.L., Menezes, I.R., Souza, D.S., Costa, J.G. and Coutinho, H.D. 2017. Phenolic composition and medicinal usage of *Psidium guajava* Linn.: Antifungal activity or inhibition of virulence? *Saudi Journal of Biological Sciences*. 24(2): 302–313.

Ojewole, J.A.O. 2005. Hypoglycemic and hypotensive effects of *Psidium guajava* Linn. (Myrtaceae) leaf aqueous extract. *Methods and Findings in Experimental and Clinical Pharmacology*. 27(10): 689–696.

Okuda, T., Yoshida, T., Hatano, T., Yazaki, K., Ikegami, Y. and Shingu, T. 1987. Guavins A, C and D, complex tannins from *Psidium guajava*. *Chemical and Pharmaceutical Bulletin*. 35(1): 443–446.

Olajide, O.A., Awe, S.O. and Makinde, J.M. 1999. Pharmacological studies on the leaf of *Psidium guajava*. *Fitoterapia*. 70(1): 25–31.

Rajendran, C., Begam, M., Kumar, D., Baruah, I., Gogoi, H.K., Srivastava, R.B. and Veer, V. 2014. Antiplasmodial activity of certain medicinal plants against chloroquine resistant *Plasmodium berghei* infected white albino BALB/c mice. *Journal of Parasitic Diseases*. 38(2): 148–152.

Sriwilaijaroen, N., Fukumoto, S., Kumagai, K., Hiramatsu, H., Odagiri, T., Tashiro, M. and Suzuki, Y. 2012. Antiviral effects of *Psidium guajava* Linn. (guava) tea on the growth of clinical isolated H1N1 viruses: Its role in viral hemagglutination and neuraminidase inhibition. *Antiviral Research*. 94(2): 139–146.

Tanaka, T., Ishida, N., Ishimatsu, M., Nonaka, G.I. and Nishioka, I. 1992. Tannins and related compounds. CXVI. Six new complex tannins, guajavins, psidinins and psiguavin from Bark: Of *Psidium guajava* L. *Chemical and Pharmaceutical Bulletin*. 40(8): 2092–2098.

Tangpu, T.V. and Yadav, A.K. 2006. Anticestodal efficacy of *Psidium guajava* against experimental Hymenolepis diminuta infection in rats. *Indian Journal of Pharmacology*. 38(1): 29.

Ugbogu, E.A., Emmanuel, O., Uche, M.E., Dike, E.D., Okoro, B.C., Ibe, C., Ude, V.C., Ekweogu, C.N. and Ugbogu, O.C. 2022. The ethnobotanical, phytochemistry and pharmacological activities of *Psidium guajava* L. *Arabian Journal of Chemistry*. 103759.

Voravuthikunchai, S.P. and Kitoipit, L. 2005. Activity of medicinal plant extracts against hospital isolates of methicillin-resistant Staphylococcus aureus. *Clinical Microbiology and Infection*. 11(6): 510–512.

4.1.11.4 Order Sapindales Juss. ex Bercht. & J. Presl (1820)

4.1.11.4.1 Family Anacardiaceae R.Br. (1818), the Cashew Family

The family Anacardiaceae consists of about 70 genera and 600 species of trees, shrubs, or climbing plants. Stems: often exude a vesicant sap or resin once incised. Leaves: simple or compound, alternate or spiral, exstipulate. Blades: often coriaceous, with straight secondary nerves. Inflorescences: panicles of tiny flowers. Calyx: tubular, 3-5-lobed. Corolla: 3–5 petals or tubular. Disc: present. Stamens: 5–10. Carpels: 1–5 forming a 1–5 locular ovary, each locule with 1 ovule. Fruits: drupes, often asymmetrical

Mangifera caesia **Jack**

[From Tamil *manga* = mango and Latin *caesium* = sky blue]

Published in *Flora Indica; or descriptions of Indian Plants* 2: 441. 1824

Local names: *beluno, binjai* (Bajau); *mangga wani* (Murut); *pahu* (Dusun)
Common name: white mango
Habitat: forests near streams and swamps, cultivated
Geographical distribution: Malaysia, Indonesia
Botanical description: this tree reaches a height of about 45 m. Bark: brown, deeply fissured, secretes a sap once incised. Stems: stout, rough, striated. Leaves: simple, spiral, exstipulate. Petioles: stout, 2–5 cm long, flattened above. Blades: broadly obovate to oblanceolate, coriaceous, smooth, glossy, 2.5–6.5 cm × 16.5–18 cm, tapering at base, rounded at apex, with 10–22 pairs of secondary nerves. Panicles: terminal, up to about 30 cm long. Calyx: tubular, tiny, 5-lobed. Petals: 5, tiny, purplish inside. Stamens: 5. Ovary: sub-globose. Styles: 1 Drupes: potato-shaped to pear- or avocado-shaped, brownish, up to 20 cm long, with an edible, sour, and pure white pulp.
Traditional therapeutic indication: medicinal (Dusun)
Pharmacology and phytochemistry: antibacterial, anticandidal (Ardiningsih et al. 2011), cytotoxic (Dwidhanti et al. 2018).
Toxicity, side effects, and drug interaction: the sap is extremely vesicant. Burning the stems produces a vesicant smoke (Wood & Calnan 1976).

REFERENCES

Ardiningsih, P., Nofiani, R. and Jayuska, A. 2011. Antimicrobial activity of leaves, stems, and barks of palasu (*Mangifera caesia* Jack) against microorganisms associated with fish spoilage. *Bioscience and Biotechnology (ICBB)*. 2011.

Dwidhanti, F., Taufiqurrahman, I. and Sukmana, B.I. 2018. Cytotoxicity test of binjai leaf (*Mangifera caesia*) ethanol extract in relation to Vero cells. *Dental Journal (Majalah Kedokteran Gigi)*. 51(3): 108–113.

Woods, B. and Calnan, C.D. 1976. Toxic woods. *British Journal of Dermatology*. 94: 1–1.

Mangifera pajang **Kosterm.**

[From Tamil *manga* = mango, Latin *ferre* = bearer, and the Malay name of the plant = *asem pajan*]

Published in *Reinwardtia* 7: 20. 1965

Local names: *bambangan* (Dusun; Murut); *embawang* (Bajau)
Common name: wild mango
Habitat: swampy forests, on the banks of rivers, villages
Geographical distribution: Borneo
Botanical description: this tree reaches a height of about 30 m. Bark: dark brown, fissured, secretes a creamy sap once incised. Stems: stout, rough, angular. Leaves: simple, spiral, exstipulate. Petioles: somewhat swollen at base, 3.5–9 cm long, stout. Blades: oblanceolate, coriaceous, smooth, glossy, narrowly obovate, 5.5–12.5 cm × 15–28 cm, tapering at base, obtuse to somewhat acuminate at apex, with 13–25 pairs of secondary nerves. Panicles: terminal, up to about 30 cm long. Calyx: tubular, dark purple, 5-lobed. Petals: 5, tiny, purplish. Stamens: 5. Drupes: of a peculiar kind of yellowish-brown color (almost like a potato), rugose, somewhat globose, about 10 cm across, containing an edible orange pulp.
Traditional therapeutic indications: medicinal (Dusun)
Pharmacology and phytochemistry: toxic to breast cancer cells (Bakar et al. 2010). Consumption of fruit juice in healthy volunteers increased the antioxidant capacity of their plasma (Ibrahim et al. 2013).
Phenolic acids (methyl gallate), triterpenes (lupeol, lupenone, 3*β*-hydroxy-cycloart-24-ene-26-oic acid, 3*β*,23-dihydroxy-cycloart-24-ene-26-oic acid) (Ahmad et al. 2015).

Comments:

(i) The Dusun and Kadazan use the fruits of this plant in a local salad dish called "*hinava*" made of raw fish, lemon juice, onions, chilli pepper, and ginger.
(ii) The Murut eat the unripe pickled fruits of *Mangifera indica* L. (local name: *longgom*).

REFERENCES

Ahmad, S., Sukari, M.A., Ismail, N., Ismail, I.S., Abdul, A.B., Abu Bakar, M.F., Kifli, N. and Ee, G.C. 2015. Phytochemicals from *Mangifera pajang* Kosterm and their biological activities. *BMC Complementary and Alternative Medicine*. 15(1): 1–8.

Bakar, M.F.A., Mohamad, M., Rahmat, A., Burr, S.A. and Fry, J.R. 2010. Cytotoxicity, cell cycle arrest, and apoptosis in breast cancer cell lines exposed to an extract of the seed kernel of *Mangifera pajang* (bambangan). *Food and Chemical Toxicology*. 48(6): 1688–1697.

Ibrahim, M., Ismail, A., Al-Sheraji, S.H., Azlan, A. and Hamid, A.A. 2013. Effects of *Mangifera pajang* Kostermans juice on plasma antioxidant status and liver and kidney function in normocholesterolemic subjects. *Journal of Functional Foods*. 5(4): 1900–1908.

Pegia sarmentosa **(Lecomte) Hand.-Mazz.**

[From Latin *pege* = a source and *sarmentosus* = sarmentose]

Published in *Sinensia* 3: 187. 1933

Synonyms: *Pegia bijuga* Hand.-Mazz.; *Phlebochiton sarmentosum* Lecomte; *Phlebochiton sinense* Diels
Local name: *ampan* (Dusun)
Habitat: forests
Geographical distribution: Southeast Asia, South China.
Botanical description: this woody climbing plant grows up to a length of about 20 m. Leaves: spiral, imparipinnate, exstipulate. Rachis: channeled. Petiolules: up to about 4–8 mm long. Blades: about 15–30 cm long, with 11–15 pairs of opposite folioles. Folioles: ovoid to oblong, rounded at base, acuminate at apex, serrate, 2.5–4 cm × 4–8 cm, with 6–8 pairs of secondary nerves. Panicles: up to 20 cm long, thin, terminal, lax, hairy, with tiny flowers. Sepals: 5, triangular. Petals: 5, ovate, white, 2 mm long. Disc: lobed. Stamens: 10. Drupes: oblong, red, about 1.3 cm long. Seeds: curved.
Traditional therapeutic indications: cuts (Dusun)
Pharmacology and phytochemistry: not known
Toxicity, side effects, and drug interaction: not known

Semecarpus cuneiformis **Blanco**

[From Greek *sema* = a mark, *karpos* = fruit, and Latin *cuneiformis* = with attenuate leaves]

Published in *Flora de Filipinas* 220. 1837

Synonyms: *Semecarpus elmeri* Perkins; *Semecarpus ferrugineus* Merr.; *Semecarpus lanceolatus* Ridl.; *Semecarpus megabotrys* Merr.; *Semecarpus merrillianus* Perkins; *Semecarpus micranthus* Perkins; *Semecarpus obtusifolius* Merr.; *Semecarpus perrottetii* Marchand; *Semecarpus philippinensis* Engl.; *Semecarpus pilosus* Merr.; *Semecarpus ridleyi* Merr.; *Semecarpus taftianus* Perkins; *Semecarpus thyrsoideus* Elmer; *Semecarpus whitfordii* Merr.
Local name: *kutang* (Murut)
Common names: marking nut tree, Oriental cashew nut
Habitat: roadsides, open lands
Geographical distribution: Borneo, the Philippines, Indonesia, Taiwan
Botanical description: this tree reaches a height of about 20 m. Bark: secretes a vesicant resin once incised. Leaves: simple, spiral, exstipulate. Petioles: 1–3 cm long. Blades: elliptic, oblong, to obovate, 2–9 cm × 8–35 cm, dark green, coriaceous, attenuate to acute at base, acute, rounded, to acuminate at apex, with a stout midrib and 11–25 pairs of secondary nerves. Panicles: 15–30 cm long, many-flowered, terminal. Sepals: 5, tiny. Petals: 5, whitish, somewhat fleshy, up to 3 mm long, revolute. Disc: lobed. Stamens: 5, about 4 mm long, spreading with straight filaments. Ovary: hairy. Styles: 3. Drupes: ovoid, asymmetrical, black, glossy, longitudinally striated, 1–2 cm long, on a bright red, fleshy, glossy, and about 3 cm long hypocarp.
Traditional therapeutic indications: wounds (Murut)
Pharmacology and phytochemistry: toxic to Lewis lung carcinoma cells, Walker carcinoma cells, human epidermoid carcinoma cells, and leukemia cells (Spjut 2005).
Toxicity, side effects, and drug interaction: poisonous

Comments:

(i) In the Philippines, the resin was used to blacken the teeth for cosmetic purposes.
(ii) *Lannea coromandelica* (Houtt.) Merr. is used by the Bugis for wounds (local name: *aju Jawa*).
(iii) In the family Meliaceae Juss. (1789), *Lansium domesticum* Corrêa is used by the Dusun and Kadazan for stomach aches (local name: *langsat*), whilst the Murut use it for diarrhea.
(iv) In the family Meliaceae Juss. (1789), *Xylocarpus granatum* J. Koenig is used by the Bajau for dysentery (local name: *terbigit*). Parts of this plant are sold as medicine in the wet markets of Sabah (local name: *tampan merah*).
(v) In the family Burseraceae Kunth (1824), *Canarium littorale* Bl., *Canarium pilosum* A.W. Benn. (local name: *adal*), and *Dacryodes incurvata* (Engl.) H.J. Lam. (local name: *nguluon*) are used as food by the Dusun.

REFERENCE

Spjut, R.W. 2005. Relationships between plant folklore and antitumor activity: An historical review. *SIDA, Contributions to Botany*: 2205–2241.

4.1.11.4.2 Family Rutaceae Juss. (1789), the Citrus Family

The family Rutaceae consists of about 150 genera and 900 species of trees, shrubs, climbing plants, or herbaceous plants. Leaves: simple or compound, spiral, exstipulate. Blades (and fruits): often with oil glands. Inflorescences: cymes, racemes, fascicles, or solitary. Calyx: 5 sepals or tubular. Corolla: 4–5 petals or tubular, often pure white, somewhat fleshy, with revolute lobes. Disc: present. Stamens: 2–60, somewhat merged at base. Carpels: 3–5, forming a 4–5 locular ovary, each locule with 1–many ovules. Styles: 1, often stout. Stigma: often clavate and humid. Fruits: berries (hesperidia), follicles, capsules.

Clausena excavata **Burm.f.**

[After the 16th century Norwegian priest Peder Claussen and Latin *excavata* = hollowed]

Published in *Flora Indica. . . nec non Prodromus Florae Capensis* 89, pl. 29, f. 2. 1768

Synonyms: *Amyris punctata* Roxb.; *Clausena forbesii* Engl.; *Clausena lunulata* Hayata; *Clausena moningerae* Merr.; *Clausena punctata* (Roxb.) Wight & Arn.; *Clausena tetramera* Hayata

Local names: *untuk paranok* (Rungus); *alab layat* (Lundayeh)

Habitat: villages, gardens, at the edges of forests

Geographical distribution: from India to the Philippines

Botanical description: this treelet reaches a height of about 15 m. Bark: somewhat smooth, light brown. Leaves: spiral, compound, exstipulate, up to about 30 cm long. Folioles: 5–31 pairs, alternate, asymmetrical, serrate, with oil glands and 9 pairs of secondary nerves, slightly aromatic, 1–2.5 cm × 1.7–4 cm. Panicles: about 15–45 cm long, with tiny flowers. Calyx: tiny. Petals: 4, lanceolate to oblong, greenish. Stamens: 8, about 3 mm long. Drupes: 1–2 cm long, oblong, of a kind of heavenly light pink.

Traditional therapeutic indications: venereal diseases (Lundayeh); swellings, headaches (Dusun, Kadazan); toothaches (Dusun, Kadazan, Rungus)

Pharmacology and phytochemistry this plant has been the subject of numerous pharmacological studies (Arbab et al. 2011) which have highlighted, among other things, hypoglycemic (Thant et al. 2019), wound healing (Albaayit et al. 2020), anti-inflammatory (Albaayit et al. 2020), and analgesic activity (Rahman et al. 2002).

Antiviral (HIV) limonoids (Sunthitikawinsakul et al. 2003). Antiviral (HIV) carbazole alkaloids and pyranocoumarins (Kongkathip et al. 2005). Antifungal and antimycobacterial carbazole alkaloids and coumarins (Sunthitikawinsakul et al. 2003a). Antifungal coumarins (Kumar et al. 2012). Antiplasmodial and cytotoxic carbazoles (Huang et al. 2017).

Toxicity, side effects, and drug interaction: methanol extract of leaves administered orally to rats at the dose of 5 g/kg/day was found to cause liver and kidney damages (Albaayit et al. 2014).

Comments:

(i) *Citrus limon* (L.) Osbeck is used by the Dusun to prevent sweating (local name: *limau*). They use a plant in the genus *Citrus* L. (1753) for sore throats (local name: *kolopis*).

(ii) *Citrus microcarpa* Bunge is used for coughs, fever, sore throats, and smoking addictions by the Kedayan and Bajau (local name: *limau kasturi*).

(iii) *Micromelum minatum* Wight & Arn. is medicinal for the Murut (local name: *kimamansak*). A plant of the genus *Micromelum* Bl. (1825) (local name: *paw*) is used for flatulence by the Dusun.

(iv) *Murraya koenigii* (L.) Spreng. is used by the Kedayan for bruises, diabetes, dyspepsia, nausea, and swellings (local name: *daun kari*).
(v) The Dusun and Kadazan use a plant of the genus *Melicope* J.R. Forst. & G. Forst. (1775) (local name: *pau*) for beriberi.
(vi) The Dusun use *Luvunga motleyi* Oliver (local name: *tiga tiga*) for pancreatitis and they smoke the leaves as tobacco.

REFERENCES

Albaayit, S.F.A., Abba, Y., Abdullah, R. and Abdullah, N. 2014. Evaluation of antioxidant activity and acute toxicity of *Clausena excavata* leaves extract. *Evidence-Based Complementary and Alternative Medicine*. 2014: 1–10.

Albaayit, S.F.A., Rasedee, A., Abdullah, N. and Abba, Y. 2020. Methanolic extract of *Clausena excavata* promotes wound healing via antiinflammatory and anti-apoptotic activities. *Asian Pacific Journal of Tropical Biomedicine*. 10(5): 232–238.

Arbab, I.A., Abdul, A.B., Aspollah, M., Abdullah, R., Abdelwahab, S.I., Mohan, S. and Abdelmageed, A.H.A. 2011. *Clausena excavata* Burm.f. (Rutaceae): A review of its traditional uses, pharmacological and phytochemical properties. *Journal of Medicinal Plants Research*. 5(33): 7177–7184.

Huang, L., Zhe-Ling, F.E.N.G., Yi-Tao, W.A.N.G. and Li-Gen, L.I.N. 2017. Anticancer carbazole alkaloids and coumarins from *Clausena* plants: A review. *Chinese Journal of Natural Medicines*. 15(12): 881–888.

Kongkathip, B., Kongkathip, N., Sunthitikawinsakul, A., Napaswat, C. and Yoosook, C. 2005. Anti-HIV-1 constituents from *Clausena excavata*: Part II. carbazoles and a pyranocoumarin. *Phytotherapy Research: An International Journal Devoted to Pharmacological and Toxicological Evaluation of Natural Product Derivatives*. 19(8): 728–731.

Kumar, R., Saha, A. and Saha, D. 2012. A new antifungal coumarin from *Clausena excavata*. *Fitoterapia*. 83(1): 230–233.

Rahman, M.T., Alimuzzaman, M., Shilpi, J.A. and Hossain, M.F. 2002. Antinociceptive activity of *Clausena excavata* leaves. *Fitoterapia*. 73(7–8): 701–703.

Sunthitikawinsakul, A., Kongkathip, N., Kongkathip, B., Phonnakhu, S., Daly, J.W., Spande, T.F., Nimit, Y., Napaswat, C., Kasisit, J. and Yoosook, C. 2003. Anti-HIV-1 limonoid: First isolation from *Clausena excavata*. *Phytotherapy Research: An International Journal Devoted to Pharmacological and Toxicological Evaluation of Natural Product Derivatives*. 17(9): 1101–1103.

Sunthitikawinsakul, A., Kongkathip, N., Kongkathip, B., Phonnakhu, S., Daly, J.W., Spande, T.F., Nimit, Y. and Rochanaruangrai, S. 2003a. Coumarins and carbazoles from *Clausena excavata* exhibited antimycobacterial and antifungal activities. *Planta Medica*. 69(2): 155–157.

Thant, T.M., Aminah, N.S., Kristanti, A.N., Ramadhan, R., Aung, H.T. and Takaya, Y. 2019. Antidiabetes and antioxidant agents from *Clausena excavata* root as medicinal plant of Myanmar. *Open Chemistry*. 17(1): 1339–1344.

4.1.11.4.3 Family Sapindaceae Juss. (1789), the Soapberry Family

The family Sapindaceae consists of about 140 genera and 2,000 species of trees, shrubs, or climbing plants. Leaves: compound, spiral, exstipulate. Folioles: often somewhat coriaceous with a peculiar kind of dull green color. Inflorescences: cymes, panicles. Calyx: 4–5 sepals or tubular. Petals: 4–5, free or merged at base. Disc: present. Stamens: 7–many. Carpels: 3, forming a trilocular ovary, each locule with 1–5 ovules. Styles: 1. 1. Fruits: drupes, berries, capsules. Seeds: often embedded in a fleshy aril.

Guioa pleuropteris **(Blume) Radlk.**

[From Greek *pleuron* = lateral outgrowth and *pteron* = wing]

Published in *Actes du IIIme Congrès international de botanique, Bruxelles* 1910 reimpr.: 10. 1879

Synonym: *Cupania pleuropteris* Bl.
Local names: *andipatan, gulambir ayam, gurujod, kanawit, kangi runok, mata pait, munggulan ayam, pengkul* (Dusun); *piri manok, tingir manok, tangkit manok, saasa* (Bajau)
Habitat: forests, roadsides, seashores
Geographical distribution: from Myanmar to the Philippines
Botanical description: this tree reaches a height of about 30 m. Bark: smooth, the inner bark white. Leaves: spiral, pinnate, exstipulate. Petioles: 5 mm–10 cm long. Rachis: winged, up to about 25 cm long. Petiolules: about 3 mm long, stout, curved. Folioles: 2–5 pairs, opposite, shorter at base of leaves, asymmetrical, elliptic to obovate, 1–18 cm × 5–10 cm, acute, dull green, attenuate and asymmetrical at base, with about 5–7 pairs of secondary nerves, acuminate at apex. Cymes: axillary, up to about 20 cm long, hairy. Sepals: 5, tiny, of unequal length. Petals: 5, about 3 mm long. Stamens: 8. Capsules: about 2.5 cm across, deeply bilobed, dull red, dehiscent. Seeds: 1 per lobe, about 1 cm long, in in an orangish aril.
Traditional therapeutic indications: fatigue, flatulence, postpartum, stomach aches thrush (Dusun, Kadazan)
Pharmacology and phytochemistry: not known
Toxicity, side effects, and drug interaction: not known

Lepisanthes amoena **(Hassk.) Leenh**

[from Greek *lepis* = scale, *anthos* = flower, and Latin *amoenus* = charming]

Published in *Blumea* 17: 71.1969

Synonyms: *Capura spectabilis* Teijsm. & Binn.; *Otolepis amoena* Kuntze; *Otolepis cordigera* Kuntze; *Otolepis imbricata* Kuntze; *Otolepis pubescens* Kuntze; *Otolepis spectabilis* Kuntze; *Otophora amoena* Blume; *Otophora confinis* Blume; *Otophora cordigera* Radlk.
Local name: *kuinin* (Dusun)
Habitat: forests
Geographical distribution: Thailand, Malaysia, Borneo
Botanical identification: this handsome tree reaches a height of about 6 m. Leaves: imparipinnate, spiral, stipulate. Petioles: 1–9 cm long. Pseudostipules: about 6 cm across, cordate. Folioles: 4–12 pairs, opposite, dull green, somewhat wavy, membranous, oblong, of a kind of heavenly pinkish purple when young, sessile, asymmetrical, wavy, 1.2–5 cm × 7–22.5 cm, obtuse at apex, with numerous pairs of secondary nerves. Panicles: axillary, up to 40 cm long.

Sepals: 4–5, orbicular, red, about 3 mm long. Petals: 4–5, tiny, white, hairy. Stamens: 7–9, hairy. Ovary: trilobed. Drupes: somewhat irregularly globose, up to 3 cm across, dull orange.

Traditional therapeutic indications: malaria (Dusun)

Pharmacology and phytochemistry: antibacterial (Batubara et al. 2009).

Toxicity, side effects, and drug interaction: not known

REFERENCE

Batubara, I., Mitsunaga, T. and Ohashi, H. 2009. Screening antiacne potency of Indonesian medicinal plants: Antibacterial, lipase inhibition, and antioxidant activities. *Journal of Wood Science*. 55(3): 230–235.

Mischocarpus pentapetalus **(Roxb.) Radlk.**

[From the Greek *mischos* = a stalk, *karpos* = a fruit, *penta* = five, and *petalon* = petals]

Published in *Sitzungsberichte der Mathematisch-Physikalischen Classe (Klasse) der K. B. Akademie der Wissenschaften zu München* 43(113). 1879

Synonyms: *Cupania pentapetala* (Roxb.) Wight & Arn.; *Mischocarpus fuscescens* Bl.; *Mischocarpus productus* H.L. Li; *Schleichera pentapetala* Roxb.
Local name: *tokingkid* (Rungus)
Habitat: forests
Geographical distribution: India, Southeast Asia, South China
Botanical description: this tree reaches a height of about 15 m. Leaves: paripinnate, alternate, exstipulate. Petioles: up to 25 cm long. Petiolules: 3 mm–1.2 cm long. Folioles: alternate or opposite, oblong to lanceolate, 2–6 cm × 6–20 cm, rounded at base, acute to acuminate at apex, coriaceous, with about 13–16 pairs of secondary nerves looping at the margin, as well as intercostal veinations. Panicles: pseudoterminal, up to 40 cm long, with numerous tiny flowers. Calyx: 5-lobed. Petals: 5. Disc: annular. Stamens: 7–8, with hairy filaments. Ovary: hairy. Stigma: trifid. Capsules: meliaceous, fleshy, obovate, bright reddish-orange, up to about 2 cm long, dehiscent. Seeds: 1, magnificent, black, glossy.
Traditional therapeutic indications: medicinal (Rungus)
Pharmacology and phytochemistry: anti-inflammatory (Le et al. 2021).
Toxicity, side effects, and drug interaction: not known

Comments:

(i) The Murut eat the fruits of *Lepisanthes fruticosa* (Roxb.) Leenh. (local name: *talikasan*) whilst the Dusun use the plant for fatigue (local name: *banculuk*). A plant of the genus *Lepisanthes* Bl. (1825) is used by the Murut (local name: *bolilingasan*) as medicine. The Dusun and Kadazan use a plant of this genus (local name: *boyongo*) for the treatment of mumps.
(ii) The fruits of *Guioa bijuga* (Hiern) Radlk. are used as food by the Rungus (local name: *anggil*).
(iii) *Nephelium uncinatus* Radlk. ex Leenh. is used as food by the Dusun (local name: *kamanggis*).
(iv) *Nephelium macrophyllum* Radlk. is endemic to Borneo and used as food and for magic rituals by the Dusun (local name: *mbokot*).

REFERENCE

Le, T.H., Duong, T.T., Tran, P.T., Pham, V.C., Nguyen, H.D. and Lee, J.H. 2021. Anti-osteoclastogenesis potential agents from plants naturalized in Vietnam. *Pharmacognosy Magazine*. 17(75): 525.

4.1.11.4.4 Family Simaroubaceae DC. (1811), the Quassia Family

The family Simaroubaceae consists of about 20 genera and 50 species of handsome trees or shrubs. Bark: often secretes some resins after incision. Wood: sometimes bitter. Leaves: simple or compound, alternate or spiral, stipulate or exstipulate. Stipules: caducous, tiny, or leafy. Folioles: often somewhat asymmetrical. Inflorescences: racemes, panicules, cymes of tiny flowers. Calyx: 3–5 sepals or tubular. Petals: 3–5 Disc: annular, cupular. Stamens: often 10. Carpels: 1–5, free or forming a unilocular ovary, each carpel or locule with 1 ovule. Styles: 2–5. Fruits: drupes, capsules, samaras.

***Brucea javanica* (L.) Merr.**

[After the 18th century British explorer James Bruce and Latin *javanica* = from Java]

Published in *Journal of the Arnold Arboretum* 9(1): 3. 1928

Synonyms: *Brucea sumatrana* Roxb.; *Gonus amarissimus* Lour.; *Rhus javanica* L.
Local names: *garakat, gompoit, riringit* (Dusun); *pazaskoloruk* (Rungus); *kuinin, magapas, tinimug di ikus, monomopuru* (Dusun, Kadazan)
Common name: Java brucea
Habitat: forests
Geographical distribution: from India to Papua New Guinea
Botanical description: this tree reaches a height of about 10 m. Leaves: compound, 20–50 cm long, exstipulate, spiral. Folioles: 3–15 pairs, oblong, lanceolate or ovate, 1.5–5 cm × 3.5–11 cm, hairy, serrate. Petiolules: about 9 mm long. Panicles: axillary, greenish-white or greenish red or purplish. Calyx: tubular, 4-lobed. Petals: 4. Disc: 4-lobed. Stamens: 4. Carpels: 4, free. Drupes: black, glossy, ovoid, about 5 mm long.
Traditional therapeutic indications: dandruff, flu, lice, gastritis, skin disease, malaria, intestinal worms, stomach aches (Dusun, Kadazan); dysentery (Rungus)
Pharmacology and phytochemistry: antimalarial (O'Neill et al. 1987), antidiabetic, antibacterial (Simamora et al. 2019), anticandidal (Harun & Razak 2013), antiviral (Yan et al. 2010), toxic to pancreatic adenocarcinoma cells (Lau et al. 2008).
Quassinoids toxic for pancreatic cancer cells (Xiang et al. 2017), antiplasmodial (Chumkaew & Srisawat 2017), antiamoebic (bruceantin) (Wright et al. 1988), gastroprotective (Zheng et al. 2023).
Toxicity, side effects, drug interaction: the median lethal dose (LD_{50}) of methanol extract of seeds administered to rats was 281.7 mg/kg (Shahida et al. 2011).

REFERENCES

Chumkaew, P. and Srisawat, T. 2017. Antimalarial and cytotoxic quassinoids from the roots of *Brucea javanica*. *Journal of Asian Natural Products Research*. 19(3): 247–253.
Harun, W.H.A.W. and Razak, F.A. 2013. Antifungal susceptibility and growth inhibitory response of oral Candida species to Brucea javanica Linn. extract. *BMC Complementary and Alternative Medicine*. 13(1): 1–8.
Lau, S.T., Lin, Z.X., Zhao, M. and Leung, P.S. 2008. Brucea javanica fruit induces cytotoxicity and apoptosis in pancreatic adenocarcinoma cell lines. *Phytotherapy Research: An International Journal Devoted to Pharmacological and Toxicological Evaluation of Natural Product Derivatives*. 22(4): 477–486.
O'Neill, M.J., Bray, D.H., Boardman, P., Chan, K.L., Phillipson, J.D., Warhurst, D.C. and Peters, W. 1987. Plants as sources of antimalarial drugs, Part 4: Activity of *Brucea javanica* fruits against chloroquine-resistant *Plasmodium falciparum in vitro* and against Plasmodium berghei *in vivo*. *Journal of Natural Products*. 50(1): 41–48.
Shahida, N.A., Choo, Y.C. and Wong, W.T. 2011. Acute/subcronic oral toxicity *of Brucea javanica* seeds with hypoglicemic activity. *Journal of Natural Remedies*. 11(1): 60–68.

Simamora, A., Timotius, K.H. and Santoso, A.W. 2019. Antidiabetic, antibacterial and antioxidant activities of different extracts from *Brucea javanica* (L.) merr seeds. *Pharmacognosy Journal*. 11(3).

Wright, C.W., O'neill, M.J., Phillipson, J.D. and Warhurst, D.C. 1988. Use of microdilution to assess *in vitro* antiamoebic activities of *Brucea javanica* fruits, *Simarouba amara* stem, and a number of quassinoids. *Antimicrobial Agents and Chemotherapy*. 32(11): 1725–1729.

Xiang, Y., Ye, W., Huang, C., Lou, B., Zhang, J., Yu, D., Huang, X., Chen, B. and Zhou, M. 2017. Brusatol inhibits growth and induces apoptosis in pancreatic cancer cells via JNK/p38 MAPK/NF-κb/Stat3/Bcl-2 signaling pathway. *Biochemical and Biophysical Research Communications*. 487(4): 820–826.

Yan, X.H., Chen, J., Di, Y.T., Fang, X., Dong, J.H., Sang, P., Wang, Y.H., He, H.P., Zhang, Z.K. and Hao, X.J. 2010. Anti-tobacco mosaic virus (TMV) Quassinoids from Brucea javanica (L.) Merr. *Journal of Agricultural and Food Chemistry*. 58(3): 1572–1577.

Zheng, X., Mai, L., Xu, Y., Wu, M., Chen, L., Chen, B., Su, Z., Chen, J., Chen, H., Lai, Z. and Xie, Y. 2023. *Brucea javanica* oil alleviates intestinal mucosal injury induced by chemotherapeutic agent 5-fluorouracil in mice. *Frontiers in Pharmacology*. 14: 1136076.

***Eurycoma longifolia* Jack**

[From Greek *eurus* = broad, *kome* = hair, from Latin *longus* = long, and *folium* = leaf]

Published in *Malayan Miscellanies* V11.45. 1822

Local names: *duli, ruli* (Murut); *timuh* (Dusun, Murut); *monompuru, tombuid* (Dusun, Kadazan); *inurod mondu* (Rungus); *tongkat Ali* (Bajau, Dusun, Kadazan); *taratus* (Dusun)

Habitat: at the edges of forests

Geographical distribution: Cambodia, Vietnam, Thailand, Malaysia, Indonesia

Botanical description: this treelet reaches a height of about 2 m. Tape root: elongated. Wood: whitish, very bitter. Leaves: spiral, terminal, exstipulate, pinnate, up to 1 m long. Folioles: about 20 pairs, subopposite, narrowly elliptic, acute at base and apex, shorter at base of blade, coriaceous, glossy, asymmetrical, about 1.5–3 cm × 6–9 cm, sessile. Panicles: axillary, tomentose, pendulous, with tiny flowers. Calyx: tubular, 5-6-lobed. Petals: 5–6, hairy, about 5 mm long, burgundy. Stamens: 5–6. Disc: present. Carpels: 5–6, free. Follicles: 5, reddish-green turning black, ellipsoid, glossy, pointed at apex, 1–1.7 cm long.

Traditional therapeutic indications: asthma, malaria, stomach aches (Dusun, Kadazan); aphrodisiac (Bajau, Dusun, Kadazan, Murut); hypertension (Kedayan, Iranun); diabetes (Iranum); coughs (Dusun)

Pharmacology and phytochemistry: antiviral (dengue virus) (George et al. 2019), male aphrodisiac (Ang et al. 1995; Ang & Sim 1997), analgesic (Han et al. 2016).

Quassinoids (14,15-β-dihydroxyklaineanone) toxic to liver cancer cells (HepG2) (Pei et al. 2020). Antiplasmodial quassinoids (eurycomanol, eurycomanol 2-*O*-β-*D*-glucopyranoside, 13β,18-dihydroeurycomanol) (Kardono et al. 1991; Ang et al. 1995).

Indole alkaloids (9,10-dimethoxycanthin-6-one, 10-hydroxy-9-methoxycanthin-6-one) and triterpene (dihydroniloticin) toxic to fibrosarcoma cells (HT-1080) (Miyake et al. 2010). Anti-inflammatory indole alkaloid (canthin-6-one) (Zhang et al. 2020).

Toxicity, side effects, and drug interaction: in rats, administration of an aqueous extract of roots induced an increase in prostate weight (Ezzat et al. 2019) and at a dose of 10 mg/kg twice a day, abnormal proliferation of prostate epithelium (Faisal et al. 2013). All this suggests that daily consumption of "extracts" of this plant could induce prostate cancer.

Comments:

(i) There is a dangerous trend in the region to consume aphrodisiac products of questionable quality supposedly containing the root of this plant. Traditionally, this plant was used by the Malays who drank light decoctions of the roots from time to time but never daily and never in the form of extracts.

(ii) A plant of the genus *Eurycoma* Jack (1822) is used to treat back pain by the Dusun and Kadazan (local name: *mumud mondu*).

REFERENCES

Ang, H.H., Chan, K.L. and Mak, J.W. 1995. *In vitro* antimalarial activity of quassinoids from *Eurycoma longifolia* against Malaysian chloroquine-resistant Plasmodium falciparum isolates. *Planta Medica*. 61(2): 177–178.

Ang, H.H. and Sim, M.K. 1997. *Eurycoma longifolia* Jack enhances libido in sexually experienced male rats. *Experimental Animals*. 46(4): 287–290.

Ezzat, S.M., Ezzat, M.I., Okba, M.M., Hassan, S.M., Alkorashy, A.I., Karar, M.M., Ahmed, S.H. and Mohamed, S.O. 2019. Brain cortical and hippocampal dopamine: A new mechanistic approach for *Eurycoma longifolia* well-known aphrodisiac activity and its chemical characterization. *Evidence-Based Complementary and Alternative Medicine*. 2019: 1–13.

Faisal, G.G., Alahmad, B.E., Mustafa, N.S., Najmuldeen, G.F., Althunibat, O. and Azzubaidi, M.S. 2013. Histopathological effects of *Eurycoma longifolia* Jack extract (Tongkat Ali) on the prostate of rats. *Journal of Asian Scientific Research*. 3(8): 843–851.

George, A., Zandi, K., Biggins, J., Chinnappan, S., Hassandarvish: and Yusof, A. 2019. Antiviral activity of a standardized root water extract of *Eurycoma longifolia* (Physta®) against dengue virus. *Tropical Biomedicine*. 36: 412–421.

Han, Y.M., Woo, S.U., Choi, M.S., Park, Y.N., Kim, S.H., Yim, H. and Yoo, H.H. 2016. Antiinflammatory and analgesic effects of *Eurycoma longifolia* extracts. *Archives of Pharmacal Research*. 39(3): 421–428.

Kardono, L.B., Angerhofer, C.K., Tsauri, S., Padmawinata, K., Pezzuto, J.M. and Kinghorn, A.D. 1991. Cytotoxic and antimalarial constituents of the roots of *Eurycoma longifolia*. *Journal of Natural Products*. 54(5): 1360–1367.

Miyake, K., Tezuka, Y., Awale, S., Li, F. and Kadota, S. 2010. Canthin-6-one alkaloids and a tirucallanoid from *Eurycoma longifolia* and their cytotoxic activity against a human HT-1080 fibrosarcoma cell line. *Natural Product Communications*. 5(1): 17–22.

Pei, X.D., He, S.Q., Shen, L.Q., Wei, J.C., Li, X.S., Wei, Y.Y., Zhang, Y.M., Wang, X.Y., Lin, F., He, Z.L. and Jiang, L.H. 2020. 14, 15β-dihydroxyklaineanone inhibits HepG2 cell proliferation and migration through p38MAPK pathway. *Journal of Pharmacy and Pharmacology*. 72(9): 1165–1175.

Zhang, Y., Zhao, W., Ruan, J., Wichai, N., Li, Z., Han, L., Zhang, Y. and Wang, T. 2020. Anti-inflammatory canthin-6-one alkaloids from the roots of Thailand *Eurycoma longifolia* Jack. *Journal of Natural Medicines*. 74(4): 804–810.

4.1.12 Superorder Caryophyllanae Takht. (1967), the Malvids

4.1.12.1 Order Caryophyllales Juss. ex Bercht. & J. Presl (1820)

4.1.12.1.1 Family Amaranthaceae Juss. (1789), the Amaranth Family

The family Amaranthaceae consists of about 75 genera and 850 species of herbaceous plants. Leaves: simple, alternate, spiral, or opposite, exstipulate. Inflorescences: spikes, heads, racemes, panicles. Flowers: tiny. Tepals: 3–5, free or merged at base. Stamens: 3–5. Carpels: 2–3, forming a unilocular ovary with 1–5 ovules. Stigma: capitate, bifid, or trifid. Fruits: achenes, utricles, capsules, berries.

Alternanthera sessilis **(L.) R.Br. ex DC.**

[From Latin *alternus* = alternate, Greek *anthos* = flower, and Latin *sessilis* = sitting]

Published in *Catalogus plantarum horti botanici monspeliensis* 77. 1813

Synonyms: *Alternanthera nodiflora* R.Br.; *Alternanthera repens* Gmel.; *Gomphrena sessilis* L.; *Illecebrum sessile* (L.) L.

Local name: *lalambi* (Dusun)

Common name: sessile joyweed

Habitat: wastelands, roadsides, drains

Geographical distribution: tropical Asia

Botanical description: this prostrate herbaceous plant grows up to a length of about 50 cm. Stems: terete, hairy at nodes, thin, somewhat reddish. Leaves: simple, opposite, subsessile, exstipulate. Blades: 1–6 cm × 0.6–1.5 cm, acute to somewhat acuminate at apex, finely and irregularly crenate, wavy, membranous, spathulate, with 6–8 pairs of secondary nerves, dull green. Flower heads: axillary, globose, about 5 mm across, sessile, white. Flowers: tiny. Perianth: 5-lobed, each lobe about 2 mm long and ovate. Stamens: 3. Ovary: ovoid. Utricles: obcordate.

Traditional therapeutic indications: medicinal (Dusun)

Pharmacology and phytochemistry: vasorelaxant (Saqib & Janbaz 2016), anti-inflammatory (Rayees et al. 2013), antimicrobial (Sivakumar & Sunnathi. 2016), hepatoprotective (Bhuyan et al. 2018).

Ellagic acid (Mondal et al. 2015).

Toxicity, side effects, and drug interaction: not known

REFERENCES

Bhuyan, B., Baishya, K. and Rajak, P. 2018. Effects of *Alternanthera sessilis* on Liver function in carbon tetra chloride induced hepatotoxicity in wister rat model. *Indian Journal of Clinical Biochemistry*: 1–6.

Mondal, H., Hossain, H., Awang, K., Saha, S., Mamun-Ur-Rashid, S., Islam, M.K., Rahman, M., Jahan, I.A., Rahman, M.M. and Shilpi, J.A. 2015. Anthelmintic activity of ellagic acid, a major constituent of *Alternanthera sessilis* against haemonchus contortus. *Pakistan Veterinary Journal*. 35(1).

Rayees, S., Kumar, A., Rasool, S., Kaiser, P., Satti, N.K., Sangwan, P.L., Singh, S., Johri, R.K. and Singh, G. 2013. Ethanolic extract of *Alternathera sessilis* (AS-1) inhibits IgE-mediated allergic response in RBL-2H3 cells. *Immunological Investigations*. 42(6): 470–480.

Saqib, F. and Janbaz, K.H. 2016. Rationalizing ethnopharmacological uses of *Alternanthera* sessilis: A folk medicinal plant of Pakistan to manage diarrhea, asthma and hypertension. *Journal of Ethnopharmacology*. 182: 110–121.

Sivakumar, R. and Sunnathi, D. 2016. Phytochemical screening and antimicrobial activity of ethanolic leaf extract of *Alternanthera sessilis* (L.) R.Br. ex Dc and *Althernantheras philoxeroides* (Mart.) Griseb. *European Journal of Pharmaceutical Sciences*. 3(3): 409–412.

***Amaranthus spinosus* L.**

[From Greek *amarantos* = unfading and Latin *spinosus* = spiny]

Published in *Species Plantarum* 2: 991. 1753

Synonym: *Galliaria spinosa* (L.) Nieuwl.

Local names: *bayam hutan, sansam kuda* (Bajau); *buam berduri* (Kedayan); *bayam hutan, sansam sau* (Murut); *samsam lodut, samsai namatai* (Dusun, Kadazan)

Common name: spiny amaranth

Habitat: wastelands, dry and sandy village roadsides, gardens

Geographical distribution: tropical

Botanical description: this herbaceous plant reaches a height of about 50 cm. Stems: fleshy, sometimes purplish, thorny at nodes, somewhat terete. Leaves: simple, spiral, exstipulate. Petioles: 1–8 cm long, sometimes purplish, thorny at base. Blades: ovate, rhombic, spathulate or lanceolate, wavy, 1–6 cm × 3–12 cm, dull green, with about 10 pairs of secondary nerves sunken above, somewhat fleshy, attenuate to tapering at base, obtuse at apex. Spikes: terminal or axillary, 8–25 cm long. Tepals: 5, membranous, somewhat glossy and scalelike, about 2 mm long, light green. Stamens: 5. Stigma 3. Utricles: globose, tiny, circumscissile, included in the perianth. Seeds: numerous, black, globose.

Traditional therapeutic indications: dysentery, flatulence, swellings, thrush, urinary tract infections, diuretic (Dusun, Kadazan); epilepsy (Murut); diuretic (Dusun)

Pharmacology and phytochemistry: hepatoprotective (Zeashan et al. 2008), anti-inflammatory, anti-diarrhea (Olajide et al. 2004), antifungal (Yusnawan 2015), anxiolytic (Abid et al. 2017), diuretic (Amuthan et al. 2012), antiplasmodial (Hilou et al. 2006).

Long chain alkanes (14*E*, 18*E*, 22*E*, 26*E*)-methyl nonacosa-14,18,22,26 tetraenoate) toxic to liver cancer cells (HepG2) (Mondal et al. 2015; Mondal et al. 2016) and antibacterial (Mondal & Maity 2016).

Toxicity, side effects, and drug interaction: the plant growing on fertilized soils is poisonous (kidney failure) (Peixoto et al. 2003).

REFERENCES

Abid, M., Gosh, A.K. and Khan, N.A. 2017. *In vivo* psychopharmacological investigation of *Delphinium denudatum* and *Amaranthus spinosus* extracts on Wistar rats. *Basic and Clinical Neuroscience*. 8(6): 503.

Amuthan, A., Chogtu, B., Bairy, K.L. and Prakash, M. 2012. Evaluation of diuretic activity of *Amaranthus spinosus* Linn. aqueous extract in Wistar rats. *Journal of Ethnopharmacology*. 140(2): 424–427.

Hilou, A., Nacoulma, O.G. and Guiguemde, T.R. 2006. *In vivo* antimalarial activities of extracts from *Amaranthus spinosus* L. and *Boerhaavia erecta* L. in mice. *Journal of Ethnopharmacology*. 103(2): 236–240.

Mondal, A., Guria, T. and Maity, T.K. 2015. A new ester of fatty acid from a methanol extract of the whole plant of *Amaranthus spinosus* and its α-glucosidase inhibitory activity. *Pharmaceutical Biology*. 53(4): 600–604.

Mondal, A., Guria, T., Maity, T.K. and Bishayee, A. 2016. A novel tetraenoic fatty acid isolated from *Amaranthus spinosus* inhibits proliferation and induces apoptosis of human liver cancer cells. *International Journal of Molecular Sciences*. 17(10): 1604.

Mondal, A. and Maity, T.K. 2016. Antibacterial activity of a novel fatty acid (14E, 18E, 22E, 26E)-methyl nonacosa-14, 18, 22, 26 tetraenoate isolated from *Amaranthus spinosus*. *Pharmaceutical Biology*. 54(10): 2364–2367.

Olajide, O.A., Ogunleye, B.R. and Erinle, T.O. 2004. Anti-inflammatory properties of *Amaranthus spinosus* leaf extract. *Pharmaceutical Biology*. 42(7): 521–525.

Peixoto, P.V., Brust, L.A.C., Brito, M.D.F., França, T.D.N., Cunha, B.R.M.D. and Andrade, G.B.D. 2003. Intoxicação natural por *Amaranthus spinosus* (Amaranthaceae) em ovinos no Sudeste do Brasil. *Pesquisa Veterinária Brasileira*. 23: 179–184.

Yusnawan, E. 2015. Inhibition of spore germination of *Phakopsora pachyrhizi* using crude extracts of *Amaranthus spinosus*. *Procedia Food Science*. 3: 340–347.

Zeashan, H., Amresh, G., Singh, S. and Rao, C.V. 2008. Hepatoprotective activity of *Amaranthus spinosus* in experimental animals. *Food and Chemical Toxicology*. 46(11): 3417–3421.

***Cyathula prostrata* (L.) Bl.**

[From Greek *cyathos* = a cup and from Latin *sternere* = lay flat]

Published in *Bijdragen tot de flora van Nederlandsch Indië* (11): 549. 1825

Synonyms: *Achyranthes prostrata* L.; *Desmochaeta prostrata* (L.) DC.; *Pupalia prostrata* Mart.
Local name: *samsam bawi* (Murut)
Common name: cyathula
Habitat: roadsides, around villages
Geographical distribution: tropical
Botanical description: this prostrate, hairy herbaceous plant grows up to a length of about 50 cm. Stems: terete to angular, articulate, swollen at nodes. Leaves: simple, opposite, exstipulate. Petioles: 1–7 mm long. Blades: lanceolate, rhombic, to ovate, asymmetrical, 0.8–1.5 cm × 2.5–5 cm, wavy, dull green, with 5–6 pairs of secondary nerves. Spikes: linear, which can make one think of a snake, yellowish-green, up to about 25 cm long. Flowers: about 4 mm across. Tepals: 5, membranous. Stamens: 5. Ovary: obovoid. Stigma: capitulate. Utricules: globose, tiny.
Traditional therapeutic indications: insect bites (Murut); colic, headaches (Dusun, Kadazan)
Pharmacology and phytochemistry: toxic to cervical cancer cells (HeLa) (Sowemimo et al. 2009), anti-inflammatory, analgesic (Ibrahim et al. 2012), gastroprotective (Richard et al. 2017), hypotensive (Ojekale et al. 2016).
Anti-inflammatory and vasoprotective steroid (ecdysterone) (Shah & De Souza 1971; Zhang et al. 2014; Akopova et al. 2023).
Toxicity, side effects, and drug interaction: not known

Comments:

(i) The shape of the spikes may have encouraged the use of this plant for the treatment of insect bites (centipedes). It is used for that purpose in India (Kamble et al. 2015). Is ecdysterone of value as an insect bites antidote?
(ii) *Celosia cristata* L. is used by the Kedayan for abscesses, dysentery, coughs, cuts, hemorrhoids, and swellings (local name: *balung ayam*).
(iii) In the family Nyctaginaceae Juss. (1789), a plant of the genus *Bougainvillea* Comm. ex Juss. (1789) is used by the Bajau for abscesses, fever, body pains, flu, wounds, and headaches (local name: *bunga kertas*).
(iv) In the family Nepenthaceae Dumort. (1829), *Nepenthes ampullaria* Jack is used by the Lundayeh for respiratory diseases (local name: *telungau becuk*). A plant of the genus *Nepenthes* L. (1753) is used by the Dusun and Kadazan (local name: *kukuanga*) to treat syphilis, whilst "*periuk pera*" is employed for diarrhea and fever.
(v) In the family Basellaceae Raf. (1837), *Anredera cordifolia* (Ten) Steenis is used by the Murut for gastritis (local name: *bina*).

REFERENCES

Akopova, O.V., Korkach, Y.P., Nosar, V.I. and Sagach, V.F. 2023, Ecdysterone treatment restores constitutive NO synthesis and alleviates oxidative damage in heart tissue and mitochondria of streptozotocin-induced diabetic rats. *Фізіол. Журн*: 13–24.

Ibrahim, B., Sowemimo, A., van Rooyen, A. and Van de Venter, M. 2012. Antiinflammatory, analgesic and antioxidant activities of *Cyathula prostrata* (Linn.) Blume (Amaranthaceae). *Journal of Ethnopharmacology*. 141(1): 282–289.

Kamble, S.R., Deokar, R.R., Patil, S.R. and Mane, S.R. 2015. Herbal antidotes used for antivenum treatment from 32-Shirala Tahasil, Dist. Sangli (MS), India. *International Journal of Innovative Research in Science, Engineering and Technology*. 4.

Ojekale, A.B., Lawal, O.A. and Lasisi, M.O. 2016. *Cyathula prostrata*: A potential herbal hope for hypertensives, an animal model study and its secondary metabolites assessment via GC-MS. *European Journal of Medicinal Plants*. 14(2).

Richard, D.K., Jauro, A., Nvau, J.B. and Dabun, L.J. 2017. Ethanol-induced gastric ulceration in rats: Protective roles of methanol and water extracts of *Cyathula prostrata* Linn Blume. *Journal of Pharmacognosy and Phytochemistry*. 6(5): 1515–1517.

Shah, V.C. and De Souza, N.J. 1971. Amaranthaceae: Ecdysterone from *Cyathula prostrata*. *Phytochemistry*. 10(6): 1398–1399.

Sowemimo, A., van de Venter, M., Baatjies, L. and Koekemoer, T. 2009. Cytotoxic activity of selected Nigerian plants. *African Journal of Traditional, Complementary and Alternative Medicines*. 6(4): 526–528.

Zhang, X., Xu, X., Xu, T. and Qin, S. 2014. β-Ecdysterone suppresses interleukin-1β-induced apoptosis and inflammation in rat chondrocytes via inhibition of NF-κB signaling pathway. *Drug Development Research*. 75(3): 195–201.

4.1.12.1.2 Family Polygonaceae Juss. (1789), the Buckwheat Family

The family Polygonaceae consists of about 45 genera and 800 species of herbaceous plants and shrubs. Stems: often hollowed, ribbed, swollen at nodes, sappy, sour in flavor. Leaves: simple, alternate, stipulate. Stipules: form a membranous sheath around the stems (ochreas). Inflorescences: cymes of tiny flowers. Tepals: 2–6, free or merged at base. Stamens: 2–9. Carpels: 3, forming a unilocular ovary with 1 ovule. Styles: 3. Fruits: trigonous achenes.

***Polygonum odoratum* Lour.**

[From Greek *polu* = many, *gonia* = angles, and Latin *odoratum* = fragrant]

Published in *Flora Cochinchinensis* 1: 243. 1790

Synonym: *Persicaria odorata* (Lour.) Soják
Local name: *daun kesum* (Dusun)
Common names: laksa leaves, Vietnamese coriander
Habitat: cultivated
Geographical distribution: Southeast Asia
Botanical description: this herbaceous plant reaches a height of about 30 cm. Stems: angular, reddish. Leaves: simple, alternate, fragrant, stipulate. Ochreas: tubular, membranous. Petioles: up to 5 mm long. Blades: lanceolate, 5–12 cm × 3–6 cm, tapering at base, acuminate at apex, with about 10–18 pairs of secondary nerves. Spikes: axillary or terminal, filiform, up to about 15 cm long. Perianth: white to pinkish, tubular, 5-lobed, the lobes about 3 mm long. Stamens: 8. Styles: 3, about 2 mm long. Achenes: tiny, glossy, orbicular, in persistent perianth.
Traditional therapeutic indications: medicinal (Dusun)
Pharmacology and phytochemistry: antibacterial, cytotoxic (Chansiw et al. 2018). Anti-inflammatory flavonol glycosides (scutellarein-7-glucoside, quercitrin) (Okonogi et al. 2016).
Toxicity, side effects, and drug interaction: not known

Comment: The Dusun use this plant to flavor their dishes.

REFERENCES

Chansiw, N., Paradee, N., Chotinantakul, K. and Srichairattanakool, S. 2018. Anti-hemolytic, antibacterial and anti-cancer activities of methanolic extracts from leaves and stems of *Polygonum odoratum*. *Asian Pacific Journal of Tropical Biomedicine*. 8(12): 580.

Okonogi, S., Kheawfu, K., Holzer, W., Unger, F.M., Viernstein, H. and Mueller, M. 2016. Anti-inflammatory effects of compounds from *Polygonum odoratum*. *Natural Product Communications*. 11(11): 1651–1654.

***Polygonum orientale* L.**

[From Greek *polu* = many, *gonia* = angles, and Latin *orientale* = from the East]

Published in *Species Plantarum* 1: 362. 1753

Synonyms: *Amblygonum orientale* (L.) Nakai ex T. Mori; *Lagunea cochinchinensis* Lour.; *Lagunea orientalis* (L.) Nakai; *Persicaria cochinchinensis* (Lour.) Kitag.; *Persicaria orientalis* E. Vilm.; *Persicaria pilosa* (Roxb. ex Meisn.) Kitag.; *Polygonum amoenum* Blume; *Polygonum cochinchinense* (Lour.) Meisn.; *Polygonum cordobense* Lindau; *Polygonum pilosum* Roxb. ex Meisn.; *Polygonum torquatum* Bruijn

Local names: *kasum, waying* (Dusun); *daun wangi* (Lundayeh); *kesum* (Bajau)

Common name: Oriental pepper

Habitat: roadsides, wastelands

Geographical distribution: tropical Asia, Pacific Islands

Botanical description: this pretty herbaceous plant reaches a height of about 1 m. Stems: terete, hairy. Leaves: simple, alternate, stipulate. Ochreas: tubular, membranous, truncate, ciliate, with expanded margin, up to about 2 cm long. Petioles: up to 10 cm long. Blades: ovate to elliptic, 10–20 cm × 5–12 cm, hairy, rounded at base, acuminate at apex. Spikes: terminal or cylindrical, up to about 7 cm long. Perianth: pink, tubular, 5-lobed, the lobes elliptic and about 4 mm long. Stamens: 7. Styles: 2. Achenes: tiny, glossy, orbicular in persistent perianth.

Traditional therapeutic indications: dandruff (Dusun)

Pharmacology and phytochemistry: antibacterial (Cai et al. 2019), cytotoxic (Islam et al. 2016), anti-inflammatory (Gou et al. 2017), antidiabetic (Nigam 2013), hepatoprotective effects (Chiu et al. 2018).

Lignans (arctiin, lappaol B, orientalin) (Zheng et al. 1998), flavonols (exoticin), flavans (epicatechin-3-*O*-gallate), benzoic acids (gallic acid, 3,4-dihydroxy benzoic acid) (Masum et al. 2019).

Toxicity, side effects, and drug interaction: the median lethal dose (LD_{50}) of ethanol extract of fruits administered to mice was 10 g/kg (Chiu et al. 2018).

Comments:

(i) *Polygonum minus* Huds is used by the Bajau for colds, indigestion, and dysmenorrhea (local name: *kesum*). The plant is sold fresh in the wet markets of Sabah as a medicinal food, or "*ulam*."

(ii) In the Superorder Santalanae Thorne ex Reveal (1992), family Loranthaceae Juss. (1808), a plant of the genus *Scurrula* L. (1753) is used by the Dusun and Kadazan for toothaches (local name: *tongom la'an*). A plant in this family is used for fatigue by the Dusun and Kadazan (local name: *surni sohod*).

(iii) In the Superorder Santalanae Thorne ex Reveal (1992), family Olacaceae Juss. (1789) ex R. Brown (1818), the seeds and leaves of *Scorodocarpus borneensis* (Baill.) Becc. are used as spice (local name: *sambawang*). The fruits are sold in the wet markets of Sabah.

REFERENCES

Cai, J., Gao, Y., Wang, M., Zhang, J., Zhang, Y., Wang, Q. and Liu, J. 2019. Antibacterial activity of polygonum orientale extracts against *Clavibacter michiganensis* subsp. *michiganensis*, the agent of bacterial canker of tomato disease. *Brazilian Archives of Biology and Technology*. 62: 1–18.

Chiu, Y.J., Chou, S.C., Chiu, C.S., Kao, C.P., Wu, K.C., Chen, C.J., Tsai, J.C. and Peng, W.H. 2018. Hepatoprotective effect of the ethanol extract of *Polygonum orientale* on carbon tetrachloride-induced acute liver injury in mice. *Journal of Food and Drug Analysis*. 26(1): 369–379.

Gou, K.J., Zeng, R., Dong, Y., Hu, Q.Q., Hu, H.W.Y., Maffucci, K.G., Dou, Q.L., Yang, Q.B., Qin, X.H. and Qu, Y. 2017. Anti-inflammatory and analgesic effects of *Polygonum orientale* L. extracts. *Frontiers in Pharmacology*. 8: 562.

Islam, M.T., Priyanka, A.K., Sultana, T., Kawsar, M.H., Sumon, M.H.U. and Sohel, M.D. 2016. *In vitro* antimicrobial, antioxidant and cytotoxic activities of *Polygonum orientale* (Bishkatali). *Journal of Pharmacy and Nutrition Sciences*. 6(3): 112–119.

Masum, M.N., Choodej, S., Yamauchi, K. and Mitsunaga, T. 2019. Isolation of phenylpropanoid sucrose esters from the roots of *Persicaria orientalis* and their potential as inhibitors of melanogenesis. *Medicinal Chemistry Research*. 28(5): 623–632.

Nigam, V. 2013. Antihyperglycaemic activity on flower of *Polygonum orientale* Linn. using steptozotocin induced diabetic mice model. *International Journal of Pharmacy Teaching & Practices*. 4(3): 1–7.

Zheng, S., Wang, D., Meng, J. and Shen, X. 1998. Studies on the lignans of *Polygonum orientale*. *Acta Botanica Sinica*. 40(5): 466–469.

4.1.13 Superorder Asteranae Takht. (1967), the Asterids

4.1.13.1 Order Ericales Bercht. & J. Presl (1820)

4.1.13.1.1 Family Actinidiaceae Engl. & Gilg. (1824), the Kiwi Fruit Family

The family Actinidiaceae consists of about 3 genera and 350 species of trees, shrubs, or climbing plants. Leaves: simple, alternate, exstipulate. Inflorescences: fascicles, cymes, panicles. Sepals: 5. Petals: 5, free or merged at base. Stamens: often 18–many. Carpels: 5–20, forming a 5–30 locular ovary, each locule having 15–50 ovaries. Styles: 3–5. Fruits: berries, capsules.

***Saurauia fragrans* Hoogland**

[After the 19th century Austrian statesman Franz Joseph von Saurau and Latin *fragrans* = fragrant]

Published in *Gard. Bull. Singapore* 30: 117. 1977

Local name: *longugan taragang* (Dusun)
Habitat: forests
Geographical distribution: Malaysia, Borneo
Botanical description: this tree reaches a height of about 12 m. Leaves: simple, alternate, exstipulate. Petioles: 2–4 cm long. Blades: oblong to obovate, coriaceous, dark green, glossy, 6–11 cm × 18–30 cm, with 14–17 pairs of secondary nerves, acute to acuminate at apex, acute to obtuse at apex. Flowers: fragrant, axillary, solitary or in fascicles, all along the stems. Sepals: 5, broadly lanceolate, about 5 mm across. Corolla: tubular, 5-lobed about 8 mm long, whitish, revolute. Stamens: about 20–30, about 5 mm long. Ovary: somewhat globose. Styles: 3, about 4 mm long. Capsules: about 1 cm across.
Traditional therapeutic indications: ulcers (Dusun)
Pharmacology and phytochemistry: not known
Toxicity, side effects, and drug interaction: not known

Comments:

(i) Plants in the family Actinidiaceae Engl. & Gilg. (1824) are medicinal in Sabah. The Dusun and Kadazan use *"longugan"* for scalds, whilst *"longugan taragan"* and *"lonlugan totomu"* are employed for swellings. The Dusun use *"kebong"* for beriberi.
(ii) *Saurauia longistyla* Merr. is medicinal for the Murut (local name: *usod usod).*
(iii) A plant of the genus *Actinidia* Lindl. (1836) (local name: *longugan*) is medicinal for the Dusun and Kadazan, whilst the Dusun employ *"kebong"* to make a bath to treat beriberi.
(iv) Dusun and Kadazan often refer to different plant species with the same name.

4.1.13.1.2 Family Balsaminaceae Bercht. & J. Presl (1820), the Touch-me-not Family

The family Balsaminaceae consists of about 4 genera and 600 species of herbaceous plants. Leaves: simple, opposite, verticillate, spiral, exstipulate. Blades: often serrate. Inflorescences: racemes, fascicles. Calyx: 5 sepals or tubular. Petals: 5, membranous. Stamens: 5. Carpels: 5, forming a 5-locular ovary, each locule with 2–50 ovules. Styles: 1. Stigma: 1-6-lobed Fruits: berries, explosively dehiscent capsules. Seeds: numerous, tiny.

Impatiens balsamina **L.**

[From Latin *impatientem* = intolerant and the Hebrew word *basam* = balsam]

Published in *Species Plantarum* 2: 938. 1753

Synonyms: *Balsamina hortensis* Desp.; *Impatiens eriocarpa* Launert; *Impatiens stapfiana* Gilg

Common name: garden balsam

Habitat: gardens, roadsides

Geographical distribution: subtropical and tropical Asia, Pacific Islands

Botanical description: this beautiful herbaceous plant reaches a height of about 60 cm. Stems: fleshy, light green, longitudinally striated, sappy. Leaves: simple, alternate or opposite, exstipulate. Petioles: 1–3 cm long, channeled. Blades: elliptic to lanceolate, 4–12 cm × 1.5–3 cm, with 4–7 pairs of secondary nerves, base attenuate, serrate, fleshy, apex acuminate. Flower fascicles: axillary with few showy flowers on 2–2.5 cm long and pubescent peduncles. Lateral sepals: 2, ovate, tiny. Lower sepal: up to 2 cm long, narrowed into an incurved spur with a length of 1–2.5 cm. Upper petal: orbicular. Lateral petals: clawed, bilobed, up to 2.5 cm long. Stamens: 5, linear. Ovary: tiny, fusiform, pubescent. Capsules: fusiform, 1–2 cm long, hairy. Seeds: numerous.

Traditional therapeutic indications: snakebites, sprains, itchiness (Lundayeh)

Pharmacology and phytochemistry: analgesic (Imam et al. 2012), nematicidal (Jiang et al. 2017), antibacterial (Voravuthikunchai & Kitoipit 2005; Niyomkam et al. 2010; Szewczyk et al. 2016).

Antibacterial naphthoquinones (lawsone methyl ether, 5-hydroxy-1,4-naphthoquinone, methylene-3,3′-bilawsone) (Macé et al. 2017; Yang et al. 2001; Sakunphueak & Panichayupakaranant 2012; Panichayupakaranant et al. 2019). Neuroprotective biflavonoid glycosides (Kim et al. 2017). Coumarin (scopoletin) (Panjchayupakaranant et al. 1995). Cytotoxic naphthoquinones (plumbagin, shikonin) (Fujii et al. 1992). Antiplasmodial and nematicidal naphthoquinone (lawsone) (Dama 2002).

Toxicity, side effects, and drug interaction: the median lethal dose (LD_{50}) of ethylacetate extract from stems and leaves administered to rats was above 2 g/kg (Anzini et al. 2014).

Comments:

(i) The Chinese people of Sabah use this plant to treat bee stings and grow it in their gardens to ward off snakes.

(ii) In the Family Lecythidaceae A. Rich. (1825), a plant of the genus *Barringtonia* J.R. Forst. & G. Forst. (1776) is used by the Dusun and Kadazan as poison antidote and for stomach aches (local name: *tampalang*).

(iii) *Barringtonia lanceolata* (Ridl.) Payens, *Barringtonia reticulata* (Blume) Miq., and *Barringtonia sarcostachys* (Merr.) Payens are employed by the Dusun as insecticides and to stupefy fish (local name: *mpalang*).

REFERENCES

Anzini, N.I.A., Kusharyanti, I. and Nurbaeti, S.N. 2014. Uji Toksisitas Akut Fraksi Etil asetat batang dan daun pacar air (*Impatiens balsamina* Linn) terhadap tikus putih betina galur Sprague Dawley. *Journal of Tropical Pharmacy and Chemistry*. 2(4): 235–247.

Dama, L.B. 2002. Effect of naturally occurring napthoquinones on root-knot nematode *Meioidogyne javanica*. *Indian Phytopathology*. 55(1): 67–69.

Fujii, N., Yamashita, Y., Arima, Y., Nagashima, M. and Nakano, H. 1992. Induction of topoisomerase II-mediated DNA cleavage by the plant naphthoquinones plumbagin and shikonin. *Antimicrobial Agents and Chemotherapy*. 36(12): 2589–2594.

Imam, M.Z., Nahar, N., Akter, S. and Rana, M.S. 2012. Antinociceptive activity of methanol extract of flowers of *Impatiens balsamina*. *Journal of Ethnopharmacology*. 142(3): 804–810.

Jiang, H.F., Zhuang, Z.H., Hou, B.W., Shi, B.J., Shu, C.J., Chen, L., Shi, G.X. and Zhang, W.M. 2017. Adverse effects of hydroalcoholic extracts and the major components in stems of *Impatiens balsamina* L. on *Caenorhabditis elegans*. *Evidence-Based Complementary and Alternative Medicine*. 2017: 1–10.

Kim, C.S., Bae, M., Oh, J., Subedi, L., Suh, W.S., Choi, S.Z., Son, M.W., Kim, S.Y., Choi, S.U., Oh, D.C. and Lee, K.R. 2017. Anti-neurodegenerative biflavonoid glycosides from *Impatiens balsamina*. *Journal of Natural Products*. 80(2): 471–478.

Macé, S., Truelstrup Hansen, L. and Rupasinghe, H.P. 2017. Anti-bacterial activity of phenolic compounds against *Streptococcus pyogenes*. *Medicines*. 4(2): 25.

Niyomkam, P., Kaewbumrung, S., Kaewnpparat, S. and Panichayupakaranant, P. 2010. Antibacterial activity of Thai herbal extracts on acne involved microorganism. *Pharmaceutical Biology*. 48(4): 375–380.

Panjchayupakaranant, P., Noguchi, H., De-Eknamkul, W. and Sankawa, U. 1995. Naphthoquinones and coumarins from *Impatiens balsamina* root cultures. *Phytochemistry*. 40(4): 1141–1143.

Panichayupakaranant, P., Septama, A.W. and Sinviratoong, A. 2019. Synergistic activity of lawsone methyl ether in combination with some antibiotics and artocarpin against methicillin-resistant Staphylococcus aureus, Candida albicans, and Trychophyton rubrum. *Chinese Herbal Medicines*, 11(3): 321–325.

Sakunphueak, A. and Panichayupakaranant, P. 2012. Comparison of antimicrobial activities of naphthoquinones from *Impatiens balsamina*. *Natural Product Research*. 26(12): 1119–1124.

Szewczyk, K., Zidorn, C., Biernasiuk, A., Komsta, Ł. and Granica, S. 2016. Polyphenols from *Impatiens* (Balsaminaceae) and their antioxidant and antimicrobial activities. *Industrial Crops and Products*. 86: 262–272.

Voravuthikunchai, S.P. and Kitoipit, L. 2005. Activity of medicinal plant extracts against hospital isolates of methicillin-resistant *Staphylococcus aureus*. *Clinical Microbiology and Infection*. 11(6): 510–512.

Yang, X., Summerhurst, D.K., Koval, S.F., Ficker, C., Smith, M.L. and Bernards, M.A. 2001. Isolation of an antimicrobial compound from *Impatiens balsamina* L. using bioassay-guided fractionation. *Phytotherapy Research*. 15(8): 676–680.

4.1.13.1.3 Family Ebenaceae Gürke (1891), the Ebony Family

The family Ebenaceae consists of about 2 genera and 500 species of handsome trees or shrubs. Wood: often dark, very durable, dense, hard. Inner bark: often becomes bluish when exposed to light. Leaves: simple, alternate, opposite, exstipulate, turning black in herbarium samples. Inflorescences: cymes or solitary. Calyx: tubular, 3–7-lobed, merged at base, the lobes leafy. Corolla: tubular, 3-5-lobed, somewhat fleshy, the lobes often contorted. Stamens: 3–many. Carpels: 2–5, forming a 4–30 locular ovary, each locule with 2 ovules. Styles: 2–5. Fruit: berries on persistent calyx.

***Diospyros elliptifolia* Merr.**

[From Greek *dios* = divine, *pyros* = grain, and Latin *elliptifolia* = with elliptical leaves]

Published in *Philippine Journal of Science* 30: 421. 1926

Local name: *radtak* (Rungus)
Habitat: forests
Geographical distribution: Indonesia, Borneo, the Philippines
Botanical description: this tree reaches a height of about 20 m. Trunk: straight. Leaves: simple, spiral, subsessile, exstipulate. Petioles: 2–5 mm long. Blades: elliptic, theaceous, 4–6 cm × 6–9 cm, attenuate at base, smooth, the midrib sunken above, with about 4–7 pairs of secondary nerves, wavy, coriaceous, acuminate at apex, turning black in herbarium samples. Cymes: 3-flowered, about 5 mm long, axillary. Calyx: tubular, about 1 cm long, 4-5-lobed, the lobes leafy. Corolla: tubular, whitish, about 1 cm long, 4-lobed. Stamens: numerous. Ovary: globose. Berries: globose, green, smooth, up to 3 cm across, on persistent calyx.
Traditional therapeutic indications: medicinal (Rungus)
Pharmacology and phytochemistry: triterpenes (betulin, lupeol). Naphthoquinones (plumbagin, elliptinone) (Fallas & Thomson 1968).
Toxicity, side effects, and drug interaction: not known

Comments:

(i) *Diospyros andamanica* Bakh. is used by the Rungus to poison chickens (local name: *bokis manuk*).
(ii) *Diospyros foxworthyi* Bakh is used by the Bajau as an antidote for poisons and for stomach aches (local name: *sungkang seribu*). Parts of this plant are sold in the wet markets of Sabah as medicine.
(iii) *Diospyros wallichii* King & Gamble is used by the Lundayeh for cachexia and jaundice (local name: *lapad perurut*).

REFERENCE

Fallas, A.L. and Thomson, R.H. 1968. Ebenaceae extractives. Part III. Binaphthaquinones from *Diospyros* species. *Journal of the Chemical Society C: Organic*. 2279–2282.

4.1.13.1.4 Family Primulaceae Batsch ex Borkh (1797), the Primrose Family

The family Primulaceae consists of about 20 genera and 1,000 species of herbaceous plants. Leaves: simple, spiral or in basal rosettes, exstipulate. Inflorescences: panicles, racemes, spikes, fascicles, heads. Calyx: tubular, 5-lobed. Corolla: tubular, 5-lobed. Stamens: 5. Carpels: 5, forming a unilocular ovary with 7–many ovules. Styles: 1. Stigma: capitate. Fruits: capsules or berries.

Embelia philippinensis **A. DC**

[From Latin *philippinense* = from the Philippines]

Published in *Prodromus Systematis Naturalis Regni Vegetabilis* 8: 83. 1844

Synonyms: *Ribesiodes philippinse* (A. DC.) Kuntze; *Samara philippensis* S. Vidal
Local name: *papaling* (Murut)
Habitat: forests
Geographical distribution: Borneo, the Philippines
Botanical description: this climbing plant grows up to a length of about 4 m. Stems: spiny when old. Leaves: simple, alternate, exstipulate. Petioles: stout, about 1 cm long. Blades: elliptic to obovate, glossy, acute at base and apex (somewhat *Ficus*-like), 3–5 cm × 7–14 cm. Panicles: axillary, thin, showy, up to about 30 cm long, with numerous tiny flowers. Calyx: tubular, 5-lobed, the lobes hairy, ciliate, ovate. Petals: 5, ciliate. Stamens: 5. Stigma: capitate. Berries: tiny, globose, purplish (grape-colored), apiculate (Figure 4.41).

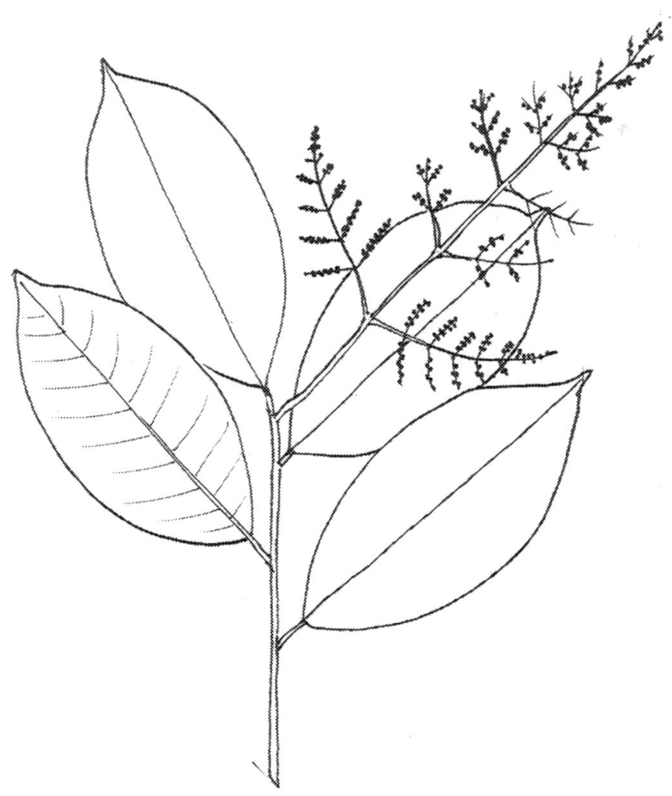

FIGURE 4.41 *Embelia philippinensis* A. DC

Traditional therapeutic indications: medicinal salads (Murut)
Pharmacology and phytochemistry: not known
Toxicity, side effects, and drug interaction: not known

Comments:

(i) Plants of the family Primulaceae Batsch ex Borkh (1797) are medicinal for the Dusun and Kadazan: *"tonsom onsom"* for rashes, *"bangkou bangkou"* for pancreatitis and diarrhea, *"suput"* for cuts and wounds, and *"tolonsi"* for feverish colds. The Dusun use *"rangup"* for fatigue and for magic rituals, *"sumping sumping"* for headaches, and *"sowolikan"* for stomach aches.

(ii) *Embelia dasytithyrsa* Miq. is used for fever by the Dusun (local names: *tarikan*, *sowolikan*). They also use this plant as food.

(iii) In the the genus *Ardisia* Sw. (1788), the Dusun and Kadazan use *"tolonsi"* for feverish cold, whilst the Dusun use *"rangup"* for fatigue and for magic rituals.

(iv) The Dusun and Kadazan use a plant of the genus *Maesa* Forssk. (1775) for itchiness (local name: *tonsom onsom*) whilst the Dusun employ *"sumping sumping"* for headaches.

(v) In the Philippines the natives used to employ *Embelia philippinensis* A. DC for mummifications (local name: *sipudak*).

(vi) *Labisia pumila* (Bl.) Fern.-Vill. is used by the Bajau for postpartum (local name: *kacip Fatimah*).

(vii) In the family Symplocaceae Desf. (1820), *Symplocos odoratissima* Choisy ex Zoll. is used by the Lundayeh for fever and malaria (local name: *lobo*).

4.1.14 Superorder Asteranae Takht. (1967), the Lamiids

4.1.14.1 Order Gentianales Juss. ex Bercht. & J. Presl (1820)

4.1.14.1.1 Family Apocynaceae Juss. (1789), the Dogbane Family

The family Apocynaceae consists of about 200 genera and 2,000 species of trees, shrubs, climbing plants, or herbaceous plants. Stems: often exude a milky and poisonous latex once incised. Leaves: simple, opposite, or whorled, exstipulate. Inflorescences: cymes, racemes. Calyx: tubular, 5-lobed. Corolla: tubular, 5-lobed, the lobes contorted. Stamens: 5. Carpels: 2, forming a unilocular or bilocular ovary, each locule with 2–many. Fruits: pairs of more or less elongated follicles drupes, berries. Seeds: often winged or comose.

***Alstonia angustifolia* Wall.**

[After the 18th century Scottish physician Charles Alston and Latin *angustifolia* = narrow leaf]

Published in *A Numerical List of Dried Specimens* n. 1650. 1828

Local names: *tembiros, pulai penipu paya* (Bajau); *tombirog, tambailik* (Murut)
Habitat: swampy forests
Geographical distribution: Malaysia, Indonesia
Botanical description: this tree reaches a height of about 25 m. Bark: brown, fissured. Stems: exude a white latex once incised. Leaves: simple, exstipulate, in whorls of 3. Petioles: 1.5 cm long. Blades: elliptic to lanceolate, 2.5–6 cm × 11.5–16 cm, with 15–20 pairs of secondary nerves. Calyx: tubular, 5-lobed, tiny, hairy. Corolla: tubular, hairy, caducous, 5-lobed, the lobes broadly rounded, 2.5 mm long and contorted. Stamens: 5, adnate to the corolla tube. Styles: 1, thin, tiny. Follicles: in pairs, linear, 25–70 cm × 3 mm. Seeds: numerous, comose.
Traditional therapeutic indications: putrefied wounds, gastritis (Dusun, Kadazan); gastritis (Dusun, Kadazan, Murut)
Pharmacology and phytochemistry: antiplasmodial (Said et al. 1992), antibacterial (Ab Rahim et al. 2019). Indole alkaloids (Kam & Choo 2004; Pan et al. 2014). Antiplasmodial indole alkaloids (macralstonine acetate, villastonine) (Wright et al. 1992; Keawpradub et al. 1999), anti-inflammatory indole alkaloids (*N*(4)-methyltalpinine) (Pan et al. 2014).
Toxicity, side effects, and drug interaction: not known

REFERENCES

Ab Rahim, N., Zakaria, N., Dzulkarnain, S.M., Othman, N. and Abdulla, M.A. 2019. Antibacterial activity of *Alstonia angustifolia* leaf extract against Staphyloccoccal and Bacilli. *Biomedical Research.* 30(1): 11–15.

Kam, T.S. and Choo, Y.M. 2004. Alkaloids from *Alstonia angustifolia. Phytochemistry.* 65(5): 603–608.

Keawpradub, N., Kirby, G.C., Steele, J.C.P. and Houghton, P.J. 1999. Antiplasmodial activity of extracts and alkaloids of three *Alstonia* species from Thailand. *Planta Medica.* 65(8): 690–694.

Pan, L., Terrazas, C., Acuña, U.M., Ninh, T.N., Chai, H., de Blanco, E.J.C., Soejarto, D.D., Satoskar, A.R. and Kinghorn, A.D. 2014. Bioactive indole alkaloids isolated from *Alstonia angustifolia. Phytochemistry Letters.* 10: liv–lix.

Said, I.M., Din, L.B., Yusoff, N.I., Wright, C.W., Cai, Y. and Phillipson, J.D. 1992. A new alkaloid from the roots of *Alstonia angustifolia. Journal of Natural Products.* 55(9): 1323–1324.

Wright, C.W., Allen, D., Cai, Y., Phillipson, J.D., Said, I.M., Kirby, G.C. and Warhurst, D.C. 1992. *In vitro* antiamoebic and antiplasmodial activities of alkaloids isolated from *Alstonia angustifolia* roots. *Phytotherapy Research.* 6(3): 121–124.

***Alstonia angustiloba* Miq.**

[After the 18th century Scottish physician Charles Alston and Latin *angustiloba* = narrow lobe]

Published in *Flora van Nederlandsch Indië* 2: 438.1857

Synonym: *Paladelpha angustiloba* (Miq.) Pichon
Local names: *tambailik, tombirog* (Murut); *tombirong* (Kadazan, Dusun); *pulai lilin* (Bajau)
Habitat: forests
Geographical distribution: Thailand, Malaysia, Indonesia
Botanical description: this magnificent tree reaches a height of about 40 m. Trunk: straight, buttressed. Bark: rough, grayish to brownish, fissured, secretes a milky latex. Stems: smooth, terete. Leaves: simple, exstipulate, pendulous, in terminal whorls of 4–7. Petioles: 1–2 cm long, sometimes channeled above. Blades: coriaceous, narrowly elliptic to obovate, 2–7 cm × 4–20 cm, obtuse at base, obtuse or shortly acuminate at apex, finely marked with numerous pairs of secondary nerves. Cymes: on 2–5 cm long peduncles, with many whitish and caducous flowers. Calyx: tiny, tubular, 5-lobed, the lobes ovate and somewhat hairy. Corolla: tubular, about 8 mm long, 5-lobed, the lobes about 6 mm long, twisted, and contorted. Stamens: 5, adnate to the corolla tube. Styles: 1, thin, tiny. Follicles: in pairs, linear, 30–60 cm × 3 mm, cylindrical, light green, somewhat twisted, pendulous. Seeds: numerous, comose, about 7 mm long.
Traditional therapeutic indications: gastritis (Dusun, Kadazan, Murut); putrefied wounds, sprains, malaria (Dusun, Kadazan)
Pharmacology and phytochemistry: cytotoxic indole alkaloids (Koyama et al. 2008; Ku 2011), smooth muscle relaxant indole alkaloids (alstilobanines A—E) (Koyama et al. 2008).
Toxicity, side effects, and drug interaction: not known

REFERENCES

Koyama, K., Hirasawa, Y., Zaima, K., Hoe, T.C., Chan, K.L. and Morita, H. 2008. Alstilobanines A—E, new indole alkaloids from *Alstonia angustiloba. Bioorganic & Medicinal Chemistry.* 16(13): 6483–6488.

Ku, W.F. 2011. *Biologically Active Alkaloids from Alstonia Angustiloba/Ku Wai Foong* (Doctoral dissertation, University of Malaya).

Alstonia macrophylla **Wall. ex G.Don**

[After the 18th century Scottish physician Charles Alston and Latin *macrophylla* = large leaf]

Published in *A General History of the Dichlamydeous Plants* 4(1): 87. 1838

Synonyms: *Alstonia acuminata* Miq.; *Alstonia oblongifolia* Merr.; *Alstonia pangkorensis* King & Gamble; *Alstonia subsessilis* Miq.,
Local name: *mangalang* (Dusun)
Common names: batino, deviltree
Habitat: forests, cultivated
Geographical distribution: Sri Lanka, Southeast Asia, South China, Papua New Guinea
Botanical description: this magnificent tree reaches a height of about 30 m. Trunk: straight, with small buttresses. Bark: blackish brown, smooth, with small square adherent scales. Leaves: simple, exstipulate, in whorls of 3–4. Petioles: 2–3 cm long. Blades: elliptic to oblanceolate, smooth, glossy, hairy beneath, 6–15 cm × 9–25 cm, with 20–24 pairs of secondary nerves, tapering and acuminate at apex. Calyx: cup-shaped, hairy, about 2 mm long, 5-lobed, the lobes triangular. Corolla: tubular, pure white, the tube about 5 mm long, 5-lobed, the lobes broadly rounded, about 3–7 mm long, hairy at the margin, and contorted. Stamens: 5, adnate to the corolla tube. Follicles: in pairs, 20–40 cm × 1.5–2 mm. Seeds: numerous, comose, narrowly elliptic, about 2 mm long.
Traditional therapeutic indications: epilepsy (Dusun)
Pharmacology and phytochemistry: anti-inflammatory (Arunachalam et al. 2002, 2005), neuroleptic (Chattopadhyay et al. 2004).
Indole alkaloids (Wong et al. 1996 ; Hirasawa et al. 2009 ;Changwichit et al. 2011 ;Lim 2013). Vasorelaxant indole alkaloids (Arai et al. 2012). Antiplasmodial and cytotoxic indole alkaloid (macrocarpamine) (Keawpradub et al. 1999; Hirasawa et al. 2009).
Toxicity, side effects, and drug interaction: not known

REFERENCES

Arai, H., Zaima, K., Mitsuta, E., Tamamoto, H., Saito, A., Hirasawa, Y., Rahman, A., Kusumawati, I., Zaini, N.C. and Morita, H. 2012. Alstiphyllanines I—O, ajmaline type alkaloids from *Alstonia macrophylla* showing vasorelaxant activity. *Bioorganic & Medicinal Chemistry*. 20(11): 3454–3459.

Arunachalam, L.G., Chattopadhyay, D., Chatterjee, S., Mandal, A.B., Sur, T.K. and Mandal, S.C. 2002. Evaluation of anti-inflammatory activity of *Alstonia macrophylla* Wall ex A. DC. leaf extract. *Phytomedicine*. 9(7): 632–635.

Arunachalam, L.G., Rajendran, K., Mandal, A.B. and Bhattacharya, S.K. 2005. Antipyretic activity of *Alstonia macrophylla* Wall ex A. DC: An ethnomedicine of Andaman Islands. *Journal of Pharmaceutical Sciences*. 8: 558–564.

Changwichit, K., Khorana, N., Suwanborirux, K., Waranuch, N., Limpeanchob, N., Wisuitiprot, W., Suphrom, N. and Ingkaninan, K. 2011. Bisindole alkaloids and secoiridoids from *Alstonia macrophylla* Wall. ex G.Don. *Fitoterapia*. 82(6): 798–804.

Chattopadhyay, D., Arunachalam, G., Ghosh, L. and Mandal, A.B. 2004. CNS activity of *Alstonia macrophylla* leaf extracts: An ethnomedicine of Onge of Bay Islands. *Fitoterapia*. 75(7–8): 673–682.

Hirasawa, Y., Arai, H., Zaima, K., Oktarina, R., Rahman, A., Ekasari, W., Widyawaruyanti, A., Indrayanto, G., Zaini, N.C. and Morita, H. 2009. Alstiphyllanines A–D, Indole Alkaloids from *Alstonia macrophylla*. *Journal of Natural Products*. 72(2): 304–307.

Keawpradub, N., Eno-Amooquaye, E., Burke, P.J. and Houghton, P.J. 1999. Cytotoxic activity of indole alkaloids from *Alstonia macrophylla*. *Planta Medica*. 65(4): 311–315.

Keawpradub, N., Kirby, G.C., Steele, J.C.P. and Houghton, P.J. 1999. Antiplasmodial activity of extracts and alkaloids of three *Alstonia* species from Thailand. *Planta Medica*. 65(8): 690–694.

Lim, S.H. 2013. *Alkaloids from Alstonia macrophylla /Lim Siew Huah* (Doctoral dissertation, University of Malaya).

Wong, W.H., Lim, P.B. and Chuah, C.H. 1996. Oxindole alkaloids from *Alstonia macrophylla*. *Phytochemistry*. 41(1): 313–315.

***Alstonia scholaris* (L.) R.Br.**

[After the 18th century Scottish physician Charles Alston and Latin *scholaris* = a scholar]

Published in *On the Asclepiadeae* 75. 1810

Synonyms: *Echites scholaris* L., *Nerium septaparna* Jones
Local names: *tomboilik* (Dusun, Kadazan); *tombirog* (Murut)
Common name: blackboard tree
Habitat: forests, cultivated
Geographical distribution: from India to Pacific Islands
Botanical description: this sinister tree reaches a height of about 18 m. Trunk: somewhat bulging. Bark: smooth, blackish, or dark brown, secretes a milky latex. Stems: with numerous lenticels. Leaves: simple, exstipulate, pendulous, in terminal whorls of 5–8. Petioles: up to 2 cm long, channeled above, stout. Blades: coriaceous, oblong-lanceolate or obovate, 7–17 cm × 3–6 cm, obtuse or shortly acuminate, dark green above and pale bluish beneath, and finely marked with numerous pairs of secondary nerves. Cymes: terminal, many-flowered. Calyx: up to 3 mm long, tubular, 5-lobed, red. Corolla: caducous, 8 mm long, somewhat yellowish brown, 5-lobed, the lobes about 5 mm long, twisted, and contorted. Stamens: 5, adnate to the corolla tube. Styles: 1, thin, tiny. Follicles: in pairs, linear, 30–60 cm × 3 mm, cylindrical, light green, somewhat twisted, pendulous. Seeds: numerous, comose, about 5 mm long.
Traditional therapeutic indications: hypertension, diabetes, malaria (Dusun, Kadazan)
Pharmacology and phytochemistry: antibacterial (Khan et al. 2003), antiviral (influenza virus, HIV) (Sabde et al. 2011; Zhou et al. 2020).
Indole alkaloids (Dutta et al. 1976). Antiplasmodial indole alkaloids (akuammidine, echitamine) (Wright et al. 1993; Hirasawa et al. 2009). Antibacterial indole alkaloids (alstoniascholarine J, vallesamine, 5-hydroxy-19,20-Z-alschomine) (Liu et al. 2015; Qin et al. 2015). Antimycobacterial indole alkaloid (tubotaiwine) (Macabeo et al. 2008). Antifungal indole alkaloids (alstoniascholarine A, E, and J) (Qin et al. 2015). Antiviral (adenovirus, herpes simplex virus, influenza virus) indole alkaloids (strictamine, 17-nor-excelsinidine) (Zhang et al. 2014; Zhou et al. 2020).
Toxicity, side effects, and drug interaction: methanol extract of bark administered orally to rats at the dose of 100 mg/kg/day for 28 days was found to cause liver damage (Bello et al. 2016).

Comment: the bark of *Alstonia scholaris* (L.) R.Br. has been used in Western medicine (*British Pharmacopoeia*, 1914) as an antimalarial drug. Early clinical trials (Manila hospital and India) have demonstrated a drop in malarial fever and amelioration in the health of patients given an extract of the plant (Mukerji et al. 1942). In Ayurvedic medicine, this plant is highly esteemed as an antimalarial (Mukerji et al. 1942).

REFERENCES

Bello, I., Bakkouri, A.S., Tabana, Y.M., Al-Hindi, B., Al-Mansoub, M.A., Mahmud, R. and Asmawi, M. 2016. Acute and sub-acute toxicity evaluation of the methanolic extract of *Alstonia scholaris* stem bark. *Medical Sciences*. 4(1): 4.

Dutta, S.C., Bhattacharya, S.K. and Ray, A.B. 1976. Flower alkaloids of *Alstonia scholaris*. *Planta Medica*. 30(5): 86–89.

Hirasawa, Y., Miyama, S. and Kawahara, N. 2009. Indole alkaloids from the leaves of *Alstonia scholaris*. *Heterocycles*. 79(1): 1107–1112.

Khan, M.R., Omoloso, A.D. and Kihara, M. 2003. Antibacterial activity of *Alstonia scholaris* and *Leea tetramera*. *Fitoterapia*. 74(7–8): 736–740.

Liu, L., Chen, Y.Y., Qin, X.J., Wang, B., Jin, Q., Liu, Y.P. and Luo, X.D. 2015. Antibacterial monoterpenoid indole alkaloids from *Alstonia scholaris* cultivated in temperate zone. *Fitoterapia.* 105: 160–164.

Macabeo, A.P.G., Krohn, K., Gehle, D., Read, R.W., Brophy, J.J., Franzblau, S.G. and Aguinaldo, M.A.M. 2008. Activity of the extracts and indole alkaloids from *Alstonia scholaris* against *Mycobacterium tuberculosis* H37Rv. *The Philippine Agricultural Scientist.* 91(3): 348–351.

Mukerji, B., Ghosh, B.K. and Siddons, L.B. 1942. Search for an antimalarial drug in the indigenous materia medica: Part I—*Alstonia scholaris*, F. Br. *The Indian Medical Gazette.* 77(12): 723.

Qin, X.J., Zhao, Y.L., Lunga, P.K., Yang, X.W., Song, C.W., Cheng, G.G., Liu, L., Chen, Y.Y., Liu, Y.P. and Luo, X.D. 2015. Indole alkaloids with antibacterial activity from aqueous fraction of *Alstonia scholaris. Tetrahedron.* 71(25): 4372–4378.

Sabde, S., Bodiwala, H.S., Karmase, A., Deshpande, P.J., Kaur, A., Ahmed, N., Chauthe, S.K., Brahmbhatt, K.G., Phadke, R.U., Mitra, D. and Bhutani, K.K. 2011. Anti-HIV activity of Indian medicinal plants. *Journal of Natural Medicines.* 65(3–4): 662–669.

Wright, C.W., Allen, D., Phillipson, J.D., Kirby, G.C., Warhurst, D.C., Massiot, G. and Le Men-Olivier, L. 1993. *Alstonia* species: Are they effective in malaria treatment? *Journal of Ethnopharmacology.* 40(1): 41–45.

Zhang, L., Zhang, C.J., Zhang, D.B., Wen, J., Zhao, X.W., Li, Y. and Gao, K. 2014. An unusual indole alkaloid with anti-adenovirus and anti-HSV activities from *Alstonia scholaris. Tetrahedron Letters.* 55(10): 1815–1817.

Zhou, H.X., Li, R.F., Wang, Y.F., Shen, L.H., Cai, L.H., Weng, Y.C., Zhang, H.R., Chen, X.X., Wu, X., Chen, R.F. and Jiang, H.M. 2020. Total alkaloids from *Alstonia scholaris* inhibit influenza a virus replication and lung immunopathology by regulating the innate immune response. *Phytomedicine.* 77: 153272.

***Alstonia spatulata* Bl.**

[After the 18th century Scottish physician Charles Alston and Latin *spatulata* = spathulate]

Published in *Bijdr. Fl. Ned. Ind.* (16): 1037. 1826

Synonyms: *Alstonia cochinchensis* Pierre ex Pit.; *Alstonia cuneata* Wall. ex G.Don;
Local name: *tembirog* (Dusun)
Common names: hard milkwood, siamese balsa
Habitat: forests
Geographical distribution: from Myanmar to Papua New Guinea
Botanical description: this tree reaches a height of about 25 m. Bark: smooth, grayish, and yields an abundant milky latex once incised. Leaves: simple, sessile, whorled, exstipulate. Blades: spathulate, coriaceous, 5–6.5 cm × 2–2.5 cm, with about 25 pairs of secondary nerves at right angles to midrib. Inflorescences: lax and few-flowered cymes on 1 cm long peduncles. Calyx: tiny, 5-lobed. Corolla: tubular, 5-lobed, the lobes contorted, 7 mm–1.2 cm long. Stamens: 5, adnate to the corolla tube. Styles: 1. tiny. Follicles: linear, twisted, about 15 cm × 2 mm. Seeds: numerous, comose, about 5 mm long.
Traditional therapeutic indications: cuts (Dusun)
Pharmacology and phytochemistry: indole alkaloids (Tan et al. 2010).
Toxicity, side effects, and drug interaction: not known

REFERENCE

Tan, S.J., Low, Y.Y., Choo, Y.M., Abdullah, Z., Etoh, T., Hayashi, M., Komiyama, K. and Kam, T.S. 2010. Strychnan and secoangustilobine A type alkaloids from *Alstonia spatulata*. Revision of the C-20 configuration of scholaricine. *Journal of Natural Products*. 73(11): 1891–1897.

Kopsia dasyrachis Ridl.

[After the 19th century Dutch botanist Jan Kops and from the Greek dasus = hairy, and *rachis* = spine]

Published in *Bull. Misc. Inform. Kew.* 123. 1934

Local name: *sarakad* (Rungus)
Habitat: forests near rivers
Geographical distribution: Borneo
Botanical description: this noble shrub reaches a height of about 3 m. Bark: yellowish gray, smooth, secretes white latex once incised. Leaves: simple, opposite, exstipulate. Petioles: 5–7 mm long. Blades: 2–10 cm × 6.5–25 cm, acuminate at apex, acute to attenuate at base, with 9–16 pairs of secondary nerves forming an intramarginal nerve. Cymes: terminal, up to about 20 cm long. Calyx: 5-lobed, the lobes about 4 mm long. Corolla: pure white, the tube up to about 3 cm long, yellow at the throat, thin, 5-lobed, the lobes obovate, about 1.5 cm long and contorted. Stamens: 5, adnate to the corolla tube. Ovary: tiny, somewhat hairy. Styles: 1, thin, up to about 1 cm long. Follicles: somewhat hairy, moon-shaped, orange, about 1.5 cm long. Seeds: few, in a red fleshy mass (Figure 4.42).
Traditional therapeutic indications: medicinal (Rungus)
Pharmacology and phytochemistry: indole alkaloids (kopsiflorine *N*(4)-oxide, 11-methoxykopsilongine *N*(4)-oxide, kopsifine, decarbomethoxykopsifine, kopsinarine, 11,12-methylenedioxykopsine, dasyrachine, rhazinicine, (+)-19 (*R*)-hydroxyeburnamine, and (−)-19 (*R*)-hydroxyisoeburnamine) (Homberger & Hesse 1982; Kam et al. 1999). Kopsiflorine increased the toxicity of vincristine against human carcinoma cells (KB) (Rho et al. 1999).
Toxicity, side effects, and drug interaction: not known

FIGURE 4.42 *Kopsia dasyrachis* Ridl.

REFERENCES

Homberger, K. and Hesse, M. 1982. Indole alkaloids of *Kopsia dasyrachis* RIDL. 185. naturally-occurring organic-substances. *Helvetica Chimica Acta.* 65(8): 2548–2557.

Kam, T.S., Choo, Y.M., Chen, W. and Yao, J.X. 1999. Indole and monoterpene alkaloids from the leaves of Kopsia dasyrachis. *Phytochemistry.* 52(5): 959–963.

Kam, T.S., Subramaniam, G. and Chen, W. 1999. Alkaloids from *Kopsia dasyrachis. Phytochemistry.* 51(1): 159–169.

Rho, M.C., Toyoshima, M., Hayashi, M., Koyano, T., Subramaniam, G., Kam, T.S. and Komiyama, K. 1999. Reversal of multidrug resistance by kopsiflorine isolated from *Kopsia dasyrachis. Planta Medica.* 65(4): 307–310.

***Plumieria acuminata* W.T. Aiton**

[After the 17th century French botanist Charles Plumier and Latin *acuminata* = acuminate]

Published in *Hortus Kew.* 2: 70. 1811

Synonym: *Plumeria rubra* L.

Local names: *campaka* (Murut); *cempaka* (Bajau, Dusun); *bunga Kemboja, semboja* (Lundayeh); *Kemboja* (Bajau)

Common name: frangipani

Habitat: cultivated

Geographical distribution: tropical

Botanical description: this tree reaches a height of about 7 m. Trunk: not straight. Bark: pale gray, rough. Stems: terete, stout, with numerous leaf scars at apex, yield a pure white latex once incised. Leaves: simple, spiral, exstipulate, few, grouped at apex stems. Petioles: stout, channeled above, 2–7 cm long. Blades: coriaceous, glossy or not, oblong, lanceolate to spathulate, tapering at base, acute, rounded or shortly acuminate at apex, 14–30 cm × 5–10 cm, with about 13–17 pairs of secondary nerves. Cymes: terminal with few flowers. Flowers: slightly fragrant, magnificent. Calyx: tiny, 5-lobed. Corolla: tubular, contorted, 5-lobed, pure white or white tinged with pink or red or yellow, fleshy, about 5 cm across, the lobes contorted. Follicles: in pairs, fusiform, coriaceous, smooth, about 25 cm × 2 cm. Seeds: numerous, winged.

Traditional therapeutic indications: internal pain, constipation, piles (Murut)

Pharmacology and phytochemistry: antiplasmodial (Highlands et al. 2016), analgesic, anti-inflammatory (Das et al. 2013), antibacterial (Gupta et al. 2008).

Leishmanicidal, cytotoxic, and antifungal iridoids (isoplumericin, plumericin) (Castillo et al. 2007; Sharma et al. 2011 ; Sing et al. 2011, Sun et al. 2016).

Toxicity, side effects, and drug interaction: dreadfully poisonous (Eddleston & Warrell 1999).

REFERENCES

Castillo, D., Arevalo, J., Herrera, H., et al. 2007. Spirolactone iridoids might be responsible for the leishmanicidal activity of a Peruvian traditional remedy made with *Himatanthus sucuuba* (Apocynaceae). *Journal of Ethnopharmacology*. 112: 410–414.

Das, B., Ferdous, T., Mahmood, Q.A., Hannan, J.M.A., Bhattacharjee, R. and Das, B.K. 2013. Antinociceptive and anti-inflammatory activity of bark: Extract of *Plumeria rubra* on laboratory animals. *European Journal of Medicinal Plants*, 1: 114–126.

Eddleston, M. and Warrell, D.A. 1999. Management of acute yellow oleander poisoning. *QJM*. 92(9): 483–485.

Gupta, M., Mazumder, U.K., Gomathi, P. and Selvan, V.T. 2008. Antimicrobial activity of methanol extracts of *Plumeria acuminata* Ait. leaves and *Tephrosia purpurea* (Linn.) Pers. roots. *Natural Product Radiance*. 7(2): 102–105.

Highlands, H.N., Mathew, S., Jani, D.V. and George, L.B. 2016. In-vitro evidence of effective anti-plasmodium activity by *Plumeria rubra* (L) extracts. *International Journal of Pharmacognosy and Phytochemical Research*. 8(8): 1377–1384.

Sharma, U., Singh, D., Kumar, P., Dobhal, M.P. and Singh, S. 2011. Antiparasitic activity of plumericin & isoplumericin isolated from *Plumeria bicolor* against *Leishmania donovani*. *The Indian Journal of Medical Research*. 134(5): 709.

Singh, D., Sharma, U., Kumar, P., Gupta, Y.K., Dobhal, M.P. and Singh, S. 2011. Antifungal activity of plumericin and isoplumericin. *Natural Product Communications*. 6(11): 1934578X1100601101.

Sun, J., Wu, J., Zhang, L., Wan, Q.Y., Wang, L.C., Liu, Y.Y., Wang, R.M. and Sun, D.Q. 2016. Plumericin inhibits the viability of cervical cancer cells by induction of apoptosis through degradation of DNA structure. *Bangladesh Journal of Pharmacology*. 11(2): 337–341.

***Tabernaemontana macrocarpa* Jack**

[After the 16th century German botanist *J. T. Tabernaemontanus* and Latin *macrocarpa* = large fruit]

Published in *Malayan Misc.* 2(7): 80. 1822

Synonyms: *Ervatamia macrocarpa* (Jack) Merr.; *Orchipeda sumatrana* Miq; *Pagiantha macrocarpa* (Jack) Markgr.; *Voacanga plumeriifolia* Elmer; *Tabernaemontana megacarpa* Merr.

Local names: *gugutukan*, *lampada* (Dusun)

Habitat: forests

Geographical distribution: Thailand, Malaysia, Indonesia, the Philippines

Botanical description: this tree reaches a height of about 30 m. Bark: grayish to yellowish brown. Stems: yield a latex once incised. Leaves: simple, opposite, exstipulate. Petioles: 1–4 cm long. Blades: elliptic, oblong to obovate, coriaceous, 2-7 cm × 20–40 cm, acute to attenuate at base, rounded to acuminate or acute at apex, with 9–12 pairs of secondary nerves. Cymes: axillary, up to 20 cm long. Calyx: tubular, 5-lobed, the lobes ovate, up to about 6 mm long. Corolla: tubular, orangish, caducous, about 3 cm long, 5-lobed, the lobes contorted. Stamens: 5. Styles: 1., tiny. Follicles: globose to somewhat ellipsoid, smooth, minutely acuminate at apex or not, glossy, orange to tomato red, up to about 15 cm across, dehiscent. Seeds: numerous in a blood-red pulp. The open fruits look like some sorts of monstrous cockles.

Traditional therapeutic indications: boils, dermatitis (Dusun)

Pharmacology and phytochemistry: indole alkaloids (Husain et al. 1997).

Toxicity, side effects, and drug interaction: not known

Comment: According to Evans (1922) this tree was sacred and greatly feared by the Dusun. Evans writes: "According to Dusun legendry, Kenharingan has put a curse upon anyone who shall violate a tree of this species, the punishment being that the offender shall die of incurable ulcers".

REFERENCE

Husain, K., Said, I.M., Din, L.B., Takayama, H., Kitajima, M. and Aimi, N. 1997. Alkaloids from the roots of *Tabernaemontana macrocarpa* Jack. *Natural Product Sciences*. 3(1): 42–48.

***Tabernaemontana sphaerocarpa* Bl.**

[After the 16th century German botanist J. T. Tabernaemontanus and Latin *sphaerocarpa* = spherical fruit]

In *Bijdragen tot de flora van Nederlandsch Indië* (16): 1028. 1826

Synonyms: *Pagiantha sphaerocarpa* (Blume) Markgr.; *Tabernaemontana fagraeoides* Miq.; *Tabernaemontana javanica* Miq;
Local name: *lampada* (Dusun)
Habitat: forests
Geographical distribution: Malaysia, Indonesia, the Philippines
Botanical description: this tree reaches a height of about 10 m. Stems: yield a latex once incised. Leaves: simple, opposite, exstipulate. Petioles: thin, about 3 cm long. Blades: elliptic, oblong to obovate, 7–14 cm × 20–30 cm, acute at base, obtuse at apex, with 9–12 pairs of secondary nerves. Cymes: axillary, pendulous. Calyx: tubular, 5-lobed, the lobes oblong, up to about 5 mm long. Corolla: tubular, orangish, caducous, about 3 cm long, 5-lobed, the lobes contorted. Stamens: 5, adnate to the corolla tube. Follicles: globose, smooth, glossy, orange, up to 10 cm across, dehiscent. Seeds: numerous in a blood-red pulp. The open fruits look like some sorts of monstrous cockles. (Figure 4.43).
Pharmacology and phytochemistry: indole alkaloids (Chatterjee et al. 1968; Zaima et al. 2009).
Toxicity, side effects, and drug interaction: not known

FIGURE 4.43 *Tabernaemontana sphaerocarpa* Bl.

Comments:

(i) This plant with local name *lampada* and very much alike *Tabernaemontana macrocarpa* Jack could also have been greatly fear and venerated (?).
(ii) The Dusun and Kadazan use a plant of the genus *Willughbeia* Roxb. (1820) for stomach aches (local name: *kombing).*
(iii) *Allamanda cathartica* L. is medicinal for the Murut (local name: *bunga loceng).*
(iv) *Catharanthus roseus* (L.) G.Don is used by the Bajau (local name: *kemunting Cina*) for coughs, diabetes, malaria, and wounds whilst the Dusun use this plant for skin whitening.
(v) The Dusun and Kadazan use a plant of the genus *Kopsia* Bl. (1823) (local name: *lodo lodo*) for toothaches, whilst a plant of the genus *Tabernaemontana* L. (1753) is employed for wounds (local name: *lado lado).*
(vi) Plants of the family Asclepiadaceae Borkh (1797) are medicinal in Sabah: *Asclepias curassavica* L. for fever and bronchitis (Dusun, Kadazan, local name: *piak piak*), *Dischidia rafflesiana* Wall. for cancer and skin diseases (Bajau), *Hoya coronaria* Bl. for pancreatitis and foul body odor (Dusun, Kadazan; local name: *launau launau).* A plant of the genus *Hoya* R. Br. (1810) is used by the Dusun and Kadazan for cancer (local name: *bina*) whilst the Murut use *"pongkukubab"* as medicine.

REFERENCES

Chatterjee, A., Banerji, A. and Majumder, P.L. 1968. Occurrence of tabernaemontanine+ dregamine in *Tabernaemontana sphaerocarpa* BL. *Indian Journal of Chemistry*. 6(9): 545-+.

Zaima, K., Hirata, T., Hosoya, T., Hirasawa, Y., Koyama, K., Rahman, A., Kusumawati, I., Zaini, N.C., Shiro, M. and Morita, H. 2009. Biscarpamontamines A and B, an aspidosperma– iboga bisindole alkaloid and an aspidosperma– spidosperma bisindole alkaloid, from *Tabernaemontana sphaerocarpa. Journal of Natural Products*. 72(9): 1686–1690.

4.1.14.1.2 Family Loganiaceae R.Br ex Mart (1827), the Logania Family

The family Loganiaceae consists of about 30 genera and 500 species magnificent trees, shrubs, climbing plants, or herbaceous plants. Leaves: simple, opposite, stipulate. Inflorescences: cymes. Calyx: tubular, 4-5-lobed. Corolla: tubular, 4-5-lobed. Stamens: 4–6. Carpels: 2, forming a bilocular ovary, each locule with 3–many ovules. Styles: 1. Stigma: capitate. Fruits: capsules, berries, drupes.

Fagraea cuspidata **Bl.**

[After the 18th century Swedish naturalist Fagraeus and from Latin *cuspidatus* = with abrupt short point]

Published in *Musée Botanique de M. Benjamin Delessert, with an introduction by F. A. Stafleu* 1: 170. 1850

Synonyms: *Fagraea crassipes* Benth.; *Fagraea cymosa* Merr.; *Fagraea pendula* Merr.; *Fagraea robusta* Bl.; *Utania cuspidata* (Blume) K.M. Wong, Sugumaran & Sugau
Local names: *solovon, gondula talun* (Murut); *kabang penah, panaa, puok, todopon* (Dusun); *kopi hutan, sepuleh* (Bajau)
Habitat: forests
Geographical distribution: Borneo, the Philippines
Botanical description: this tree reaches a height of about 10 m. Bark: fissured, dark gray. Sapwood: yellow. Stems: smooth, terete. Leaves: simple, opposite, exstipulate. Petioles: 0.5–1.5 cm long, clasping the stem at base. Blades: broadly elliptic, somewhat wavy, coriaceous, 8–40 cm × 4–16 cm, attenuate at base, acuminate to cuspidate at apex, with 5–10 pairs of secondary nerves. Cymes: terminal, pendulous, on about 30 cm long peduncles. Calyx: cup-shaped, about 1.5 cm long, 5-lobed. Corolla: tubular, whitish green, 3.5–6 cm long, 5-lobed, the lobes ovate. Stamens: 5, about 1.5 cm long. Stigma: capitate, protuding. Berries: ovoid, greenish to pale yellow, 1–2 cm long, on persistent calyx. Seeds: few, angular.
Traditional therapeutic indications: fever, postpartum (Dusun); gastritis, chest pain, jaundice, diabetes (Dusun, Kadazan)
Pharmacology and phytochemistry: not known
Toxicity, side effects, and drug interaction: not known

Comments:

(i) *Strychnos ignatii* Berg. is used by the Murut to poison blowgun darts (local name: *tataga do sangi*).
(ii) *Fagraea racemosa* Jack (Figure 4.44) is used by the Dusun and Kadazan (local name: *todopon puok*) for chest pain.

4.1.14.1.3 Family Rubiaceae Juss. (1789), the Madder Family

The family Rubiaceae is a vast taxon consisting of about 660 genera and 12,000 species of trees, shrubs, climbing plants, or herbaceous plants. Leaves: simple, opposite, stipulate. Stipules: interpetiolar. Inflorescences: cymes, panicles, heads, verticillasters, or solitary. Calyx: tubular, 4–5-lobed. Corolla: tubular, 3–8 lobed. Stamens: 4–5. Disc: present. Carpels: 2, forming a unilocular or bilocular ovary, each locule with 1–many ovules. Styles: 1, protuding. Fruits: often capsules, berries, drupes marked at apex by a small cylindrical opening.

Gardenia tubifera **Wall.**

[After the 18th century Scottish naturalist Alexander Garden and Latin *tubifera* = bearing a tube]

FIGURE 4.44 *Fagraea cuspidata* Bl.

Published in *Flora Indica; or descriptions of Indian Plants* 2: 562. 1824

Synonyms: *Gardenia elata* Ridl., *Gardenia lobbii* Craib, *Gardenia resinifera* Korth., *Gardenia speciosa* (Hook.f.) Hook.f., *Randia speciosa* Hook.f.

Local name: *piluzung* (Rungus)

Common name: golden gardenia

Habitat: roadsides, forests, parks

Geographical distribution: India, Southeast Asia

Botanical description: this handsome tree reaches a height of about 10 m. Bark: smooth, gray. Inner bark: light brown. Stems: terete, fissured, articulate, with 1.5 cm long internodes, angular at apex. Leaves: simple, opposite, stipulate. Stipules: interpetiolar, tubular, ciliate at the margin, about 5 mm long. Petioles: about 1 cm long, flattish, hairy. Blades: spathulate, 6–15 cm × 2.5–5 cm, somewhat coriaceous, tapering at base, acuminate at apex, with about 11 pairs of secondary nerves. Flowers: terminal, solitary, white turning yellow, with a strange and heavy scent, with the shape of some kind of airplane propeller. Calyx: membranous, cup-shaped, about 9 mm long. Corolla: tubular, somewhat fleshy, about 5 cm long, 8-lobed, the lobes about 3 cm long. Berries: globose, about 3 cm across, with about 3 mm long apical tubes.

Traditional therapeutic indications: magic rituals (Rungus)

Pharmacology and phytochemistry: triterpenes toxic to human cancer cells (Nuanyai et al. 2010).

Toxicity, side effects, and drug interaction: not known

Comment: A plant of the genus *Gardenia* J. Ellis (1761) is used by the Dusun and Kadazan for fatigue (local name: *marabingo*).

REFERENCE

Nuanyai, T., Chokpaiboon, S., Vilaivan, T. and Pudhom, K. 2010. Cytotoxic 3, 4-seco-cycloartane triterpenes from the exudate of *Gardenia tubifera*. *Journal of Natural Products*. 73(1): 51–54.

Hedyotis auricularia **L.**

[From Greek *hedus* = sweet, *ous* = ear, and Latin *auriculata* = auriculate]

Published in *Species Plantarum* 1: 101. 1753

Synonyms: *Hedyotis sarmentosa* Craib.; *Hedyotis venosa* (Bl.) Korth; *Metabolus venosa* Bl.; *Oldenlandia auricularia* K.Schum.
Local name: *pipisoson tinanansad* (Rungus)
Habitat: roadsides, at the edges of forests
Geographical distribution: from India to Australia
Botanical description: this herbaceous plant grows up to a length of about 1 m. Stems: ligneous, flexuous, swollen at nodes, angular, hairy. Leaves: simple, sessile, decussate, stipulate. Stipules: less than 4 mm across, lobed. Blades: lanceolate, hairy beneath, dark green, 1.5–8 cm × 0.2–2.5 cm, acute at base, acuminate or acute at apex, the secondary nerves unclearly visible. Verticillasters: with tiny white flowers. Calyx: cupular, tiny, with 4 triangular lobes. Corolla: tubular, hairy at the throat, whitish, 4-lobed, the lobes oblong. Stamens: 4. Styles: 1, thin. Stigma: bifid. Capsules: tiny (Figure 4.45).

FIGURE 4.45 *Hedyotis auricularia* L.

Traditional therapeutic indications: medicinal (Rungus)

Pharmacology and phytochemistry: antiviral (HIV, dengue virus) (Hamidi et al. 1996; Rothan et al. 2014). Anthraquinones (Caro et al. 2012).

Toxicity, side effects, and drug interaction: not known

Comments:

(i) The leaves of *Hedyotis congesta* R.Br. (local name: *tapis apiris*) and *Hedyotis rigida* (Blume) Walp (local name: *udu lomut*) are used for wounds by the Lundayeh.

(ii) A plant of the genus *Hedyotis* L. (1753) is used by the Dusun and Kadazan for swellings (local name: *mompu ompu*).

REFERENCES

Caro, Y., Anamale, L., Fouillaud, M., Laurent, P., Petit, T. and Dufossé, L. 2012. Natural hydroxyanthraquinoid pigments as potent food grade colorants: An overview. *Natural Products and Bioprospecting*. 2(5): 174–193.

Hamidi, J.A., Ismaili, N.H., Ahmadi, F.B. and Lajisi, N.H. 1996. Antiviral and cytotoxic activities of some plants used in Malaysian indigenous medicine. *Pertanika Journal of Tropical Agricultural Science*. 19(2/3): 129–136.

Rothan, H.A., Zulqarnain, M., Ammar, Y.A., Tan, E.C., Rahman, N.A. and Yusof, R. 2014. Screening of antiviral activities in medicinal plants extracts against dengue virus using dengue NS2B-NS3 protease assay. *Tropical Biomedicine*. 31(2): 286–296.

Hydnophytum formicarum **Jack**

[From Greek *hudnon* = truffle, *phyton* = plant, and from Latin *formica* = ants]

Published in *Transactions of the Linnean Society of London* 14(1): 124. 1823

Synonym: *Hydnophytum orbiculatum* Elmer
Local names: *angang, sarang semut betina* (Lundayeh); *musalag noh kilou* (Murut)
Common name: baboon's head
Habitat: coastal forests, mangroves
Geographical distribution: from the Andaman Islands to the Solomon Islands
Botanical description: this epiphytic herbaceous plant reaches a height of about 60 cm. Tuber: monstrous, woody, with cavities inside inhabited by ants, up to about 25 cm across. Stems: terete, stout. Leaves: simple, opposite, stipulate. Petioles: about 5 mm–4 cm long, stout. Blades: coriaceous, fleshy, 0.8–15 cm × 0.5–7 cm, with about 5 pairs of secondary nerves, spathulate to elliptic, somewhat wavy, acuminate at apex. Cymes: axillary, sessile. Calyx: cup-shaped, about 2 mm long. Corolla: pure white, up to about 3 mm long, hairy, 4-lobed, the lobes revolute. Stamens: 4. Styles: 1, linear, protuding. Stigma: bifid. Berries: orange-red, glossy, about 7 mm long, pulpy, oblong. Seeds: few, tiny, oblong.
Traditional therapeutic indications: cancer, diabetes, hypertension (Lundayeh)
Pharmacology and phytochemistry: antidiabetic (Rachpirom et al. 2022), neuroprotective (Gay et al. 2018), toxic to human fibrosarcoma cells (HT-1080) (Ueda et al. 2002) and cervical cancer cells (HeLa) (Senawong et al. 2013).
Chalcones (isoliquiritigenin, butein), flavanones (butin) as well as benzoic acid derivatives (protocatechualdehyde) and antibacterial hydroxycinnamic acid derivatives (sinapic acid) (Prachayasittikul et al. 2008). Sinapic has hypotensive (Silambarasan et al. 2014) and hypoglycemic activity (Cherng et al. 2013). Other constituents are iridoids (asperulosidic acid, deacetylasperulosidic acid, 6α-hydroxygeniposide, 10-hydroxyloganin) (Hanh et al. 2015).
Toxicity, side effects, and drug interaction: not known

Comment: Slices of tubers are sold in the wet markets of Sabah as medicine.

REFERENCES

Cherng, Y.G., Tsai, C.C., Chung, H.H., Lai, Y.W., Kuo, S.C. and Cheng, J.T. 2013. Antihyperglycemic action of sinapic acid in diabetic rats. *Journal of Agricultural and Food Chemistry*. 61(49): 12053–12059.

Gay, N.H., Phopin, K., Suwanjang, W., Ruankham, W., Wongchitrat, P., Prachayasittikul, S. and Prachayasittikul, V. 2018. Attenuation of oxidative stress-induced neuronal cell death by *Hydnophytum formicarum* Jack. *Asian Pacific Journal of Tropical Medicine*. 11(7): 415.

Hanh, N.P., Phan, N.H.T., Thuan, N.T.D., Hanh, T.T.H., Van Thanh, N., Cuong, N.X., Nam, N.H. and Van Minh, C. 2015. Iridoid constituents from the ant plant *Hydnophytum formicarum*. *Vietnam Journal of Chemistry*. 53(2e).

Prachayasittikul, S., Buraparuangsang, P., Worachartcheewan, A., Isarankura-Na-Ayudhya, C., Ruchirawat, S. and Prachayasittikul, V. 2008. Antimicrobial and antioxidative activities of bioactive constituents from *Hydnophytum formicarum* Jack. *Molecules*. 13(4): 904–921.

Rachpirom, M., Barrows, L.R., Thengyai, S., Ovatlarnporn, C., Sontimuang, C., Thiantongin, P. and Puttarak, P. 2022. Antidiabetic activities of medicinal plants in traditional recipes and candidate antidiabetic compounds from *Hydnophytum formicarum* Jack. Tubers. *Pharmacognosy Research*. 14(1).

Senawong, T., Misuna, S., Khaopha, S., Nuchadomrong, S., Sawatsitang, P., Phaosiri, C., Surapaitoon, A. and Sripa, B. 2013. Histone deacetylase (HDAC) inhibitory and antiproliferative activities of phenolic-rich extracts derived from the rhizome of *Hydnophytum formicarum* Jack.: Sinapinic acid acts as HDAC inhibitor. *BMC Complementary and Alternative Medicine*. 13(1): 1–11.

Silambarasan, T., Manivannan, J., Krishna Priya, M., Suganya, N., Chatterjee, S. and Raja, B. 2014. Sinapic acid prevents hypertension and cardiovascular remodeling in pharmacological model of nitric oxide inhibited rats. *PLoS ONE*. 9(12): e115682.

Ueda, J.Y., Tezuka, Y., Banskota, A.H., Le Tran, Q., Tran, Q.K., Harimaya, Y., Saiki, I. and Kadota, S. 2002. Antiproliferative activity of Vietnamese medicinal plants. *Biological and Pharmaceutical Bulletin*. 25(6): 753–760.

Ixora blumei (Bl.) Zoll. & Moritzi

[The word ixora derives from the name of the Hindu deity Isvara and after the 19th century the German botanist Karl Ludwig von Blume]

Published in *Systematisches Verzeichniss der im Indischen Archipel* 65. 1846

Synonyms: *Ixora blumeana* (Bl.) Kuntze; *Ixora odorata* (Blume) Boerl.; *Ixora valetonii* (Blume) Hochr.; *Pavetta odorata* Bl.
Local name: *lapad bala* (Lundayeh)
Habitat: forests
Geographical distribution: Indonesia, Borneo
Botanical description: this tree reaches a height of about 20 m. Leaves: simple, opposite, stipulate. Petioles: 2 mm–1 cm long. Blades: elliptic, 3–7 cm × 8–15 cm, glossy, acuminate. Cymes: showy, about 10 cm long, terminal. Calyx: tiny, somewhat hairy, 4-lobed. Corolla: about 4 mm long, whitish, tubular, 4-lobed, the lobes revolute. Stamens: 4. Style: 1. linear, protruding. Stigma: bifid. Berries: reddish, about 8 mm across. Seeds: 2.
Traditional therapeutic indications: hydrocele, swollen penis (Lundayeh), intestinal worms (Rungus)
Pharmacology and phytochemistry: not known
Toxicity, side effects, and drug interaction: not known

Comments:

(i) The Lundayeh use *Ixora fucosa* Bremek (local name: *lapad lontong*) or *Ixora javanica* (Blume) DC. (local name: *busak wudan*) for anorexia.
(ii) *Ixora capillaris* Bremek is used by the Rungus as medicine (local name: *tagandap timulu*). The Murut call this plant *"angin tolunsung"*.

Morinda citrifolia L.

[From Latin *morus* = mulberry, *indicus* = from India, and *citrofolia* = with citrus leaves]

Published *Species Plantarum* 1: 176. 1753

Synonym: *Morinda bracteata* Roxb.
Local names: *babas, kekudu* (Lundayeh); *baja* (Bugis); *bingkudu* (Kadazan); *mengkudu* (Bajau, Dusun, Murut); *mongkudu* (Murut)
Common name: Indian mulberry
Habitat: seashores, cultivated
Geographical distribution: from India to Pacific Islands
Botanical description: this tree reaches a height of about 5 m. Bark: of a peculiar plain brown, somewhat smooth. Stems: glossy, angular at apex. Leaves: simple, opposite, stipulate. Stipules: interpetiolar, up to about 1 cm long, oblong. Petioles: 1–2 cm long, stout, curved. Blades: elliptic, 10–20 cm × 5–9 cm, dark green, glossy, somewhat fleshy, with 5–7 pairs of secondary nerves, attenuate at base, obtuse or acute-acuminate at apex. Flowering heads: somewhat ovoid, up to about 3 cm long, fleshy, develop one or a few flowers at a time, repulsive. Calyx: cup-shaped. Corolla: pure white, the tube about 1 cm long, 5-6-lobed, the lobes lanceolate, about 5 mm long, revolute. Stamens: 5–6, dark brown. Styles: 1, thin. Stigma: bifid, protuding. Syncarps: ovoid to potato-shaped, glossy, about 5 cm × 3 cm, with a disgusting larval white color. Seeds: numerous.

Traditional therapeutic indications: headaches (Lundayeh; Sungai); joints pain (Bajau, Dusun, Kadazan); gastric ulcers, jaundice (Dusun, Kadazan); darken gray hair, hair loss (Bugis); poison antidote (Lundayeh); diabetes, skin diseases (Kedayan); (Bugis, Kedayan); hypertension (Bajau, Kedayan, Bugis, Dusun, locals of Javanese descent, Dusun, Kadazan, Iranun, Lundayeh); yellow fever (Rungus); cancer (Kedayan); piles (Iranum); dental caries (Dusun, Kadazan, Kedayan); dysmenorrhea (Bugis, Dusun, Kadazan, Kedayan)

Pharmacology and phytochemistry: hypotensive (Yoshitomi et al. 2020), antiulcer (Srikanth & Muralidharan 2009), estrogenic activity (Chearskul et al. 2004).

Antiviral anthraquinone´ (1,3-dihydroxy-5-methoxy-6´-methoxymethyl-2-methyl-9,10-anthraquinone) (Wang et al. 2016), antifungal and antimycobacterial anthraquinone (damnacanthal) (Ali et al. 2000; Pollo et al. 2021). Antiviral (herpes simplex virus, Epstein-Barr virus) iridoid glycoside (asperuloside) (Kapadia et al. 1996, Manzione et al. 2020). Leishmanicidal anthracene dione (morindicone) (Sattar et al. 2012).

Toxicity, side effects, and drug interaction: the fruit contains high amounts of potassium, explaining, at least in part, the hypotensive properties if eaten in small amounts, but which can lead to hemorrhages and heart attacks if eaten in excess or even in small amounts for kidney failure patients (Mueller et al. 2000). Patients on anticoagulants and beta-blockers must absolutely avoid this plant. Infusion of leaves is poisonous.

Comment: *Morinda borneensis* (Baill.) K.Schum is sold as medicine in the wet markets of Sabah.

REFERENCES

Ali, A.M., Ismail, N.H., Mackeen, M.M., Yazan, L.S., Mohamed, S.M., Ho, A.S.H. and Lajis, N.H. 2000. Antiviral, cyototoxic and antimicrobial activities of anthraquinones isolated from the roots of *Morinda elliptica*. *Pharmaceutical Biology*. 38(4): 298–301.

Chearskul, S., Kooptiwut, S., Chatchawalvanit, S., Onreabroi, S., Churintrapun, M., Saralamp, P. and Soonthornchareonnon, N. 2004. *Morinda citrifolia* has very weak estrogenic activity *in vivo*. *The Journal of Physiological Sciences*. 17(1): 22–29.

Kapadia, G.J., Sharma, S.C., Tokuda, H., Nishino, H. and Ueda, S. 1996. Inhibitory effect of iridoids on Epstein-Barr virus activation by a short-term *in vitro* assay for anti-tumor promoters. *Cancer Letters*. 102(1–2): 223–226.

Manzione, M.G., Martorell, M., Sharopov, F., Bhat, N.G., Kumar, N.V.A., Fokou, P.V.T. and Pezzani, R. 2020. Phytochemical and pharmacological properties of asperuloside, a systematic review. *European Journal of Pharmacology*. 883: 173344.

Mueller, B.A., Scott, M.K., Sowinski, K.M. and Prag, K.A. 2000. Noni juice (*Morinda citrifolia*): Hidden potential for hyperkalemia? *American Journal of Kidney Diseases*. 35(2): 310–312.

Pollo, L.A., Martin, E.F., Machado, V.R., Cantillon, D., Wildner, L.M., Bazzo, M.L., Waddell, S.J., Biavatti, M.W. and Sandjo, L.P. 2021. Search for antimicrobial activity among fifty-two natural and synthetic compounds identifies anthraquinone and polyacetylene classes that inhibit Mycobacterium tuberculosis. *Frontiers in Microbiology*. 11: 622629.

Sattar, F.A., Ahmed, F., Ahmed, N., Sattar, S.A., Malghani, M.A.K. and Choudhary, M.I. 2012. Double-blind, randomized, clinical trial on the leishmanicidal activity of a *Morinda citrifolia* (Noni) stem extract and its major constituents. *Natural Product Communications*. 7(2): 1934578X1200700218.

Srikanth, J. and Muralidharan, P. 2009. Antiulcer activity of *Morinda citrifolia* Linn fruit extract. *Journal of Scientific Research*. 1(2): 345–352.

Wang, J., Qin, X., Chen, Z., Ju, Z., He, W., Tan, Y., Zhou, X., Tu, Z., Lu, F. and Liu, Y. 2016. Two new anthraquinones with antiviral activities from barks of *Morinda citrifolia* (Noni). *Phytochemistry Letters*. 15: 13–15.

Yoshitomi, H., Zhou, J., Nishigaki, T., Li, W., Liu, T., Wu, L. and Gao, M. 2020. *Morinda citrifolia* (Noni) fruit juice promotes vascular endothelium function in hypertension via glucagon-like peptide-1 receptor-CaMKKβ-AMPK-eNOS pathway. *Phytotherapy Research*. 35(9): 2341–2350.

***Mussaenda frondosa* L.**

[After the Malagasy name of a plant of the genus *Mussaenda* L. and Latin *frondosa* = leafy]

Published in *Species Plantarum* 1: 177. 1753

Synonyms: *Gardenia frondosa* (L.) Lam.; *Mussaenda sumatrensis* Roth
Local name: *boliadok* (Dusun)
Common names: wild mussaenda, dhobi tree
Habitat: cultivated
Geographical distribution: from India to Papua New Guinea
Botanical description: this climbing shrub reaches a height of about 2 m. Stems: terete. Leaves: simple, opposite, stipulate. Stipules: bifid, 5 mm–1 cm long. Petioles: 2 mm–1 cm long, hairy. Blades: elliptic to broadly lanceolate, 8–15 cm × 3–8 cm, acute at base, acuminate at apex, with 5–10 pairs of secondary nerves, glossy, membranous. Cymes: terminal or axillary, hairy. Calyx: hairy, 5-lobed, the lobes linear, 7 mm–1 cm long, one lobe enlarged into a calycophyll. Calophylls: whitish, elliptic-oblong, 6–8 cm × 2.5–5 cm. Corolla: tubular, bright orange, hairy outside, the tube about 2 cm long and 5-lobed. Berries: ovoid to oblong, about 1 cm long, hairy.
Traditional therapeutic indications: headaches (Dusun)
Pharmacology and phytochemistry: nematicidal (Siju et al. 2010), wound healing (Suhas et al. 2012; Patil et al. 2011), hepatoprotective (Sn et al. 2010), antibacterial (Shanthi & Radha 2020).
Toxicity, side effects, and drug interaction: not known

Comments:

(i) The Dusun and Kadazan use a plant of the genus *Mussaenda* L. (1753) for colds (local name: *gayoh lubah*).
(ii) In East Kalimantan, natives use a plant genus *Mussaenda* L. (1753) to soften the skin of the face (local name: *pilanggang bulan*).

REFERENCES

Patil, S.A., Joshi, V.G. and Sambrekar, S.N. 2011. Evaluation of wound healing activity of isolated compound quercetin and alcoholic extract of leaves of Mussaenda frondosa Linn. *Research Journal of Pharmacognosy and Phytochemistry.* 3(6): 266–271.

Shanthi, S. and Radha, R. 2020. Anti-microbial and phytochemical studies of Mussaenda frondosa Linn. Leaves. *Pharmacognosy Journal*. 12(3).

Siju, E.N., Rajalakshmi, G.R., Hariraj, N., Sreejith, K.R., Sudhakaran, S., Muneer, E.K. and Premalatha, K. 2010. In-vitro anthelmintic activity of Mussaenda frondosa. *Research Journal of Pharmacy and Technology.* 3(1): 151–153.

SN, S., Patil, P.A. and Kangralkar, V.A. 2010. Protective activity of *Mussaenda frondosa* leaf extracts against paracetamol induced he-patic damage in wistar rats. *Journal of Pharmacy Research.* 3(4): 711–713.

Suhas, P.A., Joshi, V.G. and Sambrekar, S.N. 2012. Wound healing effects of Mussaenda frondosa extracts on second degree superficial burned rat. *Research Journal of Pharmacology and Pharmacodynamics.* 4(3): 3.

***Myrmecodia platytyrea* Becc.**

[From the Greek *myrmekos* = ant, *platys* = flat, and *tyros* = cheese]

Published in *Malesia Raccolta* . . . 2: 115. 1884

Synonym: *Myrmecodia antoinii* Becc.
Local names: *raja ubat* (Dusun, Lundayeh); *sarang semut* (Bajau, Murut)
Habitat: forests, swamps
Geographical distribution: Borneo, Indonesia, Australia
Botanical description: this epiphytic herbaceous plant reaches a height of about 25 cm. Tubers: repugnant, irregularly pear-shaped, with cavities inside inhabited by ants, up to 20 cm across, brown. Stems: thorny, the thorns about 5 mm long. Leaves: simple, opposite, apical, stipulate. Stipules: interpetiolar, narrowly triangular, 5 mm–1.3 cm long. Petioles: 2–9 cm long. Blades: elliptic to oblanceolate, 3–4 cm × 20–25 cm, tapering at base, acuminate at apex, with 6–10 pairs of secondary nerves. Inflorescences: solitary and in alveoli. Corolla: immaculately white, 4-lobed, about 1.5 cm long. Stamens: 4. Styles: 1. Stigma: 4-lobed. Drupes: red, about 1.5 cm long.
Traditional therapeutic indications: hypertension (Bajau, Lundayeh, Dusun); poison antidote (Bajau, Dusun); tuberculosis, fever, sinusitis, kidney diseases, cancer, headaches (Bajau); diabetes (Bajau, Murut)
Pharmacology and phytochemistry: hypoglycemic, hypocholesterolemic (Hasan et al. 2020), toxic for liver cancer cells (Ibrahim 2021).
Toxicity, side effects, and drug interaction: aqueous extract of tubers administered orally to mice at the dose of 2 g/kg was found to cause no signs of toxicity (Mohd Zin et al. 2013).

Comments:

(i) The sliced tubers of this plant are sold in the wet markets of Sabah for the treatment of cancer, from which came the appellation "*raja ubat,*" meaning the king of medicines. These slices are packed in transparent plastic bags, sometimes with ants, and one could wonder if the therapeutic value could be owed, at least in part, by natural products produced by ants.
(ii) In Borneo, the Dayak use *Myrmecodia beccari* Hook.f. (local name: *sarang semut*) for tumors, asthma, piles and postpartum.

REFERENCES

Hasan, M.H., Zakaria, H., Wahab, I.A., Ponto, T. and Adam, A. 2020b. Myrmecodia platytyrea methanol tuber extract ameliorates hyperglycemia In STZ-induced diabetic Sprague-Dawley male rats. *Indonesian Journal of Pharmacy*: 338–348.

Hasan, M.N., Wahab, I.A., Mizaton, H.H. and Rasadah, M.A. 2020a. Efficacy of Myrmecodia Platytyrae (MyP) water extract in reducing cholesterol level in hypercholesterolemia induced sprague dawley rat. *Malaysian Journal of Medical and Biological Research*. 7(2): 69–80.

Ibrahim, N. 2021. *Antitumour Activity of Myrmecodia Platytyrea Methanolic Tuber Extract on Hepatocellular Carcinoma* (Doctoral dissertation, Universiti Teknologi MARA (UiTM)).

Mohd Zin, M., Hasan, M.H., Abdul Wahab, I., Ponto, T. and Adam, A. 2013, March. Acute toxicity study of myrmecodia platytyrea aqueous tuber extract. *The Open Conference Proceedings Journal*. 4(1).

***Neonauclea gigantea* (Valeton) Merr.**

[From Greek *neo* = new, *naus* = ship, *kleio* = to close, and Latin *gigantea* = giant]

Published in *Journal of the Washington Academy of Sciences* 5: 540. 1915

Synonyms: *Nauclea gigantea* Valeton; *Neonauclea cyrtopodioides* (Wernham.) Merr.
Local name: *mahitap* (Dusun)
Habitat: forests near rivers
Geographical distribution: Borneo
Botanical description: this tree reaches a height of about 30 m. Stems: angular at apex. Leaves: simple, opposite, stipulate. Stipules: interpetiolar, about 3 cm long, ovate. Petioles: about 4 cm long, stout. Blades: broadly elliptic to lanceolate, 12–26 cm × 20–40 cm, somewhat cordate at base, acute at apex, with about 10 pairs of secondary nerves. Flowering heads: terminal, about 2.5 cm across, on approximately 7 cm long peduncles. Calyx: tubular, 5-lobed. Corolla: tubular, whitish-yellow, about 1 cm long, 5-lobed. Stamens: 5, protuding. Styles: 1. protuding. Stigma: swollen. Fruiting heads: about 4 cm across and made of numerous dehiscent capsules. Seeds: winged.
Traditional therapeutic indications: diarrhea, stomach aches, thrush (Dusun)
Pharmacology and phytochemistry: anticandidal (Rosamah et al. 2017).
Toxicity, side effects, and drug interaction: not known

Comments:

(i) The Dusun and Kadazan use a plant of the genus *Neonauclea* Merr. (1915) (local name: *intap*) for diarrhea.
(ii) *Neonauclea calycina* Merr. is used as a medicine by the Murut (local name: *kembalu*).

REFERENCE

Rosamah, E., Arung, E.T. and Siahaan, F.R. 2017. Antifungal potency from walur (Neonauclea gigantea (veleton) Merr.). *Integrated Biological Sciences for Human Welfare*. 196.

***Nauclea officinalis* (Pierre ex Pitard) Merr. & Chun.**

[From Greek *naus* = ship, *kleio* = to close, and Latin *officinalis* = of medicinal value]

Published in *Sunyatsenia* 5: 188. 1940

Synonyms: *Nauclea brunnea* Craib; *Sarcocephalus officinalis* Pierre ex Pit.
Local name: *bonggkol* (Rungus)
Habitat: forests
Geographical distribution: Cambodia, Laos, Vietnam, Malaysia, Indonesia, South China
Botanical description: this tree reaches a height of about 12 m. Stems: angular at apex. Leaves: simple, opposite, stipulate. Stipules: interpetiolar, about 6 mm–1 cm long, rounded. Petioles: about 1–1.5 cm long, stout. Blades: broadly elliptic, attenuate at base, acuminate at apex, 3.5–10 cm × 7–11 cm, with about 5–7 pairs of secondary nerves. Flowering heads: terminal, about 1.5 cm across, on about 5 cm long peduncles. Calyx: tubular, 5-lobed. Corolla: tubular, whitish-yellow, about 5 mm long, 5-lobed. Stamens: 5, protuding. Styles: 1, protruding. protuding Stigma: swollen. Fruiting heads: about 1.5 cm across, fleshy.
Traditional therapeutic indications: medicinal (Rungus)
Pharmacology and phytochemistry: anti-inflammatory indole alkaloids and lignans (Liu et al. 2017; Wang et al. 2022). Antiplasmodial indole alkaloids (Sun et al. 2008). Antiviral (HIV) indole alkaloids (Liu et al. 2019).
Toxicity, side effects, and drug interaction: not known

Comment: *Nauclea orientalis* (L.) L. is used by the Dusun for painful bowel movements (local name: *bongkol*).

REFERENCES

Liu, Q.L., Chen, A.H., Tang, J.Y., Ma, Y.L., Jiang, Z.H., Liu, Y.P., Chen, G.Y., Fu, Y.H. and Xu, W. 2017. A new indole alkaloid with anti-inflammatory activity from *Nauclea officinalis*. *Natural Product Research*. 31(18): 2107–2112.

Liu, Y.P., Liu, Q.L., Zhang, X.L., Niu, H.Y., Guan, C.Y., Sun, F.K., Xu, W. and Fu, Y.H. 2019. Bioactive monoterpene indole alkaloids from *Nauclea officinalis*. *Bioorganic Chemistry*. 83: 1–5.

Sun, J., Lou, H., Dai, S., Xu, H., Zhao, F. and Liu, K. 2008. Indole alkoloids from *Nauclea officinalis* with weak antimalarial activity. *Phytochemistry*. 69(6): 1405–1410.

Wang, G., Hou, L., Wang, Y., Liu, H., Yuan, J., Hua, H. and Sun, L. 2022. Two new neolignans and an indole alkaloid from Stems: Of *Nauclea officinalis* and their biological activities. *Fitoterapia*. 105228.

Oxyceros bispinosus **(Griff.) Tirveng.**

[From Greek *oxy* = sharp, *ceros* = horn, from Latin *bi* = two, and *spinosus* = spines]

Published in *Nordic Journal of Botany* 3(4): 466. 1983

Synonyms: *Oxyceros bispinosa* (Griff.) Tirveng.; *Stylocoryna bispinosa* Griff.; *Randia bispinosa* (Griff.) Craib; *Randia fragrantissima* Ridl.; *Randia liamsii* Elmer; *Randia uncaria* Elmer;

Local name: *kovilan* (Rungus)

Habitat: forests

Geographical distribution: from India to Papua New Guinea

Botanical description: this climbing plant grows up to a length of about 6 m. Stems: woody, terete, sometimes hairy at apex, smooth, thorned at nodes. Thorns: somewhat horn-shaped, about 1 cm long. Leaves: simple, opposite, stipulate. Stipules: broadly triangular, about 3 mm long, acuminate at apex. Petioles: about 1–2 cm long, sometimes hairy. Blades: ovate, elliptic, or lanceolate, somewhat wavy, 8–20 cm × 3–7 cm, acute at base, acute to acuminate at apex, with 5–9 pairs of secondary nerves. Cymes: about 3 cm long, terminal. Calyx: tubular, about 4 mm long, 5-lobed, the lobes lanceolate. Corolla: pure white, about 1.5–3 cm long, 5-lobed, the lobes somewhat obovate, contorted and about 1 cm long. Stamens: 5, about 1 cm long, linear. Styles: 1. protuding, about 5 mm long. Stigma: bifid. Berries: somewhat globose, 1 cm across, which can make one think of blueberries.

Traditional therapeutic indications: medicinal (Rungus)

Pharmacology and phytochemistry: not known

Toxicity, side effects, and drug interaction: not known

Paedaria verticillata **Bl.**

[From *paederos*, a plant known to Pliny and Latin *verticillatus* = arranged in a whorl]

Published in *Bijdragen tot de flora van Nederlandsch Indië* (16): 968–969. 1826

Synonym: *Hondbesseion verticillatum* (Bl.) Kuntze

Local name: *taud* (Dusun, Kadazan)

Habitat: forests

Geographical distribution: from Malaysia to the Philippines

Botanical description: this climbing plant reaches a length of about 10 m and gives off a nauseating odor. Leaves: in whorls of 3–4 or opposite, stipulate. Stipules: triangular, tiny. Petioles: about 1 cm long. Blades: ovate or elliptic, 6–17 cm × 2.5–8 cm, attenuate or rounded at base, acuminate at apex, coriaceous, somewhat hairy beneath, with 6–7 pairs of secondary nerves. Panicles: axillary or terminal, thin, up to 1 m long, with many flowers. Calyx: tubular, 5-lobed, tiny. Corolla: deep dark red or purplish to greenish-red, about 1 cm long, 5-lobed, the lobes broadly triangular. Stamens: 5. Disc: present. Styles: 1, about 1 cm long. Berries: oblate, about 1 cm across. Seeds: few, flat, winged.

Traditional therapeutic indications: intestinal worms (Dusun, Kadazan)

Pharmacology and phytochemistry: not known

Toxicity, side effects, and drug interaction: not known

Comments:

(i) In Java, the leaves of this plant are eaten as a vegetable whilst *Paederia foetida* L. is used for fever, dysentery, and herpes (local name: *kahitutan*).

(ii) Plants of the genus *Paederia* L. are used as medicines in Sabah. The Dusun and Kadazan use *"kombutong"* for dental caries, *"ubat damat"* is employed to stop bleeding from cuts by the Murut, whilst *"papaid dazing"* is used as medicine and magic rituals by the Rungus.

Psychotria gyrulosa Stapf

[From Greek *psycho* = mind and Latin *gyrulosa* = gyral]

Published in *Transactions of the Linnean Society of London* II, 4: 180. 1894

Local names: *porompong bukit, siroromuk* (Dusun)
Common name: gyrulose white coffee
Habitat: forests
Geographical distribution: Borneo
Botanical description: this handsome tree reaches a height of about 12 m. Bark: whitish to yellowish. Leaves: simple, opposite, stipulate. Stipules: interpetiolar. Petioles: 2–3 cm long. Blades: narrowly oblanceolate or obovate, 8–10 cm × 18–24 cm, tapering at base, acuminate at apex, with about 20–25 pairs of secondary nerves. Cymes: terminal, on about 7 cm long peduncles. Calyx: 5-lobed. Corolla: tubular, 5-lobed. Drupes: urn- to amphora-shaped, about 5 mm long, with an opening at apex, ribbed (Figure 4.46).
Traditional therapeutic indications: headaches (Dusun)
Pharmacology and phytochemistry: not known
Toxicity, side effects, and drug interaction: not known

Comment: *Psychotria sarmentosa* Bl. is used by the Murut for itchiness (local name: *solovondo*).

FIGURE 4.46 *Psychotria gyrulosa* Stapf

Ridsdalea pseudoternifolia (Valeton) J.T.Pereira

[After the 20th century British botanist Colin Ernest Ridsdale and Latin *pseudo* = false, and *ternifolia* = 3 leaves]

Published in *Sandakania* 21: 49. 2016

Synonym: *Gardenia pseudoternifolia* Valeton
Local name: *sarakad rahat* (Rungus)
Habitat: coastal forests
Geographical distribution: Sumatra, Borneo
Botanical description: this shrub reaches a height of about 4 m. Stems: terete, smooth, zigzag-shaped, somewhat hairy at apex. Leaves: simple, opposite, stipulate. The reduction of some leaves can make one think they are verticillate by 3. Stipules: interpetiolar, acuminate, about 3 mm long. Petioles: about 3–5 mm long. Blades: somewhat lanceolate to elliptic, attenuate at base, acuminate at apex, with about 6 pairs of secondary nerves, 2–8 cm × 7–15 cm, membranous, somewhat hairy. Cymes: axillary or terminal, about 3 cm long. Calyx: hairy, 5-lobed, tubular. Corolla: white, 5-lobed. Stamens: 5. Stigma: bifid. Berries: pear-shaped or globose, green, up to about 8 cm long. Seeds: numerous.
Traditional therapeutic indications: medicinal (Rungus)
Pharmacology and phytochemistry: not known
Toxicity, side effects, and drug interaction: not known

Uncaria acida (W. Hunter) Roxb.

[From Latin *uncus* = hook and *acidus* = sour]

Published in *Hortus Bengalensis, or a catalogue. . .* 86. 1814

Synonyms: *Uncaria acida* Roxb.; *Uncaria ovalifolia* Roxb.; *Uncaria forbesii* Wernh.; *Uncaria firma* Val.
Local name: *langkawit* (Dusun)
Habitat: forests
Geographical distribution: from Bangladesh to Papua New Guinea
Botanical description: this woody climbing plant grows up to a length of about 6 m. Stems: hairy, angular, hooked at nodes. Hooks: in pairs, revolute, ligneous, about 1 cm long. Leaves: simple, opposite, stipulate. Stipules: interpetiolar, hairy, lanceolate, bilobed, about 5–8 mm long. Petioles: hairy, 3 mm–1.5 cm long. Blades: elliptic-lanceolate, hairy beneath, 7–10 cm × 6–8 cm, with about 4–7 pairs of secondary nerves, glossy above, with scalariform tertiary nerves, rounded to acute at base, acuminate at apex. Flower heads: axillary, up to about 3 cm across on about 3 cm long peduncles. Calyx: tubular, the tube 3 mm long and 5-lobed, the lobes 3 mm long. Corolla: tubular, the tube 7 mm long, whitish, 5-lobed, somewhat hairy. Stamens: 5. Styles: 1, protuding. Fruiting heads: globose. Capsules: fusiform, dehiscent.
Traditional therapeutic indications: dental caries (Dusun)
Pharmacology and phytochemistry: not known
Toxicity, side effects, and drug interaction: not known

Comments:

(i) The Dusun and Kadazan use a plant of the genus *Uncaria* Schreb (1789) they call "*kalawit*" for the treatment of gout, blood vomiting, feverish colds, coughs, and headaches. The Dusun use another plant in this genus (local name: *kalait*) for itchiness.
(ii) *Uncaria cordata* (Lour.) Merr. is used by the Dusun and Kadazan for fever, sprains, fatigue, headaches, and bloody stools.
(iii) *Uncaria ferrea* (Bl.) DC is medicinal for the Rungus (local name: *ingangit*).
(iv) *Uncaria gambier* (W. Hunter) Roxb. is used by the Bajau for joints pain, flu, and wounds (local name: *gambir*).
(v) A plant of the genus *Neolamarckia* Bosser (1984) is used by the Dusun and Kadazan for beriberi (local name: *towo*).
(vi) A plant of the genus *Pavetta* L. (1753) is medicinal for the Murut (local name: *buntungon*).
(vii) *Lasianthus inaequalis* Bl. is used by the Bajau for fever (local name: *pikolas*).
(viii) The endemic *Praravinia suberosa* (Merr.) Bremek is medicinal for the Murut (local name: *kingkimu*).
(ix) The endemic *Urophyllum nigricans* Wernham is medicinal for the Murut.
(x) *Chassalia chartacea* Craib is used by the Dusun for blurred vision (local name: *lanci*).

4.1.14.2 Order Lamiales Bromhead (1838)

4.1.14.2.1 Family Acanthaceae Juss. (1789), the Acanthus Family

The family Acanthaceae consists of about 250 genera and 3,000 species of herbaceous plants, climbing plants, or shrubs. Stems: often ligneous, smooth, swollen at nodes. Leaves: simple, opposite, exstipulate. Inflorescences: cymes, spikes, racemes, panicles. Calyx: tubular, 4-5-lobed persistent in fruits. Corolla: tubular, often bilabiate, membranous, caducous. Stamens: 2–4. Carpels: 2, forming a bilocular ovary, each locule containing 2–many ovules. Styles: 1. Fruits: capsules. Seeds: numerous, tiny, disc-shaped.

***Clinacanthus nutans* (Burm.f.) Lindau**

[From Greek *kline* = a bed, *akantha* = a thorn, and Latin *nutans* = nodding]

Published in *Botanische Jahrbücher für Systematik, Pflanzengeschichte und Pflanzengeographie* 18: 63. 1893

Synonyms: *Clinacanthus burmanni* Nees; *Justicia nutans* Burm.f.
Local name: *belalai gajah* (Bajau)
Common name: Sabah snake grass
Habitat: cultivated
Geographical distribution: Southeast Asia
Botanical description: this herbaceous plant reaches a height of about 1 m. Stems: smooth, terete, somewhat soft. Interpetiolar nodes: cup-shaped. Leaves: simple, opposite, exstipulate. Petioles: 3 mm–1 cm long. Blades: lanceolate, wavy, somewhat light green, glossy, fleshy, soft, 1–4 cm × 7–12 cm, with 4–6 pairs of secondary nerves, attenuate to rounded at base, acuminate at apex. Panicles: terminal. Calyx: tubular, about 1 cm long, 5-lobed, the lobes linear. Corolla: tubular, which can make one think of a crocodile mouth, dull light red, about 3–4 cm long, bilabiate, the lower lip trifid. Stamens: 2, 1.5 cm long. Styles: 1, thin, about 3 cm long. Capsules: about 2 cm long. Seeds: numerous, 2 mm long.
Traditional therapeutic indications: diabetes, hypertension, cancer, leukemia, piles (Bajau)
Pharmacology and phytochemistry: toxic to cervical cancer cells (HeLa) and leukemia cells (K562) (Arullappan et al. 2014). Vasoprotective (Azemi et al. 2020), antidiabetic (Umar Imam et al. 2019), and antibacterial (Lim et al. 2020).
Lignans (2-methoxy-9*β*-hydroxydiasesamin) toxic to cervical cancer cells (HeLa), breast cancer cells (MCF-7), and non-small cell lung cancer cells (A549) (Diao et al. 2019). Other cytotoxic principles in this plant are cycloartane triterpenes (Xu et al. 2021).
Toxicity, side effects, and drug interaction: the median lethal dose (LD_{50}) of a methanol extract of leaves administered orally to rats was above 1.8 g/kg (P'ng et al. 2012).

REFERENCES

Arullappan, S., Rajamanickam, P., Thevar, N. and Kodimani, C.C. 2014. *In vitro* screening of cytotoxic, antimicrobial and antioxidant activities of *Clinacanthus nutans* (Acanthaceae) leaf extracts. *Tropical Journal of Pharmaceutical Research*. 13(9): 1455–1461.

Azemi, A.K., Mokhtar, S.S. and Rasool, A.H.G. 2020. *Clinacanthus nutans* leaves extract reverts endothelial dysfunction in type 2 diabetes rats by improving protein expression of eNOS. *Oxidative Medicine and Cellular Longevity*. 2020: 1–10.

Diao, H.Z., Chen, W.H., Cao, J., Shao, T.M., Song, X.P. and Han, C.R. 2019. Furofuran lignans and alkaloids from *Clinacanthus nutans*. *Natural Product Research*. 33(9): 1317–1321.

Lim, S.H.E., Almakhmari, M.A., Alameri, S.I., Chin, S.Y., Abushelaibi, A., Mai, C.W. and Lai, K.S. 2020. Antibacterial activity of *Clinacanthus nutans* polar and non-polar leaves and stem extracts. *Biomedical and Pharmacology Journal*. 13: 1169–1175.

P'ng, X.W., Akowuah, G.A. and Chin, J.H. 2012. Acute oral toxicity study of *Clinacanthus nutans* in mice. *International Journal of Pharmaceutical Sciences and Research*. 3(11): 4202.

Umar Imam, M., Ismail, M., George, A., Chinnappan, S.M. and Yusof, A. 2019. Aqueous leaf extract *of Clinacanthus nutans* improved metabolic indices and sorbitol-related complications in type II diabetic rats (T2D). *Food Science & Nutrition*. 7(4): 1482–1493.

Xu, W., Li, J., Li, D., Tan, J., Ma, H., Mu, Y., Wen, Y., Gan, L., Huang, X. and Li, L. 2021. Chemical characterization, antiproliferative and antifungal activities of *Clinacanthus nutans*. *Fitoterapia*. 155: 105061.

***Justicia gendarussa* Burm.f.**

[After the 18th century Scottish horticulturalist James Justice and from Sanskrit *gandharasa* = sweet odor]

Published in *Flora Indica. . . nec non Prodromus Florae Capensis* 10. 1768

Synonym: *Gendarussa vulgaris* Nees

Local names: *gombizau tobilung, insasahi, solimbangan* (Murut); *tambiau taragang, sabungum* (Dusun); *gandarusa, sarimbangun hitam* (Kedayan); *sikapapar, tolonsi* (Kadazan, Dusun)

Common name: gendarussa

Habitat: cultivated, gardens, villages

Geographical distribution: from India to Papua New Guinea

Botanical description: this shrub reaches a height of about 1.2 m. Stems: terete, stiff, swollen at nodes, somewhat dark purplish and glossy. Leaves: simple, opposite, exstipulate. Petioles: 3 mm–1 cm long. Blades: narrowly lanceolate, 5–10 cm × 1–1.5 cm, with 5–8 pairs of secondary nerves, attenuate at base, somewhat coriaceous, acute at apex. Spikes: terminal and axillary, about 10 cm long. Calyx: tubular, 5 mm long, 5-lobed. Corolla: tubular, dirty white, about 1.5 cm long, bilabiate, the lower lip with purple lines and somewhat trilobed. Stamens: 2, about 5 mm long. Styles: thin, about 1 cm long. Capsules: elongated to club-shaped, about 1 cm long.

Traditional therapeutic indications: postpartum, coughs, muscle aches, flu, boils, lumbago, aphrodisiac, arthritis (Dusun, Kadazan), flatulence (Dusun), stomach aches (Murut), diarrhea, swellings (Kedayan), rheumatisms, headaches (Dusun, Kadazan, Kedayan)

Pharmacology and phytochemistry: antibacterial (Nirmalraj et al. 2015), anti-inflammatory (Saleem et al. 2011).

Anti-inflammatory lignans (Zhang et al. 2021). Antiviral (HIV) lignan (justiprocumin B) (Zhang et al. 2017). Cytotoxic coumarin glycosides (Sun et al. 2019).

Toxicity, side effects, and drug interaction: the median lethal dose (LD_{50}) of an ethanol extract of leaves administered orally to rats was 1 g/kg (Paval et al. 2009).

Comments:

(i) The Murut use the plant for magic rituals whilst the Dusun use it to eliminate nightmares.
(ii) A plant of the genus *Acanthus* L. (1753) (local name: *tahipai*) is used by the Dusun and Kadazan for ear diseases.
(iii) A plant of the genus *Hemigraphis* Nees (1847) is used by the Dusun and Kadazan for chest pain, headaches, and fatigue.
(iv) A plant of the genus *Hypoestes* Sol. ex R.Br. (1810) is medicinal for the Murut (local name: *matopait*).
(v) *Andrographis paniculata* Nees is used for hypertension by the Bajau, Dusun, Kedayan, and Murut (local name: *hempedu bumi*), for fever, cuts, insect and snakebites (Kedayan), and skin diseases (Bajau). The Chinese people of Sabah use this plant for coronavirus disease 2019 (COVID-19).
(vi) *Graptophyllum pictum* (L.) Nees ex Griff. is medicinal in Sabah (local name: *lalamih*).

REFERENCES

Nirmalraj, S., Ravikumar, M., Mahendrakumar, M., Bharath, B. and Perinbam, K. 2015. Antibacterial and anti-inflammatory activity of *Justicia gendarussa* Burm.f. Leaves. *Journal of Plant Sciences*. 10(2): 70.

Paval, J., Kaitheri, S.K., Potu, B.K., Govindan, S., Kumar, R.S., Narayanan, S.N. and Moorkoth, S. 2009. Anti-arthritic potential of the plant *Justicia gendarussa* Burm f. *Clinics*. 64: 357–362.

Saleem, T.M., Azeem, A.K., Dilip, C., Sankar, C., Prasanth, N.V. and Duraisami, R. 2011. Anti—inflammatory activity of the leaf extracts of *Gendarussa vulgaris* Nees. *Asian Pacific Journal of Tropical Biomedicine*. 1(2): 147–149.

Sun, Y., Gao, M., Chen, H., Han, R., Chen, H., Du, K., Zhang, Y., Li, M., Si, Y. and Feng, W. 2019. Six new Coumarin glycosides from the aerial parts of *Gendarussa vulgaris*. *Molecules*. 24(8): 1456.

Zhang, H.J., Rumschlag-Booms, E., Guan, Y.F., Liu, K.L., Wang, D.Y., Li, W.F., Cuong, N.M., Soejarto, D.D., Fong, H.H. and Rong, L. 2017. Anti-HIV diphyllin glycosides from *Justicia gendarussa*. *Phytochemistry*. 136: 94–100.

Zhang, H.X., Xia, Z., Xu, T.Q., Chen, Y.M. and Zhou, G.X. 2021. New compounds from the aerial parts of *Justicia gendarussa* Burm.f. and their antioxidant and anti-inflammatory activities. *Natural Product Research*. 35(20): 3478–3486.

4.1.14.2.2 Family Avicenniaceae Miq. (1845)

The family Avicenniaceae consists of the single genus *Avicennia* L. (1753)

Avicennia marina (Forssk.) Vierh.

[After the 11th century Persian physician Avicenna and Latin *marinus* = of the sea]

Published in *Denkschriften der Kaiserlichen Akademie der Wissenschaften, Wien. Mathematisch-naturwissenschaftliche Klasse* 71: 435. 1907

Synonyms: *Avicennia alba* Bl., *Avicennia officinalis* L., *Sceura marina* Forssk
Local name: *api api* (Bajau)
Habitat: mangroves
Geographical distribution: from India to the Pacific Islands
Botanical description: this shrub reaches a height of about about 3 m. Pneumatophores: about 20 cm tall. Stems: terete, somewhat striated, swollen at nodes. Leaves: simple, opposite, subsessile, exstipulate. Petioles: up to about 3 mm long. Blades: coriaceous, broadly lanceolate to elliptic, tapering at base, 2–8 cm × 1.5–3.5 cm, rounded at apex, somewhat glossy above, hairy beneath, with 5–6 pairs of secondary nerves. Spikes: terminal or axillary, on about 2 cm long peduncles. Calyx: hairy, about 3 mm long, 5-lobed. Corolla: tubular, 4-lobed, about 5 mm across, fleshy, somewhat brownish. Stamens: 4, sessile. Ovary: hairy. Styles: 1. Stigma: bifid. Capsules: smooth, which can make one think of pistachio fruits, pointed at apex, on persistent calyx, almost glaucous, about 1 cm × 2 cm.
Traditional therapeutic indications: stingray bites (Bajau)
Pharmacology and phytochemistry: anti-inflammatory (Gandomani et al. 2012), antibacterial (Okla et al. 2021), hypoglycemic (Aljaghthmi et al. 2017), anticancer (Abbas Momtazi-borojeni et al. 2013).
Naphthoquinones (Ito et al. 2000).
Toxicity, side effects, and drug interaction: the median lethal dose (LD_{50}) of an aqueous extract of fruit administered to mice was above 10 g/kg (Xiu-mei et al. 2008).

Comments:

(i) The young leaves of this plant are used as food by the Bajau.
(ii) This family has been transferred to the Acanthaceae or Verbernaceae which makes no sense phenotypically.

REFERENCES

Abbas Momtazi-borojeni, A., Behbahani, M. and Sadeghi-Aliabadi, H. 2013. Antiproliferative activity and apoptosis induction of crude extract and fractions of *Avicennia marina*. *Iranian Journal of Basic Medical Sciences*. 16(11): 1203.

Aljaghthmi, O.H., Heba, H.M. and Zeid, I.A. 2017. Antihyperglycemic properties of mangrove plants (*Rhizophora mucronata* and *Avicennia marina*): An overview. *Advances in Biological Research*. 11(4): 161–170.

Gandomani, M.Z., Molaali, E.F., Gandomani, Z.Z., Madani, H. and Moshtaghian, S.J. 2012. Evaluation of anti-inflammatory effect of hydroalcoholic extract of Mangrove (*Avicennia marina*) leaves in male rats. *Medical Journal of Tabriz University of Medical Sciences*. 35(4): 80–85.

Ito, C., Katsuno, S., Kondo, Y., Tan, H.T.W. and Furukawa, H. 2000. Chemical constituents of *Avicennia alba*. Isolation and structural elucidation of new naphthoquinones and their analogues. *Chemical and Pharmaceutical Bulletin*. 48(3): 339–343.

Okla, M.K., Alatar, A.A., Al-Amri, S.S., Soufan, W.H., Ahmad, A. and Abdel-Maksoud, M.A. 2021. Antibacterial and antifungal activity of the extracts of different parts of Avicennia marina (Forssk.) Vierh. *Plants*. 10(2): 252.

Xiu-mei, G., Wei-Dong, H. and Zen-ji, Y. 2008. The acute toxicity and bone-merrow micronucleus tests of water extract from avicennia marina fruits in mice. *Journal of Coastal Development*. 11(2): 70–74.

4.1.14.2.3 Family Bignoniaceae Juss. (1789), the Trumpet-creeper Family

The family Bignoniaceae consists of about 120 genera and 750 species of trees or shrubs. Leaves: opposite, simple, or compound, exstipulate. Calyx: tubular, 5-lobed. Corolla: tubular, 5-lobed, membranous, caducous, bilabiate. Stamens: 4–5. Carpels: 2, forming a 1–4 locular ovary, each locule with 6–many ovules. Styles: 1. Fruits: capsules, berries. Seeds: winged.

***Oroxylum indicum* (L.) Vent.**

[From *oro* = mountain, *xylon* = wood, and Latin *indicum* = from India]

Published in *Forest Flora of British Burma* 2: 237. 1877

Synonyms: *Calosanthes indica* Bl.; *Bignonia indica* L., *Bignonia pentandra* Lour.; *Spathodea indica* (L.) Pers.
Local name: *ulunan sangku* (Murut)
Common names: Indian trumpet, tree of Damocles, trumpet flowers, broken bones
Habitats: roadsides, around villages, on the banks of rivers, swamps
Geographical distribution: from India to the Philippines
Botanical description: this sinister tree reaches a height of about 12 m. Wood: light yellow. Bark: light brown, fissured, soft, secretes a green juice once incised, with numerous corky lenticels. Leaves: enormous (90 cm–1.5 m long), 2–3-pinnate, with opposite pinnae. Rachis: stout, cylindrical. Petiolules: 6 mm–1.5 cm long. Folioles: 3–6 cm × 10–12 cm, ovate, or elliptic, acuminate at apex, rounded at base. Panicles: upright, about 60 cm long, with numerous, monstrous, caducous, stenchy flowers blooming at night. Calyx: about 2 cm long, coriaceous, oblong, campanulate. Corolla: purple outside, greenish inside, about 10 cm long, fleshy, 5-lobed, the lobes incised and about 4 cm long. Stamens: 5, filaments hairy at base. Capsules: somewhat sword- to canoe-shaped, about 1.2 m × 5 cm, dark purplish brown, coriaceous, smooth. Seeds: numerous, 7 cm × 3.5 cm, with membranous and translucent wings.
Traditional therapeutic indications: swellings (Murut); cuts, wounds, sprains, vomiting, skin disease (Dusun, Kadazan)
Pharmacology and phytochemistry: this plant has been the subject of numerous pharmacological studies (Dev et al. 2010; Jagetia 2021) which have highlighted, among other things, anti-inflammatory activity (Lalrinzuali et al. 2016). Flavones (baicalein, oroxylin A), naphthoquinones (lapachol) (Ali et al. 1998; Siddiqui et al. 2012). Anti-inflammatory flavone (chrysin) (Wu et al. 2019) and naphthoquinone (lapachol) (Ali et al. 1998). Leishmanicidal flavones (Das et al. 2006).
Toxicity, side effects, and drug interaction: methanol extract of leaves administered orally to mice at the dose of 5 g/kg was found to cause no signs of toxicity (Farhan Hanif Reduan et al. 2020).

REFERENCES

Ali, R.M., Houghton, P.J., Raman, A. and Hoult, J.R.S. 1998. Antimicrobial and antiinflammatory activities of extracts and constituents of *Oroxylum indicum* (L.) Vent. *Phytomedicine*. 5(5): 375–381.

Das, B.B., Sen, N., Roy, A., Dasgupta, S.B., Ganguly, A., Mohanta, B.C., Dinda, B. and Majumder, H.K. 2006. Differential induction of Leishmania donovani bi-subunit topoisomerase I—DNA cleavage complex by selected flavones and camptothecin: Activity of flavones against camptothecin-resistant topoisomerase I. *Nucleic Acids Research*. 35(4): 1121–1132.

Dev, L.R., Anurag, M. and Rajiv, G. 2010. *Oroxylum indicum*: A review. *Pharmacognosy Journal*. 2(9): 304–310.

Farhan Hanif Reduan, M., Hamid, F.F.A., Nordin, M.L., Shaari, R., Hamdan, R.H., Teik Chung, E.L., Peng, T.L., Kamaruzaman, I.N.A. and Noralidin, N. 2020. Acute oral toxicity study of ethanol extract of *Oroxylum indicum* leaf in mice. *The Thai Journal of Veterinary Medicine*. 50(4): 573–581.

Jagetia, G.C. 2021. A review on the medicinal and pharmacological properties of traditional ethnomedicinal plant sonapatha, *Oroxylum indicum*. *Sinusitis*. 5(1): 71–89.

Lalrinzuali, K., Vabeiryureilai, M. and Jagetia, G.C. 2016. Investigation of the anti-inflammatory and analgesic activities of ethanol extract of stem bark of Sonapatha *Oroxylum indicum in vivo*. *International Journal of Inflammation*. 2016.

Siddiqui, W.A., Ahad, A., Ganai, A.A., Sareer, O., Najm, M.Z., Kausar, M.A. and Mohd, M. 2012. Therapeutic potential of *Oroxylum indicum*: A review. *Journal of Research and Opinion*. 2(10): 163–172.

Wu, B.L., Wu, Z.W., Yang, F., Shen, X.F., Wang, L., Chen, B., Li, F. and Wang, M.K. 2019. Flavonoids from the seeds of *Oroxylum indicum* and their anti-inflammatory and cytotoxic activities. *Phytochemistry Letters*. 32: 66–69.

4.1.14.2.4 Family Lamiaceae Martinov (1820), the Mint Family

The family Lamiaceae consists of about 210 genera and 3,500 species of herbaceous plants or shrubs. Stems: quadrangular, often hairy (trichomes filled with essential oils). Leaves: simple, opposite, exstipulate. Inflorescences: verticillasters, heads. Calyx: 5-lobed. Corolla: bilabiate, the upper lip bilobed and the lower lip trilobed. Stamens: 2–5. Carpels: 2, forming a 4 locular ovary, each locule with 1–2 ovules. Styles: 1, protuding. Fruits: nutlets in persistent calyx.

***Hyptis capitata* Jacq.**

[From Greek *hyptios* = bent back and Latin *capitata* = having a head]

Published in *Collectanea* 1: 102–103. 1786

Synonyms: *Clinopodium capitatum* (Jacq.) Sw.; *Hyptis celebica* Zipp. ex Koord.; *Hyptis macrochila* Mart. ex Steud.; *Hyptis pittieri* Briq.; *Hyptis rhomboidea* M. Martens & Galeotti; *Mesosphaerum capitatum* (Jacq.) Kuntze; *Thymus virginicus* Blanco

Local names: *baing baing, rumput kepala bulat* (Murut); *bala* (Dusun)

Common name: knobweed

Habitat: roadsides, wastelands, villages

Geographical distribution: from India to Pacific Islands

Botanical description: this herbaceous plant reaches a height of about 1 m. Stems: hollowed, quadrangular, somewhat ligneous and stiff, with 7–10 cm long internodes. Leaves: opposite, simple, exstipulate. Petioles: about 2–3 cm long. Blades: 0.5–2.2 cm × 4–10.5 cm, dull green, broadly lanceolate, serrate, tapering at base, acute at apex, with about 5 pairs of secondary nerves. Flowering heads: axillary, globose, 1.5–2.5 cm across, on 3–3.5 cm long and thin peduncles. Calyx: tubular, tiny, membranous, hairy at base, striated, 5-lobed, the lobes linear. Corolla: white with light purplish blotches on the upper lip, about 5 mm long, 5-lobed. Stamens: 4. Styles: 1. protuding. Nutlets: brownish black in persistent calyx.

Traditional therapeutic indications: stomach aches (Murut); fever, colds, boils (Dusun)

Pharmacology and phytochemistry: this plant has been the subject of numerous pharmacological studies (To'bungan et al. 2022) which have highlighted, among other things, gastroprotective activity (Jesus et al. 2013).

Lignans, pyrones (Almtorp et al. 1991). Gastroprotective triterpene (oleanolic acid) (Astudillo et al. 2002). Antiviral (HIV) triterpenes (pomolic acid, oleanolic acid) (Kashiwada et al. 1998). Cytotoxic triterpenes (hyptatic acid A and B) (Yamagishi et al. 1988).

Toxicity, side effects, and drug interaction: not known

REFERENCES

Almtorp, G.T., Hazell, A.C. and Torssell, K.B. 1991. A lignan and pyrone and other constituents from *Hyptis capitata*. *Phytochemistry*. 30(8): 2753–2756.

Astudillo, L., Rodriguez, J.A. and Schmeda-Hirschmann, G. 2002. Gastroprotective activity of oleanolic acid derivatives on experimentally induced gastric lesions in rats and mice. *Journal of Pharmacy and Pharmacology*. 54(4): 583–588.

Jesus, N.Z.T., Falcão, H.S., Lima, G.R.M., Caldas Filho, M.R.D., Sales, I.R.P., Gomes, I.F., Santos, S.G., Tavares, J.F., Barbosa-Filho, J.M. and Batista, L.M., 2013. Hyptis suaveolens (L.) Poit (Lamiaceae), a medicinal plant protects the stomach against several gastric ulcer models. *Journal of Ethnopharmacology*, 150(3): 982–988.

Kashiwada, Y., Wang, H.K., Nagao, T., Kitanaka, S., Yasuda, I., Fujioka, T., Yamagishi, T., Cosentino, L.M., Kozuka, M., Okabe, H. and Ikeshiro, Y. 1998. Anti-AIDS agents. 30. Anti-HIV activity of oleanolic acid, pomolic acid, and structurally related triterpenoids. *Journal of Natural Products*. 61(9): 1090–1095.

To'bungan, N., Widyarini, S., Nugroho, L.H. and Pratiwi, R., 2022. Ethnopharmacology of Hyptis capitata. *Plant Science Today*, 9(3): 593–600.

Yamagishi, T., Zhang, D.C., Chang, J.J., McPhail, D.R., McPhail, A.T. and Lee, K.H. 1988. The cytotoxic principles of Hyptis capitata and the structures of the new triterpenes hyptatic acid-A and-B. *Phytochemistry*. 27(10): 3213–3216.

***Orthosiphon stamineus* Benth.**

[from Greek *ortho* = straight, *siphon* = tube, and Latin *stamineus* = full of threads]

Published in *Plantae Asiaticae Rariores* 2: 15. 1830

Synonyms: *Clerodendrum spicatum* Thunb.; *Orthosiphon aristatus* (Blume) Miq.; *Orthosiphon spicatus* Benth.

Local names: *misai kucing* (Dusun, Murut, Bajau, Kedayan); *kumis kucing* (Lundayeh)

Common names: cat's whiskers, Java tea

Habitat: cultivated

Geographical distribution: from India to Pacific Islands

Botanical description: this beautiful herbaceous plant reaches a height of about 1.5 m. Stems: quadrangular, dark purple. Leaves: simple, opposite, exstipulate. Petioles: 0.5–1.5 cm long. Blades: fleshy, lanceolate, serrate, 1–3.5 cm × 2–5.5 cm, with about 4 pairs of secondary nerves, attenuate at base, acute at apex. Verticillasters: about 6 flowers. Calyx: tubular, about 5 mm long, hairy, somewhat bilabiate. Corolla: heavenly pure white, bilabiate, the tube up to about 2 cm long, the lower lip trilobed and about 5 mm long. Stamens: 5, filamentous, pure white, thin, curved upward, up to about 6 cm long, which can make one think of fine jets of water springing from the corolla. Nutlets: ovoid, tiny.

Traditional therapeutic indications: obesity, diuretic (Dusun); hypertension (Bajau, Murut, Sungai, Dusun, Kadazan, Lundayeh, Rungus); kidney failure (Kedayan, Dusun); diabetes (Bajau, Kedayan, Murut, Dusun, Lundayeh)

Pharmacology and phytochemistry: this plant has been the subject of numerous pharmacological studies which have highlighted, among other things, diuretic (Arafat et al. 2008), antidiabetic (Sriplang et al. 2007), and antibacterial activity (Romulo et al. 2018).

Hypotensive and bradycardiac chromene (methylripariochromene) (Matsubara et al. 1999).

Toxicity, side effects, and drug interaction: ethanol extract administered orally to rats at the dose of 5 g/kg/day for 28 days was found to cause no signs of toxicity (Mohamed et al. 2011).

Comments:

(i) *Coleus blumei* Benth. is used for abscesses, asthma, coughs, fever, headaches, and smallpox by the Kedayan (local name: *ati ati*).

(ii) *Mentha arvensis* L. is used by the Bajau for earaches (local name: *pudina*).

(iii) *Ocimum basilicum* L. is used by the Lundayeh for fever (local name: *bawing*).

(iv) *Ocimum tenuiflorum* Burm.f. is used for coughs and gastric ulcers by the Kedayan (local name: *kemangi*) whilst it is a remedy for indigestion, headaches, coughs, asthma, fever, and colds for the Bajau.

(v) The endemic *Hosea lobbii* (C.B. Clarke) Ridl. is used by the Dusun for vomiting blood (local name: *tagalap*).

(vi) A plant of the genus *Gomphostemma* Benth. (1830) is used by the Murut as styptic.

(vii) In the family Oleaceae Hoffmanns. & Link (1809), *Jasminum aculeatum* (Blanco) Walp. ex Hassk. is used for flatulence by the Murut (local name: *onsom onsom*), whilst *Jasminum bifarium* Wall is used for sore eyes by the Lundayeh (local name: *bunga melor*).

REFERENCES

Arafat, O.M., Tham, S.Y., Sadikun, A., Zhari, I., Haughton, P.J. and Asmawi, M.Z. 2008. Studies on diuretic and hypouricemic effects of *Orthosiphon stamineus* methanol extracts in rats. *Journal of Ethnopharmacology*. 118(3): 354–360.

Matsubara, T., Bohgaki, T., Watarai, M., Suzuki, H., Ohashi, K. and Shibuya, H. 1999. Hypotensive actions of methylripariochromene A from *Orthosiphon aristatus*, an Indonesian traditional medicinal plant. *Biological and Pharmaceutical Bulletin*. 22(10): 1083–1088.

Mohamed, E.A.H., Lim, C.P., Ebrika, O.S., Asmawi, M.Z., Sadikun, A. and Yam, M.F. 2011. Toxicity evaluation of a standardised 50% ethanol extract of *Orthosiphon stamineus*. *Journal of Ethnopharmacology*. 133(2): 358–363.

Romulo, A., Zuhud, E.A., Rondevaldova, J. and Kokoska, L. 2018. Screening of *in vitro* antimicrobial activity of plants used in traditional Indonesian medicine. *Pharmaceutical Biology*. 56(1): 287–293.

Sriplang, K., Adisakwattana, S., Rungsipipat, A. and Yibchok-Anun, S. 2007. Effects of *Orthosiphon stamineus* aqueous extract on plasma glucose concentration and lipid profile in normal and streptozotocin-induced diabetic rats. *Journal of Ethnopharmacology*. 109(3): 510–514.

4.1.14.2.5 Family Gesneriaceae Rich. & Juss. (1816), the Gerneriad Family

The family Gesneriaceae consists of about 140 genera and 3,000 species of herbaceous plants or shrubs. Stems: soft, often sappy. Leaves: simple, opposite, exstipulate. Inflorescences: cymes. Calyx: tubular, 5-lobed. Corolla: tubular, bilabiate, membranous, caducous. Stamens: 2–5. Carpels: 2, forming a unilocular or bilocular ovary, each locule with 2–8 ovules. Styles: 1. Fruits: capsules. Seeds: numerous.

***Cyrtandra areolata* (Stapf) B.L. Burtt**

[From Greek *kurtos* = curved, *andros* = male, and Latin *areola* = a small open space]

Published in *Notes from the Royal Botanic Garden, Edinburgh* 30: 26. 1970

Synonym: *Didymocarpus areolatus* Stapf
Local name: *pohodo* (Murut)
Habitat: forests
Geographical distribution: Borneo
Botanical description: this herbaceous plant reaches a height of about 2 m. Stems: stout, hairy. Leaves: simple, opposite, exstipulate. Petioles: about 4–7 cm long. Blades: oblong to elliptic, about 8 cm × 20 cm, tapering at base, acute at apex, sharply denticulate, with tiny pentagonal or hexagonal areoles above, hairy beneath, with about 16 pairs of secondary nerves. Cymes: axillary, hairy, sessile. Calyx: tubular, 5-lobed, the lobes linear, hairy, about 8 mm long. Corolla: white, about 2 cm long, bilabiate, the upper lip about 8 mm long, the lower lip about 1 cm long and with yellow blotches. Stamens: 4, about 1 cm long. Styles: 1, linear, about 2 cm long (Figure 4.47).

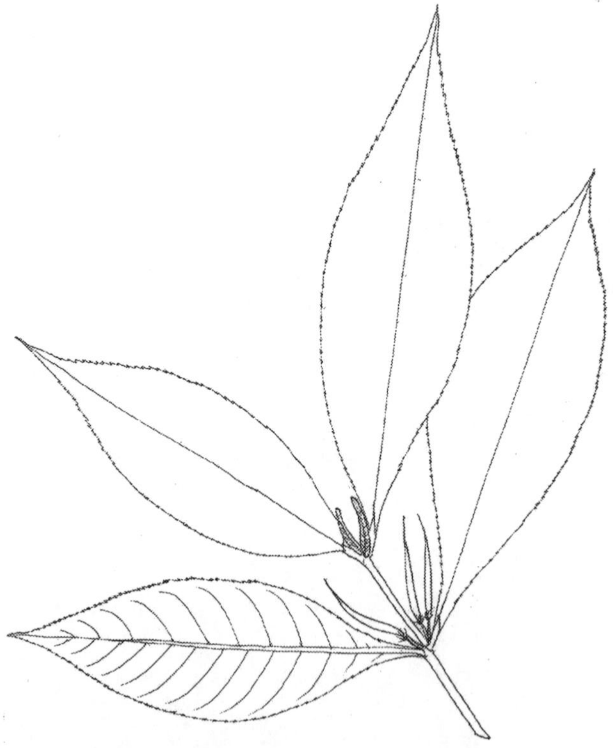

FIGURE 4.47 *Cyrtandra areolata* (Stapf) B.L. Burtt

Traditional therapeutic indications: bloody stools, skin diseases (Murut)
Pharmacology and phytochemistry: not known
Toxicity, side effects, and drug interaction: not known

Comments:

(i) A plant of the genus *Cyrtandra* J.R. Forst. & G. Forst. (1775) is used by the Dusun and Kadazan for swollen ankles (local name: *lumpoh*).
(ii) *Cyrtandromoea grandis* Ridl. is medicinal for the Murut (local name: *setawar*).
(iii) In the family Plantaginaceae Juss. (1789), *Plantago major* L. is used as a diuretic, for dysmenorrhea, hypertension, flu, coughs, and indigestion by the Bajau (local name: *ekor anjing*), anemia, cancer, wounds, and indigestion by the Lundayeh (local name: *bunga*), for urinary tract infections by the Dusun and Kadazan (local name: *kulung*), and for diabetes by the Dussun, Kadazan, and Lundayeh.

4.1.14.2.6 Family Scrophulariaceae Juss. (1789), the Figwort Family

The family Scrophulariaceae consists of about 280 genera and 3,000 species of herbaceous plants, shrubs, or trees. Leaves: simple, alternate, opposite or whorled, exstipulate. Inflorescences: racemes, spikes, cymes, or solitary. Calyx: tubular, 3–5-lobed. Corolla: bilabiate or 4–5-lobed. Carpels: 2, forming a bilocular ovary, each locule with 1–many ovary. Styles: 1. Fruits: capsules.

***Scoparia dulcis* L.**

[From Latin *scopa* = broom and *dulcis* = sweet]
Published in *Species Plantarum* 1: 116. 1753

Synonyms: *Capraria dulcis* Kuntze, *Gratiola micrantha* Nutt.; *Scoparia grandifl ora* Nash.
Local names: *ingat ingat kasar, telensi lensi* (Dusun); *ingat ingat* (Murut); *kolimpang* (Dusun, Kadazan); *mini sopan tinggi* (Kedayan)
Common names: licorice weed, sweet broomweed
Habitat: parking lots, roadsides, drains
Geographical distribution: tropical Asia, Pacific Islands
Botanical description: this beautiful herbaceous plant reaches a height of about 70 cm and has a positive aura. Stems: angular, somewhat stiff. Leaves: simple, opposite, subsessile, exstipulate. Blades: narrowly elliptic-lanceolate, up to about 2 cm × 5 mm, attenuate at bases, serrate, obtuse at apex. Flowers: solitary, axillary, tiny. Calyx: 5-lobed. Corolla: tubular, hairy at throat, pure white to light purplish, 4-lobed, the lobes about 3 mm long and ovate. Stamens: 4. Capsules: about 3 mm across.
Traditional therapeutic indications sprains (Dusun); fever, arthritis, malaise, stomach aches (Dusun, Kadazan)
Pharmacology and phytochemistry: anti-inflammatory (De Farias Freire et al. 1993), leishmanicidal (Gachet et al. 2010).
Antiviral (herpes simplex virus) diterpene (scopadulcic acid B) (Hayashi et al. 1988; Hayashi et al. 1990; Phan et al. 2006). Anti-inflammatory and cytotoxic diterpene (dulcidiol) (Phan et al. 2006; Ahsan et al. 2003). Antiplasmodial diterpene (scopadulcic acid A) (Riel et al. 2002). Anti-inflammatory triterpene (glutinol) (De Farias Freire et al. 1993).
Toxicity, side effects, and drug interaction: the median lethal dose (LD_{50}) of a methanol extract of leaves administered to mice was above 8 g/kg (Abere et al. 2015).

Comment: A plant of the genus *Scrophula* L. (1753) is used by the Dusun for asthma (local name: *mata mata*).

REFERENCES

Abere, T.A., Okoye, C.J., Agoreyo, F.O., Eze, G.I., Jesuorobo, R.I., Egharevba, C.O. and Aimator, P.O. 2015. Antisickling and toxicological evaluation of the leaves of *Scoparia dulcis* Linn (Scrophulariaceae). *BMC Complementary and Alternative Medicine*. 15(1): 1–7.

Ahsan, M., Islam, S.N., Gray, A.I. and Stimson, W.H. 2003. Cytotoxic diterpenes from *Scoparia dulcis*. *Journal of Natural Products*. 66(7): 958–961.

De Farias Freire, S.M., Da Silva Emim, J.A., Lapa, A.J., Souccar, C. and Torres, L.M.B. 1993. Analgesic and antiinflammatory properties of *Scoparia dulcis* L. extracts and glutinol in rodents. *Phytotherapy Research*. 7(6): 408–414.

Gachet, M.S., Lecaro, J.S., Kaiser, M., Brun, R., Navarrete, H., Muṇoz, R.A., Bauer, R. and Schühly, W. 2010. Assessment of anti-protozoal activity of plants traditionally used in Ecuador in the treatment of leishmaniasis. *Journal of Ethnopharmacology*. 128(1): 184–197.

Hayashi, K., Niwayama, S., Hayashi, T., Nago, R., Ochiai, H. and Morita, N. 1988. *In vitro* and *in vivo* antiviral activity of scopadulcic acid B from *Scoparia dulcis*, Scrophulariaceae, against herpes simplex virus type 1. *Antiviral Research*. 9(6): 345–354.

Hayashi, T., Kawasaki, M., Miwa, Y., Taga, T. and Morita, N. 1990. Antiviral agents of plant origin. III.: Scopadulin, a novel tetracyclic diterpene from *SScoparia dulcis* L. *Chemical and Pharmaceutical Bulletin*. 38(4): 945–947.

Phan, M.G., Phan, T.S., Matsunami, K. and Otsuka, H. 2006. Chemical and biological evaluation on scopadulane-type diterpenoids from *Scoparia dulcis* of Vietnamese origin. *Chemical and Pharmaceutical Bulletin*. 54(4): 546–549.

Riel, M.A., Kyle, D.E. and Milhous, W.K. 2002. Efficacy of *scopadulcic acid A* against Plasmodium falciparum *in vitro*. *Journal of Natural Products*. 65(4): 614–615.

4.1.14.2.7 Family Verbenaceae Martinov (1820), the Vervain Family

The family Verbenaceae consists of about 90 genera and 3,000 species of trees, shrubs, or herbaceous plants. Leaves: simple or compound, decussate, exstipulate. Inflorescences: cymes, panicles, racemes, spikes, umbels. Calyx: tubular, 4–5-lobed. Corolla: tubular, 4–5-lobed. Stamens: 2–4. Carpels: 2–5, forming a 2–10 locular ovary, each locule with 2 ovules. Styles: 1. Fruits: capsules, drupes.

***Callicarpa longifolia* Lam.**

[From Greek *kallos* = beauty, *karpos* = fruit, and Latin *longifolia* = long leaves]

Published in *Encyclopédie Méthodique, Botanique* 1(2): 563. 1785

Synonyms: *Callicarpa attenuata* Wall.; *Callicarpa lanceolaria* Roxb. ex Hornem; *Callicarpa horsfi eldii* Turcz

Local names: *katupang badak* (Lundayeh); *gongoloput, sasad* (Dusun); *indoloput* (Rungus); *meniran, tampah besi* (Bajau)

Habitat: forests, roadsides

Geographical distribution: from India to Australia

Botanical description: this shrub reaches a height of about 3 m. Stems: quadrangular, somewhat hairy. Leaves: simple, opposite, exstipulate. Petioles: thin, 1–1.5 cm long. Blades: 2–6 cm × 8–20 cm, elliptic, membranous, lanceolate, to oblanceolate, dull light green, tapering or attenuate at base, serrate, acute to somewhat acuminate at apex, with about 5–8 pairs of secondary nerves. Cymes: axillary, about 2 cm across. Calyx: tubular, hairy, and 4-lobed, tiny. Corolla: tubular, purplish, about 2 mm long, hairy outside, 4–5. Stamens: 4. Ovary: hairy. Styles: 1, thin, protuding. Stigma: capitate. Berries: about 3 mm across, white.

Traditional therapeutic indications: headaches, smallpox, shingles (Dusun); malaria, stomach aches (Dusun, Kadazan)

Pharmacology and phytochemistry: antibacterial (Kusumawati et al. 2017), anti-inflammatory (Semiawan et al. 2015), analgesic activity (Syamsul et al. 2016).

Toxicity, side effects, and drug interaction: not known

Comment: The Dusun and Kadazan use a plant of the genus *Callicarpa* L. (1753) for arthritis (local name: *subol subol*).

REFERENCES

Kusumawati, E., Apriliana, A. and Khatimah, K. 2017. Uji aktivitas antibakteri ekstrak etanol daun kerehau (*Callicarpa longifolia* Lam.) Lam. terhadap Escherichia coli dan Staphylococcus aureus. *Jurnal Ilmiah Manuntung*. 2(2): 166–172.

Semiawan, F., Ahmad, I. and Masruhim, M.A. 2015. Aktivitas antiinflamasi ekstrak daun kerehau (*Callicarpa longifolia* L.). *Jurnal Sains dan Kesehatan*. 1(1): 1–4.

Syamsul, E.S., Andani, F. and Soemarie, Y.B. 2016. Analgesic activity study of ethanolic extract of *Callicarpa longifolia* lamk. in mice. *Majalah Obat Tradisional*. 21(2): 99–103.

***Clerodendrum philippinum* Schauer**

[From Greek *kleros* = clergy, *dendron* = tree, and Latin *philippinum* = from the Philippines]

Published in *Prodromus Systematis Naturalis Regni Vegetabilis* 11: 667. 1847

Synonyms: *Clerodendron fragrans* Roxb.; *Volkameria fragrans* Vent.
Local names: *sambung tulang, bunga rayat, gutuk* (Dusun); *pangkilai* (Lundayeh); *pangi pangil, pemangil* (Bajau)
Habitat: cultivated
Geographical distribution: subtropical, tropical
Botanical description: this ornamental and fragrant shrub reaches a height of about 2 m. Stems: quadrangular, somewhat hairy. Leaves: simple, decussate, exstipulate. Petioles: 4–9 cm long, channeled, hairy. Blades: broadly triangular, 7–14 cm × 8–15 cm, hairy, somewhat dentate, truncate at base, with 4–5 pairs of secondary nerves, acute at apex. Cymes: axillary or terminal. Calyx: tubular, pubescent, about 1.5 cm long, 5-lobed. Corolla: tubular, 5-lobed, pure white, about 1.5–2 cm long, the lobes obovate. Stamens: 4, thin. Styles: 1. Stigma: bifid, protuding. Drupes: globose to somewhat lobed.
Traditional therapeutic indications: broken bones (Dusun)
Pharmacology and phytochemistry: antibacterial (Venkatanarasimman et al. 2012), hypoglycemic (Kar et al. 2015).
Anti-inflammatory diterpenes (uncinatone, clerodenone A) (Yue et al. 2015).
Toxicity, side effects, and drug interaction: not known

Comments:

(i) In Sumatra, the plant is used by the Simalungung for asthma, gastrointestinal disorders, fever, and thrush (local name: *simarbakkudu*).
(ii) *Clerodendrum laevifolium* Bl. is used by the Lundayeh for diarrhea (local name: *lilapo*).

REFERENCES

Kar, M.K., Swain, T.R. and Mishra, S.K. 2015. Antidiabetic activity of *Clerodendrum philippinum* Schauer leaves in streptozotocin induced diabetic rats. *International Journal of Pharmacy and Pharmaceutical Sciences*. 7(9): 386–389.

Venkatanarasimman, B., Rajeswari, T. and Padmapriya, B. 2012. Antibacterial potential of crude leaf extract of *Clerodendrum philippinum* Schauer. *International Journal of Pharmaceutical and Biological Science Archive*. 3(2): 307–310.

Yue, J.R., Feng, D.Q. and Xu, Y.K. 2015. A new triterpenoid bearing octacosanoate from Stems: And roots *of Clerodendrum philippinum* var. *simplex* (Verbenaceae). *Natural Product Research*. 29(13): 1228–1234.

Lantana camara **L.**

[From Latin *lantana* = viburnum plant and *camara* = arched roof]

Published in *Species Plantarum* 2: 627. 1753

Synonyms: *Camara vulgaris* Benth.; *Lantana aculeata* L.
Local names: *bunga tahi asu* (Murut); *bunga tahi ayam munai, bunga pagar* (Bajau); *ta ayam* (Dusun)
Common name: common lantana
Habitat: gardens, roadsides
Geographical distribution: subtropical, tropical
Botanical description: this herbaceous plant reaches a height of about 1 m. Stems: spiny. Leaves: simple, decussate, exstipulate. Petioles: up to 2.5 cm long. Blades: broadly lanceolate, 5–10 cm × 2–5 cm, serrate, hairy, dull green, with about 6 pairs of secondary nerves, truncate at base, acute at apex. Inflorescences: axillary umbels of heavenly colored flowers on thin and up to about 10 cm long peduncles. Calyx: tubular, about 3 mm long. Corolla: tubular, up to 1 cm long, yellow to light purple to white, membranous, wavy, 4-lobed. Stamens: 4. Drupes: glossy, up to 5 mm across, globose, black, the whole infructescence dismal.
Traditional therapeutic indications: itchiness (Dusun)
Pharmacology and phytochemistry: anti-trypanosomal, antiplasmodial (Mesia et al. 2008), antibacterial, and leishmanicidal essential oil (Deena & Thoppil 2000; Uddin et al. 2008; Machado et al. 2012; Barros et al. 2016).
Nematicidal triterpenes (lantanilic acid, camaric acid, oleanolic acid) (Qamar et al. 2005). Leishmanicidal triterpenes (Begum et al. 2014). Antibacterial triterpene (22β-acetoxylantic acid) (Barre et al. 1997), anti-inflammatory triterpenes (Wu et al. 2020). Antifungal hydroxycinnamic acid derivative (verbascoside) (Oyourou et al. 2013).
Toxicity, side effects, and drug interaction: the plant is dreadfully poisonous because of the triterpenes lantadene A and B (Wolfson & Solomons 1964; Morton 1994). The berries attract toddlers and children and therefore this plant should under no circumstances be planted in gardens and parks. Another equally poisonous Verbenaceae found in the gardens of Sabah is *Duranta repens* L.

REFERENCES

Barre, J.T., Bowden, B.F., Coll, J.C., De Jesus, J., Victoria, E., Janairo, G.C. and Ragasa, C.Y. 1997. A bioactive triterpene from *Lantana camara*. *Phytochemistry*. 45(2): 321–324.

Barros, L.M., Duarte, A.E., Morais-Braga, M.F.B., Waczuk, E.P., Vega, C., Leite, N.F., De Menezes, I.R.A., Coutinho, H.D.M., Rocha, J.B.T. and Kamdem, J.P. 2016. Chemical characterization and trypanocidal, leishmanicidal and cytotoxicity potential of Lantana camara L. (Verbenaceae) essential oil. *Molecules*. 21(2): 209.

Begum, S., Ayub, A., Qamar Zehra, S., Shaheen Siddiqui, B. and Iqbal Choudhary, M. 2014. Leishmanicidal triterpenes from Lantana camara. *Chemistry & Biodiversity*. 11(5): 709–718.

Deena, M.J. and Thoppil, J.E. 2000. Antimicrobial activity of the essential oil of Lantana camara. *Fitoterapia*. 71(4): 453–455.

Machado, R.R., Valente Júnior, W., Lesche, B., Coimbra, E.S., Souza, N.B.D., Abramo, C., Soares, G.L.G. and Kaplan, M.A.C. 2012. Essential oil from leaves of Lantana camara: A potential source of medicine against leishmaniasis. *Revista Brasileira de Farmacognosia*. 22(5): 1011–1017.

Mesia, G.K., Tona, G.L., Nanga, T.H., Cimanga, R.K., Apers, S., Cos, P., Maes, L., Pieters, L. and Vlietinck, A.J. 2008. Antiprotozoal and cytotoxic screening of 45 plant extracts from Democratic Republic of Congo. *Journal of Ethnopharmacology*. 115(3): 409–415.

Morton, J.F. 1994. Lantana, or red sage (*Lantana camara* L., [Verbenaceae]), notorious weed and popular garden flower; some cases of poisoning in Florida. *Economic Botany*. 48(3): 259–270.

Oyourou, J.N., Combrinck, S., Regnier, T. and Marston, A. 2013. Purification, stability and antifungal activity of verbascoside from *Lippia javanica* and *Lantana camara* leaf extracts. *Industrial Crops and Products*. 43: 820–826.

Qamar, F., Begum, S., Raza, S.M., Wahab, A. and Siddiqui, B.S. 2005. Nematicidal natural products from the aerial parts of *Lantana camara* Linn. *Natural Product Research*. 19(6): 609–613.

Uddin, S.J., Rouf, R., Shilpi, J.A., Alamgir, M., Nahar, L. and Sarker, S.D. 2008. Screening of some Bangladeshi medicinal plants for *in vitro* antibacterial activity. *Oriental Pharmacy and Experimental Medicine*. 8(3): 316–321.

Wolfson, S.L. and Solomons, T.W.G. 1964. Poisoning by fruit of Lantana camara: An acute syndrome observed in children following ingestion of the green fruit. *American Journal of Diseases of Children*. 107(2): 173–176.

Wu, P., Song, Z., Wang, X., Li, Y., Li, Y., Cui, J., Tuerhong, M., Jin, D.Q., Abudukeremu, M., Lee, D. and Xu, J. 2020. Bioactive triterpenoids from *Lantana camara* showing anti-inflammatory activities *in vitro* and *in vivo*. *Bioorganic Chemistry*. 101: 104004.

***Stachytarpheta jamaicensis* (L.) Vahl**

[From Greek *staphys* = spike, *tarphys* = thick, and Latin *jamaicensis* = from Jamaica]

Published in *Enumeratio Plantarum*. . . 1: 206–207. 1804

Synonyms: *Stachytarpheta indica* (L.) Vahl; *Verbena indica* L.; *Verbena jamaicensis* L.
Local names: *bandi, bunga tali, soginap, soginep, sugandap* (Dusun); *bunga tikus* (Dusun, Murut); *ekor tikus, rumput tahi babi* (Bajau); *indalupang* (Murut); *tali tali, sagandap* (Dusun, Kadazan); *selasi dandi* (Bajau, Kedayan)
Common name: blue porterweed
Habitat: wastelands, roadsides, villages
Geographical distribution: tropical
Botanical description: this herbaceous plant reaches a height of about 1 m. Stems: quadrangular at apex. Leaves: simple, opposite, subsessile, exstipulate. Blades: serrate, somewhat spathulate, 1.5–3 cm × 2.5–6 cm, tapering at base along petioles, somewhat fleshy and glossy, rounded to acute at apex, with about 4–5 pairs of secondary nerves and tertiary nerves sunken above. Spikes: terminal, somewhat shaped like a cat's tail, up to about 20 cm long, with a few flowers blossoming. Calyx: tubular, 6 mm long, 5-lobed, the lobes triangular. Corolla: heavenly purple to blue, 5-lobed, about 1 cm long, the lobes rounded, spreading. Stamens: 2. Disc: present. Ovary: bottle-shaped. Styles: 1, thin. Stigma: capitate. Capsules: dehiscent, in persistent calyx. Seeds: 1.
Traditional therapeutic indications: bleeding wounds (Dusun); snake and scorpion bites, skin diseases, diarrhea, muscle sprains, antidote for poisons (Dusun, Kadazan); abscesses, cuts, gonorrhea, swellings (Kedayan)
Pharmacology and phytochemistry: this plant has been the subject of numerous pharmacological studies (Rodríguez & Castro 1996) which have highlighted, among other things, antibacterial, antidiarrheal (Sasidharan et al. 2007), wound healing (Caluya 2017), and hypoglycemic activity (Estella et al. 2020).
Anti-inflammatory iridoid glycoside (ipolamiide). Antibacterial hydroxycinnamic derivative (verbascoside) (Agampodi et al. 2022).
Toxicity, side effects, and drug interaction: the median lethal dose (LD_{50}) of a methanol extract of leaves administered to rats was above 5 g/kg (Estella et al. 2020). Chronic ingestion of leaf powder in rats resulted in liver necrosis (Ataman et al. 2006).

REFERENCES

Agampodi, V.A., Katavic, P., Collet, C. and Collet, T. 2022. Antibacterial and anti-inflammatory activity of extracts and major constituents derived from *Stachytarpheta indica* Linn. leaves and their potential implications for wound healing. *Applied Biochemistry and Biotechnology*. 194(12): 6213–6254.
Ataman, J.E., Idu, M., Odia, E.A., Omogbai, E.K.I., Amaechina, F., Akhigbe, A.O. and Ebite, L.E. 2006. Histopathologic effects of *Stachytarpheta jamaicensis* (L.) Vahl. on wistar rats. *Planta Medica*. 72(11): P_093.
Caluya, E.D.C. 2017. Wound healing potential of the crude leaf extract of *Stachytarpheta jamaicensis* Linn. Vahl (Kandikandilaan) on induced wounds in rats. *Journal of Medicinal Plants*. 5(1): 375–381.
Estella, O.U., Obodoike, E.C. and Esua, U.E. 2020. Evaluation of the anti-diabetic and toxicological profile of the leaves of *Stachytarpheta jamaicensis* (L.) Vahl (Verbenaceae) on alloxan-induced diabetic rats. *Journal of Pharmacognosy and Phytochemistry*. 9(3): 477–484.
Rodríguez, M. and Castro, O. 1996. Pharmacological and chemical evaluation of *Stachytarpheta jamaicensis* (Verbenaceae). *Revista de Biologia Tropical*. 44(2A): 353–359.
Sasidharan, S., Latha, L.Y., Zuraini, Z., Suryani, S., Sangetha, S. and Shirley, L. 2007. Antidiarrheal and antimicrobial activities of *Stachytarpheta jamaicensis* leaves. *Indian Journal of Pharmacology*. 39(5): 245.

***Vitex pubescens* (L.) Vahl**

[From Latin *vitex* = name used by Plinius for the chaste tree and *pubescens* = pubescent]

Published in *Symbolae Botanicae*, . . . 3: 85. 1794

Synonyms: *Pistaciovitex pinnata* (L.) Kuntze; *Vitex latifolia* Lam.; *Vitex pinnata* L.; *Vitex puberula* (L.) Miq.

Local names: *kulimpapa* (Bajau, Kedayan); *bogong, kalipapa, kulimpapo* (Dusun); *kuhim papo* (Dusun, Kadazan)

Habitat: forests, roadsides

Geographical distribution: from India to the Philippines

Botanical description: this tree reaches a height of about 15 m. Stems: quadrangular, somewhat hairy. Leaves: palmately lobed, decussate, exstipulate. Petioles: 5–10 cm long, thin. Petiolules: 1–5 mm long. Folioles: 3–5, elliptic to obovate, 2.5–4 cm × 8–15 cm, acute at base, glabrous or somewhat hairy, punctuated with glands, with 10–15 pairs of secondary nerves. Panicles: somewhat triangular, 7–20 cm long. Calyx: tubular, 4–6 mm long, somewhat hairy, 5-lobed, the lobes triangular. Corolla: white, tubular, 5-lobed, about 3 mm long, the lower lobe rounded, light purple, about 4 mm long. Stamens: 4, about 1 cm long. Drupes: globose, about 1 cm across.

Traditional therapeutic indications: postpartum (Kedayan); beriberi, cuts, fever, hypertension, wounds (Dusun, Kadazan); bloody stools (Dusun); stomach aches, digestion, breathlessness (Bajau).

Pharmacology and phytochemistry: hypotensive (Al-Akwaa et al. 2020), gastroprotective (Saeed et al. 2016), antibacterial (Najiyah et al. 2021).

Toxicity, side effects, and drug interaction: not known

Comments:

(i) The young leaves of this plant used by the Dusun as food.
(ii) The young leaves of *Clerodendrum album* Ridl. are used by the Dusun as food (local name: *taum*).

REFERENCES

Al-Akwaa, A.A., Asmawi, M.Z., Dewa, A. and Mahmud, R. 2020. Hypotensive activity and vascular reactivity mechanisms of *Vitex pubescens* leaf extracts in spontaneously hypertensive rats. *Heliyon*. 6(7): e04588.

Najiyah, S.F., Edyson, E. and Khatimah, H. 2021. Aktivitas antibakteri infus kayu laban (*Vitex pubescens* Vahl.) terhadap *Escherichia coli* dan *Pseudomonas aeruginosa*. *Homeostasis*. 4(3): 559–566.

Saeed, A.L., Wajeeh, N., Halabi, M.F., Hajrezaie, M., M. Dhiyaaldeen, S., Abdulaziz Bardi, D., M. Salama, S., Rouhollahi, E., Karimian, H., Abdolmalaki, R., Azizan, A.H.S. and Mohd Ali, H. 2016. The gastroprotective effect of *Vitex pubescens* leaf extract against ethanol-provoked gastric mucosal damage in sprague-dawley rats. *PLoS ONE*. 11(9): e0157431.

4.1.14.3 Order Solanales Juss. ex Bercht. & J. Presl (1820)

4.1.14.3.1 Family Convolvulaceae Juss. (1789), the Morning Glory Family

The family Convolvulaceae consists of about 60 genera and 1,650 species of herbaceous plants, shrubs, or climbing plants. Stems: often flexuous. Leaves: simple, spiral, exstipulate. Inflorescences: cymes. Calyx: 5 sepals or tubular. Corolla: funnel-shaped, caducous, membranous. Stamens: 5. Carpels: 2, forming a unilocular or bilocular ovary, each locule with 2 ovules. Styles: 1. Fruits: capsules, berries.

Ipomoea aquatica **Forssk.**

[From Greek *ips* = worm, *homoios* = like, and from Latin *aqua* = water]

Published in *Flora Aegyptiaco-Arabica* 44. 1775

Synonym: *Ipomoea reptans* (Linn.) Poir.
Local names: *kangkong* (Dusun, Lundayeh, Bajau); *sankong* (Dusun)
Common name: water spinach
Habitat: cultivated, ditches, ponds, swamps, drains
Geographical distribution: Asia, Pacific Islands
Botanical description: this soft creeping herbaceous plant grows up to a length of about 50 cm. Stems: glabrous, rooting at nodes, fleshy. Leaves: simple, spiral, exstipulate. Petioles: 3–18 cm long. Blades: aristate, dull green, 5–9.5 cm × 7–14.5 cm, finely marked with about 12 pairs of secondary nerves. Cymes: axillary, on 1.5 cm–10 cm long peduncles, few-flowered. Sepals: 5, 6–8 mm long, oblong-lanceolate. Corolla: funnel-shaped, 4.5–5 cm long, membranous, pure white and pinkish at the throat. Stamens: 5. Ovary: conical. Styles: 1. Stigma: bifid. Capsules: 8 mm long, somewhat globose, dehiscent. Seeds: 1–4, angular.
Traditional therapeutic indications: constipation (Lundayeh)
Pharmacology and phytochemistry: antibacterial (Padmavathy et al. 2017), hypoglycemic (Malalavidhane et al. 2001; Malalavidhane et al. 2003; Sokeng et al. 2007; Sajak et al. 2017).
Amide alkaloid (*N*-feruloyltyramine) inhibitor the synthesis of prostaglandins and leukotrienes (Tseng et al. 1992).
Toxicity, side effects, and drug interaction: ethanol extract of leaves administered to rats at the dose of 5 g/kg was found to cause no signs of toxicity (Alkiyumi et al. 2012).

REFERENCES

Alkiyumi, S.S., Abdullah, M.A., Alrashdi, A.S., Salama, S.M., Abdelwahab, S.I. and Hadi, A.H.A. 2012. Ipomoea aquatica extract shows protective action against thioacetamide-induced hepatotoxicity. *Molecules*. 17(5): 6146–6155.

Malalavidhane, T.S., Wickramasinghe, S.M.D.N. and Jansz, E.R. 2001. An aqueous extract of the green leafy vegetable Ipomoea aquatica is as effective as the oral hypoglycaemic drug tolbutamide in reducing the blood sugar levels of Wistar rats. *Phytotherapy Research*. 15(7): 635–637.

Malalavidhane, T.S., Wickramasinghe, S.M.D.N., Perera, M.S.A. and Jansz, E.R. 2003. Oral hypoglycaemic activity of Ipomoea aquatica in streptozotocin-induced, diabetic wistar rats and Type II diabetics. *Phytotherapy Research: An International Journal Devoted to Pharmacological and Toxicological Evaluation of Natural Product Derivatives*. 17(9): 1098–1100.

Padmavathy, A., Rasny, M.R.M., Reyadh, R. and Khan, J. 2017. Evaluation of antibacterial activity of different extracts of Ipomoea aquatica leaves against Escherichia coli and Salmonella typhi. *Journal of Management and Science*. 15(2).

Sajak, A.A.B., Mediani, A., Dom, N.S.M., Machap, C., Hamid, M., Ismail, A., Khatib, A. and Abas, F. 2017. Effect of *Ipomoea aquatica* ethanolic extract in streptozotocin (STZ) induced diabetic rats via 1H NMR-based metabolomics approach. *Phytomedicine*. 36: 201–209.

Sokeng, S.D., Rokeya, B., Hannan, J.M.A., Junaida, K., Zitech, P., Ali, L., Ngounou, G., Lontsi, D. and Kamtchouing, P. 2007. Inhibitory effect of Ipomoea aquatica extracts on glucose absorption using a permerged rat intestinal preparation. *Fitoterapia*. 78(7–8): 526–529.

Tseng, C.F., Iwakami, S., Mikajiri, A., Shibuya, M., Hanaoka, F., Ebizuka, Y., Padmawinata, K. and Sankawa, U. 1992. Inhibition of in vitro prostaglandin and leukotriene biosyntheses by cinnamoyl-β-phenethylamine and N-acyldopamine derivatives. *Chemical and Pharmaceutical Bulletin*. 40(2): 396–400.

***Ipomoea batatas* (L.) Lam.**

[From Greek *ips* = worm, *homoios* = like, and from ancient Central American name of the plant = *batata*]

Published in *Tableau Encyclopédique et Methodique. . . Botanique* 1(2): 465. 1791

Synonyms: *Batatas edulis* (Thunb.) Choisy; *Convolvulus batatas* L.; *Convolvulus edulis* Thunb.

Local names: *kasou* (Dusun); *ubi keledek* (Murut); *ubi rambat* (Bajau)

Common name: sweet potato

Habitat: cultivated

Geographical distribution: subtropical, tropical

Botanical description: this creeping herbaceous plant reaches a height of about 2 m. Tubers: red, white, or yellow, smooth, elongated, up to about 25 cm long. Stems: green or purplish, somewhat angular, rooting at nodes. Leaves: simple, alternate, exstipulate. Petioles: thin, channeled, slightly enlarged at base, 5.5–10.5 cm long. Blades: ovate, cordate, or lobed, yellowish light green, 4–13 cm × 3–13 cm, with 5–10 pairs of secondary nerves. Cymes: axillary, 1–7-flowered. Sepals: 5, elliptic, shortly acuminate, hairy, finely nerved, glabrous, 0.7–1.3 cm long. Corolla: funnel-shaped, membranous, 3–4 cm long, pink, white, and pale purple to purple, darker at the throat. Ovary: subglabrous. Capsules: globose, up to about 1 cm long. Seeds: 4, angular.

Traditional therapeutic indications: boils, carbuncles (Murut)

Pharmacology and phytochemistry: nematicidal (Mackeen et al. 1997), antibacterial (Chakraborty et al. 2018), hypotensive (Ishiguro et al. 2007), hypoglycemic (Ludvik et al. 2004), renoprotective (Rafiu & Luka 2018).

Hypocholesterolemic anthocyanins (Miyazaki et al. 2008).

Toxicity, side effects, and drug interaction: moldy tubers are poisonous (Wilson et al. 1971). Vegetables treated by organophosphate pesticides are poisonous (Wu et al. 2001). The tubers contain calcium oxalate (Siener et al. 2020).

REFERENCES

Chakraborty, P., Sharma, S., Chakraborty, S., Siddapurand, A. and Abraham, J. 2018. Cytotoxicity and antimicrobial activity of *Ipomoea batatas*. *Research Journal of Pharmacy and Technology*. 11(7): 2741–2746.

Ishiguro, K., Yoshimoto, M., Tsubata, M. and Takagaki, K. 2007. Hypotensive effect of sweetpotato [Ipomoea batatas] tops. *Journal of the Japanese Society for Food Science and Technology (Japan)*. 54: 45–49.

Ludvik, B., Neuffer, B. and Pacini, G. 2004. Efficacy of *Ipomoea batatas* (Caiapo) on diabetes control in type 2 diabetic subjects treated with diet. *Diabetes Care*. 27(2): 436–440.

Mackeen, M.M., Ali, A.M., Abdullah, M.A., Nasir, R.M., Mat, N.B., Razak, A.R. and Kawazu, K. 1997. Antinematodal activity of some Malaysian plant extracts against the pine wood nematode, Bursaphelenchus xylophilus. *Pesticide Science*. 51(2): 165–170.

Miyazaki, K., Makino, K., Iwadate, E., Deguchi, Y. and Ishikawa, F. 2008. Anthocyanins from purple sweet potato *Ipomoea batatas* cultivar Ayamurasaki suppress the development of atherosclerotic lesions and both enhancements of oxidative stress and soluble vascular cell adhesion molecule-1 in apolipoprotein E-deficient mice. *Journal of Agricultural and Food Chemistry*. 56(23): 11485–11492.

Rafiu, A.A. and Luka, C.D. 2018. Effects of aqueous extract of Ipomoea batatas Leaf on blood glucose, kidney functions and hematological parameters of streptozotocin-induced diabetic rats. *Journal of Diabetes & Metabolism*. 1(4): 4–9.

Siener, R., Seidler, A. and Hönow, R. 2020. Oxalate-rich foods. *Food Science and Technology*. 41: 169–173.

Wilson, B.J., Boyd, M.R., Harris, T.M. and Yang, D.T.C. 1971. A lung oedema factor from mouldy sweet potatoes (Ipomoea batatas). *Nature*. 231(5297): 52–53.

Wu, M.L., Deng, J.F., Tsai, W.J., Ger, J., Wong, S.S. and Li, H.P. 2001. Food poisoning due to methamidophos-contaminated vegetables. *Journal of Toxicity, Side Effects, and Drug Interaction: Clinical Toxicology*. 39(4): 333–336.

***Merremia peltata* (L.) Merr.**

[After the 19th century German naturalist Blasius Merrem and Latin *peltata* = peltate]

Published in *An Interpretation of Rumphius's Herbarium Amboinense* 441. 1917

Synonyms: *Convolvulus peltatus* L.; *Ipomoea peltata* (L.) Choisy; *Merremia borneensis* Merr.
Local names: *babas, balaan* (Dusun)
Common name: big leaf rope
Habitat: roadsides, open forests
Geographical distribution: tropical Asia, Pacific Islands
Botanical description: this climbing plant grows up to a length of about 30 m. Stems: glabrous, with lenticels, yield a milky sap once incised. Leaves: simple, alternate, exstipulate. Petioles: about 8 cm long. Blades: dull light green, peltate, cordate, wavy, 13–19 cm × 10–15 cm, acuminate at apex, with 7–10 pairs of reddish secondary nerves. Cymes: large, fleshy, axillary, on 5–8 cm long conical peduncles, many-flowered. Sepals: 5, fleshy, lanceolate, 1.8–2.5 cm long. Corolla: funnel-shaped, membranous, pure white, 4–6 cm long, 5-lobed. Stamens: 5. Styles: 1. Stigma: bifid. Capsules: ovoid, 2.5–3 cm long. Seeds: 4, trigonous.
Traditional therapeutic indications: cuts, diarrhea, wounds (Dusun, Kadazan); stomach aches, flatulence, hair care (Dusun)
Pharmacology and phytochemistry: antidiabetic (Af-Idah et al. 2021), antibacterial (Perez et al. 2015), antiviral (HIV) (Yamamoto et al. 1997), cytotoxic (Djamaan 2016).
Toxicity, side effects, and drug interaction: not known

Comment:

(i) In Kalimantan, the Dayaks use *Merremia peltata* (L.) Merr. as an antidote for animal bites (local name: *akar jelayan*).
(ii) *Merremia gracilis* E.J.F. Campb. & Argent is used by the Dusun and Kadazan for asthma, diarrhea, fatigue, jaundice, and pancreatitis (local names: *gatas gatas*, *malagatas*).
(iii) In the family Convolvulaceae, a plant is used as a medicine for swellings by the Dusun (local name: *tawar tangau*).

REFERENCES

Af-Idah, B.M.A., Hanafi, M. and Elya, B. 2021. Antioxidant and alpha glucosidase inhibitor screening of *Merremia peltata* L. as potential traditional treatment for diabetes mellitus. *Pharmacognosy Journal*. 13(4): 13(4): 902–908.

Djamaan, A. 2016. Extraction, fractionation and Cytotoxicity Test of *Merremia peltata* (L.) Merr., (Fam. Convolvulaceae) leaves. *Der Pharmacia Lettre*. 8: 48–52.

Perez, K.J.B., Jose, M.A.I., Aranico, E. and Madamba, M.R.S.B. 2015. Phytochemical and antibacterial properties of the ethanolic leaf extract of *Merremia peltata* (L.) Merr. and *Rubus* spp. *Advances in Environmental Biology*. 9(19): 50–56.

Yamamoto, T., Takahashi, H., Sakai, K., Kowithayakorn, T. and Koyano, T. 1997. Screening of Thai plants for anti-HIV-1 activity. *Natural Medicines*= 生薬學雜誌. 51(6): 541–546.

4.1.14.3.2 Family Solanaceae Juss. (1789), the Nightshade Family

The family Solanaceae consists of about 95 genera with 2,000 species of herbaceous plants, shrubs, or treelets. Leaves: simple, alternate, exstipulate. Inflorescences: cymes, racemes, panicles, solitary. Calyx: tubular, 5-lobed. Corolla: tubular, membranous, star-shaped, 5-6-lobed. Stamens: 5, often merged into a narrow cone. Carpels: 2, forming a bilocular ovary, each locule with 1–50 ovules. Styles: 1. Fruit: berries, drupes, capsules.

***Capsicum frutescens* L.**

[From Greek *kapsikos* = a box and Latin *frutescens* = shrubby]

Published in *Species Plantarum* 1: 189. 1753

Synonym: *Capsicum annuum* var. *frutescens* (L.) Kuntze

Local names: *lado, lusun, penderoi* (Dusun, Kadazan); *lada padi* (Bajau); *cili padi* (Bugis); *lada* (Murut)

Common name: chili pepper

Habitat: cultivated, wastelands

Geographical distribution: tropical

Botanical description: this herbaceous plant reaches a height of about 1 m and has somewhat the appearance of a dwarf tree. Leaves: simple, spiral, exstipulate. Petioles: 4–6 cm long, terete, smooth, fleshy. Blades: lanceolate to elliptic, wavy, fleshy, pendulous, 4–12 cm × 1.5–5 cm. Flowers: solitary, axillary, nodding, solitary, or arranged in few-flowered axillary clusters. Calyx: tubular, tiny, 5-lobed. Corolla: tubular, about 1 cm across, 5–6-lobed, white, the lobes broadly lanceolate. Stamens: 5, bluish. Ovary: tiny. Styles: 1., thin. Berries: irregularly fusiform, erect, glossy, with a magnificent palette of yellow, red, or purple, pungent to burning, up to about 10 cm long. Seeds: numerous, tiny.

Traditional therapeutic indications: burns (Dusun); fever, sore eyes, ringworms (Bajau); canker sores, abscesses, rheumatisms, wounds (Kedayan); ringworms, itchiness, ailments associated with pregnancy (Dusun, Kadazan)

Pharmacology and phytochemistry: this plant has been the subject of numerous pharmacological studies (Srinivasan et al. 2016) which have highlighted, among other things, antifungal (Soumya & Nair 2012), antiviral (herpes simplex virus, measles virus, poliovirus) (Akanacaa et al. 2002; Hafiz et al. 2017; Ordaz-Trinidad et al. 2018), and antibacterial activity (Koffi-Nevry et al. 2012). Amide alkaloid (capsaicin) (Estrada et al. 2002) with antitumor (Surh 2002), antidiabetic, anti-inflammatory, and analgesic activity (Mazhar et al. 2021).

Toxicity, side effects, and drug interaction: the fruits are edible and almost daily ingredient of Asian dishes. They are, however, deadly poisonous to toddlers (Snyman et al. 2001).

REFERENCES

Akanaca, P., Kurokawa, M., Tewtrakul, S., Pramyon, P., Sripandulcha, B., Shirak, K. and Hattor, M. 2002. Inhibitory activities of Thai medicinal plants against herpes simplex type 1, poliovirus type 1, and measles virus. *Journal of Traditional and Medicine*. 19: 174–180.

Estrada, B., Bernal, M.A., Díaz, J., Pomar, F. and Merino, F. 2002. Capsaicinoids in vegetative organs of *Capsicum annuum* L. in relation to fruiting. *Journal of Agricultural and Food Chemistry*. 50(5): 1188–1191.

Hafiz, T.A., Mubaraki, M., Dkhil, M. and Al-Quraishy, S. 2017. Antiviral activities of *Capsicum annuum* methanolic extract against herpes simplex virus 1 and 2. *Pakistan Journal of Zoology*. 49(1): 251.

Koffi-Nevry, R., Kouassi, K.C., Nanga, Z.Y., Koussémon, M. and Loukou, G.Y. 2012. Antibacterial activity of two bell pepper extracts: *Capsicum annuum* L. and *Capsicum frutescens*. *International Journal of Food Properties*. 15(5): 961–971.

Mazhar, N., Baig, S.G., Ahmed, S., Palla, A. and Ishrat, G. 2021. Analgesic and anti-inflammatory potential of four varieties of bell pepper (*Capsicum annum* L.) in rodents. *Pakistan Journal of Pharmaceutical Sciences*. 35(4).

Ordaz-Trinidad, N., Dorantes-Álvarez, L., Salas-Benito, J., Barrón-Romero, B.L., Salas-Benito, M. and Nova-Ocampo, M.D. 2018. Cytotoxicity and antiviral activity of pepper extracts (Capsicum spp). *Polibotánica*. 46: 273–285.

Snyman, T., Stewart, M.J. and Steenkamp, V. 2001. A fatal case of pepper poisoning. *Forensic Science International*. 124(1): 43–46.

Soumya, S.L. and Nair, B.R. 2012. Antifungal efficacy of *Capsicum frutescens* L. extracts against some prevalent fungal strains associated with groundnut storage. *International Journal of Agricultural Technology*. 8(2): 739–750.

Srinivasan, K. 2016. Biological activities of red pepper (*Capsicum annuum*) and its pungent principle capsaicin: A review. *Critical Reviews in Food Science and Nutrition*. 56(9): 1488–1500.

Surh, Y.J. 2002. More than spice: Capsaicin in hot chili peppers makes tumor cells commit suicide. *Journal of the National Cancer Institute*. 94(17): 1263–1265.

***Nicotiana tabacum* L.**

[After the 17th century French diplomat Jean Nicot de Villemain and Latin *tabacum* = tobacco]

Published in *Species Plantarum* 1: 180. 1753

Synonym: *Nicotiana chinensis* Fisch. ex Lehm.
Local names: *sigup* (Bajau, Lundayeh, Murut, Dusun); *tembakau* (Kedayan)
Common name: tobacco
Habitat: cultivated
Geographical distribution: tropical
Botanical description: this herbaceous plant reaches a height of about 2 m. Stems: hairy, terete. Leaves: simple, spiral, sessile, exstipulate. Blades: lanceolate, 8–20 cm × 30–45 cm, soft, fleshy, hairy, tapering at base, clasping the stem at base, acute at apex, with about 7–15 pairs of secondary nerves. Panicles: terminal, with a few flowers. Calyx: tubular, hairy, about 2 cm long, 5-lobed. Corolla: tubular, 3.5–5 cm long, white to light pink, membranous, 5-lobed, the lobes broadly acuminate. Stamens: 5. Disc: present. Ovary: ovoid. Styles: 1, thin. Stigma: bifid. Capsules: ovoid, about 2 cm long, dehiscent. Seeds: numerous.
Traditional therapeutic indications: ringworms (Lundayeh); insect and animal bites (Dusun, Kadazana); cuts, wounds (Bajau, Dusun, Kadazan)
Pharmacology and phytochemistry: nematicidal (Nouri et al. 2016).
The plant is well known to produce the tropane alkaloid nicotine, which accounts for tobacco addiction (Benowitz 2009). Antifungal flavonol (kaempferol) (Yang et al. 2021). Antifungal sesquiterpenes and diterpenes (Xu et al. 2002).
Toxicity, side effects, and drug interaction: nicotine promotes cardiovascular diseases, cancer, and diabetes, and induces fetus malformation (Benowitz 2009).

REFERENCES

Benowitz, N.L. 2009. Pharmacology of nicotine: Addiction, smoking-induced disease, and therapeutics. *Annual Review of Pharmacology and Toxicology*. 49: 57–71.
Nouri, F., Nourollahi-Fard, S.R., Foroodi, H.R. and Sharifi, H. 2016. *In vitro* anthelmintic effect of Tobacco (*Nicotiana tabacum*) extract on parasitic nematode, *Marshallagia marshalli*. *Journal of Parasitic Diseases*. 40(3): 643–647.
Xu, K., Wang, J., Liu, J., Ni, L., Du, Y. and Wei, X. 2022. Sesquiterpenoids and diterpenoids from the flowers of *Nicotiana tabacum* L. and their antifungal activity. *Records of Natural Products*. 16: 5 483–487.
Yang, C., Xie, S.N., Ni, L., Du, Y.M., Liu, S., Li, M.Y. and Xu, K. 2021. Chemical constituents from nicotiana tabacum L. and their antifungal activity. *Natural Product Communications*. 16(11): 1934578X211059578.

***Physalis minima* L.**

[From Greek *phusallis* = bladder and Latin *minimus* = smallest]

Published in *Species Plantarum* 1: 183–184. 1753

Synonyms: *Physalis angulata* var. *villosa* Bonati; *Physalis lagascae* Roem. & Schult.; *Physalis parviflora* R.Br.

Local names: *lapak lapak* (Dusun); *pilanus* (Murut); *letup letup* (Bajau); *tulapak* (Dusun, Kadazan)

Common name: pygmy groundcherry

Habitat: wastelands, roadsides

Geographical distribution: tropical

Botanical description: this weird herbaceous plant reaches a height of about 50 m and has somewhat the appearance of a dwarf tree. Stems: fleshy, striated. Leaves: simple, spiral, exstipulate. Petioles: about 1.5 cm long. Blades: ovate to lanceolate, membranous, 1–2 cm × 2–3.5 cm, attenuate at base, with about 5 pairs of secondary nerves, loosely serrate, acuminate at apex. Flowers: solitary, axillary, on thin and about 1.5 cm long peduncles. Calyx: tubular, about 3 mm long, pubescent, 5-lobed. Corolla: somewhat convolvulaceous, dull light yellow, slightly 5-lobed, about 5 mm long. Stamens: 5. Styles: 1. Stigma: bifid. Berries: globose, bright red, glossy, dangling, about 5 mm across, encapsulated in a lantern-like, ribbed, polygonal, accrescent, and about 1.5 cm long enlarged calyx with fine purple nerves.

Traditional therapeutic indications: hydrocele (Dusun, Murut); hypertension, diabetes, malaria (Dusun, Kadazan); sore throats (Kedayan)

Pharmacology and phytochemistry: antibacterial (Patel et al. 2011), diuretic (Tammu et al. 2012), hypotensive (Saifurrohman 2015), diuretic (Tammu et al. 2012), hypoglycemic (Daud et al. 2016), hepatoprotective (Tammu & Ramana 2014), anti-Alzheimer (Joseph & Ravi 2022), and antifertility activity (Sudhakaran et al. 1999).

Steroids (withanolide E, 4β-hydroxywithanolid) toxic to liver cancer cells (HepG2), lung adenocarcinoma cells (SK-LU-1), and breast cancer cells (MCF-7) (Le Canh et al. 2021). Leishmanicidal (Choudhary et al. 2007) and trypanocidal steroids (Bravo et al. 2001; Chataing et al. 2009).

Toxicity, side effects, and drug interaction: not known

Comment: The Dayak of Borneo use *Physalis angulata* L. to treat lung diseases and influenza (local name: *lolotup*).

REFERENCES

Bravo B, J.A., Sauvain, M., Gimenez T, A., Balanza, E., Serani, L., Laprevote, O., Massiot, G. and Lavaud, C. 2001. Trypanocidal withanolides and withanolide glycosides from Dunalia brachyacantha. *Journal of Natural Products*. 64(6): 720–725.

Chataing, B., Hocquette, A., Diaz, S., Valentin, A. and Usubillaga, A. 2009. Activity of acnistins against Leishmania mexicana, Trypanosoma cruzi and Plasmodium falciparum. *Ciencia*. 17(1).

Choudhary, M.I., Yousuf, S., Samreen, Ahmed, S. and Atta-Ur-Rahman. 2007. New leishmanicidal physalins from *Physalis minima*. *Natural Product Research*. 21(10): 877–883.

Daud, D., Elias, S.F., Hassan, F.S.M., Jalil, M.N. and Tawang, A. 2016. *Physalis minima* Linn methanolic extract reduces blood glucose level without compromising sperm quality in normoglycaemic mice. *Journal of Applied Pharmaceutical Science*. 6(6): 008–011.

Joseph, L. and Ravi, C. 2022. *Physalis minima* L. fruit-a promising approach to Alzheimer's disease. *Pharmacological Research-Modern Chinese Medicine*. 2: 100038.

Le Canh, V.C., Le Ba, V., Thi Hai Yen, P., Le Thi, L., Thi Thuy Hoai, P., Huu Dat, T.T., Thao, D.T., Bach, L.G., Kim, Y.H. and Tuan Anh, H.L. 2021. Identification of potential cytotoxic inhibitors from *Physalis minima*. *Natural Product Research*. 35(12): 2082–2085.

Patel, T., Shah, K., Jiwan, K. and Shrivastava, N. 2011. Study on the antibacterial potential of *Physalis minima* Linn. *Indian Journal of Pharmaceutical Sciences*. 73(1): 111.

Saifurrohman, N.P.D.N. 2015. Effect of ethanolic extract of *Physalis minima* on lowering blood pressure in DOCA-salts hypertensive rat models. 中国药理学与毒理学杂志. 1.

Sudhakaran, S., Ramanathan, B. and Ganapathi, A. 1999. Antifertility effects of the petroleum ether extract of *Physalis minima* on female albino rats. *Pharmaceutical Biology*. 37(4): 269–272.

Tammu, J. and Ramana, K.V. 2014. Hepatoprotective effect of hydroalcoholic extract of *Physalis minima* leaves on rifampicin-isoniazid induced rats. *International Journal of Pharmacology and Biological Sciences*. 8(3): 55.

Tammu, J., Ramana, K.V., Thalla, S. and Narasimha Raju, B.H. 2012. Diuretic activity of methanolic extract of *Physalis minima* leaves. *Der Pharmacia Lettre*. 4(6): 1832–1834.

***Solanum erianthum* D.Don**

[From Latin *solanum* = nightshade, from Greek *erion* = wool, and *anthos* = flower]

Published in *Prodromus Florae Nepalensis* 96. 1825

Synonym: *Solanum verbascifolium* C.B. Wright
Local name: *limbasong* (Rungus)
Common name: potato tree
Habitat: cultivated
Geographical distribution: tropical Asia, Australia
Botanical description: this tree reaches a height of about 5 m. Stems: hairy, terete. Leaves: simple, spiral, exstipulate. Petioles: hairy, 1–5 cm long. Blades: elliptic to broadly lanceolate, 8–29 cm × 4–12 cm, attenuate at base, somewhat wavy, of a sinister sort of green, acute to acuminate at apex, hairy. Cymes: terminal, hairy, many-flowered. Calyx: tubular, hairy, 5-lobed, the lobes triangular and about 3 mm long. Corolla: about 2 cm long, 5-lobed, white. Stamens: 5, about 5 mm long, anthers bright yellow. Styles: 1. 5 mm long. Berries: globose, 8–10 mm across, dull green turning yellowish brown, on persistent calyx. Seeds: numerous, tiny, flat
Traditional therapeutic indications: medicinal (Rungus)
Pharmacology and phytochemistry: antibacterial (Alawode et al. 2018), antifungal activity (Sirajudeen & Muneer Ahamath 2014). Anti-inflammatory sesquiterpene (solavetivone) (Chen et al. 2013).
Toxicity, side effects, and drug interaction: the plant is dreadfully poisonous (Huang et al. 2009).

Comment: The Dusun and Kadazan use plants of the genus *Solanum* L. (1753) as medicines: *"tonsisiyah"* for dark urine and *"mansimang"* for swollen gums.

REFERENCES

Alawode, T.T., Lajide, L., Owolabi, B.J., Olaleye, M.T. and Ogunyemi, B.T. 2018. Antimicrobial studies on leaf and stem extracts of *Solanum erianthum*. *Microbiology Research Journal International*. 23(3): 1–6.

Chen, Y.C., Lee, H.Z., Chen, H.C., Wen, C.L., Kuo, Y.H. and Wang, G.J. 2013. Anti-inflammatory components from the root of Solanum erianthum. *International Journal of Molecular Sciences*. 14(6): 12581–12592.

Huang, S.T., Su, Y.J., Chien, D.K., Li, E.J. and Chang, W.H. 2009. *Solanum erianthum* intoxication mimicking an acute cerebrovascular disease. *The American Journal of Emergency Medicine*. 27(2): 249.e1–2.

Sirajudeen, J. and Muneer Ahamath, J. 2014. Evaluation of antibacterial and antifungal activity of medicinal plant *Solanum erianthum*. *Drug Discovery*. 9(22): 35–39.

***Solanum ferox* L.**

[From Latin *solanum* = nightshade and *ferox* = cruel]

Published in *Species Plantarum, Editio Secunda* 1: 267. 1762

Synonyms: *Solanum immane* Hance ex Walp.; *Solanum lasiocarpum* Dunal
Local names: *bintarung tondu* (Rungus); *ricontom* (locals of Javanese descent)
Common name: hairy-fruited eggplant
Habitat: wastelands, roadsides
Geographical distribution: India, Southeast Asia
Botanical description: this hideous herbaceous plant reaches a height of about 1.5 m. Stems: hairy, terete, thorny. Thorns: revolute, 8 mm long. Leaves: simple, spiral, exstipulate. Petioles: hairy, thorny, 3–8 cm long. Blades: ovate, 10–20 cm × 10–18 cm, thorny, truncate to somewhat cordate at base, hairy beneath, lobed, with 5–6 pairs of secondary nerves, acute at apex. Cymes: cauliforous, many-flowered. Calyx: tubular, hairy, 5-lobed, the lobes ovate and about 1 cm long. Corolla: about 2 cm long, 5-lobed, white. Stamens: 5, about 8 mm long and bright yellow. Styles: about 9 mm long. Berries: globose, about 2 cm across, orange, hairy, on persistent calyx. Seeds: numerous.
Traditional therapeutic indications: medicinal (Rungus)
Pharmacology and phytochemistry: antibacterial (Hazimah et al. 2019).
Toxicity, side effects, and drug interaction: not known

Comment: Locals of Javanese descent use the fruits of this plant as food (local name: *ricontom*) as well as natives of Sarawak (local names: *terung Dayak, terung asam*).

REFERENCE

Hazimah, H., Azharman, Z., Yuharmen, Y., Rahyuti, V. and Afriliani, A. 2019. Antibacterial activity of the hexane extract from leaf of plant solanum ferox L. In *AISTSSE 2018*, October 18–19, Medan, Indonesia. Available at https://eudl.eu/doi/10.4108/eai.18-10-2018.2287176

Solanum melongena **L.**

[From Latin *solanum* = nightshade and Persian *badinjan* = eggplant]

Published in *Species Plantarum* 1: 186. 1753

Synonyms: *Solanum esculentum* Dunal; *Solanum insanum* L.
Local names: *biterung* (Dusun, Lundayeh); *bontorung* (Dusun); *terung* (Bajau, Murut)
Common name: eggplant
Habitat: cultivated
Geographical distribution: temperate and tropical Asia
Botanical description: this shrubby herbaceous plant reaches a height of about 60 cm. Stems: hairy, terete, armed with revolute thorns. Leaves: simple, spiral, exstipulate. Petioles: hairy, 2–4.5 cm long. Blades: ovate to oblong, 6–18 cm × 5–11 cm, somewhat truncate and asymmetrical at base, hairy beneath, lobed, with about 4 pairs of secondary nerves, rounded at apex. Flowers: solitary, on 1–1.8 cm long peduncles. Calyx: tubular, hairy, prickly, 5-lobed. Corolla: about 3 cm long, 5-lobed, purplish. Stamens: 5, anthers about 8 mm long and bright yellow. Styles: 1, about 7 mm long. Berries: somewhat curved, oblong, smooth, of a delightful kind of dark purple, glossy, up to about 25 cm long. Seeds: numerous, tiny, in a yellowish and spongy flesh.
Traditional therapeutic indications: swollen gums, ulcers (Lundayeh). The Bajau use the plant as medicinal food or *"ulam"*.
Pharmacology and phytochemistry: anti-inflammatory (Han et al. 2003).
Anti-inflammatory lignans (Sun et al. 2014; Yang et al. 2019) and bisflavonol glycoside (solanoflavone) (Shen et al. 2005).

REFERENCES

Han, S.W., Tae, J., Kim, J.A., Kim, D.K., Seo, G.S., Yun, K.J., Choi, S.C., Kim, T.H., Nah, Y.H. and Lee, Y.M. 2003. The aqueous extract of *Solanum melongena* inhibits PAR2 agonist-induced inflammation. *Clinica Chimica Acta*. 328(1–2): 39–44.

Shen, G., Kiem, P.V., Cai, X.F., Li, G., Dat, N.T., Choi, Y.A., Lee, Y.M., Park, Y.K. and Kim, Y.H. 2005. Solanoflavone, a new biflavonol glycoside from *Solanum melongena*: Seeking for anti-inflammatory components. *Archives of Pharmacal Research*. 28(6): 657–659.

Sun, J., Gu, Y.F., Su, X.Q., Li, M.M., Huo, H.X., Zhang, J., Zeng, K.W., Zhang, Q., Zhao, Y.F., Li, J. and Tu, P.F. 2014. Anti-inflammatory lignanamides from the roots of *Solanum melongena* L. *Fitoterapia*. 98: 110–116.

Yang, B.Y., Yin, X., Liu, Y., Ye, H.L., Zhang, M.L., Guan, W. and Kuang, H.X. 2019. Bioassay-guided isolation of lignanamides with potential anti-inflammatory effect from the roots of *Solanum melongena* L. *Phytochemistry Letters*. 30: 160–164.

***Solanum nigrum* L.**

[From Latin *solanum* = nightshade and *nigrum* = black]

Published in *Species Plantarum* 1: 186. 1753

Synonym: *Solanum humile* Lam.
Local names: *tutan* (Dusun); *tutan pura* (Dusun, Kadazan)
Common name: black nightshade
Habitat: gardens, wastelands, roadsides
Geographical distribution: tropical Asia, Pacific Islands
Botanical description: this herbaceous plant reaches a height of about 60 cm and emits a kind of dismal aura. Stems: fleshy, purplish when young. Leaves: simple, spiral exstipulate, arranged in groups of 2–3. Petioles: up to 4 cm long. Blades: membranous, 2.5–10 cm × 2–5.5 cm, lanceolate, with 6 pairs of secondary nerves, tapering at base. Cymes: 3 cm long, axillary or cauliflorous, with a few nodding flowers. Calyx: tubular, 5-lobed, the lobes 2 mm long. Corolla: tubular, membranous, 5-lobed, the lobes yellowish at base, triangular and up to 6 mm long. Stamens: 5, merged into a cone. Berries: globose, reddish-black, glossy, about 8 mm across.
Traditional therapeutic indications: flu, increase appetite of children (Dusun); hypertension, intestinal worms (Dusun, Kadazan)
Pharmacology and phytochemistry: antibacterial (Zubair et al. 2011), anticancer (Lai et al. 2016), leishmanicidal (Singh et al. 2011; Mutoro et al. 2018).
Antiplasmodial steroidal alkaloids (solamargine, solasonine) (Chen et al. 2010).
Toxicity, side effects, and drug interaction: the plant and especially berries are dreadfully poisonous because of steroidal alkaloids (solanine).
The berries attract toddlers and children and therefore this plant should under no circumstances be planted in gardens and parks (Alexander et al. 1948).

Comment: The leaves and stems are used as food by the Dusun.

REFERENCES

Alexander, R.F., Forbes, G.B. and Hawkins, E.S. 1948. A fatal case of solanine poisoning. *British Medical Journal*. 2(4575): 518.

Chen, Y., Li, S., Sun, F., Han, H., Zhang, X., Fan, Y., Tai, G. and Zhou, Y. 2010. *In vivo* antimalarial activities of glycoalkaloids isolated from Solanaceae plants. *Pharmaceutical Biology*. 48(9): 1018–1024.

Lai, Y.J., Tai, C.J., Wang, C.W., Choong, C.Y., Lee, B.H., Shi, Y.C. and Tai, C.J. 2016. Anti-cancer activity of *Solanum nigrum* (AESN) through suppression of mitochondrial function and epithelial-mesenchymal transition (EMT) in breast cancer cells. *Molecules*. 21(5): 553.

Mutoro, C.N., Kinyua, J.K., Kariuki, D.W., Ingonga, J.M. and Anjili, C.O. 2018. *In vitro* study of the efficacy of *Solanum nigrum* against *Leishmania major*. *F1000Research*. 7(1329): 1329.

Singh, S.K., Bimal, S., Narayan, S., Jee, C., Bimal, D., Das, P. and Bimal, R. 2011. *Leishmania donovani*: Assessment of leishmanicidal effects of herbal extracts obtained from plants in the visceral leishmaniasis endemic area of Bihar, India. *Experimental Parasitology*. 127(2): 552–558.

Zubair, M., Rizwan, K., Rasool, N., Afshan, N., Shahid, M. and Ahmed, V.U. 2011. Antimicrobial potential of various extract and fractions of leaves of Solanum nigrum. *International Journal of Phytomedicine*. 3(1): 63.

Solanum torvum **Sw.**

[From Latin *solanum* = nightshade and *torvum* = dreadful]

Published in *Nova Genera et Species Plantarum seu Prodromus* 47. 1788

Local names: *bintorung talun* (Dusun); *lintahun* (Murut); *terung hutan, terung pipit* (Bajau, Kedayan); *ulom* (Lundayeh)
Common names: Turkey berry, pea eggplant, susumber berries
Habitat: roadsides, villages, wastelands
Geographical distribution: tropical Asia
Botanical description: this shrub reaches a height of about 2 m. Stems: spiny, hairy. Leaves: simple, alternate, exstipulate. Petioles: tomentose, 1.5–2.5 cm long, channeled. Blades: lanceolate or palmately lobed, somewhat asymmetrical, dull green, membranous, 6–16 cm × 4–11 cm long, hairy beneath, with 4–6 pairs of secondary nerves. Cymes: cauliflorous, about 5 cm long. Calyx: green, tomentose, membranous, 5-lobed, the lobes lanceolate and about 3 mm long. Corolla: white, membranous, 5-lobed, the lobes triangular and about 1 cm long. Stamens: 5, bright yellow, merged into a cone. Berries: dull green, globose, smooth, about 1.7 cm across.
Traditional therapeutic indications: fever (Dusun); hypertension (Kedayan); medicinal (Murut)
Pharmacology and phytochemistry: anti-inflammatory, analgesic (Ndebia et al. 2007), hypotensive (Mohan et al. 2009).
Antiviral (herpes simplex virus) isoflavonoid sulfate and steroidal saponin (torvanol A, torvoside H) (Arthan et al. 2002). Anti-inflammatory steroidal saponins (Lee et al. 2013). Cytotoxic hydroxycinnamic acid derivative (methyl caffeate) (Balachandran et al. 2015).
Toxicity, side effects, and drug interaction: berries administered daily to mice at the dose of 100 mg for a year were found to cause hepatic hemangiomas (Balachandran & Sivaramkrishnan 1995). The plant is poisonous because of cardiotoxic steroidal saponins (Smith et al. 2008; Glover et al. 2016).

Comments:

(i) In West Kalimantan, this plant is used for fever (local name: *terong pipit*).
(ii) In the Philippines, a decoction of roots is drunk as an antidote for poisons and to stop bleeding after childbirth.

REFERENCES

Arthan, D., Svasti, J., Kittakoop, P., Pittayakhachonwut, D., Tanticharoen, M. and Thebtaranonth, Y. 2002. Antiviral isoflavonoid sulfate and steroidal glycosides from the fruits *of Solanum torvum. Phytochemistry*. 59(4): 459–463.

Balachandran, B. and Sivaramkrishnan, V.M. 1995. Induction of tumours by Indian dietary constituents. *Indian Journal of Cancer*. 32(3): 104–109.

Balachandran, C., Emi, N., Arun, Y., Yamamoto, Y., Ahilan, B., Sangeetha, B., Duraipandiyan, V., Inaguma, Y., Okamoto, A., Ignacimuthu, S. and Al-Dhabi, N.A. 2015. *In vitro* anticancer activity of methyl caffeate isolated from Solanum torvum Swartz. fruit. *Chemico-Biological Interactions*. 242: 81–90.

Glover, R.L., Connors, N.J., Stefan, C., Wong, E., Hoffman, R.S., Nelson, L.S., Milstein, M., Smith, S.W. and Swerdlow, M. 2016. Electromyographic and laboratory findings in acute *Solanum torvum* poisoning. *Clinical Toxicology*. 54(1): 61–65.

Lee, C.L., Hwang, T.L., He, W.J., Tsai, Y.H., Yen, C.T., Yen, H.F., Chen, C.J., Chang, W.Y. and Wu, Y.C. 2013. Anti-neutrophilic inflammatory steroidal glycosides from Solanum torvum. *Phytochemistry*. 95: 315–321.

Mohan, M., Jaiswal, B.S. and Kasture, S. 2009. Effect of Solanum torvum on blood pressure and metabolic alterations in fructose hypertensive rats. *Journal of Ethnopharmacology*. 126(1): 86–89.

Ndebia, E.J., Kamgang, R. and Nkeh-Chungaganye, B.N. 2007. Analgesic and anti-inflammatory properties of aqueous extract from leaves of *Solanum torvum* (Solanaceae). *African Journal of Traditional, Complementary and Alternative Medicines*. 4(2): 240–244.

Smith, S.W., Giesbrecht, E., Thompson, M., Nelson, L.S. and Hoffman, R.S. 2008. Solanaceous steroidal glycoalkaloids and poisoning by Solanum torvum, the normally edible susumber berry. *Toxicon*. 52(6): 667–676.

4.1.15 Superorder Asteranae Takht. (1967), the Campanulids

4.1.15.1 Order Asterales Link (1829)

4.1.15.1.1 Family Asteraceae Martinov (1820), the Aster Family

The family Asteraceae is a vast taxon consisting of 1,500 genera and 13,000 species of herbaceous plants, shrubs, or climbing plants. Stems: often hairy. Leaves: simple, spiral, opposite, exstipulate. Blades: simple or dissected, often hairy. Inflorescences: aggregates of tiny flowers (florets) forming flower-like structures (capitula). Calyx: tubular, 5-lobed. Corolla: tubular, 1-5-lobed. Stamens: 5. Carpels: 2, forming a unilocular ovary with 1 ovule. Styles: 1. Stigma: bifid. Fruits: cypselae.

***Ageratum conyzoides* L.**

[From Greek *a* = without, *geros* = old age, and Latin *conyzoides* = Conyza-like]

Published in *Species Plantarum* 2: 839. 1753

Local names: *kambing kambing* (Dusun); *udu amek* (Lundayeh)
Common names: goat weed, billy goat weed, tropical white weed
Habitat: roadsides, wastelands, villages
Geographical distribution: tropical Asia, Pacific Islands
Botanical description: this common herbaceous plant reaches a height of about 50 cm. Stems: hairy, terete. Leaves: simple, opposite, exstipulate. Petioles: 1–3 cm long, hairy. Blades: hairy, dull green, soft, broadly lanceolate, 3–8 cm × 2–5 cm, attenuate or acute at base, serrate, acute at apex. Capitula: about 5 mm across, with a kind of beautiful, almost supernatural bluish to almost fluorescent color, especially at dawn and arranged in terminal and hairy corymbs. Florets: tubular, about 2 mm long, 5-lobed. Stamens: 5, elongated. Styles: 1, thin. Stigma: bifid. Cypselae: pentagonal and about 2 mm long achenes with 5 linear scales at apex.
Traditional therapeutic indications: cuts (Dusun); gastritis, skin injuries, venereal diseases (Lundayeh)
Pharmacology and phytochemistry: this plant has been the subject of numerous pharmacological studies which have highlighted, among other things, anti-inflammatory, wound healing, antimicrobial, antiparasitic, and anticancer activity (Marina & Chingakham 2013).
Chromenes (González et al. 1991). Pyrrolidine alkaloids (Wiedenfeld & Röder 1991).
Toxicity, side effects, and drug interaction: hepatotoxic pyrrolidine alkaloids (Wiedenfeld & Röder 1991).

Comment: Archaeological remains show that the use of this plant dates back to very ancient times in the Philippines (Paz 2005).

REFERENCES

González, A.G., Aguiar, Z.E., Grillo, T.A., Luis, J.G., Rivera, A. and Calle, J. 1991. Chromenes from *Ageratum conyzoides*. *Phytochemistry*. 30(4): 1137–1139.
Marina, A. and Chingakham, B.S. 2013. Ethnobotany, phytochemistry and pharmacology of *Ageratum conyzoides* Linn (Asteraceae). *Journal of Medicinal Plants Research*. 7(8): 371–385.
Paz, V. 2005. Rock shelters, caves, and archaeobotany in Island Southeast Asia. *Asian Perspectives*: 107–118.
Wiedenfeld, H. and Röder, E. 1991. Pyrrolizidine alkaloids from *Ageratum conyzoides*. *Planta Medica*. 57(6): 578–579.

***Blumea balsamifera* DC.**

[After the 19th century German botanist Karl Ludwig von Blume and Latin *balsamifera* = containing basalm]

Published in *Prodromus Systematis Naturalis Regni Vegetabilis* 5: 447. 1836

Synonyms: *Baccharis salvia* Lour.; *Conyza balsamifera* L.; *Pluchea balsamifera* (L.) Less.
Local names: *sambung* (Kedayan); *tawawoh* (Bajau, Dusun, Kadazan, Murut, Rungus); *daun sambong* (Dusun); *telinga kerbau* (Bajau); *bunga sapah, ipong* (Lundayeh)
Habitat: wastelands, roadsides
Geographical distribution: From India to South China
Botanical description: this robust and upright herbaceous plant reaches a height of about 3 m. Stems: terete, hairy, woody at base. Leaves: simple, with a slight camphorous odor when crushed, spiral, exstipulate. Petioles: about 1 cm long. Blades: oblong, 9–40 cm × 2–20 cm, hairy beneath, auriculate at base, soft, serrate, acuminate at apex, with 10–12 pairs of secondary nerves. Panicles: terminal with numerous capitula with a length of about 1 cm. Peripheral florets: filiform, 6 mm long, 2–4-lobed. Inner florets: yellow, tubular, 6 mm, long. Cypselae: brown, terete, oblong, achenes with a 4–6 mm long pappus.
Traditional therapeutic indications: postpartum (Dusun, Kadazan, Lundayeh, Murut); gastritis, stomach aches, flu (Murut); tapeworms (Rungus); fever (Dusun, Kadazan, Lundayeh, Murut); cuts, pancreatitis, wounds, colic, fatigue, insect bites (Dusun, Kadazan); flatulence (Dusun, Lundayeh); colds (Bajau); flu, gastritis, diarrhea, laxative (Murut). Other uses include asthma, body odor, and nosebleeds.
Pharmacology and phytochemistry: this plant has been the subject of numerous pharmacological studies which have highlighted, among other things, anticancer, antibacterial, anti-obesity, antiplasmodial, and anti-inflammatory activity. Monoterpenes (borneol) (Pang et al. 2014).
Toxicity, side effects, and drug interaction: not known

Comment: The leaves of this plant are placed under the stilt houses of the Suluk living in the northern islands of Sabah for magic rituals for childbirth (Dr Pauline Yong, personal communication).

REFERENCE

Pang, Y., Wang, D., Fan, Z., Chen, X., Yu, F., Hu, X., Wang, K. and Yuan, L. 2014. *Blumea balsamifera*—A phytochemical and pharmacological review. *Molecules*. 19(7): 9453–9477.

***Blumea riparia* DC.**

[After the 19th century German botanist Karl Ludwig von Blume and Latin *riparius* = riparian]

Published in *Prodromus Systematis Naturalis Regni Vegetabilis* 5: 444. 1836

Synonym: *Conyza riparia* Bl.
Habitat: forests near streams
Geographical distribution: from India to Pacific Islands
Botanical description: this shrub reaches a height of about 2.5 m. Stems: hairy at apex. Leaves: simple, spiral, exstipulate. Petioles: about 6–8 mm long. Blades: ovate-lanceolate, somewhat asymmetrical, 1.5–4 cm × 5–13 cm, rounded to acute at base, loosely serrate, acuminate at apex, with about 4–6 pairs of secondary nerves. Panicles: terminal with numerous capitula with a diameter of about 8 mm. Florets: filiform, about 5 mm long, 2–4-lobed, yellow. Cypselae: ribbed achenes with a 4–6 mm long and white pappus (Figure 4.48).
Traditional therapeutic indications: hypertension (Murut)
Pharmacology and phytochemistry: antifungal (Ninh et al. 2022).
Pro-coagulant phenolic acids (vanillic acid, syringic acid, *p*-coumaric acid, caffeic acid, protocatechuic acid) (Huang et al. 2010). Xanthenes (Cao et al. 2007). Essential oil (Ninh et al. 2022).
Toxicity, side effects, and drug interaction: not known

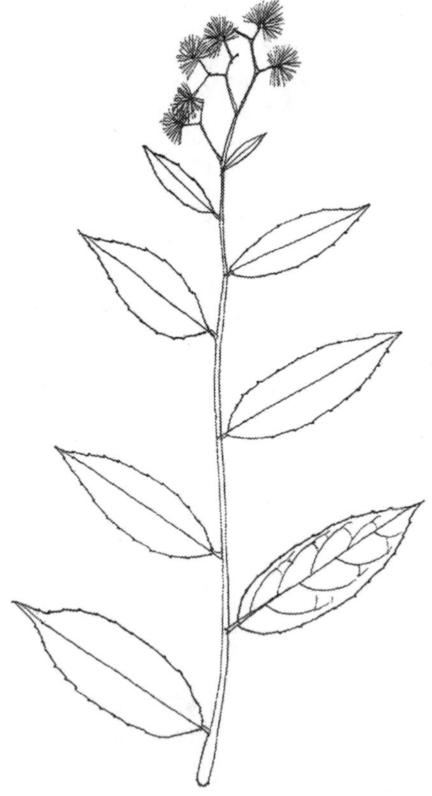

FIGURE 4.48 *Blumea riparia* DC.

REFERENCES

Cao, J.Q., Yao, Y., Chen, H., Qiao, L., Zhou, Y.Z. and Pei, Y.H. 2007. A new xanthene from *Blumea riparia*. *Chinese Chemical Letters*. 18(3): 303–305.

Huang, L., Lin, C., Li, A., Wei, B., Teng, J. and Li, L. 2010. Pro-coagulant activity of phenolic acids isolated from *Blumea riparia*. *Natural Product Communications*. 5(8): 1934578X1000500824.

Ninh The, S., Le Tuan, A., Dinh Thi Thu, T., Dinh Luyen, N. and Tran Thi, T. 2022. Essential oils of the Asteraceae plants *Blumea riparia* DC. and Pluchea *pteropoda* Hemsl. ex Hemsl. growing in Vietnam. *Natural Product Communications*. 17(6): 1–6.

***Crassocephalum crepidioides* (Benth.) S. Moore**

[From Latin *crassus* = thick and Greek *kephale* = head]

Published in *Journal of Botany, British and Foreign* 50(595): 211–212. 1912

Synonyms: *Crassocephalum diversifolium* Hiern; *Gynura crepidioides* Benth.; *Gynura polycephala* Benth.; *Senecio diversifolius* A. Rich.
Local name: *kinsau* (Murut)
Common name: thickhead
Habitat: roadsides, wastelands
Geographical distribution: tropical
Botanical description: this herbaceous plant reaches a height of about 1 m. Stems: hairy. Leaves: simple, spiral, exstipulate. Petioles: 2–4 cm long. Blades: oblanceolate, serrate, membranous, 2–7 cm × 1–3 cm, tapering at base, acute at apex, dull dark green, with about 10 pairs of secondary nerves visible. Capitula: reddish or white, arranged in cymes. Florets: tubular, up to 1 cm long. Cypselae: narrowly oblong achenes with a 1 cm long pappus.
Traditional therapeutic indications: cancer (Murut)
Pharmacology and phytochemistry: hepatoprotective (Aniya et al. 2005), anticoagulant (Ayodele et al. 2019), anticancer, anti-inflammatory (Tomimori et al. 2012), antibacterial (Owokotomo & Owokotomo 2018).
Toxicity, side effects, and drug interaction: hepatotoxic pyrrolizidine alkaloids (Asada et al. 1985). The median lethal dose (LD_{50}) of an aqueous extract of leaves administered orally to rats was 5 g/kg (Nguemfo et al. 2021).

Comments:

(i) This plant is used as food by the Dusun.
(ii) Plants of the genus *Crassocephalum* Moench (1794) are used by the Dusun and the Kadazan for broken bones "*lombon*" and "*lokop*."

REFERENCES

Aniya, Y., Koyama, T., Miyagi, C., Miyahira, M., Inomata, C., Kinoshita, S. and Ichiba, T. 2005. Free radical scavenging and hepatoprotective actions of the medicinal herb, *Crassocephalum crepidioides* from the Okinawa Islands. *Biological and Pharmaceutical Bulletin*. 28(1): 19–23.

Asada, Y., Shiraishi, M., Takeuchi, T., Osawa, Y. and Furuya, T. 1985. Pyrrolizidine alkaloids from *Crassocephalum crepidioides*. *Planta Medica*. 51(6): 539–540.

Ayodele, O.O., Onajobi, F.D. and Osoniyi, O. 2019. *In vitro* anticoagulant effect of *Crassocephalum crepidioides* leaf methanol extract and fractions on human blood. *Journal of Experimental Pharmacology*. 11: 99.

Nguemfo, E.L., Mbock, A.J., Bogning, C.Z., Fongang, A.L.M., Kedi, P.B.E. and Dongmo, A.B. 2021. Acute and sub-acute toxicity assessment of aqueous leaves extract of *Crassocephalum crepidioides* (Asteraceae) in Wistar rats. *Journal of Complementary and Integrative Medicine*. 18(2): 295–302.

Owokotomo, I.A. and Owokotomo, E.P. 2018. Anti-bacterial and brine shrimps lethality studies of the essential oils of *Crassocephalum crepidioides* (Benth S. More) grown in south west Nigeria. *African Journal of Pure and Applied Chemistry*. 12(1): 1–7.

Tomimori, K., Nakama, S., Kimura, R., Tamaki, K., Ishikawa, C. and Mori, N. 2012. Antitumor activity and macrophage nitric oxide producing action of medicinal herb, *Crassocephalum crepidioides*. *BMC Complementary and Alternative Medicine*. 12(1): 1–11.

***Elephantopus mollis* Kunth**

[From Greek *elephanto* = elephant and Latin *mollis* = soft]

Published in *Nova Genera et Species Plantarum* (folio ed.) 4: 20–21. 1820

Synonyms: *Elephantopus scaber* var. *tomentosus* (L.) Sch. Bip. ex Baker; *Elephantopus tomentosus* L.
Local names: *salaman, saraman* (Dusun, Kadazan); *honsigup, lambrunai* (Murut)
Habitat: jungle paths, villages, roadsides, lawns
Geographical distribution: tropical
Botanical description: this herbaceous plant reaches a height of about 40 cm. Roots: straight, thin. Stems: terete, thin, stiff, straight, dichotomous, hairy. Leaves: simple, exstipulate, in basal rosettes. Petioles: short, hairy. Blades: obovate, oblong, to obtuse, tapering at base, acute at apex, crenulate, hairy, finely marked with about 10 pairs of secondary nerves, 5–10 cm x 1.2–3 cm. Capitula: up to about 8 mm across, terminal, with a pair of leaflike bracts at base. Bracts: oblong, acute, green, hairy. Florets: 4-lobed, whitish-pink to purple, up to about 8 mm long. Calyx: prickly. Cypselae: 10-ribbed and about 4 mm long achenes and 5 whitish bristles.
Traditional therapeutic indications: bloody stools (Dusun, Kadazan, Murut); diarrhea, boils, coughs (Dusun); asthma, cuts, flatulence, wounds (Dusun, Kadazan)
Pharmacology and phytochemistry: hepatoprotective (Phan & Nguyen 2021), antibacterial (Alain et al. 2020), toxic to murine melanoma cells (Hasegawa et al. 2010) and liver cancer cells (HepG2) (Ooi et al. 2014).
Sesquiterpenes (But et al. 1996) toxic to myeloid leukemia cells (Li et al. 2016). Triterpenes toxic to neuroblastoma cells (Tabopda et al. 2007). Leishmanicidal sesquiterpenes (Fuchino et al. 2001). Anti-inflammatory sesquiterpenes (Wu et al. 2017).
Toxicity, side effects, and drug interaction: not known

Comments:

(i) The Dusun and Kadazan employ a plant of the genus *Elephantopus* L. (1753) as styptic.
(ii) *Elephantopus scaber* L. is used by the Dusun and Kadazan for asthma, wounds, and bloody stools.

REFERENCES

Alain, A.O.J., Le Doux, K.E., Thierry, O.A.M., Lazare, S.S., Nadia, A.H., Nga, N., Joseph, N. and Claudine, T.E. 2020. Phytochemical screening and *in-vitro* evaluation of antimicrobial and antioxidant activities of ethanolic extracts of *Elephantopus mollis* Kunth. (Asteraceae). *Journal of Pharmacognosy and Phytochemistry*. 9(1): 1711–1715.

But, P.P.H., Hon, P.M., Cao, H. and Che, C.T. 1996. A new sesquiterpene lactone from *Elephantopus mollis*. *Planta Medica*. 62(5): 474–476.

Fuchino, H., Koide, T., Takahashi, M., Sekita, S. and Satake, M. 2001. New sesquiterpene lactones from *Elephantopus mollis* and their leishmanicidal activities. *Planta Medica*. 67(7): 647–653.

Hasegawa, K., Furuya, R., Mizuno, H., Umishio, K., Suetsugu, M. and Sato, K. 2010. Inhibitory effect of *Elephantopus mollis* HB and K. extract on melanogenesis in B16 murine melanoma cells by down-regulating microphthalmia-associated transcription factor expression. *Bioscience, Biotechnology, and Biochemistry*. 74(9): 1908–1912.

Li, H., Li, M., Wang, G., Shao, F., Chen, W., Xia, C., Wang, S., Li, Y., Zhou, G. and Liu, Z. 2016. EM23, A natural sesquiterpene lactone from *Elephantopus mollis*, induces apoptosis in human myeloid leukemia cells through thioredoxin-and reactive oxygen species-mediated signaling pathways. *Frontiers in Pharmacology*. 7: 77.

Ooi, K.L., Tengku Muhammad, T.S., Lam, L.Y. and Sulaiman, S.F. 2014. Cytotoxic and apoptotic effects of ethyl acetate extract of *Elephantopus mollis* Kunth. In human liver carcinoma HepG2 cells through caspase-3 activation. *Integrative Cancer Therapies*. 13(3): NP1–NP9.

Phan, D.T. and Nguyen, P.T. 2021. Acute toxicity and liver protective effects of *Elephantopus mollis* HBK. *Tropical Journal of Natural Product Research (TJNPR)*. 5(3): 559–563.

Tabopda, T.K., Liu, J., Ngadjui, B.T. and Luu, B. 2007. Cytotoxic triterpene and sesquiterpene lactones from *Elephantopus mollis* and induction of apoptosis in neuroblastoma cells. *Planta Medica*. 73(4): 376–380.

Wu, Z.N., Zhang, Y.B., Chen, N.H., Li, M.J., Li, M.M., Tang, W., Zhuang, L., Li, Y.L. and Wang, G.C. 2017. Sesquiterpene lactones from *Elephantopus mollis* and their anti-inflammatory activities. *Phytochemistry*. 137: 81–86.

***Eupatorium odoratum* L.**

[From Greek *eu* = well, *epat* = liver, and Latin *odoratum* = fragrant]

Published in *Systema Naturae, Editio Decima* 2: 1205. 1759

Synonym: *Chromolaena odorata* (L.) R. M. King & H. Rob.
Local names: *rumput Siam* (Bajau); *gonol* (Rungus); *lambaian* (Murut); *pokok kapal terbang* (Bajau); *rumput Malaysia* (Dusun)
Common name: butterfly weed
Habitat: roadsides, wastelands
Geographical distribution: tropical Asia, Pacific Islands
Botanical description: this herbaceous plant reaches a height of about 2 m. Stems: terete, hairy. Leaves: simple, opposite, exstipulate. Petioles: 1–2.5 cm long. Blades: triangular, 4–10 cm × 1.5–5 cm, acute at apex, attenuate or acute at base, with 3 pairs of secondary nerves, serrate. Capitula: about 1 cm long, somewhat conical, organized into terminal cymes. Florets: light bluish-white, tubular, 5 mm long, 5-lobed. Styles: 1, protuding. Stigma: bifid. Cypselae: blackish brown, ribbed achenes with about 4 mm long pappus.
Traditional therapeutic indications: cuts, wounds (Dusun, Murut, Rungus); scalds (Sungai, Bajau, Kadazan, Dusun, Kedayan, Iranun, Murut, Rungus); stings (Sungai)
Pharmacology and phytochemistry: anti-inflammatory, antipyretic, antispasmodic (Taiwo et al. 2000), pro-coagulant (Triratana et al. 1991) antibacterial activity (Okwu et al. 2015).
Antimycobacterial flavonols (acacetin, luteolin) (Suksamram et al. 2004).
Toxicity, side effects, and drug interaction: hepatotoxic pyrrolizidine alkaloids (Biller et al. 1994).

REFERENCES

Biller, A., Boppré, M., Witte, L. and Hartmann, T. 1994. Pyrrolizidine alkaloids in *Chromolaena odorata*. Chemical and chemoecological aspects. *Phytochemistry*. 35(3): 615–619.

Okwu, M.U., Okorie, T.G. and Agba, M.I. 2015. *In vitro* anti-MRSA (Methicillin Resistant Staphylococcus aureus) activities of the partitions and fractions of the crude aqueous leaf extract of *Chromolaena odorata*. *IOSR Journal of Pharmacy and Biological Sciences*. 10(1): 136–141.

Suksamrarn, A., Chotipong, A., Suavansri, T., Boongird, S., Timsuksai, P., Vimuttipong, S. and Chuaynugul, A. 2004. Antimycobacterial activity and cytotoxicity of flavonoids from the flowers of *Chromolaena odorata*. *Archives of Pharmacal Research*. 27(5): 507–511.

Taiwo, O.B., Olajide, O.A., Soyannwo, O.O. and Makinde, J.M. 2000. Anti-inflammatory, antipyretic and antispasmodic properties of *Chromolaena odorata*. *Pharmaceutical Biology*. 38(5): 367–370.

Triratana, T., Suwannuraks, R. and Naengchomnong, W.A. 1991. Effect of Eupatorium odoratum on blood coagulation. *Journal of the Medical Association of Thailand*. 74(5): 283–287.

Synedrella nodiflora **(L.) Gaertn.**

[From Greek *syn* = together, *hedra* = chair, and Latin *nodiflora* = with flowers at nodes]

Published in *Fruct. Sem. Pl.* 2: 456. 1791

Synonyms: *Blainvillea gayana* Cass.; *Ucacou nodiflorum* (L.) Hitchc.; *Verbesina nodiflora* L.; *Wedelia gossweileri* S. Moore

Common names: cinderella weed, node weed

Habitat: roadsides, wastelands

Geographical distribution: tropical

Botanical description: this herbaceous plant reaches a height of about 80 cm. Stems: thin, hairy, angular. Leaves: simple, opposite, sessile, exstipulate. Blades: lanceolate, somewhat fleshy, often dark green, somewhat glossy, 3–10 cm × 2–5 cm, hairy, with a single pair of secondary nerves, attenuate to rounded at base, acuminate at apex, serrate. Capitula: sessile, axillary. Ray florets: somewhat golden yellow, about 4 mm long, bifid. Disc florets: yellow, 4-lobed, 4 mm long. Cypselae: achenes dentate, pappus made of a pair of pointed scales.

Traditional therapeutic indications: fatigue (Murut)

Pharmacology and phytochemistry: analgesic, antidepressant, antipsychotic, neuroprotective (Woode et al. 2009; Amoateng et al. 2017; Amoateng et al. 2018, Amoateng et al. 2021).

Toxicity, side effects, and drug interaction: hydroalcoholic extract administered orally to rats at a dose of 1 g per day for 14 days was found to cause no signs of toxicity (Adjei et al. 2014).

Comments:

(i) *Bidens pilosa* L. is used for teething by the Dusun and Kadazan (local name: *tondiokot*).

(ii) *Erechtites valerianaefolia* C.E.C. Fisch. is medicinal for the Murut (local name: *sumay-on*).

(iii) *Gynura procumbens* (Lour.) Merr. is used for fever and gastritis by the Murut (local names: *daun sambung, tawawo*).

(iv) *Mikania cordata* (Burm.f.) B.L. Rob. is a remedy for stomach aches and dysentery for the Bajau (local name: *selaput tunggul*).

(v) A plant of the genus *Chromolaena* DC. (1836) is used by the Dusun and Kadazan (local name: *nonokot*) as a veterinary medicine for wounded chickens.

(vi) *Cosmos caudatus* Kunth is used to purify the blood, to make the bones stronger (Kedayan), for colds, blood circulation, fatigue, indigestion, blood diseases, and aging by the Bajau (local names: *ulam rajah, ransa ransa*).

REFERENCES

Adjei, S., Amoateng, P., Safo, D.O., Sasu, C., N'guessan, B.B., Addo, P. and Gyekye, J.I.A. 2014. Sub-acute toxicity of a hydro-ethanolic whole plant extract of *Synedrella nodiflora* (L) Gaertn in rats. *International Journal of Green Pharmacy (IJGP)*. 8(4).

Amoateng, P., Adjei, S., Osei-Safo, D., Kukuia, K.K., Bekoe, E.O., Karikari, T.K. and Kombian, S.B. 2017a. Extract of *Synedrella nodiflora* (L) Gaertn exhibits antipsychotic properties in murine models of psychosis. *BMC Complementary and Alternative Medicine*. 17: 1–14.

Amoateng, P., Adjei, S., Osei-Safo, D., Kukuia, K.K.E., Kretchy, I.A., Sarkodie, J.A. and N'Guessan, B.B. 2017b. Analgesic effects of a hydro-ethanolic whole plant extract of *Synedrella nodiflora* (L.) Gaertn in paclitaxel-induced neuropathic pain in rats. *BMC Research Notes*. 10: 1–7.

Amoateng, P., Kukuia, K.K.E., Mensah, J.A., Osei-Safo, D., Adjei, S., Eklemet, A.A., Vinyo, E.A. and Karikari, T.K. 2018. An extract of *Synedrella nodiflora* (L) Gaertn exhibits antidepressant properties through monoaminergic mechanisms. *Metabolic Brain Disease*. 33: 1359–1368.

Amoateng, P., Tagoe, T.A., Karikari, T.K., Kukuia, K.K.E., Osei-Safo, D., Woode, E., Frenguelli, B.G. and Kombian, S.B. 2021. Synedrella nodiflora extract depresses excitatory synaptic transmission and chemically-induced *in vitro* seizures in the rat hippocampus. *Frontiers in Pharmacology*. 12: 610025.

Woode, E., Amoateng, P., Ansah, C. and Duwiejua, M. 2009. Anti-nociceptive effects of an ethanolic extract of the whole plant of *Synedrella nodiflora* (L.) Gaertn in mice: Involvement of adenosinergic mechanisms. *Journal of Pharmacology and Toxicology*. 4(1): 17–29.

4.1.15.2 Order Apiales Nakai (1930)

4.1.15.2.1 Family Apiaceae Lindley (1836), the Parsley Family

The family Apiaceae consists of about 440 genera and 3,000 species of herbaceous plants. Stems: often hollowed, ribbed. Leaves: simple, often deeply incised, compound, spiral, exstipulate. Petioles: forming a sheath at base. Inflorescences: umbels. Calyx: 5 sepals or tubular, or none. Petals: 5, often of unequal length. Stamens: 5. Carpels: 2, forming a unilocular or by bilocular ovary, each locule with 1–2 ovules. Styles: 2. Fruits: schizocarps.

Centella asiatica **(L.) Urb.**

[From Greek *kenteo* = to pierce and Latin *asiatica* = from Asia]

Published in *Flora Brasiliensis* 11(1): 287. 1879

Synonym: *Hydrocotyle asiatica* L.

Local names: *kulung kulung, pagago, tolingo ralan* (Dusun); *pegago* (Dusun, Kadazan, Lundayeh); *salapid* (Kadazan); *pegaga* (Bajau)

Common names: centella, Indian pennywort

Habitat: cultivated, lawns, wet and grassy places

Geographical distribution: tropical

Botanical description: this creeping herbaceous plant grows up to a length of about 80 cm. Stolons: terete, rooting at node. Leaves: simple, in basal rosettes, exstipulate. Petioles: thin, flexuous, up to about 10 cm long. Blades: somewhat kidney-shaped, up to about 2 cm × 4 cm, crenate, dull green, fleshy, with 2–3 pairs of secondary nerves. Umbels: few-flowered, the flowers tiny. Calyx: none. Petals: 5, white to pinkish, ovate. Stamens: 5. Ovary: pinkish. Schizocarps: 4 mm across.

Traditional therapeutic indications: aging, ear pain, diarrhea (Dusun); cough, diuretic, urinary tract infections, stomach aches, tachycardia (Dusun, Kadazan); canker sores, wounds (Kedayan); yellow fever (Sungai, Bajau, Iranun); Alzheimer's disease, jaundice, diabetes, anemia, abscesses, blood circulation, blood purification, cancer, fever, indigestion, rashes (Bajau); hypertension (Bajau, Dusun); fatigue (Lundayeh)

Pharmacology and phytochemistry: this plant has been the subject of numerous pharmacological studies (Sun et al. 2020) which have highlighted, among other things, anti-inflammatory, analgesic (Saha et al. 2013), and hypoglycemic activity (Masola et al. 2018). Triterpenes saponins (asiaticoside, madecassoside) (Wu et al. 2020).

Toxicity, side effects, and drug interaction: liver injuries (Jorge & Jorge 2005).

REFERENCES

Jorge, O.A. and Jorge, A.D. 2005. Hepatotoxicity associated with the ingestion of *Centella asiatica*. *Revista Espanola de Enfermedades Digestivas*. 97(2): 115–124.

Masola, B., Oguntibeju, O.O. and Oyenihi, A.B. 2018. *Centella asiatica* ameliorates diabetes-induced stress in rat tissues via influences on antioxidants and inflammatory cytokines. *Biomedicine & Pharmacotherapy*. 101: 447–457.

Saha, S., Guria, T., Singha, T. and Maity, T.K. 2013. Evaluation of analgesic and anti-inflammatory activity of chloroform and methanol extracts of *Centella asiatica* Linn. *International Scholarly Research Notices*. 2013.

Sun, B., Wu, L., Wu, Y., Zhang, C., Qin, L., Hayashi, M., Kudo, M., Gao, M. and Liu, T. 2020. Therapeutic potential of *Centella asiatica* and its triterpenes: A review. *Frontiers in Pharmacology*. 11: 568032.

Wu, Z.W., Li, W.B., Zhou, J., Liu, X., Wang, L., Chen, B., Wang, M.K., Ji, L., Hu, W.C. and Li, F. 2020. Oleanane-and ursane-type triterpene saponins from *Centella asiatica* exhibit neuroprotective effects. *Journal of Agricultural and Food Chemistry*. 68(26): 6977–6986.

Eryngium foetidum **L.**

[From Greek *eryngion* = sea holy and Latin *foetidum* = fetid]

Published in *Species Plantarum* 1: 232. 1753

Synonyms: *Eryngium antihystericum* Rottb.; *Eryngium molleri* Gand.
Local names: *rumput tabug* (Dusun); daun sup (Murut)
Common name: spiny coriander
Habitat: cultivated
Geographical distribution: subtropical, tropical
Botanical description: this erect herbaceous plant reaches a height of about 30 cm with a somewhat strange kind of dichotomous architecture. Stems: green, fleshy, striated. Leaves: in basal rosettes, spiral, opposite, sessile. Blades: oblong to oblanceolate to deeply incised, membranous, glossy, serrate, 5–25 cm × 1.2–5 cm, rounded at apex. Umbels: elongated, axillary, or terminal, about 3 cm long, with lanceolate and leafy bracts with a length of up to 2 cm long, with tiny flowers. Calyx: tiny, 5-lobed. Petals: 5, white or pale yellow. Disc: present. Stamens: 5. Styles: 2. Schizocarps: with a pair of ovoid, tiny, and tuberculate mericarps.
Traditional therapeutic indications: medicinal (Dusun)
Pharmacology and phytochemistry: anticancer (Promtes et al. 2016), antibacterial (Dalukdeniya & Rathnayaka 2017).
Nematicidal alkane (trans-2-dodecenal) (Forbes et al. 2014).
Toxicity, side effects, and drug interaction: leaves administered to mice at 3.2% of the diet was found to cause tubulonephrosis and chronic interstitial nephritis (Janwitthayanuchit et al. 2016).

Comments:

(i) The Dusun use the leaves of this plant as food.
(ii) A plant of the genus *Eryngium* L. (1753) is used by the Dusun and Kadazan for sprains (local name: *kosur*).
(iii) *Apium graveolens* L. is used for gout by the Bugis (local name: *daung sup*).

REFERENCES

Dalukdeniya, D.A.C.K. and Rathnayaka, R.M.U.S.K. 2017. Comparative study on antibacterial and selected antioxidant activities of different *Eryngium foetidum* extracts. *Journal of Applied Life Sciences International*. 12: 1–7.

Forbes, W.M., Gallimore, W.A., Mansingh, A., Reese, P.B. and Robinson, R.D. 2014. Eryngial (trans-2-dodecenal), a bioactive compound from *Eryngium foetidum*: Its identification, chemical isolation, characterization and comparison with ivermectin *in vitro*. *Parasitology*. 141(2): 269–278.

Janwitthayanuchit, K., Kupradinun, P., Rungsipipat, A., Kettawan, A. and Butryee, C. 2016. A 24-weeks toxicity study of *Eryngium foetidum* Linn. leaves in mice. *Toxicological Research*. 32(3): 231–237.

Promtes, K., Kupradinun, P., Rungsipipat, A., Tuntipopipat, S. and Butryee, C. 2016. Chemopreventive effects of *Eryngium foetidum* L: Leaves on COX-2 reduction in mice induced colorectal carcinogenesis. *Nutrition and Cancer*. 68(1): 144–153.

4.1.15.2.2 Family Araliaceae Juss. (1789), the Ginseng Family

The family Araliaceae consists of about 50 genera and 1,000 species of climbing plants or treelets. Leaves: palmate, spiral, sometimes stipulate. Petioles: often straight, clasping the stem at base. Inflorescences: globose umbels, solitary or arranged into spikes or panicles. Sepals: 3–5 or none. Petals: 5. Disc: present. Stamens: 5. Carpels: 2–5 forming a 2–5 locular ovary, each locule with 1–2 ovules. Styles: 2. Fruits: berries or drupes marked at apex with a disc.

***Schefflera nervosa* (King) R. Vig.**

[After the 19th century Polish botanist Johann Peter Ernst von Scheffler and Latin *nervosa* = with nerves]

Published in *Annales des Sciences Naturelles; Botanique, série* 9 9: 353. 1909

Synonym: *Heptapleurum nervosum* King
Local name: *miang palat* (Dusun)
Habitat: forests
Geographical distribution: Malaysia, Borneo
Botanical description: this shrub reaches a height of about 6 m. Stems: grayish, glabrous, rough, terete. Leaves: palmate, spiral, stipulate. Petioles: about 5 cm long. Petiolules: about 1 cm long. Folioles: 6–7, coriaceous, 2.5–9 cm × 1.1–6 cm, lanceolate or elliptic, attenuate at base, acute or acuminate at apex, with 6–10 pairs of secondary nerves. Panicles: about 10 cm long, hairy, with umbels of tiny flowers. Calyx: tubular, 5-lobed. Petals: 5, about 5 mm long, triangular, yellowish-green. Drupes: striated, about 7 mm long (Figure 4.49).

FIGURE 4.49 *Schefflera nervosa* (King) R.Vig

Traditional therapeutic indications: paralytic babies (Dusun)
Pharmacology and phytochemistry: not known
Toxicity, side effects, and drug interaction: not known

Comments:

(i) This plant is consumed by the very few orangutans left in Sabah.
(ii) The Dusun and Kadazan use a plant of the genus *Schefflera* J.R. Forst. & G. Forst. (1776) for pancreatitis (local name: *malad palad*), whilst a plant of the genus *Aralia* (1753) (local name: *rupa*) is a remedy for malaise and to promote lactation.
(iii) *Aralia montana* Bl. is used by the Dusun and Kadazan as poison antidote (local name: *golungang*).
(iv) *Polyscias scutellaria* (Burm.f.) Fosberg is used by the Lundayeh for anorexia (local name: *polibas*).
(v) *Schefflera petiolosa* (Miq.) Harms is used for asthma, bone aches, skin diseases, post-partum, and treatment of wounds by the Dusun and Kadazan.

4.1.15.2.3 Family Pittosporaceae R.Br. (1814), the Cheesewood Family

The family Pittosporaceae consists of about 10 genera and 200 species of shrubs, climbing plants, or trees. Leaves: simple, alternate, opposite, exstipulate. Blades: coriaceous. Inflorescences: umbels, cymes, panicles, or solitary. Calyx: 5 sepals or tubular. Petals: 5 petals or tubular. Stamens: 5. Carpels: 2 forming a unilocular or bilocular ovary, each locules with 5–many. Fruits: capsules, berries.

***Pittosporum ferrugineum* W.T. Aiton**

[From Greek *pitta* = pitch, *spora* = seed, and Latin *ferrugineum* = of the color of iron rust]

Published in *Hortus Kewensis; or, a Catalogue of the Plants Cultivated in the Royal Botanic Garden at Kew. London* (ed. 2.) 2: 27. 1811

Local names: *mensaipang, saipang* (Dusun)
Common name: rusty pittosporum
Habitat: seashores, hills, dry lands, forests
Geographical distribution: from Malaysia to Papua New Guinea
Botanical description: this tree reaches a height of about 20 m. Stems: hairy at apex. Leaves: simple, alternate, exstipulate. Petioles: 1.5–3 cm long, hairy. Blades: dark green, glossy, somewhat asymmetrical, wavy, hairy beneath, elliptic, attenuate at base, acuminate at apex, with about 5–7 pairs of secondary nerves. Panicles: hairy, terminal. Sepals: 5, hairy, linear, 1.5 mm long. Corolla: tubular, 5-lobed, yellowish white, the lobes revolute. Stamens: 5, about 5 mm long. Ovary: bottle-shaped, glossy. Styles: 1. Stigma: capitate. Capsules: light orangish, elliptic to obovate, about 1 cm long. Seeds: numerous, in a red and glossy pulp.
Traditional therapeutic indications: back and bone aches (Dusun)
Pharmacology and phytochemistry: antibacterial (Farhan et al. 2013).
Toxicity, side effects, and drug interaction: not known

Comment: In Papua New Guinea, this plant is used as poison antidote (local name: *boedobu*).

REFERENCE

Farhan, A.M., Lee, P.C., How, S.E. and Jualang, A.G. 2013. Antibacterial activities of Agave angustifolia and Pittosporum ferrugineum. *Journal of Environmental Microbiology and Toxicology*. 1(1): 15–17.

Bibliography

Ahmad, F.B. and Holdsworth, D.K. 1994. Medicinal plants of Sabah, Malaysia, part II. The Muruts. *International Journal of Pharmacognosy*. 32(4): 378–383.

Ahmad, F.B. and Holdsworth, D.K. 2003. Medicinal plants of Sabah, East Malaysia–Part I. *Pharmaceutical Biology*. 41(5): 340–346.

Ahmad, F.B. and Ismail, G. 2003. Medicinal plants used by Kadazandusun communities around Crocker Range. *ASEAN Review of Biodiversity and Environmental Conservation (ARBEC)*. 1(1): 1–10.

Alan, R., Tunung, R., Saupi, N. and Lepun, P. 2022. Wild pepper species consumed as green leafy vegetables among Orang Ulu groups in Asap-Koyan Belaga, Sarawak. *Food Research*. 6(2): 166–171.

Alex, A. 1996. *Vegetative, Floral and Fruit Characters in Mangosteen (Garcinia mangostana L.)* (Doctoral dissertation, Department of Pomology and Floriculture, College of Horticulture, Vellanikkara).

Allerton, C. 2017. Contested statelessness in Sabah, Malaysia: Irregularity and the politics of recognition. *Journal of Immigrant & Refugee Studies*. 15(3): 250–268.

Andersen, J., Nilsson, C., de Richelieu, T., Fridriksdottir, H., Gobilick, J., Mertz, O. and Gausset, Q. 2003. Local use of forest products in Kuyongon, Sabah, Malaysia. *ASEAN Review of Biodiversity and Environmental Conservation (ARBEC)*. 2: 1–18.

Andreas, C.H. and Prop, N. 1954. Florae Malesianae Precursores VIII. The genus Cnestis (Connaraceae) in Indo-Malaysia. *Blumea: Biodiversity, Evolution and Biogeography of Plants*. 7(3): 602–616.

Ara, H. and Hassan, M.A. 2006. Three new records of Aroids (Araceae) for Bangladesh. *Bangladesh Journal of Plant Taxonomy*. 13(2): 83.

Asma, J., Rinchen, G. and Namrata, S. 2018. Floral phenology of *Centella asiatica* (L.) Urban: A predominantly autogamous taxon of Apiaceae. *The International Journal of Plant Reproductive Biology*. 10(2): 189–191.

Aththorick, T.A. and Berutu, L. 2018, December. Ethnobotanical study and phytochemical screening of medicinal plants on Karonese people from North Sumatra, Indonesia. In *Journal of Physics: Conference Series* (1116, No. 5: 052008). IOP Publishing.

Austin, D.F. 2006. Fox-tail millets (Setaria: Poaceae)—Abandoned food in two hemispheres. *Economic Botany*. 60(2): 143–158.

Awang-Kanak, F., Bakar, M.F.A. and Mohamed, M. 2018, August. Ethnobotanical survey on plants used as traditional salad food (ulam) in Kampung Taun Gusi, Kota Belud Sabah, Malaysia. In *AIP Conference Proceedings* (2002, No. 1). AIP Publishing.

Awang-Kanak, F., Matawali, A., Jumat, N.R. and Bakri, S.N.S. 2021. A preliminary survey on edibles and medicinal plants used by Dusun of Kampung Pinolobu, Kadamaian, Kota Belud, Sabah, Malaysia. *Journal of Tropical Biology & Conservation (JTBC)*. 18: 21–30.

Az-Zahra, F.R., Sari, N.L.W., Saputry, R., Nugroho, G.D., Pribadi, T., Sunarto, S. and Setyawan, A.D. 2021. Traditional knowledge of the Dayak Tribes (Borneo) in the use of medicinal plants. *Biodiversitas Journal of Biological Diversity*. 22(10).

Bakker, K. and Van Steenis, C.G.G.J. 1955. Pittosporaceae. *Flora Malesiana-Series 1, Spermatophyta*. 5(1): 345–362.

Benggon, C.J.J. 2008. *Preliminary Screening of Secondary Metabolites from Selected Medicinal Plants from Kg. Pulutan, Sabah* (Bachelor of Science with Honours. Faculty of Resource Science and Technology, University Malaysia Sarawak).

Boyce, P. 1998. The genus *Epipremnum* Schott (Araceae-Monsteroideae-Monstereae) in West and Central Malesia. *Blumea: Biodiversity, Evolution and Biogeography of Plants*. 43(1): 183–213.

Boyce, P.C. 1999. The genus *Rhaphidophora* Hassk. (Araceae-Monsteroideae-Monstereae) in Peninsular Malaysia and Singapore. *Gardens' Bulletin Singapore*. 51(2): 183–256.

Bramley, G.L. 2009. The genus *Callicarpa* (Lamiaceae) on Borneo. *Botanical Journal of the Linnean Society*. 159(3): 416–455.

Cahen, D., Rickenback, J. and Utteridge, T.M. 2021. A revision of *Ziziphus* (Rhamnaceae) in Borneo. *Kew Bulletin*. 76: 767–804.

Cantoria, M.C. 1986. The identification of *Zingiber purpureum* rosc. *Transactions of the National Academy of Science and Technology Philippines*. 8: 139–150.

Chai, P.P. ed. 2000. *A Check-list of Flora, Fauna, Food and Medicinal Plants*. Forest Department Sarawak, International Tropical Timber Organization Japan.

Chantaranothai, P. 2011. A revision of the genus *Vitex* (Lamiaceae) in Thailand. *Tropical Natural History*. 11(2): 91–118.

Chase, M.W. and Reveal, J.L. 2009. A phylogenetic classification of the land plants to accompany APG III. *Botanical Journal of the Linnean Society*. 161(2): 122–127.

Chew, W.L. 1972. The genus *Piper* (Piperaceae) in New Guinea, Solomon Islands, and Australia, 1. *Journal of the Arnold Arboretum*. 53(1): 1–25.

Chin, J. 2017. Malay Muslim first: The politics of bumiputeraism in East Malaysia. In Sophie Lemiere (ed.), *Illusions of Democracy: Malaysian Politics and People* (pp. 201–220). Amsterdam University Press.

Chung, A.Y., Khen, C.V., Unchi, S. and Binti, M. 2000. Ethnoentomological uses of ants in Sabah, Malaysia. In *Proceedings of the 2nd ANeT Workshop in Kota Kinabalu, Sabah. Forest Research Centre, Forestry Department, Sabah, Malaysia* (pp. 1–4). Available at https://www.researchgate.net/profile/Arthur-Chung-3/publication/328886885_Ethnoentomological_uses_of_ants_in_Sabah_Malaysia/links/5be98ea592851c6b27b940a3/Ethnoentomological-uses-of-ants-in-Sabah-Malaysia.pdf

Chung, R.C.K. and Soepadmo, E. 2011. Taxonomic revision of the genus *Microcos* (Malvaceae-Grewioideae) in Peninsular Malaysia and Singapore. *Blumea-Biodiversity, Evolution and Biogeography of Plants*. 56(3): 273–299.

Coode, M.J.E. 1996. *Elaeocarpus* for Flora Malesiana: The 'Polystachyus group'. *Kew Bulletin*. 649–666.

Cordero, C. and Alejandro, G.J.D. 2021. Medicinal plants used by the indigenous Ati tribe in Tobias Fornier, Antique, Philippines. *Biodiversitas Journal of Biological Diversity*. 22(2): 521–536.

Damayanto, I.P.G.P., Dalimunthe, S.H. and Megawati, M. 2021. *Dinochloa scandens* (Poaceae-Bambusoideae): Distribution, habitat preference, and notes on synonymy. *Jurnal Biodjati*. 6(2): 174–189.

Davies, S.J. 2002. Ethnobotany of *Macaranga* (euphorbiaceae) among the kedayan of Brunei Darussalam. *Harvard Papers in Botany*. 7–12.

De Kok, R.P.J. 2007. The genus *Vitex* L. (Lamiaceae) in New Guinea and the south pacific islands. *Kew Bulletin*. 587–603.

De Kok, R.P.J. 2021. A revision of *Litsea* (Lauraceae) in Peninsular Malaysia and Singapore. *Gardens' Bulletin Singapore*. 73(1): 81–178.

De Winter, W.P. and Amoroso, V.B. 2003. Plant resources of South-East Asia. *Cryptogams: Ferns and Ferns Allies*. 15(2): 61–63.

Dransfield, S. 1981. The genus *Dinochloa* (Gramineae-Bambusoideae) in Sabah. *Kew Bulletin*. 613–633.

Duistermaat, H. 2012. A taxonomic revision of *Amischotolype* (Commelinaceae) in Asia. *Gardens' Bulletin Singapore*. 64: 51–131.

Evans, I.H.N. 1922. *Among Primitive Peoples in Borneo: A Description of the Lives, Habits & Customs of the Piratical Head-Hunters of North Borneo, with an Account of Interesting Objects of Prehistoric Antiquity Discovered in the Island*. Seeley, Service and Company Limited.

Evans, I.H.N. 1953. *The Religion of the Tempasuk Dusun of North Borneo*. Cambridge University Press.

Exell, A.W. 1948. Combretaceae. *Flora Malesiana-Series 1, Spermatophyta*. 4(1): 533–589.

Falah, F. and Hadiwibowo, N. 2017. Species identification of traditional medicine plants for women's health in East Kalimantan: Lesson learned from local wisdom. *Indonesian Journal of Forestry Research*. 4(1): 49–67.

Foo, J. 2018. Penglibatan komuniti tempatan dalam pasaran tumbuhan ubatan di tamu pantai barat Sabah. *Akademika*. 88(1): 35–47.

Foo, J. and Jaafar, K. 2019. Women's role in the medicinal plants market in the Tamu of West Coast, Sabah. *Jurnal Kinabalu*. 25: 13–20.

Foo, J., Mohamad, A.L., Omar, M. and Amir, A.A. 2016. Ethnobotanical survey of medicinal plants traded at Tamu in Sabah urban area. *International Journal of the Malay World and Civilisation*. 4(1): 79–87.

Forman, L.L. 1968. The menispermaceae of Malesia: V. Tribe Cocculeae Hook. F. & Thoms. *Kew Bulletin*. 22(3): 349–374.

Fukuoka, N. 1971. Contributions to the flora of Southeast Asia III. *Hedyotis* (Rubiaceae) of Thailand. *Japanese Journal of Southeast Asian Studies*. 8(3): 305–336.

Gan, C.Y. 1995. Smokeless tobacco use among rural Kadazan women in Sabah, Malaysia. *Southeast Asian Journal of Tropical Medicine and Public Health*. 26: 291–296.

Gan, C.Y. 1998. Tobacco usage among rural Bajaus in Sabah, Malaysia. *Southeast Asian Journal of Tropical Medicine and Public Health*. 29: 643–648.

Ganesan, S.K., Lim, R.C.J., Leong, P.K.F. and Ng, X.Y. 2020. *Microcos antidesmifolia* (Malvaceae-Grewioideae), a poorly known species in Singapore. *Gardens' Bulletin Singapore*. 72(2): 159–164.

Gollin, L.X. 2004. Subtle and profound sensory attributes of medicinal plants among the Kenyah leppo'ke of east Kalimantan, Borneo. *Journal of Ethnobiology*. 24(2): 173–201.

Gunggut, H., Mohd, D.S.N.S.A., Zaaba, Z. and Liu, M.S.M. 2014. Where have all the forests gone? Deforestation in land beneath the wind. *Procedia-Social and Behavioral Sciences*. 153: 363–369.

Hapid, A., Napitupulu, M. and Zubair, M.S. 2021. Ethnopharmacology and antioxidant activity studies of woody liana original Wallacea. *International Journal of Design and Nature and Ecodynamics*. 16(5): 495–503.

Haris, A., Nawan, N.A., Mei, C.A.L., Sani, S.A. and Najm, S.U.F.S. 2023. Medicinal plant applications as traditional and complementary medicine by Sabah ethnicities and the regulations and economic view in Malaysia's healthcare industry: A mini review. *Pharmacognosy Reviews*. 17(33): 1–10.

Hoogland, R.D. 1952. A revision of the genus *Dillenia*. *Blumea: Biodiversity, Evolution and Biogeography of Plants*. 7(1): 1–145.

Hoogland, R.D. 1953. The genus *Tetracera* (Dilleniaceae) in the eastern old world. *Reinwardtia*. 2(2): 185–224.

Hoogland, R.D. 1977. Sauraniae gerontogeae. I. Notes on Malayan species. *Gardens Bulletin, Singapore*. XXX.

Hou, D. 1960. Thymelaeaceae. *Flora Malesiana-Series 1, Spermatophyta*. 6(1): 1–48.

Hou, D. 1983. Florae Malesianae Praecursores LXIII. New species of Malesian Aristolochiaceae. *Blumea: Biodiversity, Evolution and Biogeography of Plants*. 28(2): 343–352.

Hou, D. 1984. Aristolochiaceae. *Flora Malesiana-Series 1, Spermatophyta*. 10(1): 53–108.

Hourt, K.E. 2008. *A Field Guide of the Rattans of Cambodia*. WWF Cambodia.

Huxley, C.R. and Jebb, M. 1993. The tuberous epiphytes of the Rubiaceae 5: A revision of *Myrmecodia*. *Blumea: Biodiversity, Evolution and Biogeography of Plants*. 37(2): 271–334.

Jarrett, F.M. 1959. Studies in *Artocarpus* and allied genera, III. A revision of Artocarpus subgenus Artocarpus. *Journal of the Arnold Arboretum*. 40(2): 113–155.

Jaya, A.M., Musa, Y., Iswoyo, H., Asmi, N. and Siregar, L.F. 2019, October. Ethnobotanical study and identification of medicinal plants based on local knowledge. In *IOP Conference Series: Earth and Environmental Science* (343, No. 1: 012028). IOP Publishing.

Kartonegoro, A. and Veldkamp, J.F. 2010. Revision of dissochaeta (melastomataceae) in Java, Indonesia. *Reinwardtia*. 13(2): 125–145.

Kessler, P.J. and Sidiyasa, K. 1999. *Trees of the Balikpapan-Samarinda Area, East Kalimantan, Indonesia: A Manual to 280 Selected Species*. The Tropenbos Foundation.

Khan, K.K. and Hemalatha, E. 2015. A review on the genus *Bambusa* and one particular species Bambusa vulgaris in Sabah (Malaysia). *International Research Journal of Pharmacy*. 6(9): 580–584.

Khaw, S.H. 2001. The genus *Etlingera* (Zingiberaceae) in Peninsular Malaysia including a new species. *Gard Bull Singapore*. 53(1–2): 191–239.

King, J.K. and King, J.W. 1984. *Languages of Sabah, a Survey Report*. Dept. of Linguistics, Research School of Pacific Studies, The Australian National University.

Kodoh, J. 2005. Surveys of non-timber forest products traded in Tamu, Sabah, Malaysia. *Sepilok Bulletin*. 3: 27–36.

Kodoh, J., Mojiol, A.R., Lintangah, W., Gisiu, F., Maid, M. and Liew, K.C. 2017. Traditional knowledge on the uses of medicinal plants among the ethnic communities in Kudat, Sabah, Malaysia. *International Journal of Agricultural, Forestry & Plantation*. 5: 79–85.

Koster, J. 1958. Notes on Malay compositae IV. *Blumea. Supplement*. 4(1): 170–177.

Kulip, J. 1997. A preliminary survey of traditional medicinal plants in the west coast and interior of Sabah. *Journal of Tropical Forest Science*. 10(2): 271–274.

Kulip, J. 2003a. An ethnobotanical survey of medicinal and other useful plants of Muruts in Sabah, Malaysia. *Telopea*. 10(1): 81–98.

Kulip, J. 2003b. Similarity of medicinal plants used by two native communities in Sabah, Malaysia. In *III WOCMAP Congress on Medicinal and Aromatic Plants-Volume 1: Bioprospecting and Ethnopharmacology 675* (pp. 81–85). Available at https://www.researchgate.net/profile/Arthur-Chung-3/publication/328886885_Ethnoentomological_uses_of_ants_in_Sabah_Malaysia/links/5be98ea592851c6b27b940a3/Ethnoentomological-uses-of-ants-in-Sabah-Malaysia.pdf

Kulip, J. 2007. Gingers in Sabah and their traditional uses. *Sepilok Bulletin*. 7(1): 23–44.

Kulip, J. 2014. The ethnobotany of Dusun people in Tikolod village, Tambunan district, Sabah, Malaysia. *Reinwardtia*. 14(1): 101–121.

Kulip, J., Fan, L.N., Manshoor, N., Julius, A., Said, I.M., Gisil, J., Joseph, J.A. and Tukin, W.F. 2010. Medicinal plants in Maliau Basin, Sabah, Malaysia. *Journal of Tropical Biology & Conservation (JTBC)*. 6: 21–33.

Kulip, J., Indu, J.P. and Mision, R. 2005. Ethnobotanical survey of medical plants in the village of Kaingaran in Sabah, Malaysia. *Journal of Tropical Biology & Conservation (JTBC)*. 1.

Kulip, J., Majawat, G. and Kulik, J. 2000. Medicinal and other useful plants of the Lundayeh community of Sipitang, Sabah, Malaysia. *Journal of Tropical Forest Science*. 12(4): 810–816.

Lemmens, R.H.M.J. 2003. Eurycoma Jack. Di dalam: Lemmens. In R.H.M.J. dan N. Bunyapraphatsara (Ed.), *Medicinal and Poisonous Plants 3. Plants Resources of South East Asia. No. 12 (3)*. Kew Bulletin.

Leong, C.E. 2009. *Lest We Forget (Security and Sovereignty of Sabah)*. Self published.

Linares, V., Jakoel, E., Be'eri, R., Lipschits, O., Neumann, R. and Gadot, Y., 2022. Opium trade and use during the Late Bronze Age: Organic residue analysis of ceramic vessels from the burials of Tel Yehud, Israel. Archaeometry.1-18

Mandia, E.H. 2004. The Alangan Mangyan of Mt. Halcon, Oriental Mindoro: Their ethnobotany. *Philippine Quarterly of Culture and Society*. 32(2): 96–117.

Merrill, E.D. 1954. Miscellaneous Malaysian notes. *Journal of the Arnold Arboretum*. 35(2): 134–157.

Middleton, D.J. 2004. A revision of *Kopsia* (Apocynaceae: Rauvolfioideae). *Harvard Papers in Botany*. 9(1): 89–142.

Mojiol, A.R., Adella, A., Kodoh, J., Lintangah, W. and Wahab, R. 2010. Common medicinal plants species found at burned and unburned areas of Klias peat swamp forest, Beaufort, Sabah Malaysia. *Journal of Sustainable Development.* 3(1): 109.

Mojiol, A.R., Lintangah, W., Ismenyah, M., Alamjuri, R.H. and Jaafar, C.S.Z. 2016. Mangroves forest produce (MFP): Importance and contribution to the local communities at Banggi Island Malaysia using free listing. *International Journal of Agriculture, Forestry and Plantation*. 3: 89–94.

Mols, J.B. and Keßler, P. 2000. Revision of the genus Phaeanthus (Annonaceae). *Blumea: Biodiversity, Evolution and Biogeography of Plants*. 45(1): 205–233.

Mood, J.D., Tanaka, N., Aung, M.M. and Murata, J. 2016. The genus *Boesenbergia* (Zingiberaceae) in Myanmar with two new records. *Gardens' Bulletin Singapore*. 68(2): 299–318.

Muhammed, N. and Muthu, T.A. 2015. Indigenous people and their traditional knowledge on tropical plant cultivation and utilization: A case study of Murut communities of Sabah, Borneo. *Journal of Tropical Resources and Sustainable Science (JTRSS)*. 3(1): 117–128.

Mulyaningsih, T. and Yamada, I. 2008. Notes on some species of agarwood in Nusa Tenggara, Celebes and West Papua. In *Natural Resource Management and Socio-economic Transformation under the Decentralization in Indonesia: Toward Sulawesi Area Studies* (pp. 365–372). CSEAS, Kyoto University .Available at http://eprints.unram.ac.id/36368/

Nahdi, M.S., Martiwi, I.N.A. and Arsyah, D.C. 2016. The ethnobotany of medicinal plants in supporting the family health in Turgo, Yogyakarta, Indonesia. *Biodiversitas*. 17(2): 900–906.

Naive, M.A.K., Pabillaran, R.O. and Escrupulo, I.G. 2018. *Etlingera coccinea* (Blume) S. Sakai and Nagam. (Zingibearaceae–Alpinieae): An addition to the Flora of the Philippines, with notes on its distribution, phenology and ecology. *Bioscience Discovery*. 9(1): 107–110.

Norsaengsri, M. and Chantaranothai, P. 2008. A revised taxonomic account of *Paspalum* L. (Poaceae) in Thailand. *Tropical Natural History*. 8(2): 99–119.

Obico, J.J.A. and Ragragio, E.M. 2014. A survey of plants used as repellents against hematophagous insects by the Ayta people of Porac, Pampanga province, Philippines. *Philippine Science Letters*. 7(1): 179–186.

Olowa, L. and Demayo, C.G. 2015. Ethnobotanical uses of medicinal plants among the Muslim Maranaos in Iligan City, Mindanao, Philippines. *Advances in Environmental Biology*. 9(27): 204–215.

On, L.K. and Ishak, S. 2016. Beliefs in the Komburongo (Acorus Calamus) and its spiritual healing among the Dusunic people of Sabah, Malaysia. *Advanced Science Letters*. 22(5–6): 1336–1339.

Ong, H.G. and Kim, Y.D. 2014. Quantitative ethnobotanical study of the medicinal plants used by the Ati Negrito indigenous group in Guimaras island, Philippines. *Journal of Ethnopharmacology*. 157: 228–242.

Othman, A.S.B., Boyce, P.C. and Keng, C.L. 2010. Studies on Monstereae (Araceae) of Peninsular Malaysia III: *Scindapsus lucens*, a New Record for Malaysia, and a Key to Peninsular Malaysian *Scindapsus*. *Gardens' Bulletin Singapore*. 62(1): 9–15.

Panoff, F. 1970. Maenge remedies and conception of disease. *Ethnology*. 9(1): 68–84.

Pathan, A.R., Vadnere, G.P. and Sabu, M. 2013. *Curcuma caesia* almost untouched drug: An updated ethnopharmacological review. *Inventi Rapid: Planta Activa*. 4: 1–4.

Phengklai, C. 1973. Studies in flora of Thailand: Gnetaceae. *Thai Forest Bulletin (Botany)*. 7: 21–38.

Priyadi, H., Takao, G., Rahmawati, I., Supriyanto, B., Nursal, W.I. and Rahman, I. 2010. *Five Hundred Plant Species in Gunung Halimun Salak National Park, West Java: A Checklist Including Sundanese Names, Distribution, and Use*. CIFOR.

Pugh-Kitingan, J. 2015. Cultural and religious diversity in Sabah and relationships with surrounding areas. In *Islam and Cultural Diversity in Southeast Asia* (pp. 269–264). ILCAA, Tokyo University of Foreign Studies.

Puri, R.K. 2001. *The Bulungan Ethnobiology Handbook: A Field Manual for Biological and Social Science Research on the Knowledge and Use of Plants and Animals Among 18 Indigenous Groups in Northern East Kalimantan, Indonesia*. CIFOR.

Rahayu, Y., Chikamawati, T. and Widjaja, E.A. 2018. Nomenclatural study of *Tetrastigma leucostaphylum* and *Tetrastigma rafflesiae* (Vitaceae): Two common hosts of Rafflesia in Reveal, J.L. and Chase, M.W. 2011. APG III: Bibliographical information and synonymy of Magnoliidae. *Phytotaxa*. 19: 71–134.

Ridley, H.N. 1897. Malay materia medica. *Journal Straits Medical Association*. 122–134.

Ridsdale, C.E. 1978. A revision of *Mitragyna* and *Uncaria* (Rubiaceae). *Blumea: Biodiversity, Evolution and Biogeography of Plants*. 24(1): 43–100.

Robson, N.K.B. 1972. Notes on Malesian species of *Hypericum* (Guttiferae). Florae Malesianae Praecursores LII. *Blumea: Biodiversity, Evolution and Biogeography of Plants*. 20(2): 251–274.

Royyani, M.F., Rahayu, M. and Susiarti, S. 2015. Pengetahuan dan pemanfaatan tumbuhan obat masyarakat Tobelo Dalam di Maluku Utara. *Media Penelitian dan Pengembangan Kesehatan*. 25(4): 20744.

Sadiq, K. 2017. When states prefer non-citizens over citizens: Conflict over illegal immigration into Malaysia. In *Immigration* (pp. 219–240). Routledge. Available at https://www.taylorfrancis.com/chapters/edit/10.4324/9781315252599-7/states-prefer-non-citizens-citizens-conflict-illegal-immigration-malaysia-kamal-sadiq

Said, I.M. 2005. A preliminary checklist of the pteridophytes of Sabah. *Journal of Tropical Biology & Conservation (JTBC)*. 1.

Samarakoon, T. 2015. *Phylogenetic Relationships of Samydaceae and Taxonomic Revision of the Species of Casearia in South-Central Asia*. The University of Southern Mississippi.

Sari, R.Y. and Wardenaar, E. 2014. Etnobotani tumbuhan obat di Dusun Serambai Kecamatan Kembayan Kabupaten Sanggau Kalimantan Barat. *Jurnal Hutan Lestari*. 2(3).

Saudah, S., Zumaidar, Z., Darusman, D., Roslim, D.I. and Ernilasari, E. 2022. Ethnobotanical knowledge of *Etlingera elatior* for medicinal and food uses among ethnic groups in Aceh Province, Indonesia. *Biodiversitas Journal of Biological Diversity*. 23(8).

Saw, L.G. 2012. A revision of *Licuala* (Arecaceae, Coryphoideae) in Borneo. *Kew Bulletin*. 67: 577–654.

Schot, A.M. 1994. A revision of *Callerya* Endl. (including Padbruggea and Whitfordiodendron) (Papilionaceae: Millettieae). *Blumea: Biodiversity, Evolution and Biogeography of Plants*. 39(1/2): 1–40.

Shaw, H.A. 1981. The Euphorbiaceae of Sumatra. *Kew Bulletin*. 36(2): 239–374.

Sidiyasa, K. 1998. Taxonomy, phylogeny, and wood anatomy of *Alstonia* (Apocynaceae). *Blumea. Supplement*. 11(1): 1–230.

Silalahi, M., Supriatna, J. and Walujo, E.B. 2015. Local knowledge of medicinal plants in sub-ethnic Batak Simalungun of North Sumatra, Indonesia. *Biodiversitas*. 16(1): 44–54.

Simpson, D.A. 2019. Flora of Singapore. In D.J. Middleton, J. Leong-Škorničková, S. Lindsay (eds.), *Cyperaceae* (7, pp. 37–211). National Parks Board.

Sinclair, J. 1955. A revision of the Malayan Annonaceae part I. *Gardens' Bulletin Straits Settlements*. 14: 149–516.

Sinha, B.K. and Srivastava, S.K. 1996. Genus *Dracaena* Vandelli Ex L. in Andaman and Nicobar Islands. *Nelumbo*. 38(1–4): 14–18.

Sleumer, H. 1955. Proteaceae. *Flora Malesiana-Series 1, Spermatophyta*. 5(1): 147–206.

Soepadmo, E., Saw, L.G. and Chung, R.C.K. 2004. *Tree Flora of Sabah and Sarawak: Volume 5*. Forest Research Institute Malaysia.

Sofiah, S. and Sulistyaningsih, L.D. 2019. The diversity of *Smilax* (Smilacaceae) in Besiq-Bermai and Bontang Forests, East Kalimantan, Indonesia. *Biodiversitas Journal of Biological Diversity*. 20(1).

Sonjit, D., Prodyut, M. and Zaman, M.K. 2013. *Curcuma caesia* Roxb. and it's medicinal uses: A review. *International Journal of Research in Pharmacy and Chemistry*. 3(2): 370–375.

Sosef, M.S.M., Hong, L.T. and Prawirohatmodjo, S. 1998. *Plant Resources of South-East Asia* (5, No. 3). Backhuys.

Suwanphakdee, C., Masuthon, S., Chantaranothai, P., Chayamarit, K. and Chansuvanich, N. 2006. Notes on the genus *Piper* L. (Piperaceae) in Thailand. *Thai Forest Bulletin (Botany)*. 34: 206–214.

Takhtajan, A., ed. 2009. *Flowering Plants*. Springer Netherlands.

Tan, A.L. and Latiff, A. 2014. A taxonomic study of *Dillenia* L. *Dilleniaceae) in Peninsular Malaysia. Malayan Nature Journal*. 66(3): 338–353.

Tan, S.Y., Chong, K.Y., Neo, L., Koh, C.Y., Lim, R.C., Loh, J.W., Ng, W.Q., Seah, W.W., Yee, A.T.K. and Tan, H.T.W. 2016. Towards a field guide to the trees of the Nee Soon Swamp Forest (IV): *Xanthophyllum* (Polygalaceae). *Nature in Singapore*. 9: 139–147.

Tangah, J., Bajau, F.E., Jilimin, W., Chan, H.T., Wong, S.K. and Chan, E.W.C. 2017. Phytochemistry and pharmacology of *Mangifera pajang*: An iconic fruit of Sabah, Malaysia. *Systematic Reviews in Pharmacy*. 8(1): 86.

Teo, L.L., Lum, S.K.Y. and Loo, A.H.B. 2011. *Plectocomiopsis geminiflora* (griff.) Becc. (arecaceae)—a new record for Singapore. *Nature in Singapore*. 4: 1–4.

Tindowen, D.J.C., Bangi, J.C. and Mendezabal, M.J.N. 2017. Ethnopharmacology of medicinal plants in a rural community in Northern Philippines. *Journal of Biodiversity and Environmental Sciences*. 11: 296–303.

Trias-Blasi, A., Dee, R., Jimbo, T., Jackes, B. and Parmar, G. 2022. Taxonomic revision of *Causonis* (vitaceae) in New Guinea. *Edinburgh Journal of Botany*. 79: 1–10.

Turner, I.M. 2012. Annonaceae of Borneo: A review of the climbing species. *The Gardens' Bulletin Singapore*. 64(2): 371–479.

Uddin, M.Z., Khan, M.S. and Hassan, M.A. 1999. *Curculigo latifolia* Dryand. (Hypoxidaceae). – A new angiospermic record for Bangladesh. *Bangladesh Journal of Plant Taxonomy,* 6(2): 105–107.

van Welzen, P.C. 2003. Revision of the Malesian and Thai species of *Sauropus* (Euphorbiaceae: Phyllanthoideae). *Blumea-Biodiversity, Evolution and Biogeography of Plants*. 48(2): 319–391.

Veldkamp, J.F., Duistermaat, H., Wong K.M. and Middleton, D.J. 2019. Flora of Singapore. *Poaceae (Gramineae)*. 7: 219–501.

Veldkamp, J.F., Teerawatananon, A. and Sungkaew, S. 2015. A revision of *Garnotia* (Gramineae) in Malesia and Thailand. *Blumea-Biodiversity, Evolution and Biogeography of Plants*. 59(3): 229–237.

Voeks, R.A. and bin Nyawa, S. 2006. Dusun ethnobotany: Forest knowledge and nomenclature in northern Borneo. *Journal of Cultural Geography*. 23(2): 1–31.

Wan Zakaria, W.N.F., Ahmad Puad, A.S., Geri, C., Zainudin, R. and Latiff, A. 2016. Tetrastigma diepenhorstii (Miq.) Latiff (Vitaceae), a new host of *Rafflesia tuan-mudae* Becc. (Rafflesiaceae) in Borneo. *Journal of Botany*. 2016. Article ID 3952323: 6 pages.

Watson, L. and Dallwitz, M.J. 1992 onwards. *The Families of Flowering Plants: Descriptions, Illustrations, Identification, and Information Retrieval*. 13th January 2024. delta-intkey.com

Wee-Lek, C. 1963. A revision of the genus *Poikilospermum* (Urticaceae). *Gardens Bulletin, Singapore*. 20: 1–104.

Wegener, M. 2020, February. Evaluation and identification of the native Zingiberaceae specie in Mijen, Central Java, Indonesia. In *IOP Conference Series: Earth and Environmental Science* (457, No. 1: 012025). IOP Publishing.

Wong, K.M. and Pereira, J.T. 2016. A taxonomic treatment of the Asiatic allies of *Rothmannia* (Rubiaceae: Gardenieae), including the new genera Ridsdalea and Singaporandia. *Sandakania*. 21: 21–64.

World Health Organization. 2009. *Medicinal Plants in Papua New Guinea: Information on 126 Commonly Used Medicinal Plants in Papua New Guinea*. WHO Regional Office for the Western Pacific.

Wulandari, I., Iskandar, B.S., Parikesit, P., Hudoso, T., Iskandar, J., Megantara, E.N., Gunawan, E.F. and Shanida, S.S. 2021. Ethnoecological study on the utilization of plants in Ciletuh-Palabuhanratu Geopark, Sukabumi, West Java, Indonesia. *Biodiversitas Journal of Biological Diversity*. 22(2).

Yandre, R. 2023. Spesies Tumbuhan Dalam Makanan Tradisi Masyarakat Bugis di Pulau Sebatik, Malaysia: Plant species used in the traditional cuisine of the Bugis community in Sebatik Island, Malaysia. *Jurnal Komunikasi Borneo (JKoB)*. 11: 134–145.

Yeng, W.S. 2016. Araceae of peat swamp forests. *Biodiversity of Tropical Peat Swamp Forests of Sarawak*. 35.

Yusli, N.A., Saupi, N., Ramaiya, S.D. and Lirong, Y.A. 2021. An ethnobotanical study on indigenous food flavourings and aromatic enhancing plants used by the native communities of the central region of Sarawak. *Malaysian Applied Biology*. 50(3): 105–115.

Zakaria, W., Puad, A.S.A., Zainudin, R. and Latiff, A. 2017. A revision of *Tetrastigma* (Miq.) Planch. (Vitaceae) in Sarawak, Borneo. *Malayan Nature Journal*. 69: 71–90.

Zumbroich, T.J. 2009. The ethnobotany of teeth blackening in Southeast Asia. *Ethnobotany Research and Applications*. 7: 381–398.

Index

Note: Page numbers in *italics* indicate a figure on the corresponding page.

A

abdominal pain
- *Baccaurea lanceolata* for, 206
- *Curculigo latifolia* for, 83

Abelmoschus esculentus (lady's fingers), 259–260
abscesses
- *Blechnum orientale* for, 13
- *Capsicum frutescens* for, 354
- *Carica papaya* for, 249
- *Centella asiatica* for, 374
- *Jatropha curcas* for, 215
- *Mimosa pudica* for, 192
- *Peperomia pellucida* for, 39
- *Piper betle* for, 41
- *Stachytarpheta jamaicensis* for, 348

acne, *Aloe vera* for, 84
Acorus calamus (sweet flag), 70–71
Agathis borneensis (Borneo kauri), 31, *32*
Ageratum conyzoides (goat weed, billy goat weed, tropical white weed), 365
aging, *Centella asiatica* for, 374
Aglaonema oblongifolium, 72
Airyantha borneensis, 189–190
Alocasia macrorrhizos (giant taro, elephant's ears), 73
Aloe vera (Barbado aloe, true aloe), 84
alopecia, *Lycopodium phlegmaria* for, 4
Alphitonia incana (cooper's wood), 242, *243*
Alpinia galanga (greater galanga), 135
Alstonia angustifolia, 303
Alstonia angustiloba, 304
Alstonia macrophylla (batino, deviltree), 305
Alstonia scholaris (blackboard tree), 306
Alstonia spatulata (hard milkwood, siamese balsa), 308
Alternanthera sessilis (weed, sessile joyweed), 289
Alzheimer's disease, *Centella asiatica* for, 374
Amaranthus spinosus (spiny amaranth), 290
Amydrium medium, 74
Ananas comosus (pineapple), 104
anemia
- *Areca catechu* for, 91
- *Centella asiatica* for, 374

anemia in pregnant women, *Manihot esculenta* for, 223
Anisophyllea disticha (leechwood), 181
Annona muricata (soursop), 55
anticandidal properties of
- *Blechnum orientale*, 12
- *Brucea javanica*, 285
- *Cymbopogon citratus*, 114
- *Eichhornia crassipes*, 102
- *Eleusine indica*, 119
- *Lycopodium cernuum*, 1
- *Mangifera caesia*, 276
- *Neonauclea gigantea*, 324

anticancer properties of
- *Ageratum conyzoides*, 365
- *Avicennia marina*, 334
- *Blumea balsamifera*, 366
- *Capsicum frutescens*, 354
- *Ceiba pentandra*, *255*
- *Crassocephalum crepidioides*, 368
- *Durio zibethinus*, 245
- *Eryngium foetidum*, 375
- *Jatropha curcas*, 215
- *Luffa cylindrica*, 183
- *Momordica charantia*, 184
- *Solanum nigrum*, 362

antidote for poisons
- *Sansevieria trifasciata* for, 81
- *Stachytarpheta jamaicensis* for, 348
- *Tetracera indica* for, 172

aphrodisiacs
- *Dianella ensifolia*, 85
- *Dichapetalum gelonioides*, 203
- *Eurycoma longifolia*, 287
- *Justicia gendarussa*, 332

Areca catechu (betel palm, areca palm; betel nut palm), 91
Aristolochia minutiflora, 37
Aristolochia papillifolia, 37–38, *38*
Artabotrys roseus, 57
arthritis
- *Jatropha curcas* for, 215
- *Justicia gendarussa* for, 332
- *Scoparia dulcis* for, 342

Artocarpus elasticus (wild breadfruit), 236
Artocarpus tamaran (elephant Jack), 237
asthma
- *Aloe vera* for, 84
- *Annona muricata* for, 55
- *Boesenbergia stenophylla* for, 140
- *Cassia alata* for, 191
- *Costus speciosus* for, 129
- *Curculigo latifolia* for, 83
- *Drynaria sparsisora* for, 21
- *Eichhornia crassipes* for, 102
- *Elephantopus mollis* for, 369
- *Eleusine indica* for, 119
- *Euphorbia hirta* for, 210
- *Eurycoma longifolia* for, 287
- *Lycopodium cernuum* for, 1
- *Mimosa pudica* for, 192
- *Piper betle* for, 41
- *Selaginella argentea* for, 6

Avicennia marina, 334

B

Baccaurea lanceolata, 206
back and bone aches, *Pittosporum ferrugineum* for, 377
bad breath
- *Piper betle* for, 41
- *Rhizophora apiculata* for, 232

Bambusa vulgaris (golden bamboo), 109–110
Benincasa hispida (ash gourd, wax gourd), 182

beriberi
Cassia alata for, 191
Etlingera punicea for, 150
Vitex pubescens for, 349
Bischofia javanica (bishopwood tree), *207*, 207–208
Bixa orellana (achiote, annatto, lipstick tree), 252
bladder stones, *Commelina communis* for, 98
Blechnum orientale (centipede fern), *12*, 12–13
bleeding
Boesenbergia stenophylla for, 140
Dinochloa sublaevigata for, 118
Melastoma malabathricum for, 272
Stachytarpheta jamaicensis for, 348
blemishes, *Melastoma malabathricum* for, 272
blisters, *Peperomia pellucida* for, 39
bloating, *Areca catechu* for, 91
blood circulation
Bischofia javanica for, 208
Centella asiatica for, 374
Sansevieria trifasciata for, 81
blood detoxification
Etlingera coccinea for, 146
Melastoma malabathricum for, 272
blood diseases, *Peperomia pellucida* for, 39
blood purification, *Centella asiatica* for, 374
blood vomiting
Bombax ceiba for, 253
Garcinia mangostana for, 201
bloody stools
Cyrtandra areolata for, 342
Dillenia excelsa for, 168
Elephantopus mollis for, 369
Etlingera coccinea for, 146
Phyllanthus niruri for, 224
Vitex pubescens for, 349
blowguns darts, *Dissochaeta monticola* for, 272
Blumea balsamifera, 366
Blumea riparia, 367
Boesenbergia pulchella, 137
Boesenbergia rotunda (fingerroot), 137–138
Boesenbergia stenophylla, *139*, 139–140
boils
Aglaonema oblongifolium for, 72
Blechnum orientale for, 13
Elephantopus mollis for, 369
Euphorbia hirta for, 210
Hyptis capitata for, 338
Imperata cylindrica for, 120
Ipomoea batatas for, 352
Justicia gendarussa for, 332
Piper betle for, 41
Tabernaemontana macrocarpa for, 311
Urena lobata for, 261
boils in the ears, *Piper umbellatum* for, 48
Bombax ceiba (red silk cotton tree), 253
bone aches
Baccaurea lanceolata for, 206
Costus speciosus for, 129
Curculigo latifolia for, 83
Eleusine indica for, 119
Homalanthus populneus for, 213
Litsea odorifera for, 53
Octomeles sumatrana for, 188
Paspalum conjugatum for, 124
bone pains, *Piper betle* for, 41
breast milk, *Carica papaya* for, 249
breathlessness
Boesenbergia stenophylla for, 140
Cymbopogon citratus for, 114
Vitex pubescens for, 349
Bridelia stipularis, 209
broken bones
Clerodendrum philippinum for, 345
Euphorbia hirta for, 210
Brucea javanica (Java Brucea), 285
Bruguiera parviflora (thin-fruit orange mangrove), 230
burns
Aloe vera for, 84
Capsicum frutescens for, 354

C

Callicarpa longifolia, 344
cancer
Centella asiatica for, 374
Clinacanthus nutans for, 330
Crassocephalum crepidioides for, 368
Helminthostachys zeylanica for, 9
Hydnophytum formicarum for, 318
Morinda citrifolia for, 321
Myrmecodia platytyrea for, 323
canker sores
Capsicum frutescens for, 354
Centella asiatica for, 374
Durio zibethinus for, 256
Lycopodium cernuum for, 1
Canna indica (Indian shot), 127
Capsicum frutescens (tabasco pepper, bird pepper), 354
carbuncles, *Ipomoea batatas* for, 352
Carica papaya (papaya; melon tree), 249
Caryota mitis (fishtail palm), 93
Casearia grewiifolia, 228, *229*
Cassia alata (candle bush), *190*, 190–191
Casuarina sumatrana (Sumatran ru), 200
Ceiba pentandra (kapok tree), 255
Centella asiatica (centella, Indian pennywort), 374
chest pain
Antidesma montanum for, 205
Costus speciosus for, 129
Dillenia excelsa for, 168
Fagraea cuspidata for, 314
Fibraurea tinctoria for, 162
Lycopodium cernuum for, 1
Octomeles sumatrana for, 188
Tetrastigma leucostaphylum for, 178
chicken pox
Imperata cylindrica for, 120
Lygodium salicifolium for, 27
Psidium guajava for, 274
childbirth pain, *Imperata cylindrica* for, 120
child rashes, *Manihot esculenta* for, 223
Cinnamomum iners (wild cinnamon), 49
Clausena excavata, 280–281
Cleome chelidonii (elandine spider flower), 251
Clerodendrum philippinum, 345
Clidemia hirta (Koster's curse, soap bush), 270
Clinacanthus nutans (Sabah snake grass), 330
Cnestis platantha, 233

Coix lacryma-jobi (Job's tears), 112
colds
Alpinia galanga for, 135
Blumea balsamifera for, 366
Cymbopogon citratus for, 114
Dichapetalum gelonioides for, 203
Diplazium cordifolium for, 18
Ficus deltoidea for, 237
Hyptis capitata for, 338
Kaempferia galanga for, 152
Zingiber officinale for, 154
colds and shivers in children, *Uvaria grandiflora* for, 67
colic
Blumea balsamifera for, 366
Cyathula prostrata for, 292
Fissistigma latifolium for, 59
Lindera pipericarpa for, 50
Combretum nigrescens, 268
Commelina communis (Asiatic dayflower), 98
Commelina nudiflora (doveweed), 99
constipation
Carica papaya for, 249
Ipomoea aquatica for, 350
Plumieria acuminata for, 310
Urena lobata for, 261
convulsions, *Jatropha curcas* for, 215
cooling
Globba francisci for, 151
Miscanthus floridulus for, 123
Zingiber officinale for, 154
Cordyline fruticosa (ti plant), 77–78
Coscinium fenestratum (false calumba, knotted plant, tree turmeric), 159
Costus speciosus (spiral ginger), 129
coughing up blood
Cyperus rotundus for, 105
Lygodium flexuosum for, 27
coughs
Annona muricata for, 55
Boesenbergia stenophylla for, 140
Centella asiatica for, 374
Cinnamomum iners for, 49
Cnestis platantha for, 233
Coix lacryma-jobi for, 112
Costus speciosus for, 129
Curculigo latifolia for, 83
Curcuma caesia for, 141
Curcuma longa for, 142
Curcuma xanthorrhiza for, 143
Cymbopogon citratus for, 114
Donax canniformis for, 132
Eichhornia crassipes for, 102
Elephantopus mollis for, 369
Eurycoma longifolia for, 287
Flagellaria indica for, 107
Justicia gendarussa for, 332
Kaempferia galanga for, 152
Lawsonia inermis for, 269
Lycopodium cernuum for, 1
Lygodium flexuosum for, 27
Momordica charantia for, 184
Peperomia pellucida for, 39
Phyllanthus niruri for, 224
Piper betle for, 41
Piper sarmentosum for, 47
Polyalthia insignis for, 65
Polygala paniculata for, 196
Tetracera indica for, 172
Tetracera scandens for, 174
Urena lobata for, 261
Crassocephalum crepidioides (thickhead), 368
Curculigo latifolia (weevil lily), *82*, 82–83
Curcuma caesia (black turmeric), 141
Curcuma longa (turmeric), 142
Curcuma xanthorrhiza (false turmeric, Javanese ginger), 143
cuts
Ageratum conyzoides for, 365
Aloe vera for, 84
Alstonia spatulata for, 308
Areca catechu for, 91
Baccaurea lanceolata for, 206
Blumea balsamifera for, 366
Combretum nigrescens for, 268
Curculigo latifolia for, 83
Curcuma longa for, 142
Elephantopus mollis for, 369
Eleusine indica for, 119
Etlingera brevilabrum for, 145
Eupatorium odoratum for, 371
Imperata cylindrica for, 120
Jatropha curcas for, 215
Kalanchoe pinnata for, 175
Mallotus paniculatus for, 222
Melastoma malabathricum for, 272
Merremia peltata for, 353
Mimosa pudica for, 192
Nicotiana tabacum for, 356
Octomeles sumatrana for, 188
Oroxylum indicum for, 336
Pegia sarmentosa for, 278
Phyllanthus niruri for, 224
Piper betle for, 41
Sauropus androgynus for, 226
Stachytarpheta jamaicensis for, 348
Vitex pubescens for, 349
Cyathula prostrata (cyathula), 292
Cymbopogon citratus (citronella grass, fever grass, lemon grass), 114
Cyperus rotundus, 105
Cyrtandra areolata, *341*, 341–342

D

dandruff
Ananas comosus for, 104
Brucea javanica for, 285
Kaempferia galanga for, 152
Polygonum orientale for, 295
darken gray hair, *Morinda citrifolia* for, 321
deafness, *Dichapetalum gelonioides* for, 203
dehydration, *Licuala spinosa* for, 94
Dendrobium umbellatum (umbeled cadetia), *86*, 86–87
Dendrocnide elliptica, 246
dengue, *Carica papaya* for, 249
dental caries
Morinda citrifolia for, 321
Uncaria acida for, 328

dermatitis, *Tabernaemontana macrocarpa* for, 311
Desmos teysmannii, 58
diabetes
Alstonia scholaris for, 306
Boesenbergia stenophylla for, 140
Bridelia stipularis for, 209
Carica papaya for, 249
Centella asiatica for, 374
Clinacanthus nutans for, 330
Dichapetalum gelonioides for, 203
Eurycoma longifolia for, 287
Fagraea cuspidata for, 314
Hydnophytum formicarum for, 318
Melastoma malabathricum for, 272
Momordica charantia for, 184
Morinda citrifolia for, 321
Myrmecodia platytyrea for, 323
Orthosiphon stamineus for, 339
Peperomia pellucida for, 39
Phyllanthus niruri for, 224
Physalis minima for, 357
Piper betle for, 41
Tinospora crispa for, 164
Dianella ensifolia (umbrella dracaena), 85
diarrhea
Acorus calamus for, 71
Annona muricata for, 55
Aristolochia papillifolia for, 38
Bischofia javanica for, 208
Blumea balsamifera for, 366
Boesenbergia stenophylla for, 140
Centella asiatica for, 374
Elephantopus mollis for, 369
Eleusine indica for, 119
Justicia gendarussa for, 332
Litsea odorifera for, 53
Macaranga gigantea for, 219
Macaranga gigantifolia for, 220
Manihot esculenta for, 223
Melastoma malabathricum for, 272
Merremia peltata for, 353
Mimosa pudica for, 192
Neonauclea gigantea for, 324
Phyllanthus niruri for, 224
Psidium guajava for, 274
Stachytarpheta jamaicensis for, 348
Dichapetalum gelonioides (gelonium poison-leaf), 203
digestion, *Vitex pubescens* for, 349
Dillenia excelsa, 168
Dillenia grandifolia, 169
Dillenia indica (elephant apple), 169–170, *170*
Dinochloa scabrida, 117
Dinochloa scandens, 116, *116*
Dinochloa sublaevigata, 117–118
Dinochloa trichogona, 117
Diospyros elliptifolia, 300
Diplazium cordifolium, 18, *18*
Dissochaeta monticola, *271*, 271–272
diuretic
Amaranthus spinosus as, 290
Centella asiatica as, 374
Drymoglossum piloselloides as, 20
Orthosiphon stamineus as, 339
dizziness, *Mimosa pudica* for, 192
Donax canniformis, *131*, 131–132
Dracaena elliptica, 79, *79*
Dracaena umbratica, 80
Drymoglossum piloselloides (dragon scale), *19*, 19–20
Drynaria sparsisora, *21*, 21–22
Durio zibethinus (durian tree), 256
dysentery
Amaranthus spinosus for, 290
Brucea javanica for, 285
Psidium guajava for, 274
dysmenorrhea
Dichapetalum gelonioides for, 203
Kaempferia galanga for, 152
Morinda citrifolia for, 321

E

earache
Piper caducibracteum for, 45
Sansevieria trifasciata for, 81
ear pain, *Centella asiatica* for, 374
eczema, *Fibraurea tinctoria* for, 162
Eichhornia crassipes (common water hyacinth, water lily), 102
Elaeocarpus clementis, 235, *235*
Elephantopus mollis, 369
Eleusine indica (Indian goose grass), 119
Embelia philippinensis, *301*, 301–302
epilepsy
Alstonia macrophylla for, 305
Amaranthus spinosus for, 290
epileptic seizures, *Polyalthia bullata* for, 64
Eryngium foetidum (spiny coriander), 375
Etlingera brevilabrum, 145
Etlingera coccinea, 146
Etlingera elatior (kantan flower, Philippine waxflower, torch ginger), 147
Etlingera littoralis, 149
Etlingera punicea, 150
Eupatorium odoratum (butterfly weed), 371
Euphorbia hirta (hairy spurge, asthma-plant), 210
Euphorbia prostrata (ground spurge; prostrate sandmat), 212
Eurycoma longifolia, 287
excessive menses,
Mimosa pudica for, 192
Piper caducibracteum for, 45

F

Fagraea cuspidata, 314, *315*
fainting, *Poikilospermum cordifolium* for, 247
fatigue
Acorus calamus for, 71
Anisophyllea disticha for, 181
Areca catechu for, 91
Blumea balsamifera for, 366
Boesenbergia stenophylla for, 140
Bombax ceiba for, 253
Bruguiera parviflora for, 230
Cassia alata for, 191
Centella asiatica for, 374
Cnestis platantha for, 233
Cymbopogon citratus for, 114

Dracaena elliptica for, 79
Etlingera punicea for, 150
Fibraurea tinctoria for, 162
Ficus lepicarpa for, 239
Ficus septica for, 240
Fissistigma fulgens for, 58
Fissistigma manubriatum for, 60
Gnetum macrostachyum for, 34
Goniothalamus roseus for, 62
Guioa pleuropteris for, 282
Hanguana malayana for, 101
Homalanthus populneus for, 213
Kaempferia galanga for, 152
Leea indica for, 177
Litsea odorifera for, 53
Macaranga tanarius for, 221
Metroxylon sagu for, 95
Octomeles sumatrana for, 188
Phyllanthus niruri for, 224
Shorea parvistipulata for, 259
Synedrella nodiflora for, 372
Tetrastigma leucostaphylum for, 178
Uvaria grandiflora for, 67
Uvaria sorzogonensis for, 69
Wikstroemia ridleyi for, 267
Zingiber purpureum for, 156
Zingiber zerumbet for, 158
festering wounds, *Manihot esculenta* for, 223
fever
Abelmoschus esculentus for, 260
Acorus calamus for, 71
Airyantha borneensis for, 190
Aloe vera for, 84
Alpinia galanga for, 135
Annona muricata for, 55
Blumea balsamifera for, 366
Boesenbergia stenophylla for, 140
Bridelia stipularis for, 209
Capsicum frutescens for, 354
Carica papaya for, 249
Centella asiatica for, 374
Clidemia hirta for, 270
Cnestis platantha for, 233
Coix lacryma-jobi for, 112
Commelina nudiflora for, 99
Costus speciosus for, 129
Curculigo latifolia for, 83
Curcuma longa for, 142
Cymbopogon citratus for, 114
Diplazium cordifolium for, 18
Etlingera elatior for, 147
Etlingera littoralis for, 149
Euphorbia prostrata for, 212
Fagraea cuspidata for, 314
Fibraurea tinctoria for, 162
Ficus lepicarpa for, 239
Ficus septica for, 240
Goniothalamus roseus for, 62
Hedychium longicornutum for, 152
Homalanthus populneus for, 213
Hyptis capitata for, 338
Imperata cylindrica for, 120
Kalanchoe pinnata for, 175
Lycopodium cernuum for, 1
Lygodium flexuosum for, 27
Manihot esculenta for, 223
Melastoma malabathricum for, 272
Musa paradisiaca for, 133
Myrmecodia platytyrea for, 323
Phyllanthus niruri for, 224
Piper betle for, 41
Polygala paniculata for, 196
Psidium guajava for, 274
Sauropus androgynus for, 226
Scoparia dulcis for, 342
Selaginella argentea for, 6
Selaginella plana for, 7
Solanum torvum for, 363
Stenochlaena palustris for, 15
Tetrastigma leucostaphylum for, 178
Thysanolaena latifolia for, 125
Urena lobata for, 261
Uvaria cuneifolia for, 67
Vitex pubescens for, 349
feverish children
Ceiba pentandra for, 255
Vigna unguiculata for, 194
feverish colds
Glochidion macrostigma for, 213
Homalomena propinqua for, 74
Fibraurea tinctoria (yellow roots), *161*, 161–162
Ficus deltoidea (mistletoe fig, mistletoe rubber-plant), 237
Ficus elliptica, 239
Ficus lepicarpa (saraca fig), 239
Ficus septica (hauli tree), 240, *241*
Fissistigma fulgens, 58
Fissistigma latifolium, 59
Fissistigma manubriatum, 60
Flagellaria indica (false rattan, whip vine), 107
flatulence
Amaranthus spinosus for, 290
Blechnum orientale for, 13
Blumea balsamifera for, 366
Carica papaya for, 249
Cordyline fruticosa for, 78
Curcuma longa for, 142
Cymbopogon citratus for, 114
Dinochloa sublaevigata for, 118
Elephantopus mollis for, 369
Etlingera elatior for, 147
Ficus septica for, 240
Guioa pleuropteris for, 282
Imperata cylindrica for, 120
Justicia gendarussa for, 332
Lindera pipericarpa for, 50
Merremia peltata for, 353
Miscanthus floridulus for, 123
Piper betle for, 41
Polygala paniculata for, 196
Urena lobata for, 261
Zingiber officinale for, 154
flu
Abelmoschus esculentus for, 260
Amydrium medium for, 74
Blumea balsamifera for, 366
Brucea javanica for, 285
Clidemia hirta for, 270
Cnestis platantha for, 233

Coix lacryma-jobi for, 112
Costus speciosus for, 129
Curcuma longa for, 142
Eleusine indica for, 119
Flagellaria indica for, 107
Justicia gendarussa for, 332
Peperomia pellucida for, 39
Piper sarmentosum for, 47
Solanum nigrum for, 362
Thysanolaena latifolia for, 125
food poisoning
Boesenbergia stenophylla for, 140
Eleusine indica for, 119
Kaempferia galanga for, 152
Shorea macroptera for, 258
Forrestia griffithii, 100
foul body odor
Dillenia excelsa for, 168
Piper betle for, 41
fractures, *Curcuma longa* for, 142
fungal infection
Curcuma longa for, 142
Hanguana malayana for, 101

G

gallstones, *Drymoglossum piloselloides* for, 20
Garcinia mangostana (mangosteen), 200–201, *201*
Gardenia tubifera (golden gardenia), 314–315
Garnotia acutigluma, 120
gastric ulcers, *Morinda citrifolia* for, 321
gastritis
Acorus calamus for, 71
Ageratum conyzoides for, 365
Alstonia angustifolia for, 303
Alstonia angustiloba for, 304
Areca catechu for, 91
Bixa orellana for, 252
Blumea balsamifera for, 366
Brucea javanica for, 285
Cymbopogon citratus for, 114
Dichapetalum gelonioides for, 203
Fagraea cuspidata for, 314
Ficus septica for, 240
Hanguana malayana for, 101
Jatropha curcas for, 215
Litsea odorifera for, 53
Polygala paniculata for, 196
Salacca zalacca for, 96
Urena lobata for, 261
Xanthophyllum excelsum for, 199
gastroenteritis, *Psidium guajava* for, 274
Gleichenia truncata (creepy fingers fern), *16*, 16–17
Globba francisci, 151
Globba propinqua (mazoloso), 151
Glochidion macrostigma, 213
Gnetum macrostachyum, *33*, 33–34
Goniothalamus roseus, *61*, 61–62
gonorrhea
Carica papaya for, 249
Stachytarpheta jamaicensis for, 348
gout
Carica papaya for, 249
Melastoma malabathricum for, 272
Piper betle for, 41
Rhaphidophora korthalsii for, 76
Guioa pleuropteris, 282
gynecological disorders, *Melastoma malabathricum* for, 272

H

hair care, *Merremia peltata* for, 353
hair loss
Aloe vera for, 84
Casuarina sumatrana for, 200
Morinda citrifolia for, 321
Hanguana malayana, 101
headaches
Alphitonia incana for, 242
Blechnum orientale for, 13
Callicarpa longifolia for, 344
Carica papaya for, 249
Clausena excavata for, 280
Coix lacryma-jobi for, 112
Costus speciosus for, 129
Cyathula prostrata for, 292
Cymbopogon citratus for, 114
Desmos teysmannii for, 58
Dianella ensifolia for, 85
Euphorbia hirta for, 210
Fibraurea tinctoria for, 162
Ficus septica for, 240
Fissistigma fulgens for, 58
Homalanthus populneus for, 213
Justicia gendarussa for, 332
Kaempferia galanga for, 152
Kalanchoe pinnata for, 175
Leea indica for, 177
Manihot esculenta for, 223
Morinda citrifolia for, 321
Mussaenda frondosa for, 322
Myrmecodia platytyrea for, 323
Psychotria gyrulosa for, 327
Selaginella argentea for, 6
Selaginella plana for, 7
Tetracera indica for, 172
Thysanolaena latifolia for, 125
Vigna unguiculata for, 194
Wikstroemia androsaemifolia for, 266
heartburn, *Clidemia hirta* for, 270
heart diseases
Boesenbergia stenophylla for, 140
Drynaria sparsisora for, 21
Tinospora crispa for, 164
Hedychium longicornutum (hornbill's ginger), 151–152
Hedyotis auricularia, *316*, 316–317
Helicia serrata, *166*, 166–167
Helminthostachys zeylanica, 9–10, *10*
hematuria nosebleed, *Cyperus rotundus* for, 105
hemorrhoids
Carica papaya for, 249
Luffa cylindrica for, 183
Momordica charantia for, 184

hepatitis
Aristolochia papillifolia for, 38
Imperata cylindrica for, 120
high cholesterol
Momordica charantia for, 184
Pandanus amaryllifolius for, 90
Homalanthus populneus, 213
Homalomena propinqua, 74, *75*
hot body, *Kalanchoe pinnata* for, 175
Hydnophytum formicarum (baboon's head), 318
hydrocele
Ixora blumei for, 320
Physalis minima for, 357
hypertension
Airyantha borneensis for, 190
Alstonia scholaris for, 306
Annona muricata for, 55
Areca catechu for, 91
Blumea riparia for, 367
Boesenbergia rotunda for, 137
Boesenbergia stenophylla for, 140
Carica papaya for, 249
Centella asiatica for, 374
Clinacanthus nutans for, 330
Dichapetalum gelonioides for, 203
Drymoglossum piloselloides for, 20
Etlingera coccinea for, 146
Euphorbia prostrata for, 212
Eurycoma longifolia for, 287
Fibraurea tinctoria for, 162
Hydnophytum formicarum for, 318
Lycopodium cernuum for, 1
Melastoma malabathricum for, 272
Mimosa pudica for, 192
Momordica charantia for, 184
Morinda citrifolia for, 321
Myrmecodia platytyrea for, 323
Orthosiphon stamineus for, 339
Peperomia pellucida for, 39
Phyllanthus niruri for, 224
Physalis minima for, 357
Piper betle for, 41
Polygala paniculata for, 196
Solanum nigrum for, 362
Solanum torvum for, 363
Tinospora crispa for, 164
Vitex pubescens for, 349
Hypolytrum nemorum, 107
Hyptis capitata (knobweed), 338

I

Impatiens balsamina (garden balsam), 298
Imperata cylindrica (cogon grass), 120–121
indications sprains, *Scoparia dulcis* for, 342
indigestion
Alpinia galanga for, 135
Centella asiatica for, 374
Psidium guajava for, 274
insect bites
Blumea balsamifera for, 366
Combretum nigrescens for, 268
Curcuma longa for, 142
Cyathula prostrata for, 292
Mimosa pudica for, 192
Nicotiana tabacum for, 356
Piper betle for, 41
Tinospora crispa for, 164
insect stings
Acorus calamus for, 71
Blechnum orientale for, 13
internal injuries, *Combretum nigrescens* for, 268
internal pain, *Plumieria acuminata* for, 310
intestinal worms
Brucea javanica for, 285
Coix lacryma-jobi for, 112
Ixora blumei for, 320
Paedaria verticillata for, 326
Solanum nigrum for, 362
Tinospora crispa for, 164
Uvaria grandiflora for, 67
Ipomoea aquatica (water spinach), 350
Ipomoea batatas (sweet potato), 352
itchiness
Alocasia macrorrhizos for, 73
Aloe vera for, 84
Alphitonia incana for, 242
Capsicum frutescens for, 354
Curcuma longa for, 142
Cymbopogon citratus for, 114
Euphorbia hirta for, 210
Ficus deltoidea for, 237
Impatiens balsamina for, 298
Lantana camara for, 346
Lawsonia inermis for, 269
Theobroma cacao for, 264
itchy skin
Piper caducibracteum for, 45
Sansevieria trifasciata for, 81
Ixora blumei, 320

J

Jatropha curcas (physic nut, Barbados nut), 215
Jatropha podagrica (Buddha belly plant, bottle euphorbia, gout plant), 217
jaundice
Alphitonia incana for, 242
Aristolochia papillifolia for, 38
Cassia alata for, 191
Centella asiatica for, 374
Coscinium fenestratum for, 159
Curcuma longa for, 142
Cymbopogon citratus for, 114
Fagraea cuspidata for, 314
Fibraurea tinctoria for, 162
Flagellaria indica for, 107
Jatropha podagrica for, 217
Morinda citrifolia for, 321
Phyllanthus niruri for, 224
Tetrastigma leucostaphylum for, 178
Urena lobata for, 261
Uvaria grandiflora for, 67
jaundiced babies, *Benincasa hispida* for, 182
joint dislocation, *Litsea garciae* for, 51
Justicia gendarussa (gendarussa), 332

K

Kadsura borneensis, 35, *36*
Kaempferia galanga (East Indian galangal, resurrection lily), 152
Kalanchoe pinnata (air plant), 175–176
kidney diseases
    Carica papaya for, 249
    Myrmecodia platytyrea for, 323
kidney failure, *Orthosiphon stamineus* for, 339
kidney stones
    Aloe vera for, 84
    Boesenbergia stenophylla for, 140
    Sansevieria trifasciata for, 81
Kopsia dasyrachis, 309

L

lactation
    Carica papaya for, 249
    Caryota mitis for, 93
Lantana camara (Lantana), 346
Lawsonia inermis (henna), 268–269
laxative
    Annona muricata as, 55
    Blumea balsamifera as, 366
    Etlingera coccinea as, 146
    Psidium guajava as, 274
    Rhizophora apiculata for, 232
Leea indica (bandicoot berry), 177
Lepisanthes amoena, 282–283
leukemia, *Clinacanthus nutans* for, 330
lice, *Brucea javanica* for, 285
Licuala spinosa (mangrove fan palm, spiny licuala palm), 94
Lindera pipericarpa, 50
Litsea accedens, 51
Litsea garciae (Borneo avocado), 51–52, *52*
Litsea odorifera (trawas oil tree), 53
Litsea umbellata (hairy medang), 54
Lophatherum delicate (bamboo grass; bamboo leaf), 122
Luffa cylindrica (sponge gourd), 183
lumbago, *Justicia gendarussa* for, 332
Lycopodium cernuum (stag's-horn clubmoss), 1–2, *2*
Lycopodium phlegmaria (common tassel fern), *4*, 4–5
Lygodium circinnatum, *25*, 25–26
Lygodium flexuosum, 27
Lygodium salicifolium, 27

M

Macaranga gigantea (elephant's ear, giant mahang), *218*, 218–219
Macaranga gigantifolia, 220
Macaranga tanarius (macaranga, parasol tree), 221
magic rituals
    Gardenia tubifera for, 315
    Microcos antidesmifolia for, 264
malaise, *Scoparia dulcis* for, 342
malaria
    Alstonia angustiloba for, 304
    Alstonia scholaris for, 306
    Boesenbergia stenophylla for, 140
    Brucea javanica for, 285
    Callicarpa longifolia for, 344
    Carica papaya for, 249
    Euphorbia prostrata for, 212
    Eurycoma longifolia for, 287
    Fibraurea tinctoria for, 162
    Lepisanthes amoena for, 283
    Mimosa pudica for, 192
    Panicum palmifolium for, 123
    Phyllanthus niruri for, 224
    Physalis minima for, 357
    Piper sarmentosum for, 47
    Tinospora crispa for, 164
Mallotus paniculatus (turn-in-the-wind), 222
Mangifera caesia (white mango), 276
Mangifera pajang (wild mango), 277
Manihot esculenta (cassava, tapioca), 223
measles
    Cnestis platantha for, 233
    Imperata cylindrica for, 120
    Manihot esculenta for, 223
    Melastoma malabathricum for, 272
medicinal for the Dusun, *Dinochloa scandens* for, 116
medicinal for the Murut
    Dendrobium umbellatum for, 87
    Dinochloa scabrida for, 117
    Forrestia griffithii for, 100
    Helicia serrata for, 167
    Hypolytrum nemorum for, 107
    Plectocomiopsis geminiflora for, 96
    Scleria bancana for, 109
    Vitis trifolia for, 179
medicinal for the Rungus, *Xylopia dehiscens* for, 70
medicinal salads, *Embelia philippinensis* for, 302
Melastoma malabathricum (Malabar melastome, white melastome), 272–273
menorrhagia, *Globba propinqua* for, 151
menstruation
    Carica papaya for, 249
    Curcuma longa for, 142
Merremia peltata (big leaf rope), 353
metrorrhagia, *Canna indica* for, 127
Metroxylon sagu (sago palm), 95
Microcos antidesmifolia, *263*, 263–264
Millettia nieuwenhuis, 192
Mimosa pudica (sensitive mimosa, sensitive plant), 192–193
Miscanthus floridulus (Pacific Island silvergrass), 123
Mischocarpus pentapetalus, 284
Momordica charantia (bitter gourd), 184
Morinda citrifolia (Indian mulberry), 320–321
mouthwash, *Dillenia indica* as, 169
Musa paradisiaca (plantain), 133
muscle aches, *Justicia gendarussa* for, 332
muscle cramps
    Hanguana malayana for, 101
    Homalanthus populneus for, 213
muscle sprains, *Stachytarpheta jamaicensis* for, 348
muscular pains, *Kadsura borneensis* for, 35
Mussaenda frondosa (wild mussaenda, dhobi tree), 322
Myrmecodia platytyrea, 323

N

nails, *Combretum nigrescens* for, 268
Nauclea officinalis, 325
nausea, *Annona muricata* for, 55
Neonauclea gigantea, 324
Nicotiana tabacum (tobacco), 356
nosebleeds, *Piper betle* for, 41

O

obesity, *Orthosiphon stamineus* for, 339
Octomeles sumatrana (benuang, ilimo tree), 188, *189*
Oroxylum indicum (Indian trumpet, tree of Damocles, trumpet flowers, broken bones), 336
Orthosiphon stamineus (cat's whiskers, Java tea), 339
Oxyceros bispinosus, 326

P

Paedaria verticillata, 326–327
pain, *Bixa orellana* for, 252
painful joints
  Anisophyllea disticha for, 181
  Carica papaya for, 249
  Cinnamomum iners for, 49
  Morinda citrifolia for, 321
pancreatitis
  Alphitonia incana for, 242
  Aristolochia papillifolia for, 38
  Blumea balsamifera for, 366
  Casearia grewiifolia for, 228
  Lophatherum delicate for, 122
Pandanus amaryllifolius (pandan), 90
Panicum palmifolium (palm grass), 123–124
paralysis, *Flagellaria indica* for, 107
paralytic babies, *Schefflera nervosa* for, 376
Paspalum conjugatum (carabao grass, hilo grass), 124
Pegia sarmentosa, 278
Peperomia pellucida (man to man), 39
Phaeanthus ophthalmicus, 62–63
phlegm, *Phyllanthus niruri* for, 224
Phyllanthus niruri (niruri, stonebreaker), 224
Physalis minima (pygmy groundcherry), 357
piles
  Clinacanthus nutans for, 330
  Eleusine indica for, 119
  Jatropha curcas for, 215
  Morinda citrifolia for, 321
  Plumieria acuminata for, 310
pimples
  Aloe vera for, 84
  Curcuma longa for, 142
  Peperomia pellucida for, 39
  Psidium guajava for, 274
  Pycnarrhena tumefacta for, 163
Piper betle (betel pepper, betel vine), 41–42
Piper caducibracteum, *44*, 44–45
Piper caninum, 46
Piper sarmentosum, 47
Piper umbellatum (cow-foot leaf), 48
Pittosporum ferrugineum (rusty Pittosporum), 377
Plectocomiopsis geminiflora, 96
Plumieria acuminata (frangipani), 310
Poikilospermum cordifolium, *247*, 247–248
Poikilospermum suaveolens, 248
poison antidote
  Acorus calamus as, 71
  Areca catechu as, 91
  Aristolochia papillifolia as, 38
  Bambusa vulgaris as, 110
  Curcuma longa as, 142
  Hanguana malayana as, 101
  Morinda citrifolia as, 321
  Myrmecodia platytyrea as, 323
  Stephania corymbosa as, 165
Polyalthia bullata, 64
Polyalthia insignis, 65
Polyalthia sumatrana, 66
Polygala paniculata, 196
Polygonum odoratum (laksa leaves, Vietnamese coriander), 294
Polygonum orientale (oriental pepper), 295
postpartum
  Aloe vera for, 84
  Blumea balsamifera for, 366
  Boesenbergia rotunda for, 137
  Bridelia stipularis for, 209
  Carica papaya for, 249
  Cordyline fruticosa for, 78
  Costus speciosus for, 129
  Curcuma longa for, 142
  Curcuma xanthorrhiza for, 143
  Dichapetalum gelonioides for, 203
  Eleusine indica for, 119
  Fagraea cuspidata for, 314
  Ficus septica for, 240
  Gnetum macrostachyum for, 34
  Guioa pleuropteris for, 282
  Justicia gendarussa for, 332
  Kaempferia galanga for, 152
  Lophatherum delicate for, 122
  Melastoma malabathricum for, 272
  Momordica charantia for, 184
  Poikilospermum suaveolens for, 248
  Psidium guajava for, 274
  Stenochlaena palustris for, 15
  Vitex pubescens for, 349
  Zingiber officinale for, 154
pregnancy, *Capsicum frutescens* for, 354
Pronephrium asperum, 29, *30*
Prunus arborea (current laurel), 244–245, *245*
Psidium guajava (guava tree), 274–275
Psychotria gyrulosa (gyrulose white coffee), 327, *327*
putrefied wounds
  Alstonia angustifolia for, 303
  Alstonia angustiloba for, 304
Pycnarrhena tumefacta, 163
Pyrrosia lanceolata (lance leaf tongue fern), *23*, 23–24

R

rashes
  Centella asiatica for, 374
  Urena lobata for, 261
respiratory problems, *Costus speciosus* for, 129

Rhaphidophora korthalsii, 76
rheumatisms
Capsicum frutescens for, 354
Homalanthus populneus for, 213
Imperata cylindrica for, 120
Jatropha curcas for, 215
Justicia gendarussa for, 332
Peperomia pellucida for, 39
Zingiber officinale for, 154
Rhizophora apiculata (mangroves, tall-stilted mangrove), *231*, 231–232
Ridsdalea pseudoternifolia, 328
ringworms
Capsicum frutescens for, 354
Cassia alata for, 191
Ficus lepicarpa for, 239
Nicotiana tabacum for, 356
Piper sarmentosum for, 47

S

Salacca zalacca (salak palm, snake fruit), 96–97
Sansevieria trifasciata (mother-in-law's tongue; snake plant), 81
Saurauia fragrans, 297
Sauropus androgynus (Sabah vegetable), 226
scabies
Areca catechu for, 91
Cassia alata for, 191
Mallotus paniculatus for, 222
Piper betle for, 41
Tinospora crispa for, 164
scalds
Eupatorium odoratum for, 371
Psidium guajava for, 274
Schefflera nervosa, 376
Scindapsus perakensis, 77
Scleria bancana (windged scleria), 109
Scoparia dulcis (licorice weed, sweet broomweed), 342
Selaginella argentea, 6, *6*
Selaginella plana (Asian spikemoss), 7, *7*
Semecarpus cuneiformis (marking nut tree, Oriental cashew nut), 278–279
Shorea macroptera, 258
Shorea parvistipulata, 259
sinusitis
Carica papaya for, 249
Myrmecodia platytyrea for, 323
skin diseases
Acorus calamus for, 71
Aloe vera for, 84
Alphitonia incana for, 242
Alpinia galanga for, 135
Boesenbergia pulchella for, 137
Boesenbergia stenophylla for, 140
Brucea javanica for, 285
Cassia alata for, 191
Curculigo latifolia for, 83
Curcuma longa for, 142
Cyrtandra areolata for, 342
Litsea odorifera for, 53
Mimosa pudica for, 192
Morinda citrifolia for, 321
Oroxylum indicum for, 336
Piper betle for, 41
Stachytarpheta jamaicensis for, 348
Stenochlaena palustris for, 15
Theobroma cacao for, 264
Urena lobata for, 261
skin injuries, *Ageratum conyzoides* for, 365
skin itchiness, *Piper betle* for, 41
smallpox
Callicarpa longifolia for, 344
Costus speciosus for, 129
Imperata cylindrica for, 120
Lygodium salicifolium for, 27
Momordica charantia for, 184
Smilax odoratissima, *88*, 88–89
snake and scorpion bites
Impatiens balsamina for, 298
Lindera pipericarpa for, 50
Stachytarpheta jamaicensis for, 348
Solanum erianthum (potato tree), 359
Solanum ferox (hairy-fruited eggplant), 360
Solanum melongena (eggplant), 361
Solanum nigrum (black nightshade), 362
Solanum torvum (Turkey berry, pea eggplant, susumber berries), 363
sore eyes
Capsicum frutescens for, 354
Dinochloa sublaevigata for, 118
Dinochloa trichogona for, 117
Euphorbia hirta for, 210
Ficus deltoidea for, 237
Peperomia pellucida for, 39
Phaeanthus ophthalmicus for, 62
Sauropus androgynus for, 226
Tinospora crispa for, 164
sore throats
Aloe vera for, 84
Dinochloa sublaevigata for, 118
Kaempferia galanga for, 152
Musa paradisiaca for, 133
Physalis minima for, 357
Smilax odoratissima for, 89
spitting blood, *Mimosa pudica* for, 192
sprains
Alphitonia incana for, 242
Alstonia angustiloba for, 304
Boesenbergia rotunda for, 137
Cnestis platantha for, 233
Curcuma longa for, 142
Hanguana malayana for, 101
Homalanthus populneus for, 213
Impatiens balsamina for, 298
Leea indica for, 177
Litsea garciae for, 51
Oroxylum indicum for, 336
Tetrastigma diepenhorstii for, 179
Zingiber officinale for, 154
Stachytarpheta jamaicensis (blue porterweed), 348
Stenochlaena palustris (climbing ferns), *14*, 14–15
Stephania corymbosa, 165
stingray bites, *Avicennia marina* for, 334
stings
Aloe vera for, 84
Eupatorium odoratum for, 371

stomach aches
Acorus calamus for, 71
Aloe vera for, 84
Annona muricata for, 55
Areca catechu for, 91
Baccaurea lanceolata for, 206
Bischofia javanica for, 208
Blumea balsamifera for, 366
Boesenbergia rotunda for, 137
Boesenbergia stenophylla for, 140
Brucea javanica for, 285
Callicarpa longifolia for, 344
Carica papaya for, 249
Cassia alata for, 191
Centella asiatica for, 374
Cleome chelidonii for, 251
Cnestis platantha for, 233
Costus speciosus for, 129
Curcuma longa for, 142
Cymbopogon citratus for, 114
Dillenia grandifolia for, 169
Dinochloa sublaevigata for, 118
Durio zibethinus for, 256
Eichhornia crassipes for, 102
Etlingera littoralis for, 149
Eurycoma longifolia for, 287
Fibraurea tinctoria for, 162
Ficus septica for, 240
Guioa pleuropteris for, 282
Hyptis capitata for, 338
Imperata cylindrica for, 120
Justicia gendarussa for, 332
Kaempferia galanga for, 152
Kalanchoe pinnata for, 175
Litsea odorifera for, 53
Melastoma malabathricum for, 272
Merremia peltata for, 353
Mimosa pudica for, 192
Neonauclea gigantea for, 324
Piper betle for, 41
Psidium guajava for, 274
Scoparia dulcis for, 342
Urena lobata for, 261
Uvaria grandiflora for, 67
Vigna unguiculata for, 194
Vitex pubescens for, 349
stroke
Flagellaria indica for, 107
Lycopodium cernuum for, 1
swellings
Amaranthus spinosus for, 290
Carica papaya for, 249
Clausena excavata for, 280
Costus speciosus for, 129
Euphorbia hirta for, 210
Justicia gendarussa for, 332
Mimosa pudica for, 192
Oroxylum indicum for, 336
Phyllanthus niruri for, 224
Stachytarpheta jamaicensis for, 348
Urena lobata for, 261
swollen feet, *Homalanthus populneus* for, 213
swollen gums, *Solanum melongena* for, 361
swollen legs, *Amydrium medium* for, 74
swollen pancreas, *Pyrrosia lanceolata* for, 24
swollen penis, *Ixora blumei* for, 320
Synedrella nodiflora (cinderella weed, node weed), 372

T

Tabernaemontana macrocarpa, 311
Tabernaemontana sphaerocarpa, *312*, 312–313
tachycardia, *Centella asiatica* for, 374
tapeworms, *Blumea balsamifera* for, 366
teething, *Phyllanthus niruri* for, 224
Tetracera akara, 171
Tetracera indica, 172
Tetracera scandens, 174
Tetrastigma diepenhorstii, 179
Tetrastigma leucostaphylum (Indian chestnut vine), 178
Theobroma cacao (cacao), 264–265, *265*
thrush
Amaranthus spinosus for, 290
Bridelia stipularis for, 209
Guioa pleuropteris for, 282
Imperata cylindrica for, 120
Macaranga gigantea for, 219
Macaranga gigantifolia for, 220
Millettia nieuwenhuis for, 192
Neonauclea gigantea for, 324
Polyalthia insignis for, 65
Urena lobata for, 261
Thysanolaena latifolia (bamboo grass, broom grass, tiger grass), 125
Tinospora crispa (Chinese tinospora), 164
Toothaches
Airyantha borneensis for, 190
Areca catechu for, 91
Clausena excavata for, 280
Curcuma xanthorrhiza for, 143
Dillenia indica for, 169
Eichhornia crassipes for, 102
Fissistigma fulgens for, 58
Jatropha curcas for, 215
Piper betle for, 41
Piper sarmentosum for, 47
Polygala paniculata for, 196
Sansevieria trifasciata for, 81
Urena lobata for, 261
Trichosanthes cucumerina (snake gourd), 186
tuberculosis, *Myrmecodia platytyrea* for, 323

U

ulcers
Blechnum orientale for, 13
Melastoma malabathricum for, 272
Phyllanthus niruri for, 224
Piper betle for, 41
Saurauia fragrans for, 297
Solanum melongena for, 361
Uncaria acida, 328–329
Urena lobata (aramina, caesar weed, burr mallow), 261–262
urinary diseases, *Imperata cylindrica* for, 120
urinary tract infections
Amaranthus spinosus for, 290
Centella asiatica for, 374

Uvaria cuneifolia, 67
Uvaria grandiflora, 67
Uvaria sorzogonensis, *68*, 68–69

V

venereal diseases
Ageratum conyzoides for, 365
Clausena excavata for, 280
Garnotia acutigluma for, 120
Lygodium circinnatum for, 26
veterinary medicine, *Prunus arborea* for, 245
Vigna unguiculata (black-eyed pea, cowpea, catjang), 194–195
Vitex pubescens, 349
Vitis trifolia (bush grape, fox grape, three-leaved cayratia), 179–180, *180*
vomiting
Cymbopogon citratus for, 114
Flagellaria indica for, 107
Oroxylum indicum for, 336

W

warts, *Carica papaya* for, 249
weak uterus, *Carica papaya* for, 249
white hair, *Bixa orellana* for, 252
Wikstroemia androsaemifolia (male gaharu, red gaharu), 266
Wikstroemia ridleyi, 267, *267*
womb diseases, *Lygodium circinnatum* for, 26
wounds
Aloe vera for, 84
Blechnum orientale for, 13
Blumea balsamifera for, 366
Capsicum frutescens for, 354
Centella asiatica for, 374
Combretum nigrescens for, 268
Curculigo latifolia for, 83
Elephantopus mollis for, 369
Eleusine indica for, 119
Eupatorium odoratum for, 371
Fibraurea tinctoria for, 162
Imperata cylindrica for, 120
Kalanchoe pinnata for, 175
Melastoma malabathricum for, 272
Merremia peltata for, 353
Mimosa pudica for, 192
Nicotiana tabacum for, 356
Octomeles sumatrana for, 188
Oroxylum indicum for, 336
Phyllanthus niruri for, 224
Sansevieria trifasciata for, 81
Semecarpus cuneiformis for, 278
Tinospora crispa for, 164
Vitex pubescens for, 349
Zingiber zerumbet for, 158

X

Xanthophyllum excelsum, *198*, 198–199
Xylopia dehiscens, 70

Y

yellow fever
Centella asiatica for, 374
Euphorbia prostrata for, 212
Momordica charantia for, 184
Morinda citrifolia for, 321

Z

Zingiber officinale (ginger), 154
Zingiber purpureum (Bengal ginger, cassumunar ginger, Javanese ginger), 156
Zingiber zerumbet (bitter ginger, shampoo ginger), 158
Ziziphus horsfieldii, 244

Printed in the United States
by Baker & Taylor Publisher Services